T0319557

Matrix and Tensor Decompositions in Signal Processing

Matrices and Tensors with Signal Processing Set

coordinated by
Gérard Favier

Volume 2

Matrix and Tensor Decompositions in Signal Processing

Gérard Favier

　　　WILEY

First published 2021 in Great Britain and the United States by ISTE Ltd and John Wiley & Sons, Inc.

ISTE Ltd
27-37 St George's Road
London SW19 4EU
UK

www.iste.co.uk

John Wiley & Sons, Inc.
111 River Street
Hoboken, NJ 07030
USA

www.wiley.com

Library of Congress Control Number: 2021938218

British Library Cataloguing-in-Publication Data
A CIP record for this book is available from the British Library
ISBN 978-1-78630-155-0

MIX
Paper from
responsible sources
FSC® C013604

Contents

Introduction

The first book of this series was dedicated to introducing matrices and tensors (of order greater than two) from the perspective of their algebraic structure, presenting their similarities, differences and connections with representations of linear, bilinear and multilinear mappings. This second volume will now study tensor operations and decompositions in greater depth.

In this introduction, we will motivate the use of tensors by answering five questions that prospective users might and should ask:

– What are the advantages of tensor approaches?

– For what uses?

– In what fields of application?

– With what tensor decompositions?

– With what cost functions and optimization algorithms?

Although our answers are necessarily incomplete, our aim is to:

– present the advantages of tensor approaches over matrix approaches;

– show a few examples of how tensor tools can be used;

– give an overview of the extensive diversity of problems that can be solved using tensors, including a few example applications;

– introduce the three most widely used tensor decompositions, presenting some of their properties and comparing their parametric complexity;

– state a few problems based on tensor models in terms of the cost functions to be optimized;

– describe various types of tensor-based processing, with a brief glimpse of the optimization methods that can be used.

I.1. What are the advantages of tensor approaches?

In most applications, a tensor $\mathcal{X}$ of order N is viewed as an array of real or complex numbers. The current element of the tensor is denoted $x_{i_1,\cdots,i_N}$, where each index $i_n \in \langle I_n \rangle \triangleq \{1, \cdots, I_n\}$, for $n \in \langle N \rangle \triangleq \{1, \cdots, N\}$, is associated with the nth mode, and I_n is its dimension, i.e. the number of elements for the nth mode. The order of the tensor is the number N of indices, i.e. the number of modes. Tensors are written with calligraphic letters[1]. An Nth-order tensor with entries $x_{i_1,\cdots,i_N}$ is written $\mathcal{X} = [x_{i_1,\cdots,i_N}] \in \mathbb{K}^{I_1 \times \cdots \times I_N}$, where $\mathbb{K} = \mathbb{R}$ or $\mathbb{C}$, depending on whether the tensor is real-valued or complex-valued, and $I_1 \times \cdots \times I_N$ represents the size of $\mathcal{X}$.

In general, a mode (also called a way) can have one of the following interpretations: (i) as a source of information (user, patient, client, trial, etc.); (ii) as a type of entity attached to the data (items/products, types of music, types of film, etc.); (iii) as a tag that characterizes an item, a piece of music, a film, etc.; (iv) as a recording modality that captures diversity in various domains (space, time, frequency, wavelength, polarization, color, etc.). Thus, a digital image in color can be represented as a three-dimensional tensor (of pixels) with two spatial modes, one for the rows (width) and one for the columns (height), and one channel mode (RGB colors). For example, a color image can be represented as a tensor of size $1024 \times 768 \times 3$, where the third mode corresponds to the intensity of the three RGB colors (red, green, blue). For a volumetric image, there are three spatial modes ($width \times height \times depth$), and the points of the image are called voxels. In the context of hyperspectral imagery, in addition to the two spatial dimensions, there is a third dimension corresponding to the emission wavelength within a spectral band.

Tensor approaches benefit from the following advantages over matrix approaches:

– the essential uniqueness property[2], satisfied by some tensor decompositions, such as PARAFAC (parallel factors) (Harshman 1970) under certain mild conditions; for matrix decompositions, this property requires certain restrictive conditions on the factor matrices, such as orthogonality, non-negativity, or a specific structure (triangular, Vandermonde, Toeplitz, etc.);

– the ability to solve certain problems, such as the identification of communication channels, directly from measured signals, without requiring the calculation of

---

[1] Scalars, vectors, and matrices are written in lowercase, bold lowercase, and bold uppercase letters, respectively: $a, \mathbf{a}, \mathbf{A}$.

[2] A decomposition satisfies the essential uniqueness property if it is unique up to permutation and scaling factors in the columns of its factor matrices.

high-order statistics of these signals or the use of long pilot sequences. The resulting deterministic and semi-blind processings can be performed with signal recordings that are shorter than those required by statistical methods, based on the estimation of high-order moments or cumulants. For the blind source separation problem, tensor approaches can be used to tackle the case of underdetermined systems, i.e. systems with more sources than sensors;

– the possibility of compressing big data sets via a data tensorization and the use of a tensor decomposition, in particular, a low multilinear rank approximation;

– a greater flexibility in representing and processing multimodal data by considering the modalities separately, instead of stacking the corresponding data into a vector or a matrix. This allows the multilinear structure of data to be preserved, meaning that interactions between modes can be taken into account;

– a greater number of modalities can be incorporated into tensor representations of data, meaning that more complementary information is available, which allows the performance of certain systems to be improved, e.g. wireless communication, recommendation, diagnostic, and monitoring systems, by making detection, interpretation, recognition, and classification operations easier and more efficient. This led to a generalization of certain matrix algorithms, like SVD (singular value decomposition) to MLSVD (multilinear SVD), also known as HOSVD (higher order SVD) (de Lathauwer et al. 2000a); similarly, certain signal processing algorithms were generalized, like PCA (principal component analysis) to MPCA (multilinear PCA) (Lu et al. 2008) or TRPCA (tensor robust PCA) (Lu et al. 2020) and ICA (independent component analysis) to MICA (multilinear ICA) (Vasilescu and Terzopoulos 2005) or tensor PICA (probabilistic ICA) (Beckmann and Smith 2005).

It is worth noting that, with a tensor model, the number of modalities considered in a problem can be increased either by increasing the order of the data tensor or by coupling tensor and/or matrix decompositions that share one or several modes. Such a coupling approach is called data fusion using a coupled tensor/matrix factorization. Two examples of this type of coupling are presented later in this introductory chapter. In the first, EEG signals are coupled with functional magnetic resonance imaging (fMRI) data to analyze the brain function; in the second, hyperspectral and multispectral images are merged for remote sensing.

The other approach, namely, increasing the number of modalities, will be illustrated in Volume 3 of this series by giving a unified presentation of various models of wireless communication systems designed using tensors. In order to improve system performance, both in terms of transmission and reception, the idea is to employ multiple types of diversity simultaneously in various domains (space, time, frequency, code, etc.), each type of diversity being associated with a mode of the tensor of received signals. Coupled tensor models will also be presented in the context of cooperative communication systems with relays.

I.2. For what uses?

In the big data[3] era, digital information processing plays a key role in various fields of application. Each field has its own specificities and requires specialized, often multidisciplinary, skills to manage both the multimodality of the data and the processing techniques that need to be implemented. Thus, the "intelligent" information processing systems of the future will have to integrate representation tools, such as tensors and graphs, signal and image processing methods, with artificial intelligence techniques based on artificial neural networks and machine learning.

The needs of such systems are diverse and numerous – whether in terms of storage, visualization (3D representation, virtual reality, dissemination of works of art), transmission, imputation, prediction/forecasting, analysis, classification or fusion of multimodal and heterogeneous data. The reader is invited to refer to Lahat *et al.* (2015) and Papalexakis *et al.* (2016) for a presentation of various examples of data fusion and data mining based on tensor models.

Some of the key applications of tensor tools are as follows:

– decomposition or separation of heterogeneous datasets into components/factors or subspaces with the goal of exploiting the multimodal structure of the data and extracting useful information for users from uncertain or noisy data or measurements provided by different sources of information and/or types of sensor. Thus, features can be extracted in different domains (spatial, temporal, frequential) for classification and decision-making tasks;

– imputation of missing data within an incomplete database using a low-rank tensor model, where the missing data results from defective sensors or communication links, for example. This task is called tensor completion and is a higher order generalization of matrix completion (Candès and Recht 2009; Signoretto *et al.* 2011; Liu *et al.* 2013);

– recovery of useful information from compressed data by reconstructing a signal or an image that has a sparse representation in a predefined basis, using compressive sampling (CS; also known as compressed sensing) techniques (Candès and Wakin 2008; Candès and Plan 2010), applied to sparse, low-rank tensors (Sidiropoulos and Kyrillidis 2012);

– fusion of data using coupled tensor and matrix decompositions;

– design of cooperative multi-antenna communication systems (also called MIMO (multiple-input multiple-output); this type of application, which led to the

---

3 Big data is characterized by 3Vs (Volume, Variety, Velocity) linked to the size of the data set, the heterogeneity of the data and the rate at which it is captured, stored and processed.

development of several new tensor models, will be considered in the next two volumes of this series;

– multilinear compressive learning that combines compressed sensing with machine learning;

– reduction of the dimensionality of multimodal, heterogeneous databases with very large dimensions (big data) by solving a low-rank tensor approximation problem;

– multiway filtering and tensor data denoising.

Tensors can also be used to tensorize neural networks with fully connected layers by expressing the weight matrix of a layer as a tensor train (TT) whose cores represent the parameters of the layer. This considerably reduces the parametric complexity and, therefore, the storage space. This compression property of the information contained in layered neural networks when using tensor decompositions provides a way to increase the number of hidden units (Novikov *et al.* 2015). Tensors, when used together with multilayer perceptron neural networks to solve classification problems, achieve lower error rates with fewer parameters and less computation time than neural networks alone (Chien and Bao 2017). Neural networks can also be used to learn the rank of a tensor (Zhou *et al.* 2019), or to compute its eigenvalues and singular values, and hence the rank-one approximation of a tensor (Che *et al.* 2017).

I.3. In what fields of application?

Tensors have applications in many domains. The fields of psychometrics and chemometrics in the 1970s and 1990s paved the way for signal and image processing applications, such as blind source separation, digital communications, and computer vision in the 1990s and early 2000s. Today, there is a quantitative explosion of big data in medicine, astronomy, meteorology, with fifth-generation wireless communications (5G), for medical diagnostic aid, web services delivered by recommendation systems (video on demand, online sales, restaurant and hotel reservations, etc.), as well as for information searching within multimedia databases (texts, images, audio and video recordings) and with social networks. This explains why various scientific communities and the industrial world are showing a growing interest in tensors.

Among the many examples of applications of tensors for signal and image processing, we can mention:

– blind source separation and blind system identification. These problems play a fundamental role in signal processing. They involve separating the input signals (also called sources) and identifying a system from the knowledge of only the output signals and certain hypotheses about the input signals, such as statistical independence in the case of independent component analysis (Comon 1994), or the assumption of a finite alphabet in the context of digital communications. This type of processing is, in particular, used to jointly estimate communication channels and information symbols emitted by a transmitter. It can also be used for speech or music separation, or to process seismic signals;

– use of tensor decompositions to analyze biomedical signals (EEG, MEG, ECG, EOG[4]) in the space, time and frequency domains, in order to provide a medical diagnostic aid; for instance, Acar *et al.* (2007) used a PARAFAC model of EEG signals to analyze epileptic seizures; Becker *et al.* (2014) used the same type of decomposition to locate sources within EEG signals;

– analysis of brain activity by merging imaging data (fMRI) and biomedical signals (EEG and MEG) with the goal of enabling non-invasive medical tests (see Table I.4);

– analysis and classification of hyperspectral images used in many fields (medicine, environment, agriculture, monitoring, astrophysics, etc.). To improve the spatial resolution of hyperspectral images, Li *et al.* (2018) merged hyperspectral and multispectral images using a coupled Tucker decomposition with a sparse core (coupled sparse tensor factorization (CSTF)) (see Table I.4);

– design of semi-blind receivers for point-to-point or cooperative MIMO communication systems based on tensor models; see the overviews by de Almeida *et al.* (2016) and da Costa *et al.* (2018);

– modeling and identification of nonlinear systems via a tensor representation of Volterra kernels or Wiener–Hammerstein systems (see, for example, Kibangou and Favier 2009a, 2010; Favier and Kibangou 2009; Favier and Bouilloc 2009, 2010; Favier *et al.* 2012a);

– identification of tensor-based separable trilinear systems that are linear with respect to (w.r.t.) the input signal and trilinear w.r.t. the coefficients of the global impulse response, modeled as a Kronecker product of three individual impulse responses (Elisei-Iliescu *et al.* 2020). Note that such systems are to be compared with third-order Volterra filters that are linear w.r.t. the Volterra kernel coefficients and trilinear w.r.t. the input signal;

– facial recognition, based on face tensors, for purposes of authentication and identification in surveillance systems. For facial recognition, photos of people to recognize are stored in a database with different lighting conditions, different facial expressions, from multiple angles, for each individual. In Vasilescu and Terzopoulos (2002), the tensor of facial images is of order five, with dimensions: $28 \times 5 \times 3 \times 3 \times 7943$, corresponding to the modes: *people* $\times$ *views* $\times$ *illumination* $\times$ *expressions* $\times$ *pixels per image*. For an overview of various facial recognition systems, see Arachchilage and Izquierdo (2020);

---

4 Electroencephalography (EEG), magnetoencephalography (MEG), electrocardiography (ECG) and electrooculography (EOG).

– tensor-based anomaly detection used in monitoring and surveillance systems.

Table I.1 presents a few examples of signal and image tensors, specifying the nature of the modes in each case.

Signals	Modes	References
Antenna processing	space (antennas) × time × sensor subnetwork space × time × polarization	(Sidiropoulos *et al.* 2000a) (Raimondi *et al.* 2017)
Digital communications	space (antennas) × time × code antennas × blocks × symbol periods × code × frequencies	(Sidiropoulos *et al.* 2000b) (Favier and de Almeida 2014b)
ECG	space (electrodes) × time × frequencies	(Acar *et al.* 2007; Padhy *et al.* 2019)
EEG	space (electrodes) × time × frequencies × subjects or trials	(Becker *et al.* 2014; Cong *et al.* 2015)
EEG + fMRI	subjects × electrodes × time + subjects × voxels (model with matrix and tensor factorizations coupled via the "subjects" mode)	(Acar *et al.* 2017)
Images	Modes	References
Color images	space (width) × space (height) × channel (colors)	
Videos in grayscale	space (width) × space (height) × time	
Videos in color	space × space × channel × time	
Hyperspectral images	space × space × spectral bands	(Makantasis *et al.* 2018)
Computer vision	people × views × illumination × expressions × pixels	(Vasilescu and Terzopoulos 2002)

Table I.1. *Signal and image tensors*

Other fields of application are considered in Table I.2.

Below, we give some details about the application concerning recommendation systems, which play an important role in various websites. The goal of these systems is to help users to select *items* from *tags* that have been assigned to each item by users. These items could, for example, be movies, books, musical recordings, webpages, products for sale on an e-commerce site, etc. A standard recommendation system is based on the three following modes: *users* × *items* × *tags*.

Collaborative filtering techniques use the opinions of a set of people, or assessments from these people based on a rating system, to generate a list of recommendations for a specific user. This type of filtering is, for example, used by websites like Netflix for renting DVDs. Collaborative filtering methods are classified into three categories, depending on whether the filtering is based on (a) history and a similarity metric; (b) a model based on matrix factorization using algorithms like SVD or non-negative matrix factorization (NMF); (c) some combination of both, known as hybrid collaborative filtering techniques. See Luo *et al.* (2014) and Bokde *et al.* (2015) for approaches based on matrix factorization.

Other so-called passive filtering techniques exploit the data of a matrix of relations between items to deduce recommendations for a user from correlations between items and the user's previous choices, without using any kind of rating system. This is known as a content-based approach.

Domains	Modes	References
Phonetics	subjects × vowels × formants	(Harshman 1970)
Chemometrics (fluorescence)	excitation × emission × samples (excitation/emission wavelengths)	(Bro 1997, 2006; Smilde *et al.* 2004)
Contextual recommendation systems	users × items × tags × context 1 × ⋯ × context N	(Rendle and Schmidt-Thieme 2010) (Symeonidis and Zioupos 2016) (Frolov and Oseledets 2017)
Transportation (speed measurements)	Space (sensors) × time (days) × time (weeks) (periods of 15s and 24h)	(Goulart *et al.* 2017) (Tan *et al.* 2013; Ran *et al.* 2016)
Music	types of music × frequencies × frequencies users × keywords × songs recordings × (audio) characteristics × segments	(Panagakis *et al.* 2010) (Nanopoulos *et al.* 2010) (Benetos and Kotropoulos 2008)
Bioinformatics	medicine × targets × diseases	(Wang *et al.* 2019)

Table I.2. *Other fields of application*

Recommendation systems can also use information about the users (age, nationality, geographic location, participation on social networks, etc.) and the items themselves (types of music, types of film, classes of hotels, etc.). This is called contextual information. Taking this additional information into account allows the relevance of the recommendations to be improved, at the cost of increasing the dimensionality and the complexity of the data representation model and, therefore, of the processing algorithms. This is why tensor approaches are so important for this type of application today. Note that, for recommendation systems, the data tensors are sparse. Consequently, some tags can be automatically generated by the system based on similarity metrics between items. This is, for example, the case for music recommendations based on the acoustic characteristics of songs (Nanopoulos *et al.* 2010). Personalized tag recommendations take into account the user's profile, preferences, and interests. The system can also help the user select existing tags or create new ones (Rendle and Schmidt-Thieme 2010).

The articles by Bobadilla *et al.* (2013) and Frolov and Oseledets (2017) present various recommendation systems with many bibliographical references. Operating according to a similar principle as recommendation systems, social network websites, such as Wikipedia, Facebook, or Twitter, allow different types of data to be exchanged and shared, content to be produced and connections to be established.

I.4. With what tensor decompositions?

It is important to note that, for an Nth-order tensor $\mathcal{X} \in \mathbb{K}^{I_1 \times \cdots \times I_N}$, the number of elements is $\prod_{n=1}^{N} I_n$, and, assuming $I_n = I$ for $n \in \langle N \rangle$, this number becomes I^N, which induces an exponential increase with the tensor order N. This is called the curse of dimensionality (Oseledets and Tyrtyshnikov 2009). For big data tensors, tensor decompositions play a fundamental role in alleviating this curse of dimensionality, due to the fact that the number of parameters that characterize the decompositions is generally much smaller than the number of elements in the original tensor.

We now introduce three basic decompositions: PARAFAC/CANDECOMP/CPD, TD and TT[5]. The first two are studied in depth in Chapter 5, whereas the third, briefly introduced in Chapter 3, will be considered in more detail in Volume 3.

Table I.3 gives the expression of the element $x_{i_1,\cdots,i_N}$ of a tensor $\mathcal{X} \in \mathbb{K}^{I_1 \times \cdots \times I_N}$ of order N and size $I_1 \times \cdots \times I_N$, either real ($\mathbb{K} = \mathbb{R}$) or complex ($\mathbb{K} = \mathbb{C}$), for each of the three decompositions cited above. Their parametric complexity is compared in terms of the size of each matrix and tensor factor, assuming $I_n = I$ and $R_n = R$ for all $n \in \langle N \rangle$.

Figures I.1–I.3 show graphical representations of the PARAFAC model $[\![\mathbf{A}^{(1)}, \mathbf{A}^{(2)}, \mathbf{A}^{(3)}; R]\!]$ and the TD model $[\![\mathcal{G}; \mathbf{A}^{(1)}, \mathbf{A}^{(2)}, \mathbf{A}^{(3)}]\!]$ for a third-order tensor $\mathcal{X} \in \mathbb{K}^{I_1 \times I_2 \times I_3}$, and of the TT model $[\![\mathbf{A}^{(1)}, \mathcal{A}^{(2)}, \mathcal{A}^{(3)}, \mathbf{A}^{(4)}]\!]$ for a fourth-order tensor $\mathcal{X} \in \mathbb{K}^{I_1 \times I_2 \times I_3 \times I_4}$. In the case of the PARAFAC model, we define $\mathbf{A}^{(n)} \triangleq \left[\mathbf{a}_1^{(n)}, \cdots, \mathbf{a}_R^{(n)}\right] \in \mathbb{K}^{I_n \times R}$ using its columns, for $n \in \{1, 2, 3\}$.

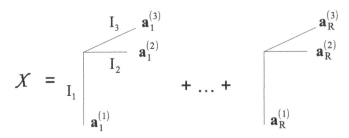

Figure I.1. *Third-order PARAFAC model*

We can make a few remarks about each of these decompositions:

– The PARAFAC decomposition (Harshman 1970), also known as CANDECOMP (Carroll and Chang 1970) or CPD (Hitchcock 1927), of a Nth-order tensor $\mathcal{X}$ is a sum

---

5 PARAFAC for parallel factors; CANDECOMP for canonical decomposition; CPD for canonical polyadic decomposition; TD for Tucker decomposition; TT for tensor train.

of R rank-one tensors, each defined as the outer product of one column from each of the N matrix factors $\mathbf{A}^{(n)} \in \mathbb{K}^{I_n \times R}$. When R is minimal, it is called the rank of the tensor. If the matrix factors satisfy certain conditions, this decomposition has the essential uniqueness property. See Figure I.1 for a third-order tensor $(N = 3)$, and Chapter 5 for a detailed presentation.

Decompositions	Notation	Element $x_{i_1, \cdots, i_N}$
PARAFAC / CPD	$[\![\mathbf{A}^{(1)}, \cdots, \mathbf{A}^{(N)}; R]\!]$	$\sum\limits_{r=1}^{R} \prod\limits_{n=1}^{N} a_{i_n, r}^{(n)}$
TD	$[\![\mathcal{G}; \mathbf{A}^{(1)}, \cdots, \mathbf{A}^{(N)}]\!]$	$\sum\limits_{r_1=1}^{R_1} \cdots \sum\limits_{r_N=1}^{R_N} g_{r_1, \cdots, r_N} \prod\limits_{n=1}^{N} a_{i_n, r_n}^{(n)}$
TT	$[\![\mathbf{A}^{(1)}, \mathcal{A}^{(2)}, \cdots \\ \cdots, \mathcal{A}^{(N-1)}, \mathbf{A}^{(N)}]\!]$	$\sum\limits_{r_1=1}^{R_1} \cdots \sum\limits_{r_{N-1}=1}^{R_{N-1}} \prod\limits_{n=1}^{N} a_{r_{n-1}, i_n, r_n}^{(n)}$

Decompositions	Parameters	Complexity
CPD	$\mathbf{A}^{(n)} \in \mathbb{K}^{I_n \times R}, \forall n \in \langle N \rangle$	$O(NIR)$
TD	$\mathcal{G} \in \mathbb{K}^{R_1 \times \cdots \times R_N} \\ \mathbf{A}^{(n)} \in \mathbb{K}^{I_n \times R_n}, \forall n \in \langle N \rangle$	$O(NIR + R^N)$
TT	$\mathbf{A}^{(1)} \in \mathbb{K}^{I_1 \times R_1}, \mathbf{A}^{(N)} \in \mathbb{K}^{R_{N-1} \times I_N} \\ \mathcal{A}^{(n)} \in \mathbb{K}^{R_{n-1} \times I_n \times R_n} \\ \forall n \in \{2, 3, \cdots, N-1\}$	$O(2IR \\ + (N-2)IR^2)$

Table I.3. *Parametric complexity of the CPD, TD, and TT decompositions*

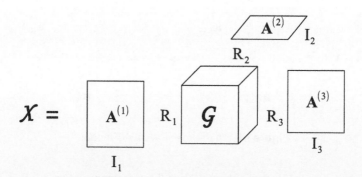

Figure I.2. *Third-order Tucker model*

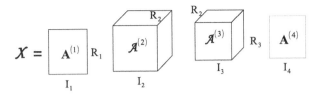

Figure I.3. *Fourth-order TT model*

– The Tucker decomposition (Tucker 1966) can be viewed as a generalization of the PARAFAC decomposition that takes into account all the interactions between the columns of the matrix factors $\mathbf{A}^{(n)} \in \mathbb{K}^{I_n \times R_n}$ via the introduction of a core tensor $\mathcal{G} \in \mathbb{K}^{R_1 \times \cdots \times R_N}$. This decomposition is not unique in general. Note that, if $R_n \leq I_n$ for $\forall n \in \langle N \rangle$, then the core tensor $\mathcal{G}$ provides a compressed form of $\mathcal{X}$. If R_n, for $n \in \langle N \rangle$, is chosen as the rank of the mode-n matrix unfolding[6] of $\mathcal{X}$, then the N-tuple $(R_1, \cdots, R_N)$ is minimal, and it is called the multilinear rank of the tensor.

Such a Tucker decomposition can be obtained using the truncated high-order SVD (THOSVD), under the constraint of column-orthonormal matrices $\mathbf{A}^{(n)}$ (de Lathauwer *et al.* 2000a). This algorithm is described in section 5.2.1.8.

See Figure I.2 for a third-order tensor, and Chapter 5 for a detailed presentation.

– The TT decomposition (Oseledets 2011) is composed of a train of third-order tensors $\mathcal{A}^{(n)} \in \mathbb{K}^{R_{n-1} \times I_n \times R_n}$, for $n \in \{2, 3, \cdots, N-1\}$, the first and last carriages of the train being matrices $\mathbf{A}^{(1)} \in \mathbb{K}^{I_1 \times R_1}$ and $\mathbf{A}^{(N)} \in \mathbb{K}^{R_{N-1} \times I_N}$, which implies $r_0 = r_N = 1$, and therefore $a_{r_0, i_1, r_1}^{(1)} = a_{i_1, r_1}^{(1)}$ and $a_{r_{N-1}, i_N, r_N}^{(N)} = a_{r_{N-1}, i_N}^{(N)}$. The dimensions R_n, for $n \in \langle N-1 \rangle$, called the TT ranks, are given by the ranks of some matrix unfoldings of the original tensor.

This decomposition has been used to solve the tensor completion problem (Grasedyck *et al.* 2015; Bengua *et al.* 2017), for facial recognition (Brandoni and Simoncini 2020) and for modeling MIMO communication channels (Zniyed *et al.* 2020), among many other applications. A brief description of the TT decomposition is given in section 3.13.4 using the mode-(p, n) product. Note that a specific SVD-based algorithm, called TT-SVD, was proposed by Oseledets (2011) for computing a TT decomposition.

This decomposition and the hierarchical Tucker (HT) one (Grasedyck and Hackbush 2011; Ballani *et al.* 2013) are special cases of tensor networks (TNs) (Cichocki 2014), as will be discussed in more detail in the next volume.

---

6 See definition [3.41], in Chapter 3, of the mode-n matrix unfolding $\mathbf{X}_n$ of a tensor $\mathcal{X}$, whose columns are the mode-n vectors obtained by fixing all but n indices.

From this brief description of the three tensor models, one can conclude that, unlike matrices, the notion of rank is not unique for tensors, since it depends on the decomposition used. Thus, as mentioned above, one defines the tensor rank (also called the canonical rank or Kruskal's rank) associated with the PARAFAC decomposition, the multilinear rank that relies on the Tucker's model, and the TT-ranks linked with the TT decomposition.

It is important to note that the number of characteristic parameters of the PARAFAC and TT decompositions is proportional to N, the order of the tensor, whereas the parametric complexity of the Tucker decomposition increases exponentially with N. This is why the first two decompositions are especially valuable for large-scale problems. Although the Tucker model is not unique in general, imposing an orthogonality constraint on the matrix factors yields the HOSVD decomposition, a truncated form of which gives an approximate solution to the best low multilinear rank approximation problem (de Lathauwer *et al.* 2000a). This solution, which is based on an *a priori* choice of the dimensions R_n of the core tensor, is to be compared with the truncated SVD in the matrix case, although it does not have the same optimality property. It is widely used to reduce the parametric complexity of data tensors.

From the above, it can be concluded that the TT model combines the advantages of the other two decompositions, in terms of parametric complexity (like PARAFAC) and numerical stability (like Tucker's model), due to a parameter estimation algorithm based on a calculation of SVDs.

To illustrate the use of the PARAFAC decomposition, let us consider the case of multi-user mobile communications with a CDMA (code-division multiple access) encoding system. The multiple access technique allows multiple emitters to simultaneously transmit information over the same communication channel by assigning a code to each emitter. The information is transmitted as symbols $s_{n,m}$, with $n \in \langle N \rangle$ and $m \in \langle M \rangle$, where N and M are the number of transmission time slots, i.e. the number of symbol periods, and the number of emitting antennas, respectively. The symbols belong to a finite alphabet that depends on the modulation being used. They are encoded with a space-time coding that introduces code diversity by repeating each symbol P times with a code $c_{p,m}$ assigned to the mth emitting antenna, $p \in \langle P \rangle$, where P denotes the length of the spreading code. The signal received by the kth receiving antenna, during the nth symbol period and the pth chip period, is a linear combination of the symbols encoded and transmitted by the M emitting antennas:

$$x_{k,n,p} = \sum_{m=1}^{M} h_{k,m} s_{n,m} c_{p,m}, \qquad [I.1]$$

where $h_{k,m}$ is the fading coefficient of the communication channel between the receiving antenna k and the emitting antenna m.

The received signals, which are complex-valued, therefore form a third-order tensor $\mathcal{X} \in \mathbb{C}^{K \times N \times P}$ whose modes are: *space* × *time* × *code*, associated with the indices (k, n, p). This signal tensor satisfies a PARAFAC decomposition $[\![\mathbf{H}, \mathbf{S}, \mathbf{C}; M]\!]$ whose rank is equal to the number M of emitting antennas and whose matrix factors are the channel ($\mathbf{H} \in \mathbb{C}^{K \times M}$), the matrix of transmitted symbols ($\mathbf{S} \in \mathbb{C}^{N \times M}$) and the coding matrix ($\mathbf{C} \in \mathbb{C}^{P \times M}$). This example is a simplified form of the DS-CDMA (direct-sequence CDMA) system proposed by (Sidiropoulos *et al.* 2000b).

I.5. With what cost functions and optimization algorithms?

We will now briefly describe the most common processing operations carried out with tensors, as well as some of the optimization algorithms that are used. It is important to first present the preprocessing operations that need to be performed. Preprocessing typically involves data centering operations (offset elimination), scaling of non-homogeneous data, suppression of outliers and artifacts, image adjustment (size, brightness, contrast, alignment, etc.), denoising, signal transformation using certain transforms (wavelets, Fourier, etc.), and finally, in some cases, the calculation of statistics of signals to be processed.

Preprocessing is fundamental, both to improve the quality of the estimated models and, therefore, of the subsequent processing operations, and to avoid numerical problems with optimization algorithms, such as conditioning problems that may cause the algorithms to fail to converge. Centering and scaling preprocessing operations are potentially problematic because they are interdependent and can be combined in several different ways. If data are missing, centering can also reduce the rank of the tensor model. For a more detailed description of these preprocessing operations, see Smilde *et al.* (2004).

For the processing operations themselves, we can distinguish between several different classes:

– supervised/non-supervised (blind or semi-blind), i.e. with or without training data, for example, to solve classification problems, or when *a priori* information, called a pilot sequence, is transmitted to the receiver for channel estimation;

– real-time (online)/batch (offline) processing;

– centralized/distributed;

– adaptive/blockwise (with respect to the data);

– with/without coupling of tensor and/or matrix models;

– with/without missing data.

It is important to distinguish batch processing, which is performed to analyze data recorded as signal and image sets, from the real-time processing required by wireless communication systems, recommendation systems, web searches and social networks. In real-time applications, the dimensionality of the model and the algorithmic complexity are predominant factors. The signals received by receiving antennas, the information exchanged between a website and the users and the messages exchanged between the users of a social network are time-dependent. For instance, a recommendation system interacts with the users in real-time, via a possible extension of an existing database by means of machine learning techniques. For a description of various applications of tensors for data mining and machine learning, see Anandkumar *et al.* (2014) and Sidiropoulos *et al.* (2017).

Tensor-based processings lead to various types of optimization algorithm as follows:

– constrained/unconstrained optimization;

– iterative/non-iterative, or closed-form;

– alternating/global;

– sequential/parallel.

Furthermore, depending on the information that is available *a priori*, different types of constraints can be taken into account in the cost function to be optimized: low rank, sparseness, non-negativity, orthogonality and differentiability/smoothness. In the case of constrained optimization, weights need to be chosen in the cost function according to the relative importance of each constraint and the quality of the *a priori* information that is available.

Table I.4 presents a few examples of cost functions that can be minimized for the parameter estimation of certain third-order tensor models (CPD, Tucker, coupled matrix Tucker (CMTucker) and coupled sparse tensor factorization (CSTF)), for the imputation of missing data in a tensor and for the estimation of a sparse data tensor with a low-rank constraint expressed in the form of the nuclear norm of the tensor.

REMARK I.1.– We can make the following remarks:

– the cost functions presented in Table I.4 correspond to data fitting criteria. These criteria, expressed in terms of tensor and matrix Frobenius norms ($\|.\|_F$), are quadratic in the difference between the data tensor $\mathcal{X}$ and the output of CPD and TD models, as well as between the data matrix $\mathbf{Y}$ and a matrix factorization model, in the case of the CMTucker model. They are trilinear and quadrilinear, respectively, with respect to the parameters of the CPD and TD models to be estimated, and bilinear with respect to the parameters of the matrix factorization model;

– for the missing data imputation problem using a CPD or TD model, the binary tensor $\mathcal{W}$, which has the same size as $\mathcal{X}$, is defined as:

$$w_{ijk} = \begin{cases} 1 & \text{if } x_{ijk} \text{ is known} \\ 0 & \text{if } x_{ijk} \text{ is missing} \end{cases} \qquad [\text{I.2}]$$

The purpose of the Hadamard product (denoted $\odot$) of $\mathcal{W}$, with the difference between $\mathcal{X}$ and the output of the CPD and TD models, is to fit the model to the available data only, ignoring any missing data for model estimation. This imputation problem, known as the tensor completion problem, was originally dealt with by Tomasi and Bro (2005) and Acar *et al.* (2011a) using a CPD model, followed by Filipovic and Jukic (2015) using a TD model. Various articles have discussed this problem in the context of different applications. An overview of the literature will be given in the next volume;

Problems Data	$\mathcal{X} \in \mathbb{K}^{I \times J \times K}, \mathbf{Y} \in \mathbb{K}^{I \times M}$
Estimation	**Cost functions**
CPD	$f(\mathbf{A}, \mathbf{B}, \mathbf{C}) = \left\| \mathcal{X} - [\![\mathbf{A}, \mathbf{B}, \mathbf{C}; R]\!] \right\|_F^2$
TD	$f(\mathcal{G}, \mathbf{A}, \mathbf{B}, \mathbf{C}) = \left\| \mathcal{X} - [\![\mathcal{G}; \mathbf{A}, \mathbf{B}, \mathbf{C}]\!] \right\|_F^2$
CMTucker	$f(\mathcal{G}, \mathbf{A}, \mathbf{B}, \mathbf{C}, \mathbf{U}) = \left\| \mathcal{X} - [\![\mathcal{G}; \mathbf{A}, \mathbf{B}, \mathbf{C}]\!] \right\|_F^2 + \left\| \mathbf{Y} - \mathbf{A}\mathbf{U}^T \right\|_F^2$
CSTF	$f(\mathcal{G}, \mathbf{W}, \mathbf{H}, \mathbf{S}) = \left\| \mathcal{X} - [\![\mathcal{G}; \mathbf{W}^*, \mathbf{H}^*, \mathbf{S}]\!] \right\|_F^2 +$ $+ \left\| \mathcal{Y} - [\![\mathcal{G}; \mathbf{W}, \mathbf{H}, \mathbf{S}^*]\!] \right\|_F^2 + \lambda \left\| \mathcal{G} \right\|_1$
Imputation	**Cost functions**
CPD	$f_{\mathcal{W}}(\mathbf{A}, \mathbf{B}, \mathbf{C}) = \left\| \mathcal{W} \odot (\mathcal{X} - [\![\mathbf{A}, \mathbf{B}, \mathbf{C}; R]\!]) \right\|_F^2$
TD	$f_{\mathcal{W}}(\mathcal{G}, \mathbf{A}, \mathbf{B}, \mathbf{C}) = \left\| \mathcal{W} \odot (\mathcal{X} - [\![\mathcal{G}; \mathbf{A}, \mathbf{B}, \mathbf{C}]\!]) \right\|_F^2$
Imputation with low-rank constraint	**Cost functions**
CPD	$f_{\mathcal{W}}(\mathbf{A}, \mathbf{B}, \mathbf{C}) = \left\| \mathcal{W} \odot (\mathcal{X} - [\![\mathbf{A}, \mathbf{B}, \mathbf{C}; R]\!]) \right\|_F^2 + \lambda \left\| \mathcal{X} \right\|_\star$
TD	$f_{\mathcal{W}}(\mathcal{G}, \mathbf{A}, \mathbf{B}, \mathbf{C}) = \left\| \mathcal{W} \odot (\mathcal{X} - [\![\mathcal{G}; \mathbf{A}, \mathbf{B}, \mathbf{C}]\!]) \right\|_F^2 + \lambda \left\| \mathcal{X} \right\|_\star$

Table I.4. *Cost functions for model estimation and recovery of missing data*

– for the imputation problem with the low-rank constraint, the term $\left\| \mathcal{X} \right\|_\star$ in the cost function replaces the low-rank constraint with the nuclear norm of $\mathcal{X}$, since the function rank($\mathcal{X}$) is not convex, and the nuclear norm is the closest convex approximation of the rank. In Liu *et al.* (2013), this term is replaced by $\sum_{n=1}^{3} \lambda_n \|\mathbf{X}_n\|_\star$, where $\mathbf{X}_n$ represents the mode-n unfolding of $\mathcal{X}$[7];

– in the case of the CMTucker model, the coupling considered here relates to the first modes of the tensor $\mathcal{X}$ and the matrix $\mathbf{Y}$ of data via the common matrix factor $\mathbf{A}$.

Coupled matrix and tensor factorization (CMTF) models were introduced in Acar *et al.* (2011b) by coupling a CPD model with a matrix factorization and using the gradient descent algorithm to estimate the parameters. This type of model was used by Acar *et al.* (2017) to merge EEG and fMRI data with the goal of analyzing brain activity. The EEG signals are modeled with a normalized CPD model (see Chapter 5), whereas the fMRI data are modeled with a matrix factorization. The data are coupled through the *subjects* mode (see Table I.1). The cost function to be minimized is therefore given by:

$$f(\mathbf{g}, \mathbf{\Sigma}, \mathbf{A}, \mathbf{B}, \mathbf{C}, \mathbf{U}) = \left\| \mathcal{X} - [\![\mathbf{g}; \mathbf{A}, \mathbf{B}, \mathbf{C}]\!] \right\|_F^2 + \left\| \mathbf{Y} - \mathbf{A}\,\mathbf{\Sigma}\,\mathbf{U}^T \right\|_F^2$$
$$+ \alpha \|\mathbf{g}\|_1 + \alpha \|\boldsymbol{\sigma}\|_1, \qquad \text{[I.3]}$$

where the column vectors of the matrix factors $(\mathbf{A}, \mathbf{B}, \mathbf{C})$ have unit norm, $\mathbf{\Sigma}$ is a diagonal matrix whose diagonal elements are the coefficients of the vector $\boldsymbol{\sigma}$ and $\alpha > 0$ is a penalty parameter that allows the importance of the sparseness constraints on the weight vectors $(\mathbf{g}, \boldsymbol{\sigma})$ to be increased or decreased, modeled by means of the l_1 norm. The advantage of merging EEG and fMRI data with the criterion [I.3] is that the acquisition and observation methods are complementary in terms of resolution, since EEG signals have a high temporal resolution but low spatial resolution, while fMRI imaging provides high spatial resolution;

– in the case of the CSTF model (Li *et al.* 2018), the tensor of high-resolution hyperspectral images (HR-HSI) is represented using a third-order Tucker model that has a sparse core ($\mathcal{X} = \mathcal{G} \times_1 \mathbf{W} \times_2 \mathbf{H} \times_3 \mathbf{S}$), with the following modes: *space* (*width*) $\times$ *space* (*height*) $\times$ *spectral bands*. The matrices $\mathbf{W} \in \mathbb{R}^{M \times n_w}$, $\mathbf{H} \in \mathbb{R}^{N \times n_h}$ and $\mathbf{S} \in \mathbb{R}^{P \times n_s}$ denote the dictionaries for the width, height and spectral modes, composed of n_w, n_h and n_s atoms, respectively, and the core tensor $\mathcal{G}$ contains the coefficients relative to the three dictionaries. The matrices $\mathbf{W}^*, \mathbf{H}^*$ and $\mathbf{S}^*$ are spatially and spectrally subsampled versions with respect to each mode. The term λ is a regularization parameter for the sparseness constraint on the core tensor, expressed in terms of the l_1 norm of $\mathcal{G}$.

The criteria listed in Table I.4 can be globally minimized using a nonlinear optimization method such as a gradient descent algorithm (with fixed or optimal step size), or the Gauss–Newton and Levenberg–Marquardt algorithms, the latter being a

---

7 See definition [3.41] of the unfolding $\mathbf{X}_n$, and definitions [1.65] and [1.67] of the Frobenius norm ($\|.\|_F$) and the nuclear norm ($\|.\|_*$) of a matrix; for a tensor, see section 3.16.

regularized form of the former. In the case of constrained optimization, the augmented Lagrangian method is very often used, as it allows the constrained optimization problem to be transformed into a sequence of unconstrained optimization problems.

The drawbacks of these optimization methods include slow convergence for gradient-type algorithms and high numerical complexity for the Gauss–Newton and Levenberg–Marquardt algorithms due to the need to compute the Jacobian matrix of the criterion w.r.t. the parameters being estimated, as well as the inverse of a large matrix.

Alternating optimization methods are therefore often used instead of a global optimization w.r.t. all matrix and tensor factors to be estimated. These iterative methods perform a sequence of separate optimizations of criteria linear in each unknown factor while fixing the other factors with the values estimated at previous iterations. An example is the standard ALS (alternating least squares) algorithm, presented in Chapter 5 for estimating PARAFAC models. For constrained optimization, the alternating direction method of multipliers (ADMM) is often used (Boyd *et al.* 2011).

To complete this introductory chapter, let us outline the key knowledge needed to employ tensor tools, whose presentation constitutes the main objective of this second volume:

– arrangement (also called reshaping) operations that express the data tensor as a vector (vectorization), a matrix (matricization), or a lower order tensor by combining modes; conversely, the tensorization and Hankelization operations allow us to construct tensors from data contained in large vectors or matrices;

– tensor operations such as transposition, symmetrization, Hadamard and Kronecker products, inversion and pseudo-inversion;

– the notions of eigenvalue and singular value of a tensor;

– tensor decompositions/models, and their uniqueness properties;

– algorithms used to solve dimensionality reduction problems and, hence, best low-rank approximation, parameter estimation and missing data imputation. This algorithmic aspect linked to tensors will be explored in more depth in Volume 3.

I.6. Brief description of content

Tensor operations and decompositions often use matrix tools, so we will begin by reviewing some matrix decompositions in Chapter 1, going into further detail on eigenvalue decomposition (EVD) and SVD, as well as a few of their applications.

The Hadamard, Kronecker and Khatri–Rao matrix products are presented in detail in Chapter 2, together with many of their properties and a few relations between them. To illustrate these operations, we will use them to represent first-order partial derivatives of a function, and solve matrix equations, such as Sylvester and Lyapunov ones. This chapter also introduces an index convention that is very useful for tensor computations. This convention, which generalizes Einstein's summation convention (Pollock 2011), will be used to represent various matrix products and to prove some matrix product vectorization formulae, as well as various relations between the Kronecker, Khatri-Rao and Hadamard products. It will be used in Chapter 3 for tensor matricization and vectorization in an original way, as well as in Chapter 5 to establish matrix forms of the Tucker and PARAFAC decompositions.

Chapter 3 presents various sets of tensors before introducing the notions of matrix and tensor slices and of mode combination on which reshaping operations are based. The key tensor operations listed above are then presented. Several links between products of tensors and systems of tensor equations are also outlined, and some of these systems are solved with the least squares method.

Chapter 4 is dedicated to introducing the notions of eigenvalue and singular value for tensors. The problem of the best rank-one approximation of a tensor is also considered.

In Chapter 5, we will give a detailed presentation of various tensor decompositions, with a particular focus on the basic Tucker and CPD decompositions, which can be viewed as generalizations of matrix SVD to tensors of order greater than two. Block tensor models and constrained tensor models will also be described, as well as certain variants, such as HOSVD and BTD (block term decomposition). CPD-type decompositions are generally used to estimate latent parameters, whereas Tucker decomposition is often used to estimate modal subspaces and reduce the dimensionality via low multilinear rank approximation and truncated HOSVD.

A description of the ALS algorithm for parameter estimation of PARAFAC models will also be given. The uniqueness properties of the Tucker and CDP decompositions will be presented, as well as the various notions of the rank of a tensor. The chapter will end with illustrations of BTD and CPD decompositions for the tensor modeling of multidimensional harmonics, the problem of source separation in an instantaneous linear mixture and the modeling and estimation of a finite impulse response (FIR) linear system, using a tensor of fourth-order cumulants of the system output.

High-order cumulants of random signals that can be viewed as tensors play a central role in various signal processing applications, as illustrated in Chapter 5. This motivated us to include an Appendix to present a brief overview of some basic results concerning the higher order statistics (HOS) of random signals, with two applications to the HOS-based estimation of a linear time-invariant system and a homogeneous quadratic system.

1

Matrix Decompositions

1.1. Introduction

The goal of this chapter is to give an overview of the most important matrix decompositions, with a more detailed presentation of the eigenvalue decomposition (EVD) and singular value decomposition (SVD), as well as some of their applications. Matrix decompositions (also called factorizations) play a key role in matrix computation, in particular, for computing the pseudo-inverse of a matrix (see section 1.5.4), the low-rank approximation of a matrix (see section 1.5.7), the solution of a system of linear equations using the least squares (LS) method (see section 1.5.9), or for parametric estimation of nonlinear models using the ALS method, as illustrated in Chapter 5 with the estimation of tensor models.

Matrix decompositions have two goals. The first is to factorize a given matrix with structured factor matrices that are easier to invert, and the second is to reduce the dimensionality, in order to reduce both the memory capacity required to store the data and the computational cost of the data processing algorithms. After giving a brief overview of the most common decompositions, we will recall a few results about the eigenvalues of a matrix, and then present the EVD decomposition of a square matrix. The use of this decomposition will be illustrated by computing the powers of a matrix, a matrix polynomial, a state transition matrix and the transfer function of a discrete-time linear system.

The $\mathbf{URV}^H$ decomposition of a rectangular matrix will then be introduced, followed by a presentation of the SVD decomposition. The latter can be viewed as a special case of the $\mathbf{URV}^H$ decomposition with $\mathbf{U}$ and $\mathbf{V}$ orthogonal (respectively, unitary) in the case of a real (respectively, complex) matrix and $\mathbf{R}$ pseudo-diagonal. The SVD can also be viewed as an extension of the EVD for diagonalizing rectangular matrices.

We will present several results relating to the SVD as the links between SVD and the fundamental spaces of a matrix and certain matrix norms. Applications of the SVD to compute the pseudo-inverse of a matrix and hence the LS estimator, as well as a low-rank matrix approximation, will also be described. Polar decomposition will be demonstrated using the SVD. The connection between SVD and principal component analysis (PCA) will be established, with an application to data compression by reducing the dimensionality of a data matrix. The use of the SVD for the blind source separation (BSS) problem will also be considered. Finally, the CUR decomposition of a matrix, which is based on selecting certain columns and rows, will be briefly described.

1.2. Overview of the most common matrix decompositions

Table 1.1 presents the most common matrix decompositions as products of matrices. These decompositions differ from one another in terms of the structural properties of their factor matrices (diagonal/pseudo-diagonal, upper/lower triangular, orthogonal/unitary).

The EVD, the SVD and the polar decomposition are presented in detail in sections 1.3 and 1.5. The $\mathbf{URV}^H$ and CUR decompositions are also described in this chapter. Full-rank decomposition is discussed in Chapter 3, in section 3.15.4. For other decompositions, see Lawson and Hanson (1974), Favier (1982, 2019), Golub and Van Loan (1983), Lancaster and Tismenetsky (1985), Horn and Johnson (1985, 1991) and Meyer (2000), among many others.

We can make the following remarks:

– A square root of a symmetric positive semi-definite square matrix $\mathbf{A} \in \mathbb{R}_S^{I \times I}$ is defined as a square matrix $\mathbf{S} \in \mathbb{R}^{I \times I}$ such that $\mathbf{A} = \mathbf{SS}^T$. We often write $\mathbf{S} = \mathbf{A}^{1/2}$. The square root is not unique, since any matrix $\mathbf{SQ}$ with $\mathbf{Q}$ orthogonal is also a square root of $\mathbf{A}$.

The Cholesky decomposition gives a square root in lower triangular ($\mathbf{L}$) or upper triangular ($\mathbf{U}$) form for a symmetric positive semi-definite matrix $\mathbf{A}$. The matrix $\mathbf{L}$ is computed row by row, from left to right and top to bottom, whereas $\mathbf{U}$ is computed column by column, from right to left and bottom to top. See Favier (1982) for a detailed presentation. The factors $\mathbf{L}$ and $\mathbf{U}$ are unique if $\mathbf{A}$ is positive definite. In the complex case, Cholesky decompositions are expressed in the form $\mathbf{A} = \mathbf{LL}^H = \mathbf{UU}^H$.

The UD decomposition is obtained by modifying the Cholesky decomposition so that the factor matrices $\mathbf{L}$ and $\mathbf{U}$ are unit lower triangular and unit upper triangular, respectively, and $\mathbf{D}$ is diagonal.

– The Schur decomposition is written as $\mathbf{A} = \mathbf{UTU}^H$ (respectively, $\mathbf{A} = \mathbf{QTQ}^T$), where $\mathbf{U}$ is unitary (respectively, $\mathbf{Q}$ is orthogonal), and $\mathbf{T}$ is upper or lower triangular with the eigenvalues of $\mathbf{A}$ along the diagonal. This decomposition, which can also be written $\mathbf{U}^H \mathbf{AU} = \mathbf{T}$ (respectively, $\mathbf{Q}^T \mathbf{AQ} = \mathbf{T}$), shows that every complex (respectively, real) square matrix is unitarily (respectively, orthogonally) similar to a triangular matrix (see Table 1.6).

– The decomposition $\mathbf{A} = \mathbf{LU}$, where $\mathbf{A} \in \mathbb{C}^{I \times I}$ is of rank R, $\mathbf{L} \in \mathbb{C}^{I \times I}$ is lower triangular and $\mathbf{U} \in \mathbb{C}^{I \times I}$ is upper triangular, satisfies the property that $\mathbf{U}$ or $\mathbf{L}$ is non-singular. This decomposition does not always exist.

In the case of a non-singular matrix $\mathbf{A} \in \mathbb{R}^{I \times I}$, a variant of the LU decomposition is $\mathbf{A} = \mathbf{LDU}$, where $\mathbf{L}$ and $\mathbf{U}$ are unit lower and upper triangular, respectively, and $\mathbf{D}$ is diagonal. In this case, the matrices $\mathbf{L}$ and $\mathbf{U}$ are unique.

– The QR decomposition can be computed using various orthogonalization methods based on Householder, Givens or Gram–Schmidt transformations[1].

– The LU and QR decompositions are used to solve systems of linear equations of the form $\mathbf{Ax} = \mathbf{b}$ using the LS method.

Using the decomposition $\mathbf{A} = \mathbf{LU}$, with $\mathbf{A} \in \mathbb{R}^{I \times I}$, the original system is replaced by two triangular systems $\mathbf{Ly} = \mathbf{b}$ and $\mathbf{Ux} = \mathbf{y}$. To solve these new systems, the square triangular matrices $\mathbf{L}$ and $\mathbf{U}$ are inverted using forward and backward substitution algorithms, respectively, instead of inverting $\mathbf{A}$.

Similarly, using the decomposition $\mathbf{A} = \mathbf{QR}$, with $\mathbf{A} \in \mathbb{R}^{I \times J}$ assumed to have full column rank ($I \geq J$), the original system of equations can be solved by inverting a triangular matrix. To see this, set:

$$\mathbf{Q}^T \mathbf{A} = \mathbf{R} = \begin{bmatrix} \mathbf{R}_1 \\ \mathbf{0} \end{bmatrix} \ , \ \mathbf{Q}^T \mathbf{b} = \begin{bmatrix} \mathbf{c} \\ \mathbf{d} \end{bmatrix}, \qquad [1.1]$$

where $\mathbf{R}_1 \in \mathbb{R}^{J \times J}$ is non-singular, $\mathbf{c} \in \mathbb{R}^J$ and $\mathbf{d} \in \mathbb{R}^{I-J}$. Using the fact that pre-multiplying a vector by an orthogonal matrix preserves its Euclidean norm ($\|\mathbf{Qy}\|_2 = \|\mathbf{y}\|_2$), the LS criterion to minimize can be rewritten as follows:

$$\|\mathbf{Ax} - \mathbf{b}\|_2^2 = \|\mathbf{Q}^T(\mathbf{Ax} - \mathbf{b})\|_2^2 = \|\mathbf{R}_1\mathbf{x} - \mathbf{c}\|_2^2 + \|\mathbf{d}\|_2^2. \qquad [1.2]$$

---

1 A matrix $\mathbf{Q} \in \mathbb{R}^{I \times J}$, with $J \leq I$, is said to be column-wise orthonormal (or simply column orthonormal) if its column vectors form an orthonormal set ($\mathbf{Q}^T \mathbf{Q} = \mathbf{I}_J$). Similarly, when $I \leq J$, the matrix $\mathbf{Q}$ is said to be row-wise orthonormal (or simply row orthonormal) if its row vectors form an orthonormal set ($\mathbf{QQ}^T = \mathbf{I}_I$).

Matrices	Decompositions	Properties
$\mathbf{A} \in \mathbb{R}_S^{I \times I}, \mathbf{A} \geq 0$	**Square root** $\mathbf{A} = \mathbf{A}^{1/2}(\mathbf{A}^{1/2})^T = \mathbf{A}^{1/2}\mathbf{A}^{T/2}$	$\mathbf{A}^{1/2} \geq 0$
$\mathbf{A} \in \mathbb{R}_S^{I \times I}, \mathbf{A} \geq 0$	**Cholesky** $\mathbf{A} = \mathbf{L}\mathbf{L}^T$ $\mathbf{A} = \mathbf{U}\mathbf{U}^T$	**L** lower triangular , $l_{ii} \geq 0$, $\forall i \in \langle I \rangle$ **U** upper triangular , $u_{ii} \geq 0$, $\forall i \in \langle I \rangle$
$\mathbf{A} \in \mathbb{R}_S^{I \times I}, \mathbf{A} > 0$	**UD** $\mathbf{A} = \mathbf{L}\mathbf{D}\mathbf{L}^T$ $\mathbf{A} = \mathbf{U}\mathbf{D}\mathbf{U}^T$	**L** unit lower triangular , $l_{ii} = 1$, $\forall i \in \langle I \rangle$ **U** unit upper triangular , $u_{ii} = 1$, $\forall i \in \langle I \rangle$ **D** diagonal , $d_{ii} > 0$, $\forall i \in \langle I \rangle$
$\mathbf{A} \in \mathbb{C}^{I \times I}$ $\mathbf{A} \in \mathbb{R}^{I \times I}$	**Schur** $\mathbf{A} = \mathbf{U}\mathbf{T}\mathbf{U}^H$ $\mathbf{A} = \mathbf{Q}\mathbf{T}\mathbf{Q}^T$	**U** unitary **T** upper triang. , $t_{ii} = \lambda_i \in \text{sp}(\mathbf{A})$, $\forall i \in \langle I \rangle$ **Q** orthogonal
$\mathbf{A} \in \mathbb{C}^{I \times I}$	**LU** $\mathbf{A} = \mathbf{L}\mathbf{U}$	**L** lower triangular **U** upper triangular
$\mathbf{A} \in \mathbb{C}^{I \times I}, r_{\mathbf{A}} = I$	**LDU** $\mathbf{A} = \mathbf{L}\mathbf{D}\mathbf{U}$	**L** unit lower triangular , $l_{ii} = 1$, $\forall i \in \langle I \rangle$ **U** unit upper triangular , $u_{ii} = 1$, $\forall i \in \langle I \rangle$ **D** diagonal
$\mathbf{A} \in \mathbb{R}^{I \times J}, r_{\mathbf{A}} = J$ $\mathbf{A} \in \mathbb{C}^{I \times I}, r_{\mathbf{A}} = I$	**QR** $\mathbf{A} = \mathbf{Q}\mathbf{R}$ $\mathbf{A} = \mathbf{Q}\mathbf{R}$	**Q** column orthonormal ($\mathbf{Q}^T\mathbf{Q} = \mathbf{I}_J$) **R** upper triang. , $r_{jj} > 0$, $\forall j \in \langle J \rangle$ **Q** unitary ($\mathbf{Q}^H\mathbf{Q} = \mathbf{Q}\mathbf{Q}^H = \mathbf{I}_I$) **R** upper triang. , $r_{ii} > 0$, $\forall i \in \langle I \rangle$
$\mathbf{A} \in \mathbb{C}^{I \times J}, r_{\mathbf{A}} = R$	**full rank** $\mathbf{A} = \mathbf{F}\mathbf{G}$	$\mathbf{F} \in \mathbb{C}^{I \times R}, \mathbf{G} \in \mathbb{C}^{R \times J}, r_{\mathbf{F}} = r_{\mathbf{G}} = R$
$\mathbf{A} \in \mathbb{R}^{I \times I}$ $\mathbf{A} \in \mathbb{C}^{I \times I}$ **Hermitian** $\mathbf{A} \in \mathbb{R}_S^{I \times I}$ **symmetric**	**EVD** $\mathbf{A} = \mathbf{P}\mathbf{D}\mathbf{P}^{-1}$ $\mathbf{A} = \mathbf{P}\mathbf{D}\mathbf{P}^H$ $\mathbf{A} = \mathbf{P}\mathbf{D}\mathbf{P}^T$	**P** matrix of eigenvectors **D** diagonal matrix of eigenvalues **P** unitary **D** real diagonal **P** orthogonal **D** real diagonal
$\mathbf{A} \in \mathbb{C}^{I \times J}$	**SVD** $\mathbf{A} = \mathbf{U}\boldsymbol{\Sigma}\mathbf{V}^H$	$\mathbf{U} \in \mathbb{C}^{I \times I}, \mathbf{V} \in \mathbb{C}^{J \times J}$ unitary $\boldsymbol{\Sigma} \in \mathbb{R}^{I \times J}$ positive pseudo-diagonal
$\mathbf{A} \in \mathbb{C}^{I \times I}$	**Polar** $\mathbf{A} = \mathbf{Q}\mathbf{P}$ or $\mathbf{A} = \mathbf{S}\mathbf{Q}$	**Q** unitary $\mathbf{P}, \mathbf{S} \geq 0$ Hermitian

Table 1.1. *The most common matrix decompositions*

The LS solution is therefore given by $\mathbf{x}_{LS} = \mathbf{R}_1^{-1}\mathbf{c}$, where $\mathbf{R}_1$ is upper triangular, and the estimation error is equal to $\|\mathbf{d}\|_2^2$.

For a discussion of solving the LS problem with SVD, see section 1.5.9.

1.3. Eigenvalue decomposition

1.3.1. *Reminders about the eigenvalues of a matrix*

Given a square matrix $\mathbf{A} \in \mathbb{K}^{I \times I}$, a scalar λ_k is said to be an eigenvalue of $\mathbf{A}$ if there exists a non-zero vector $\mathbf{v}_k \in \mathbb{K}^I$ such that:

$$\mathbf{A}\mathbf{v}_k = \lambda_k \mathbf{v}_k. \qquad [1.3]$$

The vector $\mathbf{v}_k$ is called a right eigenvector (or simply an eigenvector) of $\mathbf{A}$ associated with the eigenvalue λ_k.

This definition can also be written as follows:

$$(\mathbf{A} - \lambda_k \mathbf{I}_I)\mathbf{v}_k = \mathbf{0}_I, \quad \text{with} \quad \mathbf{v}_k \neq \mathbf{0}_I, \qquad [1.4]$$

which implies that the matrix $\mathbf{A} - \lambda_k \mathbf{I}_I$ is singular, and therefore has determinant zero. We can then conclude that the eigenvalues of $\mathbf{A}$ are the I roots of the characteristic equation:

$$p_{\mathbf{A}}(\lambda) = \det(\lambda \mathbf{I}_I - \mathbf{A}) = 0. \qquad [1.5]$$

Similarly, a left eigenvector of $\mathbf{A}$ associated with the eigenvalue μ_i is a vector $\mathbf{u}_i \neq \mathbf{0}_I$ such that[2]:

$$\mathbf{u}_i^H \mathbf{A} = \mu_i \mathbf{u}_i^H, \qquad [1.6]$$

or, equivalently:

$$\mathbf{u}_i^H(\mu_i \mathbf{I}_I - \mathbf{A}) = \mathbf{0}_I^T \iff (\mu_i^* \mathbf{I}_I - \mathbf{A}^H)\mathbf{u}_i = \mathbf{0}_I, \qquad [1.7]$$

which implies that $\det(\mu_i \mathbf{I}_I - \mathbf{A}) = 0$. We again recover the characteristic equation [1.5], which shows that the left and right eigenvectors are associated with the same eigenvalues.

Below, we recall a few results about the eigenvalues of a matrix that were presented in Volume 1 (Favier 2019).

Table 1.2 summarizes certain key results about the definitions of left and right eigenvectors, i.e. the mode-1 and mode-2 eigenvectors, respectively, of a real square matrix, as well as the characteristic equation to compute them[3]. Note that the mode-1

---

2 In the case of a real matrix ($\mathbb{K} = \mathbb{R}$), replace the transconjugation operation in [1.6] by that of transposition, and equation [1.7] becomes $(\mu_i \mathbf{I}_I - \mathbf{A}^T)\mathbf{u}_i = \mathbf{0}_I$.

3 For the definition of the mode-p product, denoted $\times_p$ see section 3.13.2

and mode-2 eigenvalues are identical, whereas the mode-1 and mode-2 eigenvectors associated with the same eigenvalue are, in general, not the same, except when $\mathbf{A}$ is Hermitian (or symmetric, in the real case). Indeed, in this case, it is easy to check that the equations defining the left and right eigenvectors are identical, and therefore the mode-1 and mode-2 eigenvectors associated with the same eigenvalue are also identical.

$\mathbf{A} \in \mathbb{R}^{I \times I}$, $\mathbf{u}, \mathbf{v} \neq \mathbf{0}_I$	Definitions, characteristic equation, and properties
Right eigenvectors (mode-2)	$\mathbf{Av} = \mathbf{A} \times_2 \mathbf{v}^T = \lambda \mathbf{v}$ or $\sum_{j=1}^{I} a_{ij} v_j = \lambda v_i$, $i \in \langle I \rangle$
Left eigenvectors (mode-1)	$\mathbf{u}^T \mathbf{A} = \mathbf{A} \times_1 \mathbf{u}^T = \mu \mathbf{u}^T$ or $\sum_{i=1}^{I} a_{ij} u_i = \mu u_j$, $j \in \langle I \rangle$
Characteristic equation	$\det(\lambda \mathbf{I} - \mathbf{A}) = \det(\mu \mathbf{I} - \mathbf{A}) = 0 \Rightarrow \lambda_i = \mu_i \ \forall i \in \langle I \rangle$

Table 1.2. *Eigenvalues of a real square matrix*

Table 1.3 recalls the links between the eigenvalues of a real symmetric matrix[4] $\mathbf{A} \in \mathbb{R}_S^{I \times I}$ and the extrema of the Rayleigh quotient $\lambda = \frac{\mathbf{x}^T \mathbf{A} \mathbf{x}}{\mathbf{x}^T \mathbf{x}}$, as well as the optimization[5] of the quadratic criterion $\mathbf{x}^T \mathbf{A} \mathbf{x}$ subject to the constraint of the unit Euclidean norm, $\|\mathbf{x}\|_2 = 1$. The notions of generalized eigenvalue and generalized eigenpair are also recalled.

Eigenvalues of $\mathbf{A} \in \mathbb{R}_S^{I \times I}$
min/max $\lambda = \frac{\mathbf{x}^T \mathbf{A} \mathbf{x}}{\mathbf{x}^T \mathbf{x}} = \frac{\sum_{i,j=1}^{I} a_{ij} x_i x_j}{\mathbf{x}^T \mathbf{x}}$, $\mathbf{x} \neq 0 \Rightarrow \mathbf{A} \mathbf{x} = \lambda \mathbf{x}$
$\Updownarrow$
min/max $\mathbf{x}^T \mathbf{A} \mathbf{x} = \sum_{i,j=1}^{I} a_{ij} x_i x_j$, $\|\mathbf{x}\|_2 = 1$
KKT optimality conditions: $\mathbf{A} \mathbf{x} = \lambda \mathbf{x}$ with $\mathbf{x}^T \mathbf{x} = 1$
Generalized eigenvalues of $\mathbf{A} \in \mathbb{R}_S^{I \times I}$, $\mathbf{A} \geq 0$ with $\mathbf{B} \in \mathbb{R}_S^{I \times I}$, $\mathbf{B} > 0$
min/max $\lambda = \frac{\mathbf{x}^T \mathbf{A} \mathbf{x}}{\mathbf{x}^T \mathbf{B} \mathbf{x}}$, $\mathbf{x} \neq 0 \Rightarrow \mathbf{A} \mathbf{x} = \lambda \mathbf{B} \mathbf{x}$

Table 1.3. *Eigenvalues and extrema of the Rayleigh quotient*

REMARK 1.1.– The generalized eigenvalues λ are found by solving the following equation:

$$\det(\lambda \mathbf{B} - \mathbf{A}) = 0 \iff \det(\lambda \mathbf{I} - \mathbf{B}^{-1} \mathbf{A}) = 0, \qquad [1.8]$$

---

4 $\mathbb{R}_S^{I \times I}$ is the subspace of symmetric square matrices of order I.

5 This optimization problem with an equality constraint is solved using the Lagrangian $L(\mathbf{x}, \lambda) = \mathbf{x}^T \mathbf{A} \mathbf{x} - \lambda(\mathbf{x}^T \mathbf{x} - 1)$. The Karush–Kuhn–Tucker (KKT) optimality conditions are $\frac{\partial L(\mathbf{x}, \lambda)}{\partial \mathbf{x}} = \mathbf{0}$, $\frac{\partial L(\mathbf{x}, \lambda)}{\partial \lambda} = 0$, which leads to $\mathbf{A} \mathbf{x} = \lambda \mathbf{x}$, with $\mathbf{x}^T \mathbf{x} = 1$.

i.e. by computing the eigenvalues of $\mathbf{B}^{-1}\mathbf{A}$, where $\mathbf{A}$ and $\mathbf{B}$ are symmetric positive definite or semi-definite and symmetric positive definite, respectively. The equation $\mathbf{A}\mathbf{x} = \lambda\mathbf{B}\mathbf{x}$ can be written as $\mathbf{A}\mathbf{P} = \mathbf{B}\mathbf{P}\mathbf{D}$, where $\mathbf{D} = \mathrm{diag}(\lambda_1, \cdots, \lambda_I)$ and $\mathbf{P}$ contain the generalized eigenvalues and the generalized eigenvectors of the pair $(\mathbf{A}, \mathbf{B})$, respectively, and $(\mathbf{P}, \mathbf{D})$ is called an eigenpair of $(\mathbf{A}, \mathbf{B})$.

The following proposition shows that any left eigenvector is orthogonal to any right eigenvector when the two vectors are associated with distinct eigenvalues.

PROPOSITION 1.2.– Given a right eigenvector $\mathbf{v}_k$ and a left eigenvector $\mathbf{u}_i$ associated with distinct eigenvalues λ_k and λ_i, respectively, the following orthogonality relation is satisfied with respect to the Hermitian scalar product if $\mathbb{K} = \mathbb{C}$ (or with respect to the Euclidean scalar product if $\mathbb{K} = \mathbb{R}$):

$$\langle \mathbf{v}_k, \mathbf{u}_i \rangle = \mathbf{u}_i^H \mathbf{v}_k = 0. \tag{1.9}$$

PROOF.– Pre-multiplying equation [1.3] by $\mathbf{u}_i^H$ and post-multiplying [1.6] by $\mathbf{v}_k$, after replacing μ_i with λ_i, yields:

$$\mathbf{u}_i^H \mathbf{A} \mathbf{v}_k = \lambda_k \mathbf{u}_i^H \mathbf{v}_k \tag{1.10}$$

$$= \lambda_i \mathbf{u}_i^H \mathbf{v}_k.$$

Therefore, by subtracting these two equations on both sides and using the hypothesis that $\lambda_i \neq \lambda_k$, we deduce the orthogonality relation $\mathbf{u}_i^H \mathbf{v}_k = 0$. $\qquad\square$

1.3.2. *Eigendecomposition and properties*

Concatenating the I columns associated with equation [1.3] for $k \in \langle I \rangle$ gives us the following matrix equation:

$$\mathbf{A} \begin{bmatrix} \mathbf{v}_1 & \cdots & \mathbf{v}_I \end{bmatrix} = \begin{bmatrix} \lambda_1 \mathbf{v}_1 & \cdots & \lambda_I \mathbf{v}_I \end{bmatrix}$$

$$= \begin{bmatrix} \mathbf{v}_1 & \cdots & \mathbf{v}_I \end{bmatrix} \begin{bmatrix} \lambda_1 & & \mathbf{0} \\ & \ddots & \\ \mathbf{0} & & \lambda_I \end{bmatrix}.$$

After defining the diagonal matrix $\mathbf{D} = \mathrm{diag}(\lambda_1, \cdots, \lambda_I)$ with the eigenvalues of $\mathbf{A}$ along the diagonal, repeated with their multiplicity, and the matrix $\mathbf{P} = [\mathbf{v}_1, \cdots, \mathbf{v}_I]$ of right eigenvectors in the same order as the corresponding eigenvalues, this equation becomes:

$$\mathbf{A}\mathbf{P} = \mathbf{P}\mathbf{D}. \tag{1.11}$$

We can therefore conclude that $\mathbf{A}$ is diagonalizable if and only if it has I linearly independent eigenvectors, i.e. if the matrix $\mathbf{P}$ is invertible. If so, the matrix $\mathbf{D} = \mathbf{P}^{-1}\mathbf{A}\mathbf{P}$ is diagonal, and $\mathbf{A}$ decomposes into:

$$\mathbf{A} = \mathbf{P}\mathbf{D}\mathbf{P}^{-1}. \qquad [1.12]$$

This decomposition is called the eigendecomposition of $\mathbf{A}$, also known as the spectral decomposition of $\mathbf{A}$, named after the "spectrum", which is the set of eigenvalues of a matrix.

Diagonalization condition: In order for a matrix $\mathbf{A} \in \mathbb{K}^{I \times I}$ to be diagonalizable, it is necessary and sufficient for all of its eigenvalues to belong to $\mathbb{K}$ and for the dimension of the eigenspace associated with each eigenvalue to be equal to the multiplicity of this eigenvalue. In other words, an eigenvalue of multiplicity q must have q linearly independent eigenvectors. This condition implies that any matrix $\mathbf{A} \in \mathbb{K}^{I \times I}$ with I distinct eigenvalues is diagonalizable.

The decomposition [1.12] can be used to show the next proposition.

PROPOSITION 1.3.– Given the matrix $\mathbf{A} \in \mathbb{K}^{I \times I}$, we have the following relations:

$$\text{tr}(\mathbf{A}) = \sum_{i=1}^{I} \lambda_i, \ \ \det(\mathbf{A}) = \prod_{i=1}^{I} \lambda_i. \qquad [1.13]$$

PROOF.– From the decomposition [1.12], we deduce the following equalities:

$$\text{tr}(\mathbf{A}) = \text{tr}(\mathbf{P}\mathbf{D}\mathbf{P}^{-1}) = \text{tr}(\mathbf{P}^{-1}\mathbf{P}\mathbf{D}) = \text{tr}(\mathbf{D}) = \sum_{i=1}^{I} \lambda_i$$

$$\det(\mathbf{A}) = \det(\mathbf{P}\mathbf{D}\mathbf{P}^{-1}) = \det(\mathbf{D}) = \prod_{i=1}^{I} \lambda_i,$$

which proves the identities [1.13].   □

The next proposition can be deduced from the orthogonality property [1.9].

PROPOSITION 1.4.– Let $\mathbf{P}$ and $\mathbf{Q}$ be the square matrices of order I formed by the right and left eigenvectors of $\mathbf{A}$:

$$\mathbf{P} = [\mathbf{v}_1, \cdots, \mathbf{v}_I], \ \ \mathbf{Q} = [\mathbf{u}_1, \cdots, \mathbf{u}_I]. \qquad [1.14]$$

If all the eigenvalues of $\mathbf{A}$ are distinct, then $\mathbf{P}$ and $\mathbf{Q}$ satisfy the following orthogonality relation:

$$\mathbf{Q}^H \mathbf{P} = \mathbf{I}_I, \qquad [1.15]$$

i.e. $\mathbf{P}$ and $\mathbf{Q}$ are bi-unitary and their inverses satisfy $\mathbf{P}^{-1} = \mathbf{Q}^H$ and $\mathbf{Q}^{-1} = \mathbf{P}^H$.

By taking the orthogonality relation [1.15] into account, the decomposition [1.12] can be rewritten in term of the matrix $\mathbf{Q}$ of left eigenvectors of $\mathbf{A}$ as follows:

$$\mathbf{A} = \mathbf{PDQ}^H, \tag{1.16}$$

or, in expanded form:

$$\mathbf{A} = \begin{bmatrix} \mathbf{v}_1 & \cdots & \mathbf{v}_I \end{bmatrix} \begin{bmatrix} \lambda_1 & & \mathbf{0} \\ & \ddots & \\ \mathbf{0} & & \lambda_I \end{bmatrix} \begin{bmatrix} \mathbf{u}_1^H \\ \vdots \\ \mathbf{u}_I^H \end{bmatrix}$$

$$= \sum_{i=1}^{I} \lambda_i \mathbf{v}_i \mathbf{u}_i^H, \tag{1.17}$$

where $\mathbf{u}_i$ and $\mathbf{v}_i$ are the left and right eigenvectors of $\mathbf{A}$ associated with the eigenvalue λ_i.

EXAMPLE 1.5.– Consider the matrix $\mathbf{A} = \begin{bmatrix} 0 & j \\ j & 0 \end{bmatrix}$, with $j^2 = -1$. This is a skew-Hermitian matrix ($\mathbf{A}^H = -\mathbf{A}$). Its eigenvalues are therefore purely imaginary; they are the solutions of the equation $\det(\lambda \mathbf{I}_2 - \mathbf{A}) = \lambda^2 + 1 = 0$, namely $\lambda_1 = j, \lambda_2 = -j$. The right eigenvectors satisfy:

$$jx_1 = jy_1 \Rightarrow \mathbf{v}_1 = \frac{1}{\sqrt{2}} \begin{bmatrix} 1 \\ 1 \end{bmatrix}$$

$$jx_2 = -jy_2 \Rightarrow \mathbf{v}_2 = \frac{1}{\sqrt{2}} \begin{bmatrix} 1 \\ -1 \end{bmatrix}.$$

It is easy to check that we can choose the left eigenvectors so that $\mathbf{u}_1 = \mathbf{v}_1, \mathbf{u}_2 = \mathbf{v}_2$. The eigenvector matrices are therefore $\mathbf{P} = \mathbf{Q} = \frac{1}{\sqrt{2}} \begin{bmatrix} 1 & 1 \\ 1 & -1 \end{bmatrix}$. These matrices are such that $\mathbf{P}^{-1} = \mathbf{P} = \mathbf{Q}$ and, hence, $\mathbf{Q}^H \mathbf{P} = \mathbf{I}_2$. The EVD of $\mathbf{A}$ is given by:

$$\mathbf{A} = \mathbf{PDP}^{-1} = \frac{1}{2} \begin{bmatrix} 1 & 1 \\ 1 & -1 \end{bmatrix} \begin{bmatrix} j & 0 \\ 0 & -j \end{bmatrix} \begin{bmatrix} 1 & 1 \\ 1 & -1 \end{bmatrix} = \begin{bmatrix} 0 & j \\ j & 0 \end{bmatrix},$$

or, equivalently:

$$\mathbf{A} = \sum_{i=1}^{2} \lambda_i \mathbf{v}_i \mathbf{u}_i^H = \frac{j}{2} \begin{bmatrix} 1 \\ 1 \end{bmatrix} \begin{bmatrix} 1 & 1 \end{bmatrix} - \frac{j}{2} \begin{bmatrix} 1 \\ -1 \end{bmatrix} \begin{bmatrix} 1 & -1 \end{bmatrix} = \begin{bmatrix} 0 & j \\ j & 0 \end{bmatrix}.$$

1.3.3. *Special case of symmetric/Hermitian matrices*

Table 1.4 recalls the eigenvalue properties of (skew-)symmetric/(skew-)Hermitian matrices, orthogonal matrices and unitary matrices.

Matrix classes	Eigenvalue properties
$\mathbf{A} \in \mathbb{C}^{I \times I}$ Hermitian $\mathbf{A} \in \mathbb{C}^{I \times I}$ skew-Hermitian	$\lambda_i \in \mathbb{R}$ $\lambda_i^* = -\lambda_i$
$\mathbf{A} \in \mathbb{R}_S^{I \times I}$ symmetric $\mathbf{A} \in \mathbb{R}^{I \times I}$ skew-symmetric	$\lambda_i \in \mathbb{R}$ $\lambda_i^* = -\lambda_i$
$\mathbf{A} \in \mathbb{R}_S^{I \times I}$ positive definite $\mathbf{A} \in \mathbb{R}_S^{I \times I}$ positive semi-definite	$\lambda_i > 0$ $\lambda_i \geq 0$
$\mathbf{A} \in \mathbb{R}^{I \times I}$ orthogonal $\mathbf{A} \in \mathbb{C}^{I \times I}$ unitary	$|\lambda_i| = 1$ $|\lambda_i| = 1$

Table 1.4. *Eigenvalue properties of certain matrix classes*

REMARK 1.6.– The eigenvalues of a symmetric or Hermitian matrix $\mathbf{A}$ are real-valued, and any right eigenvector is also a left eigenvector associated with the same eigenvalue. Therefore, any two eigenvectors associated with distinct eigenvalues are orthogonal with respect to the Euclidean inner product (real case) or the Hermitian inner product (complex case), and the columns of $\mathbf{P}$ form an orthonormal basis. Note that, if $\mathbf{A}$ is non-symmetric, its eigenvectors are not orthogonal. For more details, see Favier (2019).

The above results allow us to deduce the following proposition.

PROPOSITION 1.7.– Every real symmetric (respectively, complex Hermitian) matrix has orthogonal eigenvectors and real-valued eigenvalues, and can be diagonalized using its EVD:

$$\mathbf{A} = \mathbf{P}\mathbf{D}\mathbf{P}^T \quad (\text{respectively, } \mathbf{A} = \mathbf{P}\mathbf{D}\mathbf{P}^H), \qquad [1.18]$$

where $\mathbf{P}$ is orthogonal (respectively, unitary), and $\mathbf{D}$ is diagonal and real. We can also write $\mathbf{A}$ as follows:

$$\mathbf{A} = \sum_{i=1}^{I} \lambda_i \mathbf{u}_i \mathbf{u}_i^T \quad (\text{respectively, } \mathbf{A} = \sum_{i=1}^{I} \lambda_i \mathbf{u}_i \mathbf{u}_i^H). \qquad [1.19]$$

This expression shows that the eigenvectors are only determined up to some phase ambiguity, since, for $j^2 = -1$ and $\theta_i \in \mathbb{R}$, $\forall i \in \langle I \rangle$, we have:

$$\sum_{i=1}^{I} \lambda_i (e^{j\theta_i} \mathbf{u}_i)(e^{j\theta_i} \mathbf{u}_i)^H = \sum_{i=1}^{I} \lambda_i \mathbf{u}_i \mathbf{u}_i^H. \qquad [1.20]$$

Table 1.5 summarizes the eigenvalue properties of positive (or negative) (semi-) definite real symmetric matrices.

$\mathbf{A} \in \mathbb{R}_S^{I \times I}$	Constraints	Eigenvalue properties
$\mathbf{A} > 0$	$\mathbf{x}^T \mathbf{A} \mathbf{x} > 0 \ \forall \mathbf{x} \neq \mathbf{0}$	$\lambda_i > 0 \ , \ \forall i \in \langle I \rangle$
$\mathbf{A} \geq 0$	$\mathbf{x}^T \mathbf{A} \mathbf{x} \geq 0 \ \forall \mathbf{x}$	$\lambda_i \geq 0 \ , \ \forall i \in \langle I \rangle$
$\mathbf{A} < 0$	$\mathbf{x}^T \mathbf{A} \mathbf{x} < 0 \ \forall \mathbf{x} \neq \mathbf{0}$	$\lambda_i < 0 \ , \ \forall i \in \langle I \rangle$
$\mathbf{A} \leq 0$	$\mathbf{x}^T \mathbf{A} \mathbf{x} \leq 0 \ \forall \mathbf{x}$	$\lambda_i \leq 0 \ , \ \forall i \in \langle I \rangle$

Table 1.5. *Eigenvalues of positive/negative (semi-)definite matrices*

Table 1.6 recalls the definitions of equivalent, similar and orthogonally/unitarily similar matrices, as well as their properties, where $\mathrm{sp}(\mathbf{A})$ denotes the spectrum of $\mathbf{A}$, i.e. the set of its eigenvalues.

$\mathbf{A}, \mathbf{B}$	Relations	$\mathbf{A}$ and $\mathbf{B}$	Properties
$\mathbf{A}, \mathbf{B} \in \mathbb{K}^{I \times J}$ $\mathbf{P} \in \mathbb{K}^{J \times J}, \mathbf{Q} \in \mathbb{K}^{I \times I}$ non-singular	$\mathbf{B} = \mathbf{Q} \mathbf{A} \mathbf{P}$	Equivalent	$r(\mathbf{A}) = r(\mathbf{B})$
$\mathbf{A}, \mathbf{B} \in \mathbb{K}^{I \times I}$ $\mathbf{P} \in \mathbb{K}^{I \times I}$ non-singular	$\mathbf{B} = \mathbf{P}^{-1} \mathbf{A} \mathbf{P}$	Similar	$r(\mathbf{A}) = r(\mathbf{B})$ $\mathrm{tr}(\mathbf{A}) = \mathrm{tr}(\mathbf{B})$ $\mathrm{sp}(\mathbf{A}) = \mathrm{sp}(\mathbf{B})$
$\mathbf{A}, \mathbf{B} \in \mathbb{R}^{I \times I}$ $\mathbf{P} \in \mathbb{R}^{I \times I}$ orthogonal	$\mathbf{B} = \mathbf{P}^T \mathbf{A} \mathbf{P}$	Orthogonally similar	$r(\mathbf{A}) = r(\mathbf{B})$ $\mathrm{tr}(\mathbf{A}) = \mathrm{tr}(\mathbf{B})$ $\mathrm{sp}(\mathbf{A}) = \mathrm{sp}(\mathbf{B})$ $\|\mathbf{A}\|_F = \|\mathbf{B}\|_F$
$\mathbf{A}, \mathbf{B} \in \mathbb{C}^{I \times I}$ $\mathbf{P} \in \mathbb{C}^{I \times I}$ unitary	$\mathbf{B} = \mathbf{P}^H \mathbf{A} \mathbf{P}$	Unitarily similar	$r(\mathbf{A}) = r(\mathbf{B})$ $\mathrm{tr}(\mathbf{A}) = \mathrm{tr}(\mathbf{B})$ $\mathrm{sp}(\mathbf{A}) = \mathrm{sp}(\mathbf{B})$ $\|\mathbf{A}\|_F = \|\mathbf{B}\|_F$

Table 1.6. *Equivalent and similar matrices*

REMARK 1.8.– Setting $\mathbf{Q} = \mathbf{P}^T$, we say that $\mathbf{A}$ and $\mathbf{B}$ are orthogonally similar if the following relation is satisfied for some orthogonal matrix $\mathbf{Q} \in \mathbb{R}^{I \times I}$:

$$\mathbf{B} = \mathbf{A} \times_1 \mathbf{Q} \times_2 \mathbf{Q}. \tag{1.21}$$

For unitarily similar matrices, with $\mathbf{Q} = \mathbf{P}^H$, the relation [1.21] becomes:

$$\mathbf{B} = \mathbf{A} \times_1 \mathbf{Q} \times_2 \mathbf{Q}^*. \tag{1.22}$$

The notion of orthogonally/unitarily similar matrices is extended to tensors in Chapter 4, in section 4.2.3.

1.3.4. *Application to compute the powers of a matrix and a matrix polynomial*

If $\mathbf{A}$ is diagonalizable, then, for a given positive or negative integer k, we can deduce the following relations from the EVD $\mathbf{A} = \mathbf{P}\mathbf{D}\mathbf{P}^{-1}$:

$$\mathbf{A}^k = \mathbf{P}\mathbf{D}^k\mathbf{P}^{-1} = \mathbf{P}\begin{bmatrix} \lambda_1^k & & \mathbf{0} \\ & \ddots & \\ \mathbf{0} & & \lambda_I^k \end{bmatrix}\mathbf{P}^{-1} \qquad [1.23]$$

and

$$p(\mathbf{A}) = \sum_{k=1}^{K} \alpha_k \mathbf{A}^k = \mathbf{P}\begin{bmatrix} p(\lambda_1) & & \mathbf{0} \\ & \ddots & \\ \mathbf{0} & & p(\lambda_I) \end{bmatrix}\mathbf{P}^{-1},$$

where $p(\lambda_i) = \sum_{k=1}^{K} \alpha_k \lambda_i^k$.

From [1.23], we deduce the following relation:

$$\mathrm{tr}(\mathbf{A}^k) = \mathrm{tr}(\mathbf{P}\mathbf{D}^k\mathbf{P}^{-1}) = \mathrm{tr}(\mathbf{P}^{-1}\mathbf{P}\mathbf{D}^k) = \mathrm{tr}(\mathbf{D}^k) = \sum_{i=1}^{I} \lambda_i^k.$$

1.3.5. *Application to compute a state transition matrix*

Consider the continuous-time state-space model:

$$\dot{x}(t) = \mathbf{A}\mathbf{x}(t) + \mathbf{G}\mathbf{u}(t) \qquad [1.24]$$

$$\mathbf{y}(t) = \mathbf{C}\mathbf{x}(t), \qquad [1.25]$$

where $\mathbf{A} \in \mathbb{R}^{n \times n}, \mathbf{G} \in \mathbb{R}^{n \times p}, \mathbf{C} \in \mathbb{R}^{m \times n}$. The state transition matrix is the matrix $e^{\mathbf{A}t}$ that allows us to compute the state $\mathbf{x}(t)$ at time t from the state $\mathbf{x}_0$ at the initial time $t = 0$.

This matrix exponential also appears in the discretization of the state-space model [1.24]–[1.25], which gives:

$$\mathbf{x}(k+1) = \mathbf{F}\mathbf{x}(k) + \mathbf{B}\mathbf{u}(k) \qquad [1.26]$$

$$\mathbf{y}(k) = \mathbf{C}\mathbf{x}(k), \qquad [1.27]$$

where $\mathbf{F} = e^{\mathbf{A}T}, \mathbf{G} = \int_0^T e^{\mathbf{A}\tau}d\tau\,\mathbf{B}$ and T represents the sampling period. There are several ways to compute the exponential of a matrix. If $\mathbf{A}$ is diagonalizable, the computation can be performed using the EVD [1.12] of $\mathbf{A}$, as follows:

$$\mathbf{F} = e^{\mathbf{A}T} = \mathbf{P}\mathrm{diag}\big(e^{\lambda_1 T}, \cdots, e^{\lambda_I T}\big)\mathbf{P}^{-1}. \qquad [1.28]$$

PROOF.– Using the series expansion of the exponential, the decomposition [1.12] of $\mathbf{A}$, and the identity [1.23], we have:

$$e^{\mathbf{A}T} = \mathbf{I} + \mathbf{A}T + \frac{\mathbf{A}^2 T^2}{2!} + \frac{\mathbf{A}^3 T^3}{3!} + \cdots$$

$$= \mathbf{P}[\mathbf{I} + \mathbf{D}T + \frac{\mathbf{D}^2 T^2}{2!} + \frac{\mathbf{D}^3 T^3}{3!} + \cdots]\mathbf{P}^{-1}$$

$$= \mathbf{P} e^{\mathbf{D}T} \mathbf{P}^{-1} = \mathbf{P} \operatorname{diag}\left(e^{\lambda_1 T}, \cdots, e^{\lambda_I T}\right)\mathbf{P}^{-1},$$

which proves the expression [1.28] for the matrix exponential.   □

1.3.6. *Application to compute the transfer function and the output of a discrete-time linear system*

Consider a time-invariant discrete-time dynamic system modeled by means of the state-space model [1.26]–[1.27], with $\mathbf{F} \in \mathbb{R}^{n \times n}, \mathbf{B} \in \mathbb{R}^{n \times p}, \mathbf{C} \in \mathbb{R}^{m \times n}$. The transfer function and system output at the sampling time kT are given by:

$$H(z) = \mathbf{C}[z\mathbf{I}_n - \mathbf{F}]^{-1}\mathbf{B},$$

$$\mathbf{y}(k) = \mathbf{C}[\mathbf{F}^k \mathbf{x}(0) + \sum_{i=0}^{k-1} \mathbf{F}^{k-i-1} \mathbf{B}\mathbf{u}(i)].$$

Assuming that the state transition matrix, also called the system matrix, admits the EVD $\mathbf{F} = \mathbf{P}\mathbf{D}\mathbf{P}^{-1}$, the transfer function and the system output can be calculated using the following formulae:

$$H(z) = \mathbf{C}[\mathbf{P}(z\mathbf{I}_n - \mathbf{D})\mathbf{P}^{-1}]^{-1}\mathbf{B} = \mathbf{C}\mathbf{P}\operatorname{diag}\left(\frac{1}{z - \lambda_1}, \cdots, \frac{1}{z - \lambda_I}\right)\mathbf{P}^{-1}\mathbf{B}$$

$$\mathbf{y}(k) = \mathbf{C}\mathbf{P}[\mathbf{D}^k \mathbf{P}^{-1}\mathbf{x}(0) + \sum_{i=0}^{k-1} \mathbf{D}^{k-i-1}\mathbf{P}^{-1}\mathbf{B}\mathbf{u}(i)].$$

1.4. URV^H decomposition

Recall that any matrix $\mathbf{A} \in \mathbb{K}^{I \times J}$ has the four fundamental subspaces listed in Table 1.7[6].

The relations between the subspaces $C(\mathbf{A})$, $L(\mathbf{A})$, $\mathcal{N}(\mathbf{A})$ and $\operatorname{Im}(\mathbf{A})$ are recalled in Table 1.8, where $[C(\mathbf{A})]^{\perp}$, $[L(\mathbf{A})]^{\perp}$, and $[\mathcal{N}(\mathbf{A})]^{\perp}$ are the orthogonal complements of $C(\mathbf{A})$, $L(\mathbf{A})$ and $\mathcal{N}(\mathbf{A})$, respectively.

---

6 In the case of a real matrix ($\mathbb{K} = \mathbb{R}$), replace the transconjugation operation by that of transposition.

Fundamental subspaces	Notation
Column space	$C(\mathbf{A}) = \mathrm{Im}(\mathbf{A}) \subseteq \mathbb{K}^I$
Row space	$L(\mathbf{A}) = C(\mathbf{A}^H) \subseteq \mathbb{K}^J$
Kernel	$\mathcal{N}(\mathbf{A}) \subseteq \mathbb{K}^J$
Left kernel	$\mathcal{N}(\mathbf{A}^H) \subseteq \mathbb{K}^I$

Table 1.7. *Fundamental subspaces of* $\mathbf{A} \in \mathbb{K}^{I \times J}$

The equalities $C(\mathbf{A}) \cap \mathcal{N}(\mathbf{A}^H) = \{\mathbf{0}_I\}$ and $\mathcal{N}(\mathbf{A}) \cap C(\mathbf{A}^H) = \{\mathbf{0}_J\}$ follow from the fact that the intersection of two orthogonal subspaces is just the zero vector. From the equalities in this table, it follows that any matrix $\mathbf{A} \in \mathbb{C}^{I \times J}$ yields an orthogonal decomposition of the spaces $\mathbb{C}^I$ and $\mathbb{C}^J$ as follows:

$$\mathbb{C}^I = C(\mathbf{A}) \oplus [C(\mathbf{A})]^\perp = C(\mathbf{A}) \oplus \mathcal{N}(\mathbf{A}^H) \qquad [1.29]$$

$$\mathbb{C}^J = \mathcal{N}(\mathbf{A}) \oplus [\mathcal{N}(\mathbf{A})]^\perp = \mathcal{N}(\mathbf{A}) \oplus C(\mathbf{A}^H). \qquad [1.30]$$

Relations	
$C(\mathbf{A}) = [\mathcal{N}(\mathbf{A}^H)]^\perp = \mathrm{Im}(\mathbf{A})$,	$C(\mathbf{A}^H) = [\mathcal{N}(\mathbf{A})]^\perp = L(\mathbf{A})$
$C(\mathbf{A}) \cap \mathcal{N}(\mathbf{A}^H) = \{\mathbf{0}_I\}$,	$\mathcal{N}(\mathbf{A}) \cap C(\mathbf{A}^H) = \{\mathbf{0}_J\}$
$\dim[C(\mathbf{A})] + \dim[\mathcal{N}(\mathbf{A})] = J$	
$\dim[L(\mathbf{A})] + \dim[\mathcal{N}(\mathbf{A}^H)] = I$	

Table 1.8. *Relations between the fundamental subspaces*

The orthogonal decompositions [1.29] and [1.30] of $\mathbb{C}^I$ and $\mathbb{C}^J$ then give us the matrix decomposition described in the next proposition.

PROPOSITION 1.9.– Every matrix $\mathbf{A} \in \mathbb{C}^{I \times J}$ of rank r admits the following decomposition (Meyer 2000):

$$\mathbf{A} = \mathbf{U}\mathbf{R}\mathbf{V}^H = \mathbf{U} \begin{bmatrix} \mathbf{W} & \mathbf{0}_{r \times (J-r)} \\ \mathbf{0}_{(I-r) \times r} & \mathbf{0}_{(I-r) \times (J-r)} \end{bmatrix} \mathbf{V}^H, \qquad [1.31]$$

where $\mathbf{W} \in \mathbb{C}^{r \times r}$ is a non-singular square matrix, and $\mathbf{U} \in \mathbb{C}^{I \times I}$ and $\mathbf{V} \in \mathbb{C}^{J \times J}$ are unitary matrices partitioned as follows:

$$\mathbf{U} = \begin{bmatrix} \mathbf{U}_1 & \mathbf{U}_2 \end{bmatrix} , \quad \mathbf{V} = \begin{bmatrix} \mathbf{V}_1 & \mathbf{V}_2 \end{bmatrix}, \qquad [1.32]$$

$$\mathbf{U}_1 \in \mathbb{C}^{I \times r} , \quad \mathbf{U}_2 \in \mathbb{C}^{I \times (I-r)} , \quad \mathbf{V}_1 \in \mathbb{C}^{J \times r} , \quad \mathbf{V}_2 \in \mathbb{C}^{J \times (J-r)}. \qquad [1.33]$$

The columns of $\mathbf{U}_1$ and $\mathbf{V}_1$ form orthonormal bases for $C(\mathbf{A})$ and $C(\mathbf{A}^H) = L(\mathbf{A})$, respectively, whereas the columns of $\mathbf{U}_2$ and $\mathbf{V}_2$ form orthonormal bases for $\mathcal{N}(\mathbf{A}^H)$ and $\mathcal{N}(\mathbf{A})$, respectively. Hence, the spaces spanned by the columns of $\mathbf{U}_1$ and $\mathbf{V}_1$ are the column and row spaces of $\mathbf{A}$, respectively.

The block-diagonal structure of $\mathbf{R}$ follows from the definition of the subspaces $\mathcal{N}(\mathbf{A})$ and $\mathcal{N}(\mathbf{A}^H)$, which imply $\mathbf{A}\mathbf{V}_2 = \mathbf{0}_{I \times (J-r)}$ and $\mathbf{A}^H\mathbf{U}_2 = \mathbf{0}_{J \times (I-r)}$, or, equivalently, $\mathbf{U}_2^H \mathbf{A} = \mathbf{0}_{(I-r) \times J}$. Since the matrices $\mathbf{U}$ and $\mathbf{V}$ are unitary, we deduce that:

$$\mathbf{R} = \mathbf{U}^H \mathbf{A} \mathbf{V} = \begin{bmatrix} \mathbf{U}_1^H \mathbf{A}\mathbf{V}_1 & \mathbf{U}_1^H \mathbf{A}\mathbf{V}_2 \\ \mathbf{U}_2^H \mathbf{A}\mathbf{V}_1 & \mathbf{U}_2^H \mathbf{A}\mathbf{V}_2 \end{bmatrix} = \begin{bmatrix} \mathbf{W} & \mathbf{0}_{r \times (J-r)} \\ \mathbf{0}_{(I-r) \times r} & \mathbf{0}_{(I-r) \times (J-r)} \end{bmatrix}.$$

Furthermore, $r(\mathbf{W}) = r(\mathbf{U}_1^H \mathbf{A}\mathbf{V}_1) = r(\mathbf{A}) = r$, which implies that $\mathbf{W}$ is non-singular.

1.5. Singular value decomposition

1.5.1. *Definition and properties*

The SVD was discovered independently by both Eugenio Beltrami (1835–1900) and Camille Jordan (1838–1922), in 1873 and 1874, respectively. It is one of the most widely used matrix factorizations in signal processing and data analysis. It can be viewed as a generalization of the eigendecomposition to any rectangular matrix $\mathbf{A} \in \mathbb{C}^{I \times J}$ in the sense of a diagonalization by means of two unitary matrices (or orthogonal matrices, if $\mathbf{A} \in \mathbb{R}^{I \times J}$). The SVD plays a very important role in many applications. One example is data compression using the low-rank matrix approximation theorem, established by Eckart and Young (1936). Other applications relate to determining the rank of a matrix, solving systems of linear equations with the LS method, which requires the computation of a matrix pseudo-inverse (see Table 1.13), simplifying linear models in the sense of reducing their dimensionality, estimating missing data within matrices of incomplete measurements (i.e. the matrix completion problem), computing high-order SVD (HOSVD) for the decomposition of tensors of order greater than two and many others.

For $\mathbf{A} \in \mathbb{C}^{I \times J}$, the SVD can be written as:

$$\mathbf{A} = \mathbf{U}\mathbf{\Sigma}\mathbf{V}^H, \tag{1.34}$$

where $\mathbf{U} \in \mathbb{C}^{I \times I}$ and $\mathbf{V} \in \mathbb{C}^{J \times J}$ are unitary matrices, and $\mathbf{\Sigma} \in \mathbb{R}^{I \times J}$ is a pseudo-diagonal matrix of the same size as $\mathbf{A}$.

The derivation of the SVD is directly related to the property that, for $\mathbf{A} \in \mathbb{C}^{I \times J}$, the products $\mathbf{A}\mathbf{A}^H$ and $\mathbf{A}^H \mathbf{A}$ (or $\mathbf{A}\mathbf{A}^T$ and $\mathbf{A}^T \mathbf{A}$, in the case of a real matrix) are Hermitian matrices (or real symmetric matrices) and are therefore diagonalizable by means of their eigendecompositions (see [1.18]), namely:

$$\mathbf{A}\mathbf{A}^H = \mathbf{U}\mathbf{D}_1\mathbf{U}^H, \tag{1.35}$$

$$\mathbf{A}^H \mathbf{A} = \mathbf{V}\mathbf{D}_2\mathbf{V}^H, \tag{1.36}$$

where $\mathbf{U}$ is the matrix of eigenvectors of $\mathbf{AA}^H$, and $\mathbf{V}$ is the matrix of eigenvectors of $\mathbf{A}^H\mathbf{A}$, whose columns $(\mathbf{u}_1, \cdots, \mathbf{u}_I)$ and $(\mathbf{v}_1, \cdots, \mathbf{v}_J)$ form two orthonormal bases, which implies $\mathbf{U}^H\mathbf{U} = \mathbf{UU}^H = \mathbf{I}_I$ and $\mathbf{V}^H\mathbf{V} = \mathbf{VV}^H = \mathbf{I}_J$. The columns of $\mathbf{U}$ and $\mathbf{V}$ are called the left and right singular vectors of $\mathbf{A}$, respectively. The non-zero eigenvalues of $\mathbf{AA}^H$ and $\mathbf{A}^H\mathbf{A}$ are equal and non-negative (see Chapter 3 of Volume 1); they are ordered non-increasingly, such that:

$$\mathbf{D}_1 = \begin{bmatrix} \lambda_1 & & & \\ & \ddots & & \\ & & \lambda_r & \\ & & & \mathbf{0}_{I-r} \end{bmatrix}, \ \mathbf{D}_2 = \begin{bmatrix} \lambda_1 & & & \\ & \ddots & & \\ & & \lambda_r & \\ & & & \mathbf{0}_{J-r} \end{bmatrix}, \quad [1.37]$$

where r is the rank of $\mathbf{A}$ and hence of $\mathbf{AA}^H$ and $\mathbf{A}^H\mathbf{A}$, with $\lambda_1 \geq \lambda_2 \geq \cdots \geq \lambda_r > 0$.

Using the SVD [1.34] of $\mathbf{A}$, the decompositions [1.35] and [1.36] can be rewritten as follows:

$$\mathbf{AA}^H = (\mathbf{U\Sigma V}^H)(\mathbf{V\Sigma}^T\mathbf{U}^H) = \mathbf{U\Sigma\Sigma}^T\mathbf{U}^H, \tag{1.38}$$

$$\mathbf{A}^H\mathbf{A} = (\mathbf{V\Sigma}^T\mathbf{U}^H)(\mathbf{U\Sigma V}^H) = \mathbf{V\Sigma}^T\mathbf{\Sigma V}^H, \tag{1.39}$$

which implies $\mathbf{D}_1 = \mathbf{\Sigma\Sigma}^T$ and $\mathbf{D}_2 = \mathbf{\Sigma}^T\mathbf{\Sigma}$, and, hence, $\lambda_1 = \sigma_1^2, \cdots, \lambda_r = \sigma_r^2$. The diagonal coefficients σ_k of $\mathbf{\Sigma}$ are the positive square roots of the eigenvalues λ_k, i.e. $\sigma_k = \sqrt{\lambda_k}$; they are called the singular values of $\mathbf{A}$. Since these singular values are ordered non-increasingly along the diagonal of $\mathbf{\Sigma}$, we define $\mathbf{\Sigma}_r = \text{diag}(\sigma_1, \cdots, \sigma_r)$, where r is the rank of $\mathbf{A}$, equal to the number of non-zero singular values.

From [1.38] and [1.39], we conclude that the span of the columns of $\mathbf{U}$ is the column space of $\mathbf{A}$, whereas the span of the columns of $\mathbf{V}$ is the column space of $\mathbf{A}^H$, i.e. the row space of $\mathbf{A}$. Since $\mathbf{V}$ is unitary, we have $\mathbf{V}^{-1} = \mathbf{V}^H$, and we can rewrite the SVD as $\mathbf{AV} = \mathbf{U\Sigma}$, from which we deduce that the columns of $\mathbf{U}$ and $\mathbf{V}$ satisfy the relations:

$$\mathbf{Av}_k = \sigma_k\mathbf{u}_k \ , \ \ k \in \langle r \rangle. \tag{1.40}$$

Thus, $\mathbf{A}$ can be viewed as the matrix of the linear transformation that transforms a vector $\mathbf{v}_k$ of its row space into a vector $\mathbf{u}_k$ of its column space. Similarly, from equation [1.34], we have: $\mathbf{U}^H\mathbf{A} = \mathbf{\Sigma V}^H$, which gives $\mathbf{u}_k^H\mathbf{A} = \sigma_k\mathbf{v}_k^H$, or, after transconjugating both sides:

$$\mathbf{A}^H\mathbf{u}_k = \sigma_k\mathbf{v}_k \ , \ \ k \in \langle r \rangle. \tag{1.41}$$

From the equations [1.40] and [1.41], we deduce that:

$$\mathbf{A}^H\mathbf{Av}_k = \sigma_k\mathbf{A}^H\mathbf{u}_k = \sigma_k^2\mathbf{v}_k \tag{1.42}$$

$$\mathbf{AA}^H\mathbf{u}_k = \sigma_k\mathbf{Av}_k = \sigma_k^2\mathbf{u}_k. \tag{1.43}$$

By comparison with [1.35]–[1.37], we conclude that $\sigma_k^2 = \lambda_k$ is an eigenvalue of both $\mathbf{A}^H\mathbf{A}$ and $\mathbf{A}\mathbf{A}^H$. The above results are summarized in Table 1.9.

$\mathbf{A} \in C^{I \times J}, \mathbf{u} \in C^I, \mathbf{v} \in C^J$
$\mathbf{A}\mathbf{v} = \sigma\mathbf{u} \Leftrightarrow \sum_{j=1}^J a_{ij}v_j = \sigma u_i \ , \ i \in \langle I \rangle$
$\mathbf{A}^H\mathbf{u} = \sigma\mathbf{v} \Leftrightarrow \sum_{i=1}^I a_{ij}^* u_i = \sigma v_j \ , \ j \in \langle J \rangle$

Table 1.9. *Relations between left and right singular vectors*

1.5.2. *Reduced SVD and dyadic decomposition*

In the case where $\mathbf{A} \in \mathbb{C}^{I \times J}$ is of rank $r \leq \min(I, J)$, by partitioning as $\mathbf{U} = [\mathbf{U}_r \, \mathbf{U}_{I-r}]$ and $\mathbf{V} = [\mathbf{V}_r \, \mathbf{V}_{J-r}]$, the SVD can be written in the following partitioned form:

$$\mathbf{A} = [\mathbf{U}_r \, \mathbf{U}_{I-r}] \begin{bmatrix} \boldsymbol{\Sigma}_r & \mathbf{0}_{r \times J-r} \\ \mathbf{0}_{I-r \times r} & \mathbf{0}_{I-r \times J-r} \end{bmatrix} \begin{bmatrix} \mathbf{V}_r^H \\ \mathbf{V}_{J-r}^H \end{bmatrix}, \qquad [1.44]$$

with $\mathbf{U}_r \in \mathbb{C}^{I \times r}, \mathbf{U}_{I-r} \in \mathbb{C}^{I \times I-r}, \mathbf{V}_r \in \mathbb{C}^{J \times r}, \mathbf{V}_{J-r} \in \mathbb{C}^{J \times J-r}$, and $\boldsymbol{\Sigma}_r \in \mathbb{R}^{r \times r}$.

After simplification, we obtain:

$$\mathbf{A} = \mathbf{U}_r \boldsymbol{\Sigma}_r \mathbf{V}_r^H . \qquad [1.45]$$

This simplified form is called the reduced SVD of $\mathbf{A}$. It is also known as the compact SVD of $\mathbf{A}$. It corresponds to the special case of the $\mathbf{U}\mathbf{R}\mathbf{V}^H$ decomposition described in [1.31] where the matrix $\mathbf{R}$ is diagonal. Using [1.45], we can interpret the reduced SVD in terms of a dyadic decomposition, i.e. as a sum of rank-one matrices:

$$\mathbf{A} = \sum_{k=1}^r \sigma_k \mathbf{u}_k \mathbf{v}_k^H . \qquad [1.46]$$

In the case of a multiple singular value σ_i with multiplicity n, the SVD is not unique. Indeed, if we suppose that $\sigma_i = \sigma_{i+1} = \cdots = \sigma_{i+n-1}$, then the sum of the terms in [1.46] associated with this singular value can be written as:

$$\sigma_i \sum_{k=i}^{i+n-1} \mathbf{u}_k \mathbf{v}_k^H = \sigma_i \mathbf{U}_{i,n} \mathbf{V}_{i,n}^H, \qquad [1.47]$$

where $\mathbf{U}_{i,n} = [\mathbf{u}_i \ \cdots \ \mathbf{u}_{i+n-1}]$ and $\mathbf{V}_{i,n} = [\mathbf{v}_i \ \cdots \ \mathbf{v}_{i+n-1}]$.

For any unitary matrix $\mathbf{Q} \in \mathbb{C}^{n \times n}$, equation [1.47] is equivalent to:

$$\sigma_i (\mathbf{U}_{i,n} \mathbf{Q})(\mathbf{V}_{i,n} \mathbf{Q})^H = \sigma_i \mathbf{U}_{i,n} \mathbf{Q}\mathbf{Q}^H \mathbf{V}_{i,n}^H = \sigma_i \mathbf{U}_{i,n} \mathbf{V}_{i,n}^H, \qquad [1.48]$$

which shows that the left and right singular vectors associated with the multiple singular value σ_i are unique up to a unitary matrix.

If σ_k is a simple singular value ($\sigma_m \neq \sigma_k, \forall m \neq k$), then the singular vectors $\mathbf{u}_k$ and $\mathbf{v}_k$ are unique up to a factor $e^{j\theta}$, with $j^2 = -1$ and $\theta \in \mathbb{R}$, i.e.:

$$(e^{j\theta}\mathbf{u}_k)(e^{j\theta}\mathbf{v}_k)^H = (e^{j\theta}\mathbf{u}_k)(e^{-j\theta}\mathbf{v}_k^H) = \mathbf{u}_k\mathbf{v}_k^H .$$

Case of a full-rank matrix

In the case of a full-rank matrix, the SVD takes one of two possible forms.

If $\mathbf{A} \in \mathbb{C}^{I \times J}$ has full row rank ($r = I \leq J$), the reduced SVD can be rewritten as:

$$\mathbf{A} = \mathbf{U}\left[\boldsymbol{\Sigma}_I \; \mathbf{0}_{I \times J-I}\right]\begin{bmatrix} \mathbf{V}_I^H \\ \mathbf{V}_{J-I}^H \end{bmatrix} = \mathbf{U}\boldsymbol{\Sigma}_I\mathbf{V}_I^H, \qquad [1.49]$$

where $\mathbf{U} \in \mathbb{C}^{I \times I}$ is unitary, $\mathbf{V}_I \in \mathbb{C}^{J \times I}$ and $\mathbf{V}_{J-I} \in \mathbb{C}^{J \times (J-I)}$ are column orthonormal, and $\boldsymbol{\Sigma}_I \in \mathbb{R}^{I \times I}$ is a square matrix of order I.

If $\mathbf{A} \in \mathbb{C}^{I \times J}$ has full column rank ($r = J \leq I$), the reduced SVD simplifies as follows:

$$\mathbf{A} = [\mathbf{U}_J \; \mathbf{U}_{I-J}]\begin{bmatrix} \boldsymbol{\Sigma}_J \\ \mathbf{0}_{I-J \times J} \end{bmatrix}\mathbf{V}^H = \mathbf{U}_J\boldsymbol{\Sigma}_J\mathbf{V}^H, \qquad [1.50]$$

where $\mathbf{U}_J \in \mathbb{C}^{I \times J}$ and $\mathbf{U}_{I-J} \in \mathbb{C}^{I \times (I-J)}$ are column orthonormal, $\mathbf{V} \in \mathbb{C}^{J \times J}$ is unitary and $\boldsymbol{\Sigma}_J \in \mathbb{R}^{J \times J}$ is a square matrix of order J.

The key results about the SVD are summarized in Table 1.10.

From the reduced SVD [1.45], we deduce $\mathbf{A}^H = \mathbf{V}_r\boldsymbol{\Sigma}_r\mathbf{U}_r^H$. Noting that $\boldsymbol{\Sigma}_r$ is invertible and that $\mathbf{V}_r$ and $\mathbf{U}_r$ are column orthonormal ($\mathbf{V}_r^H\mathbf{V}_r = \mathbf{U}_r^H\mathbf{U}_r = \mathbf{I}_r$), we can deduce the relations listed in Table 1.11, which encompass the relations of Table 1.9 for the singular vectors associated with non-zero singular values. For $k \in \langle r \rangle$, the column vectors $\mathbf{u}_k$ and $\mathbf{v}_k$ of $\mathbf{U}_r$ and $\mathbf{V}_r$ form two orthonormal bases. From the relations in Table 1.11, we deduce:

$$\langle \mathbf{A}\mathbf{v}_i, \mathbf{A}\mathbf{v}_j \rangle = \langle \sigma_i\mathbf{u}_i, \sigma_j\mathbf{u}_j \rangle = \sigma_i\sigma_j\langle \mathbf{u}_i, \mathbf{u}_j \rangle = \sigma_i^2\delta_{ij} \qquad [1.51]$$

$$\langle \mathbf{A}^H\mathbf{u}_i, \mathbf{A}^H\mathbf{u}_j \rangle = \langle \sigma_i\mathbf{v}_i, \sigma_j\mathbf{v}_j \rangle = \sigma_i^2\delta_{ij}. \qquad [1.52]$$

$$\mathbf{A} \in \mathbb{C}^{I \times J} \quad (\mathbf{A} \in \mathbb{R}^{I \times J}) \quad \text{of rank } r$$

SVD

$$\mathbf{A} = \mathbf{U} \boldsymbol{\Sigma} \mathbf{V}^H \quad (\mathbf{A} = \mathbf{U} \boldsymbol{\Sigma} \mathbf{V}^T)$$
$$\mathbf{U} \in \mathbb{C}^{I \times I}, \mathbf{V} \in \mathbb{C}^{J \times J} \text{ unitary}; \boldsymbol{\Sigma} \in \mathbb{R}^{I \times J} \text{ pseudo-diagonal}$$
$$(\mathbf{U} \in \mathbb{R}^{I \times I}, \mathbf{V} \in \mathbb{R}^{J \times J} \text{ orthogonal}; \boldsymbol{\Sigma} \in \mathbb{R}^{I \times J} \text{ pseudo-diagonal})$$

Columns of $\mathbf{U}$ are eigenvectors of $\mathbf{A}\mathbf{A}^H$ $(\mathbf{A}\mathbf{A}^T)$, called left singular vectors
Columns of $\mathbf{V}$ are eigenvectors of $\mathbf{A}^H\mathbf{A}$ $(\mathbf{A}^T\mathbf{A})$, called right singular vectors
$\boldsymbol{\Sigma} = \text{diag}(\sigma_1, \cdots, \sigma_{\min(I,J)})$ with: $\sigma_1 \geq \cdots \geq \sigma_r > 0$; $\sigma_{r+1} = \cdots = \sigma_{\min(I,J)} = 0$

Reduced SVD

$$\mathbf{A} = \mathbf{U}_r \boldsymbol{\Sigma}_r \mathbf{V}_r^H \quad (\mathbf{A} = \mathbf{U}_r \boldsymbol{\Sigma}_r \mathbf{V}_r^T)$$
$$\mathbf{U}_r \in \mathbb{C}^{I \times r}, \mathbf{V}_r \in \mathbb{C}^{J \times r} \text{ column orthonormal}; \boldsymbol{\Sigma}_r = \text{diag}(\sigma_1, \cdots, \sigma_r) \in \mathbb{R}^{r \times r}$$
$$(\mathbf{U}_r \in \mathbb{R}^{I \times r}, \mathbf{V}_r \in \mathbb{R}^{J \times r} \text{ column orthonormal}; \boldsymbol{\Sigma}_r = \text{diag}(\sigma_1, \cdots, \sigma_r) \in \mathbb{R}^{r \times r})$$

Dyadic decomposition of A

$$\mathbf{A} = \sum_{k=1}^r \sigma_k \mathbf{u}_k \mathbf{v}_k^H \quad (\mathbf{A} = \sum_{k=1}^r \sigma_k \mathbf{u}_k \mathbf{v}_k^T)$$

Table 1.10. *Definition and properties of the SVD*

$\left| \mathbf{A} = \mathbf{U}_r \boldsymbol{\Sigma}_r \mathbf{V}_r^H \right|$	
$\mathbf{U}_r = \mathbf{A}\mathbf{V}_r \boldsymbol{\Sigma}_r^{-1}$	$\mathbf{V}_r = \mathbf{A}^H \mathbf{U}_r \boldsymbol{\Sigma}_r^{-1}$
$\mathbf{A}\mathbf{V}_r = \mathbf{U}_r \boldsymbol{\Sigma}_r$	$\mathbf{A}^H \mathbf{U}_r = \mathbf{V}_r \boldsymbol{\Sigma}_r$
$\mathbf{A}\mathbf{v}_k = \sigma_k \mathbf{u}_k , \ k \in \langle r \rangle$	$\mathbf{A}^H \mathbf{u}_k = \sigma_k \mathbf{v}_k , \ k \in \langle r \rangle$

Table 1.11. *Relations between the left and right singular vectors*

Case of a square matrix

For a square matrix $\mathbf{A} \in \mathbb{C}^{I \times I}$, noting that $\det(\mathbf{U}\boldsymbol{\Sigma}\mathbf{V}^H) = \det(\mathbf{U})\det(\boldsymbol{\Sigma})\det(\mathbf{V}^H)$, $\det(\mathbf{V}^H) = [\det(\mathbf{V})]^*$ and $\det(\mathbf{U}) = \det(\mathbf{V}) = \pm 1$ because $\mathbf{U}$ and $\mathbf{V}$ are unitary, we have:

$$|\det(\mathbf{A})| = |\det(\mathbf{U}\boldsymbol{\Sigma}\mathbf{V}^H)| = |\det(\boldsymbol{\Sigma})| = \prod_{i=1}^I \sigma_i. \tag{1.53}$$

From this relation, we can conclude that, if $\mathbf{A}$ is non-singular ($\det(\mathbf{A}) \neq 0$), then all the singular values are non-zero. Conversely, if $\mathbf{A}$ is singular, at least one singular value must be zero.

In the case where $\mathbf{A} \in \mathbb{C}^{I \times I}$ is a Hermitian matrix, it admits an EVD $\mathbf{PDP}^H$ that corresponds to the special case of the SVD where $\mathbf{U} = \mathbf{V} = \mathbf{P}$ and $\mathbf{\Sigma}^2 = \mathbf{D}$.

1.5.3. *SVD and fundamental subspaces associated with a matrix*

By considering the partitions $\mathbf{U} = [\mathbf{U}_r \, \mathbf{U}_{I-r}]$ and $\mathbf{V} = [\mathbf{V}_r \, \mathbf{V}_{J-r}]$ and the decompositions [1.29]–[1.30], we deduce that the span of the columns of $\mathbf{U}_r$ (respectively, $\mathbf{V}_r$) formed by the r first columns of $\mathbf{U}$ (respectively, $\mathbf{V}$) is the column space of $\mathbf{A}$ (respectively, the row space of $\mathbf{A}$), whereas the left null space of $\mathbf{A}$ (respectively, the null space of $\mathbf{A}$) is spanned by the $I - r$ last columns of $\mathbf{U}$ (respectively, the $J - r$ last columns of $\mathbf{V}$). These results are summarized in Table 1.12[7]. This leads us back to the interpretations given in proposition 1.9 for the $\mathbf{URV}^H$ decomposition.

$\mathbf{A} \in \mathbb{C}^{I \times J}$ of rank r
$C(\mathbf{A}) = C(\mathbf{AA}^H) = C(\mathbf{U}_r) = \text{Vect}(\mathbf{u}_1, \cdots, \mathbf{u}_r)$
$C(\mathbf{A}^H) = C(\mathbf{A}^H\mathbf{A}) = C(\mathbf{V}_r) = \text{Vect}(\mathbf{v}_1, \cdots, \mathbf{v}_r)$
$\mathcal{N}(\mathbf{A}) = C(\mathbf{V}_{J-r}) = \text{Vect}(\mathbf{v}_{r+1}, \cdots, \mathbf{v}_J)$
$\mathcal{N}(\mathbf{A}^H) = C(\mathbf{U}_{I-r}) = \text{Vect}(\mathbf{u}_{r+1}, \cdots, \mathbf{u}_I)$

Table 1.12. *SVD and fundamental subspaces*

1.5.4. *SVD and the Moore–Penrose pseudo-inverse*

From the expressions [1.34] and [1.45], we can deduce the formulae of the following proposition for the inverse and the Moore–Penrose pseudo-inverse of $\mathbf{A}$.

PROPOSITION 1.10.– If $\mathbf{A} \in \mathbb{C}^{I \times I}$ is invertible, its inverse may be expressed as a function of the SVD as follows:

$$\mathbf{A}^{-1} = \mathbf{V}\mathbf{\Sigma}^{-1}\mathbf{U}^H = \sum_{i=1}^{I} \frac{1}{\sigma_i} \mathbf{v}_i \mathbf{u}_i^H. \tag{1.54}$$

If $\mathbf{A} \in \mathbb{C}^{I \times I}$ is singular, or $\mathbf{A} \in \mathbb{C}^{I \times J}$ is rectangular, of rank r, then the Moore–Penrose pseudo-inverse can be expressed as a function of the reduced SVD as follows:

$$\mathbf{A}^{\dagger} = \mathbf{V}_r \mathbf{\Sigma}_r^{-1} \mathbf{U}_r^H = \sum_{k=1}^{r} \frac{1}{\sigma_k} \mathbf{v}_k \mathbf{u}_k^H \in \mathbb{C}^{J \times I}. \tag{1.55}$$

---

7 The notation $\text{Vect}(\mathbf{u}_1, \cdots, \mathbf{u}_r)$ denotes the subspace spanned by $\{\mathbf{u}_1, \cdots, \mathbf{u}_r\}$.

PROOF.– Using the expression [1.34] of the SVD and taking into account the fact that $\mathbf{U}$ and $\mathbf{V}$ are unitary ($\mathbf{U}^{-1} = \mathbf{U}^H, \mathbf{V}^{-1} = \mathbf{V}^H$), we obtain:

$$\mathbf{A}^{-1} = (\mathbf{U}\mathbf{\Sigma}\mathbf{V}^H)^{-1} = (\mathbf{V}^H)^{-1}\mathbf{\Sigma}^{-1}\mathbf{U}^{-1} = \mathbf{V}\mathbf{\Sigma}^{-1}\mathbf{U}^H.$$

Similarly, using the expression [1.45] of the reduced SVD and the orthonormality properties $\mathbf{U}_r^H\mathbf{U}_r = \mathbf{V}_r^H\mathbf{V}_r = \mathbf{I}_r$, it is easy to check that [1.55] satisfies the four relations defining the Moore–Penrose pseudo-inverse:

$$\mathbf{A}\mathbf{A}^\dagger = (\mathbf{U}_r\mathbf{\Sigma}_r\mathbf{V}_r^H)(\mathbf{V}_r\mathbf{\Sigma}_r^{-1}\mathbf{U}_r^H) = \mathbf{U}_r\mathbf{U}_r^H = (\mathbf{A}\mathbf{A}^\dagger)^H \qquad [1.56]$$

$$\mathbf{A}^\dagger\mathbf{A} = (\mathbf{V}_r\mathbf{\Sigma}_r^{-1}\mathbf{U}_r^H)(\mathbf{U}_r\mathbf{\Sigma}_r\mathbf{V}_r^H) = \mathbf{V}_r\mathbf{V}_r^H = (\mathbf{A}^\dagger\mathbf{A})^H \qquad [1.57]$$

$$\mathbf{A}\mathbf{A}^\dagger\mathbf{A} = \mathbf{U}_r\mathbf{U}_r^H(\mathbf{U}_r\mathbf{\Sigma}_r\mathbf{V}_r^H) = \mathbf{U}_r\mathbf{\Sigma}_r\mathbf{V}_r^H = \mathbf{A} \qquad [1.58]$$

$$\mathbf{A}^\dagger\mathbf{A}\mathbf{A}^\dagger = \mathbf{V}_r\mathbf{V}_r^H(\mathbf{V}_r\mathbf{\Sigma}_r^{-1}\mathbf{U}_r^H) = \mathbf{V}_r\mathbf{\Sigma}_r^{-1}\mathbf{U}_r^H = \mathbf{A}^\dagger, \qquad [1.59]$$

which concludes the proof of the proposition. □

REMARK 1.11.– If $\mathbf{A}$ has full row or column rank, then, using the expressions [1.49] and [1.50] of the reduced SVD, we can rewrite the pseudo-inverse [1.55] as follows:

$$\mathbf{A}^\dagger = (\mathbf{V}_I^H)^\dagger \mathbf{\Sigma}_I^{-1}\mathbf{U}^H \text{ and } \mathbf{A}^\dagger = \mathbf{V}\mathbf{\Sigma}_J^{-1}\mathbf{U}_J^\dagger, \qquad [1.60]$$

with $\mathbf{U}_J^\dagger = \mathbf{U}_J^H, \mathbf{V}_I^\dagger = \mathbf{V}_I^H$, and so $(\mathbf{V}_I^H)^\dagger = \mathbf{V}_I$.

1.5.5. *SVD computation*

There are several algorithms for computing the SVD. Table 1.13 describes the SVD computation based on the eigendecomposition [1.36] of $\mathbf{A}^H\mathbf{A}$ (replace $\mathbf{A}^H$ by $\mathbf{A}^T$ for the case of a real matrix) and the use of the relations in Table 1.11. A formula to compute the Moore–Penrose pseudo-inverse obtained from the reduced SVD is also recalled; this formula is very useful for computing the pseudo-inverse of a non-square matrix, as is often needed when solving systems of linear equations with the LS method.

Note that, in the case where $I < J$, the SVD can be computed from the eigendecomposition [1.35] of $\mathbf{A}\mathbf{A}^H = \mathbf{U}\mathbf{D}_1\mathbf{U}^H$, whose dimension is lower than that of $\mathbf{A}^H\mathbf{A}$, and the matrix $\mathbf{V}$ is computed using the relation $\mathbf{v}_k = \frac{1}{\sigma_k}\mathbf{A}^H\mathbf{u}_k, k \in \langle r \rangle$, given in Table 1.11. In this case, computing the SVD is numerically less expensive than if we use the eigendecomposition [1.36] of $\mathbf{A}^H\mathbf{A}$. Several different algorithms for computing the SVD are presented in Cline and Dhillon (2007).

SVD computation
$\mathbf{A} \in \mathbb{C}^{I \times J}$ of rank r
$\mathbf{A}^H \mathbf{A} = \mathbf{V} \mathbf{D}_2 \mathbf{V}^H$ $\mathbf{V} = \begin{bmatrix} \mathbf{v}_1 & \cdots & \mathbf{v}_J \end{bmatrix}, \mathbf{D}_2 = \operatorname{diag}(\lambda_1, \cdots, \lambda_r, 0, \cdots, 0) \in \mathbb{R}^{J \times J}$ with $r \leq J$
$\boldsymbol{\Sigma} = \begin{bmatrix} \boldsymbol{\Sigma}_r & \mathbf{0}_{r \times J-r} \\ \mathbf{0}_{I-r \times r} & \mathbf{0}_{I-r \times J-r} \end{bmatrix}, \ \boldsymbol{\Sigma}_r = \operatorname{diag}(\sigma_1, \cdots, \sigma_r), \ \sigma_k = \sqrt{\lambda_k}, \ k \in \langle r \rangle$
$\mathbf{U}_r = \mathbf{A} \mathbf{V}_r \boldsymbol{\Sigma}_r^{-1}, \ \mathbf{u}_k = \frac{1}{\sigma_k} \mathbf{A} \mathbf{v}_k, \ k = 1, \cdots, r$ $\mathbf{u}_{r+1}, \cdots, \mathbf{u}_I$ chosen such that the columns of $\mathbf{U}$ form an orthonormal set of vectors
$\mathbf{A} = \mathbf{U} \boldsymbol{\Sigma} \mathbf{V}^H$
Reduced SVD
$\mathbf{A} = \mathbf{U}_r \boldsymbol{\Sigma}_r \mathbf{V}_r^H$
Computation of the Moore-Penrose pseudo-inverse
$\mathbf{A}^\dagger = \mathbf{V}_r \boldsymbol{\Sigma}_r^{-1} \mathbf{U}_r^H \in \mathbb{C}^{J \times I}$

Table 1.13. *Computation of an SVD and of the Moore-Penrose pseudo-inverse*

1.5.6. *SVD and matrix norms*

Let us begin by recalling the definition of a matrix norm. We will then describe different types of matrix norm.

DEFINITION.– *Given any matrix* $\mathbf{A} \in \mathbb{K}^{I \times J}$, *a matrix norm is a function, denoted* $\|.\| : \mathbb{K}^{I \times J} \to \mathbb{R}^+$, *that satisfies the following properties:*

Positivity: $\|\mathbf{A}\| \geq 0, \ \|\mathbf{A}\| = 0 \Leftrightarrow \mathbf{A} = \mathbf{0}.$ [1.61]

Homogeneity: $\|\alpha \mathbf{A}\| = |\alpha| \|\mathbf{A}\|, \ \forall \alpha \in \mathbb{K}.$ [1.62]

Triangle inequality: $\|\mathbf{A} + \mathbf{B}\| \leq \|\mathbf{A}\| + \|\mathbf{B}\|, \ \forall \mathbf{A}, \mathbf{B} \in \mathbb{K}^{I \times J}.$ [1.63]

If we also have $\|\mathbf{AB}\| \leq \|\mathbf{A}\| \|\mathbf{B}\|$ *for* $\mathbf{A} \in \mathbb{K}^{I \times J}, \mathbf{B} \in \mathbb{K}^{J \times K}$, *then we say that the norm is submultiplicative.*

Hölder norm: For $\mathbf{A} \in \mathbb{K}^{I \times J}$, we define the Hölder norm, also known as the l_p-norm, as follows:

$$\|\mathbf{A}\|_{l_p} \triangleq \left(\sum_{i=1}^{I} \sum_{j=1}^{J} |a_{ij}|^p \right)^{1/p}, \ p \geq 1. \qquad [1.64]$$

This norm is equivalent to the l_p-norm of the vector $\operatorname{vec}(\mathbf{A})$.

Frobenius norm: For $p = 2$, we obtain the Frobenius norm, denoted $\|\mathbf{A}\|_F$:

$$\|\mathbf{A}\|_F^2 \triangleq \sum_{i=1}^I \sum_{j=1}^J |a_{ij}|^2 = \mathrm{tr}(\mathbf{A}^H \mathbf{A}) = \mathrm{tr}(\mathbf{A}\mathbf{A}^H)$$

$$= \mathrm{tr}(\mathbf{U}\mathbf{\Sigma}\mathbf{V}^H \mathbf{V}\mathbf{\Sigma}^H \mathbf{U}^H) = \mathrm{tr}(\mathbf{U}\mathbf{\Sigma}\mathbf{\Sigma}^H \mathbf{U}^H) = \mathrm{tr}(\mathbf{U}^H \mathbf{U}\,\mathbf{\Sigma}\mathbf{\Sigma}^H)$$

$$= \mathrm{tr}(\mathbf{\Sigma}\mathbf{\Sigma}^H) = \sum_{n=1}^r \sigma_n^2 = \|\boldsymbol{\sigma}(\mathbf{A})\|_2^2, \qquad [1.65]$$

where r is the rank of $\mathbf{A}$ and $\boldsymbol{\sigma}(\mathbf{A}) \triangleq [\sigma_1, \cdots, \sigma_r]^T$ denotes the vector of non-zero singular values. This norm is also equal to the Hermitian norm of a vectorized form of $\mathbf{A}$. Its square is equal to the sum of the squares of the Hermitian norms of the row vectors or column vectors of $\mathbf{A}$:

$$\|\mathbf{A}\|_F^2 = \|\mathrm{vec}(\mathbf{A})\|_2^2 = \sum_{i=1}^I \|\mathbf{A}_{i.}\|_2^2 = \sum_{j=1}^J \|\mathbf{A}_{.j}\|_2^2.$$

REMARK 1.12.–

– The Frobenius norm is submultiplicative: $\|\mathbf{A}\mathbf{B}\|_F \le \|\mathbf{A}\|_F \|\mathbf{B}\|_F$.

– This norm is left unchanged if $\mathbf{A} \in \mathbb{C}^{I \times J}$ is pre-multiplied by a unitary matrix $\mathbf{P} \in \mathbb{C}^{I \times I}$ or post-multiplied by a unitary matrix $\mathbf{Q} \in \mathbb{C}^{J \times J}$: $\|\mathbf{P}\mathbf{A}\|_F^2 = \mathrm{tr}(\mathbf{A}^H \mathbf{P}^H \mathbf{P}\mathbf{A}) = \mathrm{tr}(\mathbf{A}^H \mathbf{A}) = \|\mathbf{A}\|_F^2$. Similarly, $\|\mathbf{A}\mathbf{Q}\|_F^2 = \|\mathbf{A}\|_F^2$.

Schatten norm: The Schatten p-norm is defined as:

$$\|\mathbf{A}\|_{\sigma_p} \triangleq \|\boldsymbol{\sigma}(\mathbf{A})\|_{l_p} = \left(\sum_{n=1}^r \sigma_n^p \right)^{1/p}, \qquad [1.66]$$

i.e. the l_p-norm of the vector of singular values. We therefore have $\|\mathbf{A}\|_F = \|\mathbf{A}\|_{\sigma_2}$.

Nuclear norm: The nuclear norm, denoted $\|\mathbf{A}\|_*$, is equal to the sum of the singular values, i.e. the l_1-norm of $\boldsymbol{\sigma}(\mathbf{A})$, which also coincides with the Schatten 1-norm:

$$\|\mathbf{A}\|_* \triangleq \sum_{n=1}^r \sigma_n = \|\boldsymbol{\sigma}(\mathbf{A})\|_{l_1} = \|\mathbf{A}\|_{\sigma_1}. \qquad [1.67]$$

Induced norms: For $\mathbf{A} \in \mathbb{K}^{I \times J}$, we can also define matrix norms induced by vector norms as follows:

$$\|\mathbf{A}\| \triangleq \max_{\|\mathbf{x}\|=1} \|\mathbf{A}\mathbf{x}\| \quad \text{for} \ \mathbf{x} \in \mathbb{K}^J. \qquad [1.68]$$

This matrix norm, induced by the Euclidean vector norm and denoted $\|\mathbf{A}\|_2$, is such that:

$$\|\mathbf{A}\|_2 \triangleq \max_{\|\mathbf{x}\|_2=1} \|\mathbf{A}\mathbf{x}\|_2 = \sigma_1, \qquad [1.69]$$

where σ_1 is the largest singular value of $\mathbf{A}$. It is called the spectral norm.

If $\mathbf{A} \in \mathbb{K}^{I \times I}$ is non-singular, we have:

$$\|\mathbf{A}^{-1}\|_2 = \frac{1}{\min_{\|\mathbf{x}\|_2=1} \|\mathbf{A}\mathbf{x}\|_2} = \frac{1}{\sigma_I}, \qquad [1.70]$$

where σ_I is the smallest singular value of $\mathbf{A}$.

We also have:

$$\|\mathbf{A}\|_1 \triangleq \max_{\|\mathbf{x}\|_{l_1}=1} \|\mathbf{A}\mathbf{x}\|_{l_1} = \max_j \sum_{i=1}^{I} |a_{ij}|, \qquad [1.71]$$

$$\|\mathbf{A}\|_\infty \triangleq \max_{\|\mathbf{x}\|_\infty=1} \|\mathbf{A}\mathbf{x}\|_\infty = \max_i \sum_{j=1}^{J} |a_{ij}|. \qquad [1.72]$$

The norm $\|\mathbf{A}\|_1$ (respectively, $\|\mathbf{A}\|_\infty$) is therefore equal to the largest sum of the absolute values of the elements of a column (respectively, row) over the set of columns (respectively, rows). These norms are called the maximum absolute column sum and maximum absolute row sum norms, respectively.

The main matrix norms are summarized in Table 1.14.

Norms	Expressions
Hölder	$\|\mathbf{A}\|_{l_p} = \left(\sum_{i=1}^{I} \sum_{j=1}^{J} |a_{ij}|^p \right)^{1/p}$, $p \geq 1$
Frobenius	$\|\mathbf{A}\|_F^2 = \sum_{i=1}^{I} \sum_{j=1}^{J} |a_{ij}|^2 = \mathrm{tr}(\mathbf{A}^H \mathbf{A}) = \mathrm{tr}(\mathbf{A}\mathbf{A}^H) = \sum_{n=1}^{r} \sigma_n^2$
Schatten	$\|\mathbf{A}\|_{\sigma_p} = \|\boldsymbol{\sigma}(\mathbf{A})\|_{l_p} = \left(\sum_{n=1}^{r} \sigma_n^p \right)^{1/p}$
Nuclear	$\|\mathbf{A}\|_* = \sum_{n=1}^{r} \sigma_n = \|\boldsymbol{\sigma}(\mathbf{A})\|_{l_1} = \|\mathbf{A}\|_{\sigma_1}$
	Induced norms
	$\|\mathbf{A}\|_1 = \max_{\|\mathbf{x}\|_{l_1}=1} \|\mathbf{A}\mathbf{x}\|_{l_1} = \max_j \sum_{i=1}^{I} |a_{ij}|$
Spectral	$\|\mathbf{A}\|_2 = \max_{\|\mathbf{x}\|_2=1} \|\mathbf{A}\mathbf{x}\|_2 = \sigma_1$
	$\|\mathbf{A}\|_\infty = \max_{\|\mathbf{x}\|_\infty=1} \|\mathbf{A}\mathbf{x}\|_\infty = \max_i \sum_{j=1}^{J} |a_{ij}|$

Table 1.14. *Matrix norms*

1.5.7. *SVD and low-rank matrix approximation*

In the next proposition, we give the expression for the best rank-k approximation of a matrix of rank $r \geq k$.

PROPOSITION 1.13.– Let $\mathbf{A} \in \mathbb{K}^{I \times J}$ be a matrix of rank r. The best approximation of rank $k \leq r$ of $\mathbf{A}$ is given by:

$$\mathbf{A}_k = \mathbf{U}_k \mathbf{\Sigma}_k \mathbf{V}_k^H = \sum_{n=1}^{k} \sigma_n \mathbf{u}_n \mathbf{v}_n^H. \qquad [1.73]$$

This approximation, known as the Eckart–Young theorem (1936), is given by the SVD truncated to order k, i.e. the sum of the first k terms in the dyadic decomposition of $\mathbf{A}$, associated with the k largest singular values.

The approximation errors obtained when minimizing the Frobenius and spectral norms are given by:

$$\|\mathbf{A} - \mathbf{A}_k\|_F^2 = \sum_{n=k+1}^{r} \sigma_n^2 \;, \quad \|\mathbf{A} - \mathbf{A}_k\|_2 = \sigma_{k+1}. \qquad [1.74]$$

The square of the Frobenius norm of the rank-k approximation error is therefore equal to the sum of the squares of the $r - k$ smallest non-zero singular values, whereas the spectral norm of the approximation error is equal to the first singular value that is not taken into account in the approximation.

PROOF.– Using the best approximation $\mathbf{A}_k$ of rank $k \leq r$, the SVD can be written as:

$$\mathbf{A} = [\mathbf{U}_k \, \mathbf{U}_{r-k}] \begin{bmatrix} \mathbf{\Sigma}_k & \mathbf{0} \\ \mathbf{0} & \mathbf{\Sigma}_{r-k} \end{bmatrix} \begin{bmatrix} \mathbf{V}_k^H \\ \mathbf{V}_{r-k}^H \end{bmatrix}, \qquad [1.75]$$

where $\mathbf{U}_k \in \mathbb{K}^{I \times k}$ ($\mathbf{V}_k \in \mathbb{K}^{J \times k}$) is a column orthonormal matrix composed of the k left (right) singular vectors of $\mathbf{A}$ associated with the k largest singular values, and $\mathbf{U}_{r-k} \in \mathbb{K}^{I \times r-k}$ ($\mathbf{V}_{r-k} \in \mathbb{K}^{J \times r-k}$) is a column orthonormal matrix composed of the $r - k$ left (right) singular vectors associated with the $r - k$ smallest non-zero singular values. The diagonal elements of the diagonal matrices $\mathbf{\Sigma}_k \in \mathbb{R}^{k \times k}$ and $\mathbf{\Sigma}_{r-k} \in \mathbb{R}^{(r-k) \times (r-k)}$ are the k largest and $r - k$ smallest non-zero singular values, respectively. We can expand $\mathbf{A}$ as follows:

$$\mathbf{A} = \mathbf{A}_k + \mathbf{U}_{r-k} \mathbf{\Sigma}_{r-k} \mathbf{V}_{r-k}^H, \qquad [1.76]$$

from which we deduce the rank-k approximation error: $\mathbf{A} - \mathbf{A}_k = \mathbf{U}_{r-k}\mathbf{\Sigma}_{r-k}\mathbf{V}_{r-k}^H$, and therefore:

$$\|\mathbf{A} - \mathbf{A}_k\|_F^2 = \|\mathbf{U}_{r-k}\mathbf{\Sigma}_{r-k}\mathbf{V}_{r-k}^H\|_F^2 = \text{tr}[(\mathbf{U}_{r-k}\mathbf{\Sigma}_{r-k}\mathbf{V}_{r-k}^H)(\mathbf{U}_{r-k}\mathbf{\Sigma}_{r-k}\mathbf{V}_{r-k}^H)^H]$$

$$= \text{tr}[\mathbf{U}_{r-k}\mathbf{\Sigma}_{r-k}\mathbf{\Sigma}_{r-k}^H\mathbf{U}_{r-k}^H] = \text{tr}[\mathbf{\Sigma}_{r-k}^2] = \sum_{n=k+1}^{r} \sigma_n^2. \tag{1.77}$$

Furthermore, from the definition [1.69], we deduce that the spectral norm of the approximation error is equal to the largest singular value of $\mathbf{U}_{r-k}\mathbf{\Sigma}_{r-k}\mathbf{V}_{r-k}^H$, i.e. the first singular value not taken into account in the approximation:

$$\|\mathbf{A} - \mathbf{A}_k\|_2 = \sigma_{k+1}. \tag{1.78}$$

This concludes the proof of the approximation errors. □

Table 1.15 summarizes the approximation formula for a matrix of rank r by a matrix of rank $k \leq r$, as well as the approximation errors obtained when minimizing the spectral and Frobenius norms.

Rank-k approximation formula
$\mathbf{A} \in \mathbb{C}^{I \times J}$ of rank $r \leq \min(I, J)$
$\mathbf{A}_k = \mathbf{U}_k\mathbf{\Sigma}_k\mathbf{V}_k^H = \sum_{n=1}^{k} \sigma_n \mathbf{u}_n \mathbf{v}_n^H$, $k \leq r$
Approximation errors
$\min_{\mathbf{B}\,/\,\text{rank}(\mathbf{B})=k} \|\mathbf{A} - \mathbf{B}\|_2 = \|\mathbf{A} - \mathbf{A}_k\|_2 = \sigma_{k+1}$
$\min_{\mathbf{B}\,/\,\text{rank}(\mathbf{B})=k} \|\mathbf{A} - \mathbf{B}\|_F^2 = \|\mathbf{A} - \mathbf{A}_k\|_F^2 = \sum_{n=k+1}^{r} \sigma_n^2$

Table 1.15. *Rank-k approximation and approximation errors*

Interpretation of the approximation in terms of orthogonal projection

The previous proposition leads us to an interpretation of the best rank-k approximation in terms of orthogonal projection.

PROPOSITION 1.14.– The matrix $\mathbf{A}_k$ can be interpreted in terms of projections:

$$\mathbf{A}_k = \mathbf{P}_{\mathbf{U}_k}\mathbf{A} = (\mathbf{U}_k\mathbf{U}_k^H)\mathbf{A} = \left(\sum_{n=1}^{k} \mathbf{u}_n\mathbf{u}_n^H\right)\mathbf{A} \tag{1.79}$$

$$= \mathbf{A}\mathbf{P}_{\mathbf{V}_k} = \mathbf{A}(\mathbf{V}_k\mathbf{V}_k^H) = \mathbf{A}\left(\sum_{n=1}^{k} \mathbf{v}_n\mathbf{v}_n^H\right), \tag{1.80}$$

where $\mathbf{P}_{\mathbf{U}_k} = \mathbf{U}_k\mathbf{U}_k^H$ is an orthogonal projection matrix[8] onto the column space of $\mathbf{U}_k$, i.e. the space spanned by the k left singular vectors of $\mathbf{A}$ associated with the k largest singular values. Similarly, $\mathbf{P}_{\mathbf{V}_k} = \mathbf{V}_k\mathbf{V}_k^H$ is an orthogonal projection matrix onto the column space of $\mathbf{V}_k$, i.e. the space spanned by the k right singular vectors of $\mathbf{A}$ associated with the k largest singular values.

PROOF.– From the expression [1.75] of $\mathbf{A}$, and noting that the orthonormality of the columns of $\mathbf{U}$ implies $\mathbf{U}_k^H\mathbf{U}_k = \mathbf{I}_k$ and $\mathbf{U}_k^H\mathbf{U}_{r-k} = \mathbf{0}_{k\times(r-k)}$, we have:

$$(\mathbf{U}_k\mathbf{U}_k^H)\mathbf{A} = (\mathbf{U}_k\mathbf{U}_k^H)[\mathbf{U}_k\ \mathbf{U}_{r-k}]\begin{bmatrix} \mathbf{\Sigma}_k & \mathbf{0} \\ \mathbf{0} & \mathbf{\Sigma}_{r-k} \end{bmatrix}\begin{bmatrix} \mathbf{V}_k^H \\ \mathbf{V}_{r-k}^H \end{bmatrix}$$

$$= \mathbf{U}_k\mathbf{U}_k^H\mathbf{U}_k\mathbf{\Sigma}_k\mathbf{V}_k^H = \mathbf{U}_k\mathbf{\Sigma}_k\mathbf{V}_k^H = \mathbf{A}_k. \tag{1.82}$$

The proof of [1.80] can be established in the same way by using the orthonormality of the columns of $\mathbf{V}$. $\square$

EXAMPLE 1.15.– Let $\mathbf{A} = \begin{bmatrix} 1 & 1 \\ 1 & 0 \\ 0 & 1 \end{bmatrix}$. We have $\mathbf{A}^T\mathbf{A} = \begin{bmatrix} 2 & 1 \\ 1 & 2 \end{bmatrix}$ and $\det(\mathbf{A}^T\mathbf{A} -$

$\lambda\mathbf{I}) = \lambda^2 - 4\lambda + 3 = 0$, which gives $\lambda_1 = 3, \lambda_2 = 1$, and $\mathbf{v}_1 = \frac{1}{\sqrt{2}}\begin{bmatrix} 1 \\ 1 \end{bmatrix}, \mathbf{v}_2 = \frac{1}{\sqrt{2}}\begin{bmatrix} 1 \\ -1 \end{bmatrix}$.

Hence, we deduce that:

$$\mathbf{V}_2 = [\mathbf{v}_1\ \mathbf{v}_2] = \frac{1}{\sqrt{2}}\begin{bmatrix} 1 & 1 \\ 1 & -1 \end{bmatrix}$$

$$\sigma_1 = \sqrt{\lambda_1} = \sqrt{3},\ \sigma_2 = \sqrt{\lambda_2} = 1 \Rightarrow \mathbf{\Sigma}_2 = \begin{bmatrix} \sqrt{3} & 0 \\ 0 & 1 \end{bmatrix}$$

$$\mathbf{U}_2 = \mathbf{A}\mathbf{V}_2\mathbf{\Sigma}_2^{-1} = \frac{1}{\sqrt{6}}\begin{bmatrix} 2 & 0 \\ 1 & \sqrt{3} \\ 1 & -\sqrt{3} \end{bmatrix}.$$

The matrix $\mathbf{A}$ therefore factorizes as follows:

$$\mathbf{A} = \mathbf{U}_2\mathbf{\Sigma}_2\mathbf{V}_2^T = \frac{1}{2\sqrt{3}}\begin{bmatrix} 2 & 0 \\ 1 & \sqrt{3} \\ 1 & -\sqrt{3} \end{bmatrix}\begin{bmatrix} \sqrt{3} & 0 \\ 0 & 1 \end{bmatrix}\begin{bmatrix} 1 & 1 \\ 1 & -1 \end{bmatrix},$$

---

8 *Definition*: A square matrix $\mathbf{P} \in \mathbb{K}^{n\times n}$ is an orthogonal projector if it is idempotent and symmetric in the case $\mathbb{K} = \mathbb{R}$ (Hermitian in the case $\mathbb{K} = \mathbb{C}$), i.e.

$$\mathbf{P}^2 = \mathbf{P} \quad \text{and} \quad \mathbf{P}^T = \mathbf{P} \quad (\mathbf{P}^H = \mathbf{P} \text{ in the complex case}). \tag{1.81}$$

The matrix $\mathbf{P}^\perp = \mathbf{I}_n - \mathbf{P}$ is called the orthogonal complement of $\mathbf{P}$.

and the Moore–Penrose pseudo-inverse of $\mathbf{A}$ is given by:

$$\mathbf{A}^\dagger = \mathbf{V}_2 \mathbf{\Sigma}_2^{-1} \mathbf{U}_2^T = \frac{1}{2\sqrt{3}} \begin{bmatrix} 1 & 1 \\ 1 & -1 \end{bmatrix} \begin{bmatrix} \frac{1}{\sqrt{3}} & 0 \\ 0 & 1 \end{bmatrix} \begin{bmatrix} 2 & 1 & 1 \\ 0 & \sqrt{3} & -\sqrt{3} \end{bmatrix}$$

$$= \frac{1}{3} \begin{bmatrix} 1 & 2 & -1 \\ 1 & -1 & 2 \end{bmatrix}.$$

The best rank-one approximation is equal to $\mathbf{A}_1 = \sigma_1 \mathbf{u}_1 \mathbf{v}_1^T = \frac{1}{2} \begin{bmatrix} 2 & 2 \\ 1 & 1 \\ 1 & 1 \end{bmatrix}$.

Furthermore, the approximation error is given by: $\mathbf{A} - \mathbf{A}_1 = \frac{1}{2} \begin{bmatrix} 0 & 0 \\ 1 & -1 \\ -1 & 1 \end{bmatrix}$, which

implies that $\|\mathbf{A} - \mathbf{A}_1\|_2^2 = \|\mathbf{A} - \mathbf{A}_1\|_F^2 = 1 = \sigma_2^2$.

1.5.8. *SVD and orthogonal projectors*

The SVD of a matrix $\mathbf{A} \in \mathbb{K}^{I \times J}$ of rank r allows us to define orthogonal projectors onto the fundamental subspaces recalled in Table 1.7. These projectors, denoted $\mathbf{P}_{C(\mathbf{A})}$, $\mathbf{P}_{C(\mathbf{A}^H)}$, $\mathbf{P}_{\mathcal{N}(\mathbf{A})}$ and $\mathbf{P}_{\mathcal{N}(\mathbf{A}^H)}$, are orthogonal projections onto the column space $C(\mathbf{A})$, the row space $C(\mathbf{A}^H)$ and the null spaces $\mathcal{N}(\mathbf{A})$ and $\mathcal{N}(\mathbf{A}^H)$, respectively. Using equations [1.56] and [1.57], as well as the relations in Table 1.8, we have (Meyer 2000):

$$\mathbf{A}\mathbf{A}^\dagger = \mathbf{U}_r \mathbf{U}_r^H = \mathbf{P}_{C(\mathbf{A})} \qquad [1.83]$$

$$\mathbf{A}^\dagger \mathbf{A} = \mathbf{V}_r \mathbf{V}_r^H = \mathbf{P}_{C(\mathbf{A}^H)} \qquad [1.84]$$

$$\mathbf{P}_{\mathcal{N}(\mathbf{A})} = \mathbf{I} - \mathbf{P}_{C(\mathbf{A}^H)} = \mathbf{I} - \mathbf{A}^\dagger \mathbf{A} \qquad [1.85]$$

$$\mathbf{P}_{\mathcal{N}(\mathbf{A}^H)} = \mathbf{I} - \mathbf{P}_{C(\mathbf{A})} = \mathbf{I} - \mathbf{A}\mathbf{A}^\dagger. \qquad [1.86]$$

1.5.9. *SVD and LS estimator*

PROPOSITION 1.16.– Given the overdetermined system of linear equations $\mathbf{y} = \mathbf{A}\mathbf{x}$, where $\mathbf{A} \in \mathbb{C}^{I \times J}$ has full column rank ($r_{\mathbf{A}} = J$, $I > J$), the vector $\mathbf{x} \in \mathbb{C}^J$ that minimizes the LS criterion $J_{\text{LS}}(\mathbf{x}) = \|\mathbf{y} - \mathbf{A}\mathbf{x}\|_2^2$ is given by:

$$\mathbf{x}_{\text{LS}} = (\mathbf{A}^H \mathbf{A})^{-1} \mathbf{A}^H \mathbf{y} = \mathbf{A}^\dagger \mathbf{y}, \qquad [1.87]$$

with the following minimum criterion value:

$$J_{\text{LS}}(\mathbf{x}_{\text{LS}}) = \|\mathbf{y}\|_2^2 - \|\mathbf{A}\mathbf{x}_{\text{LS}}\|_2^2 \qquad [1.88]$$

$$= \|\mathbf{P}_{\mathbf{A}}^\perp \mathbf{y}\|_2^2 = \mathbf{y}^H \mathbf{P}_{\mathbf{A}}^\perp \mathbf{y}, \qquad [1.89]$$

where $\mathbf{P}_{\mathbf{A}} \triangleq \mathbf{P}_{C(\mathbf{A})}$ is the orthogonal projector onto the space $C(\mathbf{A})$ defined in [1.83].

PROOF.– The criterion $J_{LS}(\mathbf{x})$ is minimized by setting its gradient to zero for $\mathbf{x} = \mathbf{x}_{LS}$. We have:

$$\frac{\partial J_{LS}(\mathbf{x})}{\partial \mathbf{x}} = \frac{\partial}{\partial \mathbf{x}}\left([\mathbf{y} - \mathbf{Ax}]^H [\mathbf{y} - \mathbf{Ax}]\right)$$

$$= \frac{\partial}{\partial \mathbf{x}}(\mathbf{x}^H \mathbf{A}^H \mathbf{Ax} - \mathbf{x}^H \mathbf{A}^H \mathbf{y} - \mathbf{y}^H \mathbf{Ax} + \mathbf{y}^H \mathbf{y})$$

$$= 2(\mathbf{x}^H \mathbf{A}^H \mathbf{A} - \mathbf{y}^H \mathbf{A}). \qquad [1.90]$$

Hence, by canceling the gradient and transconjugating the right-hand side of the above equation, we obtain the LS solution $\mathbf{x}_{LS}$, satisfying the so-called normal equations:

$$\mathbf{A}^H \mathbf{A} \mathbf{x}_{LS} = \mathbf{A}^H \mathbf{y}. \qquad [1.91]$$

Since $\mathbf{A}$ is assumed to have full column rank, $\mathbf{A}^H \mathbf{A}$ is non-singular, which gives:

$$\mathbf{x}_{LS} = (\mathbf{A}^H \mathbf{A})^{-1} \mathbf{A}^H \mathbf{y} = \mathbf{A}^\dagger \mathbf{y}. \qquad [1.92]$$

We still need to check that this optimum is indeed a minimum, i.e. that the Hessian matrix is positive semi-definite: $\frac{\partial^2 J_{LS}(\mathbf{x})}{\partial \mathbf{x}^2} = 2\mathbf{A}^H \mathbf{A} \geq 0$.

The normal equations [1.91] can be rewritten as follows:

$$\mathbf{A}^H (\mathbf{y} - \mathbf{A}\mathbf{x}_{LS}) = \mathbf{0}, \qquad [1.93]$$

which shows that the estimation error $\mathbf{y} - \mathbf{A}\mathbf{x}_{LS}$ is orthogonal to the columns of $\mathbf{A}$. This result is called the orthogonality principle or condition. Using this orthogonality condition, the minimum criterion value is given by:

$$J_{LS}(\mathbf{x}_{LS}) = [\mathbf{y} - \mathbf{A}\mathbf{x}_{LS}]^H [\mathbf{y} - \mathbf{A}\mathbf{x}_{LS}] \qquad [1.94]$$

$$= \mathbf{y}^H (\mathbf{y} - \mathbf{A}\mathbf{x}_{LS}) \quad \text{(by [1.93])}$$

$$= \|\mathbf{y}\|_2^2 - (\mathbf{A}^H \mathbf{y})^H \mathbf{x}_{LS}$$

$$= \|\mathbf{y}\|_2^2 - \left[\mathbf{A}^H \left[\mathbf{A}\mathbf{x}_{LS} + (\mathbf{y} - \mathbf{A}\mathbf{x}_{LS})\right]\right]^H \mathbf{x}_{LS}$$

$$= \|\mathbf{y}\|_2^2 - (\mathbf{A}^H \mathbf{A}\mathbf{x}_{LS})^H \mathbf{x}_{LS} \quad \text{(by [1.93])}$$

$$= \|\mathbf{y}\|_2^2 - \|\mathbf{A}\mathbf{x}_{LS}\|_2^2, \qquad [1.95]$$

which corresponds to equation [1.88].

Using the expression [1.92] of $\mathbf{x}_{LS}$, the estimated value of $\mathbf{y}$ is given by:

$$\widehat{\mathbf{y}} = \mathbf{A}\mathbf{x}_{LS} = \mathbf{A}\mathbf{A}^\dagger \mathbf{y} = \mathbf{P}_{C(\mathbf{A})}\mathbf{y}, \qquad [1.96]$$

where $\mathbf{P}_{C(\mathbf{A})} \triangleq \mathbf{A}\mathbf{A}^\dagger$ is the projector onto the column space of $\mathbf{A}$ defined in [1.83].

By the expression [1.96] and the idempotence property of the orthogonal complement $\mathbf{P}_{\mathbf{A}}^\perp = \mathbf{I} - \mathbf{P}_{\mathbf{A}} = \mathbf{I} - \mathbf{A}\mathbf{A}^\dagger$, i.e. $(\mathbf{P}_{\mathbf{A}}^\perp)^2 = \mathbf{P}_{\mathbf{A}}^\perp$, the minimum criterion value [1.88] can be rewritten as follows:

$$J_{\mathrm{LS}}(\mathbf{x}_{\mathrm{LS}}) = \|(\mathbf{I} - \mathbf{A}\mathbf{A}^\dagger)\mathbf{y}\|_2^2 = \|\mathbf{P}_{\mathbf{A}}^\perp \mathbf{y}\|_2^2 = \mathbf{y}^H \mathbf{P}_{\mathbf{A}}^\perp \mathbf{y}, \qquad [1.97]$$

which proves the expression [1.89] of the minimum criterion value $J_{\mathrm{LS}}(\mathbf{x}_{\mathrm{LS}})$. $\qquad \square$

EXAMPLE 1.17.– Consider the system of equations $\mathbf{y} = \mathbf{u}\,x$, with $\mathbf{y}, \mathbf{u} \in \mathbb{C}^I$, and $x \in \mathbb{C}$. The LS estimator of x is given as:

$$\mathbf{x}_{\mathrm{LS}} = \mathbf{u}^\dagger \mathbf{y} = (\mathbf{u}^H \mathbf{u})^{-1} \mathbf{u}^H \mathbf{y} = \frac{\langle \mathbf{y}, \mathbf{u} \rangle}{\|\mathbf{u}\|_2^2} = \frac{\sum_{i=1}^{I} y_i u_i^*}{\sum_{i=1}^{I} |u_i|^2}.$$

Case of a rank-deficient matrix

It is important to note that the LS solution is unique when $\mathbf{A}$ has full column rank. However, if $\mathbf{A}$ has deficient rank r, then $\dim[\mathcal{N}(\mathbf{A})] = J - r$, and there are infinitely many solutions, as stated in the following proposition.

PROPOSITION 1.18.– In the case where $\mathbf{A} \in \mathbb{C}^{I \times J}$ is of rank r, the set of solutions of the system of equations $\mathbf{y} = \mathbf{A}\mathbf{x}$, in the LS sense, is given by the following expression:

$$\mathbf{x}_{\mathrm{LS}} = \mathbf{A}^\dagger \mathbf{y} + (\mathbf{I} - \mathbf{A}^\dagger \mathbf{A})\mathbf{z} = \mathbf{A}^\dagger \mathbf{y} + \mathbf{P}_{\mathcal{N}(\mathbf{A})}\mathbf{z}, \quad \forall \mathbf{z} \in \mathbb{K}^J, \qquad [1.98]$$

where the pseudo-inverse $\mathbf{A}^\dagger$ can be computed using the formula [1.55] of the reduced SVD of $\mathbf{A}$, and $\mathbf{I} - \mathbf{A}^\dagger \mathbf{A} = \mathbf{P}_{\mathcal{N}(\mathbf{A})}$ is the orthogonal projector onto the null space $\mathcal{N}(\mathbf{A})$ defined in [1.85].

From the property $\mathbf{A}\mathbf{A}^\dagger \mathbf{A} = \mathbf{A}$ of the Moore–Penrose pseudo-inverse, we deduce that $\mathbf{A}\mathbf{P}_{\mathcal{N}(\mathbf{A})} = \mathbf{A} - \mathbf{A}\mathbf{A}^\dagger \mathbf{A} = \mathbf{0}$. Hence, by the definition [1.83], combining the expression [1.96] of the estimate $\widehat{\mathbf{y}}$ with [1.98] gives:

$$\widehat{\mathbf{y}} = \mathbf{A}\mathbf{x}_{\mathrm{LS}} = \mathbf{A}\mathbf{A}^\dagger \mathbf{y} + \mathbf{A}(\mathbf{I} - \mathbf{A}^\dagger \mathbf{A})\mathbf{z} = \mathbf{A}\mathbf{A}^\dagger \mathbf{y} = \mathbf{P}_{C(\mathbf{A})}\mathbf{y}. \qquad [1.99]$$

Thus, we recover the same expression as in the case where $\mathbf{A}$ has full column rank.

From the expression [1.98], we can conclude that the set of solutions in the LS sense is obtained by adding an element $\mathbf{P}_{\mathcal{N}(\mathbf{A})}\mathbf{z}$ to $\mathbf{A}^\dagger \mathbf{y}$; this element corresponds to the orthogonal projection of an arbitrary vector $\mathbf{z} \in \mathbb{K}^J$ onto the null space $\mathcal{N}(\mathbf{A})$, such that $\mathbf{A}\mathbf{P}_{\mathcal{N}(\mathbf{A})}\mathbf{z} = \mathbf{0}$.

Case of an ill-conditioned matrix and regularized LS solution

If the matrix $\mathbf{A}$ is ill-conditioned, i.e. close to the singularity, we can use Tikhonov's regularization method, which introduces a regularization term into the LS criterion:

$$J_{\mathrm{LS}}(\mathbf{x}) = \|\mathbf{y} - \mathbf{A}\mathbf{x}\|_2^2 + \|\mathbf{\Gamma}\mathbf{x}\|_2^2, \qquad [1.100]$$

where $\mathbf{\Gamma}$ is a real diagonal matrix with positive diagonal elements. The LS solution is then given by:

$$\mathbf{x}_{\mathbf{\Gamma}} = (\mathbf{A}^H\mathbf{A} + \mathbf{\Gamma}^2)^{-1}\mathbf{A}^H\mathbf{y}. \qquad [1.101]$$

The regularization works by adding positive terms to the diagonal of the matrix to be inverted. The larger these terms, the further we move away from the optimal LS solution [1.92].

If we choose $\mathbf{\Gamma} = \mathbf{0}$, we recover the non-regularized LS solution. In practice, we often choose $\mathbf{\Gamma} = \mathbf{I}$, which favors a solution with small norm, since the second term of the criterion [1.100] becomes $\|\mathbf{x}\|_2^2$.

Table 1.16 summarizes the key results established in this section.

$\mathbf{y} = \mathbf{A}\mathbf{x}, \mathbf{A} \in \mathbb{C}^{I \times J}$
Case where A has full column rank
$\mathbf{x}_{\mathrm{LS}} = \mathbf{A}^\dagger\mathbf{y} = (\mathbf{A}^H\mathbf{A})^{-1}\mathbf{A}^H\mathbf{y}$
Case where A has deficient rank
$\mathbf{x}_{\mathrm{LS}} = \mathbf{A}^\dagger\mathbf{y} + (\mathbf{I} - \mathbf{A}^\dagger\mathbf{A})\mathbf{z}, \quad \forall \mathbf{z} \in \mathbb{K}^J$ $\mathbf{A}^\dagger = \mathbf{V}_r\mathbf{\Sigma}_r^{-1}\mathbf{U}_r^H = \sum_{k=1}^{r} \frac{1}{\sigma_k}\mathbf{v}_k\mathbf{u}_k^H$
Case where A is ill-conditioned
$\mathbf{x}_{\mathbf{\Gamma}} = (\mathbf{A}^H\mathbf{A} + \mathbf{\Gamma}^2)^{-1}\mathbf{A}^H\mathbf{y}, \ \mathbf{\Gamma}$ diagonal

Table 1.16. *Least squares estimator*

1.5.10. *SVD and polar decomposition*

DEFINITION.– A polar decomposition of a square matrix $\mathbf{A} \in \mathbb{C}^{I \times I}$ is a decomposition of the form:

$$\mathbf{A} = \mathbf{Q}\mathbf{P} \ \text{ or } \ \mathbf{A} = \mathbf{S}\mathbf{Q}, \qquad [1.102]$$

where $\mathbf{Q}$ is a unitary matrix, and $\mathbf{P} = (\mathbf{A}^H\mathbf{A})^{1/2}$ and $\mathbf{S} = (\mathbf{A}\mathbf{A}^H)^{1/2}$ are positive semi-definite Hermitian matrices. Such a decomposition always exists, and it is unique if $\mathbf{A}$ is non-singular. If so, the matrices $\mathbf{P}$ and $\mathbf{S}$ are positive definite.

PROOF.– The decomposition $\mathbf{A} = \mathbf{QP}$ can be shown using the SVD. Indeed, since $\mathbf{V}$ is unitary, we can write:

$$\mathbf{A} = \mathbf{U\Sigma V}^H = \mathbf{U}(\mathbf{V}^H\mathbf{V})\mathbf{\Sigma V}^H = \mathbf{QP}, \qquad [1.103]$$

where $\mathbf{Q} \triangleq \mathbf{UV}^H$ and $\mathbf{P} \triangleq \mathbf{V\Sigma V}^H$. By these definitions, and the fact that $\mathbf{U}$ and $\mathbf{V}$ are unitary, we can deduce that:

$$\mathbf{Q}^H\mathbf{Q} = (\mathbf{UV}^H)^H\mathbf{UV}^H = \mathbf{V}(\mathbf{U}^H\mathbf{U})\mathbf{V}^H = \mathbf{VV}^H = \mathbf{I}. \qquad [1.104]$$

Similarly, we have $\mathbf{QQ}^H = \mathbf{I}$, and so $\mathbf{Q}$ is unitary. Furthermore, noting that $\mathbf{A}^H\mathbf{A} = \mathbf{V\Sigma}^2\mathbf{V}^H = (\mathbf{V\Sigma V}^H)^2$, we deduce that $\mathbf{P} = (\mathbf{A}^H\mathbf{A})^{1/2} \geq \mathbf{0}$, where, by definition, $\mathbf{P}^H = (\mathbf{V\Sigma V}^H)^H = \mathbf{P}$, which implies that $\mathbf{P}$ is Hermitian positive semi-definite and equal to a square root of $\mathbf{A}^H\mathbf{A}$. Similarly,

$$\mathbf{A} = \mathbf{U\Sigma V}^H = \mathbf{U\Sigma}(\mathbf{U}^H\mathbf{U})\mathbf{V}^H = \mathbf{SQ}, \qquad [1.105]$$

where $\mathbf{Q}$ is defined as before, and $\mathbf{S} \triangleq \mathbf{U\Sigma U}^H$. It is easy to check that $\mathbf{S}$ can also be written $\mathbf{S} = (\mathbf{AA}^H)^{1/2}$, i.e. as a square root of $\mathbf{AA}^H$. □

The polar decompositions [1.103] and [1.105] are said to be right and left polar decompositions, respectively. In the case of a real square matrix $\mathbf{A}$, the matrix $\mathbf{Q}$ is orthogonal, whereas $\mathbf{P}$ and $\mathbf{S}$ are symmetric positive semi-definite.

Case of a full-rank rectangular matrix

If $\mathbf{A}$ has full column rank ($r = J$), then, following the same approach as above, the reduced SVD [1.50] gives us the following polar decomposition:

$$\mathbf{A} = \mathbf{U}_J\mathbf{\Sigma}_J\mathbf{V}^H = \mathbf{U}_J(\mathbf{V}^H\mathbf{V})\mathbf{\Sigma}_J\mathbf{V}^H = \mathbf{QP}, \qquad [1.106]$$

where $\mathbf{U}_J \in \mathbb{C}^{I \times J}$ is column orthonormal, $\mathbf{V} \in \mathbb{C}^{J \times J}$ is unitary, $\mathbf{Q} \triangleq \mathbf{U}_J\mathbf{V}^H$ and $\mathbf{P} \triangleq \mathbf{V\Sigma}_J\mathbf{V}^H$. The matrix $\mathbf{Q}$ is then column orthonormal ($\mathbf{Q}^H\mathbf{Q} = \mathbf{VU}_J^H\mathbf{U}_J\mathbf{V}^H = \mathbf{VV}^H = \mathbf{I}_J$), and $\mathbf{P} = (\mathbf{A}^H\mathbf{A})^{1/2}$ is Hermitian positive definite.

In the case where $\mathbf{A}$ has full row rank ($r = I \leq J$), then, from the reduced SVD [1.49], we obtain the following polar decomposition:

$$\mathbf{A} = \mathbf{U\Sigma}_I\mathbf{V}_I^H = \mathbf{U\Sigma}_I(\mathbf{U}^H\mathbf{U})\mathbf{V}_I^H = \mathbf{SQ}, \qquad [1.107]$$

where $\mathbf{U} \in \mathbb{C}^{I \times I}$ is unitary, $\mathbf{V}_I \in \mathbb{C}^{J \times I}$ is column orthonormal, $\mathbf{Q} \triangleq \mathbf{UV}_I^H$ and $\mathbf{S} \triangleq \mathbf{U\Sigma}_I\mathbf{U}^H$. The matrix $\mathbf{Q}$ is then row orthonormal ($\mathbf{QQ}^H = \mathbf{UV}_I^H\mathbf{V}_I\mathbf{U}^H = \mathbf{UU}^H = \mathbf{I}_I$), and $\mathbf{S} = (\mathbf{AA}^H)^{1/2}$ is Hermitian positive definite.

1.5.11. *SVD and PCA*

PCA is a widely used method for data analysis and compression. This method, also known as the Karhunen–Loève transform (KLT), is one of the techniques of factor analysis. From a statistical perspective, it decorrelates correlated random variables. This is called statistical orthogonalization, with the scalar product of two random variables x_i and x_j defined in terms of mathematical expectation and, therefore, as the correlation between variables: $\langle x_i, x_j \rangle = E[x_i x_j]$. The decorrelated variables are called principal components or, alternatively, features in the context of classification.

PCA can also be viewed as a method for reducing the dimensionality, whose goal is to find L latent variables, called factors, to represent a set of M observed random variables, where $L < M$ and often $L \ll M$.

PCA was introduced by Pearson (1901) to represent a system of points in a space with a reduced dimension. Today, it is used in many different fields of applications, such as denoising, collaborative filtering, signal separation and data visualization or classification. See Sanguansat (2012) for a review of the multidisciplinary applications of PCA. Its many extensions include independent component analysis (ICA) and multilinear PCA (MPCA). The latter technique, introduced by Kroonenberg and de Leeuw (1980), can be used to analyze tensors of multimodal data or for applications involving the reconstruction of 3D objects (Lu *et al.* 2008). Other extensions, such as kernel PCA, allow the nonlinear nature of the data to be taken into consideration, so that nonlinear features can be extracted to support data classification (Scholkopf *et al.* 1998; Mika *et al.* 1999).

1.5.11.1. *Principle of the method*

Consider a random vector $\mathbf{x} \in \mathbb{R}^M$ that characterizes a physical phenomenon observed over N sampling periods, where M might, for example, represent the number of sensors in a source separation problem or the number of objects in a classification problem. Suppose that the measurement vector $\mathbf{x}$ is centered. If not, the data can be centered by subtracting their statistical mean. The covariance matrix $\mathbf{C_x} = E[\mathbf{xx}^T]$, which is equal to the autocorrelation matrix in the case of centered data, is symmetric non-negative definite and therefore diagonalizable using its EVD $\mathbf{C_x} = \mathbf{P \Lambda P}^T$, where $\mathbf{P}$ is an orthogonal matrix formed by the eigenvectors, and $\mathbf{\Lambda}$ is a diagonal matrix whose diagonal elements are the non-negative eigenvalues $\lambda_m = \sigma_m^2 \geq 0$, with $m \in \langle M \rangle$, in decreasing order:

$$\mathbf{P} = [\mathbf{u}_1\, \mathbf{u}_2 \cdots \mathbf{u}_M], \quad \mathbf{\Lambda} = \begin{bmatrix} \lambda_1 & & & 0 \\ & \lambda_2 & & \\ & & \ddots & \\ 0 & & & \lambda_M \end{bmatrix}.$$

The principle of the PCA method is to linearly transform the vector $\mathbf{x}$ into a vector $\mathbf{y} = \mathbf{Qx}$ so as to (statistically) orthogonalize its components, i.e. decorrelate them, since decorrelating the components of $\mathbf{y}$ is equivalent to orthogonalizing them. This amounts to diagonalizing the covariance matrix $\mathbf{C_y}$. Indeed, if $[\mathbf{C_y}]_{i,j} = E[y_i y_j] = 0$ for $i \neq j$, then the components y_i and y_j are decorrelated. The diagonal elements $[\mathbf{C_y}]_{i,i} = E[y_i^2]$ are the variances of the components of $\mathbf{y}$. The diagonalization of the covariance matrix $\mathbf{C_y}$ is obtained by taking $\mathbf{Q} = \mathbf{P}^{-1} = \mathbf{P}^T$. Indeed, we have:

$$\mathbf{y} = \mathbf{P}^{-1}\mathbf{x}$$

$$\mathbf{C_y} = E[\mathbf{yy}^T]$$

$$= \mathbf{P}^{-1}E[\mathbf{xx}^T]\mathbf{P}^{-T} = \mathbf{P}^{-1}(\mathbf{P\Lambda P}^T)\mathbf{P}^{-T} = \mathbf{\Lambda}, \qquad [1.108]$$

and so $E[y_m^2] = \lambda_m$. The components of the vector $\mathbf{y} = \mathbf{P}^T\mathbf{x}$ are called the principal components. Since the eigenvalues λ_m are arranged in decreasing order, we have: $E[y_1^2] \geq E[y_2^2] \geq \cdots \geq E[y_M^2]$.

1.5.11.2. *PCA and variance maximization*

The transformation $\mathbf{y} = \mathbf{Qx} = [\mathbf{q}_1, \cdots, \mathbf{q}_M]^T\mathbf{x}$, where $\mathbf{Q}$ is expressed in terms of its rows $\mathbf{q}_m^T$, can also be determined by maximizing the variance of the principal components. Thus, the first principal component $y_1 = \mathbf{q}_1^T\mathbf{x}$ has variance:

$$E[y_1^2] = \mathbf{q}_1^T E[\mathbf{xx}^T]\mathbf{q}_1 = \mathbf{q}_1^T \mathbf{C_x}\mathbf{q}_1. \qquad [1.109]$$

Maximizing this variance subject to the constraint that $\mathbf{q}_1$ is a unit vector ($\|\mathbf{q}_1\|_2 = 1$) leads us to solve the constrained optimization problem:

$$\mathbf{u}_1 = \arg \max_{\|\mathbf{q}_1\|_2=1} \mathbf{q}_1^T \mathbf{C_x}\mathbf{q}_1, \qquad [1.110]$$

which is equivalent to maximizing the Rayleigh quotient, defined as $\frac{\mathbf{q}_1^T \mathbf{C_x}\mathbf{q}_1}{\mathbf{q}_1^T \mathbf{q}_1}$, under the constraint $\mathbf{q}_1^T \mathbf{q}_1 = 1$. The solution is given by the eigenvector associated with the largest eigenvalue of $\mathbf{C_x}$, i.e. the first column $\mathbf{u}_1$ of the matrix $\mathbf{P}$ of eigenvectors of $\mathbf{C_x}$. Furthermore, the maximum variance is equal to the largest eigenvalue $\lambda_1 = \sigma_1^2$.

Thus, we again find that the first principal component is equal to $y_1 = \mathbf{u}_1^T\mathbf{x}$, i.e. the first row of $\mathbf{y} = \mathbf{P}^T\mathbf{x}$, and its variance is equal to the first element of the diagonal matrix $\mathbf{\Lambda}$ of eigenvalues.

The second principal component y_2 is obtained similarly, by solving the following optimization problem:

$$\mathbf{u}_2 = \arg \max_{\|\mathbf{q}_2\|_2=1} E\left\{ \left[\mathbf{q}_2^T [\mathbf{x} - \mathbf{u}_1\mathbf{u}_1^T\mathbf{x}]\right]^2 \right\}, \qquad [1.111]$$

where $\mathbf{x} - \mathbf{u}_1\mathbf{u}_1^T\mathbf{x} = (\mathbf{I} - \mathbf{u}_1\mathbf{u}_1^T)\mathbf{x}$ is the error associated with reconstructing $\mathbf{x}$ from the first principal component. Since the matrix $\mathbf{I} - \mathbf{u}_1\mathbf{u}_1^T$ is the orthogonal complement of the projector onto the column space generated by $\mathbf{u}_1$, the reconstruction error is orthogonal to $\mathbf{u}_1$. Hence, maximizing the criterion [1.111] implies that the vector $\mathbf{u}_2$ must be orthogonal to $\mathbf{u}_1$.

As before, the solution is given by the eigenvector associated with the largest eigenvalue of the covariance matrix:

$$E\left\{ \left[(\mathbf{I} - \mathbf{u}_1\mathbf{u}_1^T)\mathbf{x}\right] \left[(\mathbf{I} - \mathbf{u}_1\mathbf{u}_1^T)\mathbf{x}\right]^T \right\} = [\mathbf{I} - \mathbf{u}_1\mathbf{u}_1^T]\mathbf{C_x}[\mathbf{I} - \mathbf{u}_1\mathbf{u}_1^T]^T$$

$$= \mathbf{C_x} - \lambda_1\mathbf{u}_1\mathbf{u}_1^T = \sum_{m=2}^{M} \lambda_m\mathbf{u}_m\mathbf{u}_m^T,$$

where the penultimate equality follows from the orthonormality property of the eigenvectors ($\mathbf{u}_m^T\mathbf{u}_1 = \delta_{m1}$) and the eigendecomposition of $\mathbf{C_x} = \sum_{m=1}^{M} \lambda_m\mathbf{u}_m\mathbf{u}_m^T$.

The solution is therefore given by the eigenvector $\mathbf{u}_2$ associated with the second eigenvalue of $\mathbf{C_x}$. Repeating this procedure m times yields the mth principal component by maximizing the criterion:

$$\mathbf{u}_m = \arg \max_{\|\mathbf{q}_m\|_2=1} E\left\{ \left[\mathbf{q}_m^T[\mathbf{I} - \sum_{i=1}^{m-1} \mathbf{u}_i\mathbf{u}_i^T]\mathbf{x}\right]^2 \right\}$$

$$= \arg \max_{\|\mathbf{q}_m\|_2=1} E\left\{ \left[\mathbf{q}_m^T[\mathbf{I} - \mathbf{U}_{m-1}\mathbf{U}_{m-1}^T]\mathbf{x}\right]^2 \right\}, \tag{1.112}$$

where $\mathbf{U}_{m-1}$ represents the matrix composed of the $m-1$ first columns of the matrix $\mathbf{P}$ of eigenvectors. The solution $\mathbf{u}_m$ of this maximization problem is orthogonal to the column subspace of $\mathbf{U}_{m-1}$. It is given by the eigenvector associated with the largest eigenvalue of the matrix $\mathbf{C_x} - \sum_{i=1}^{m-1} \lambda_i\mathbf{u}_i\mathbf{u}_i^T$, i.e. the eigenvector $\mathbf{u}_m$ associated with the mth eigenvalue of $\mathbf{C_x}$.

1.5.11.3. PCA and dimensionality reduction

From the perspective of reducing the dimensionality, consider the matrix $\mathbf{P}_L \in \mathbb{R}^{M \times L}$, consisting of the first L columns of $\mathbf{P}$, i.e. the L eigenvectors of $\mathbf{C_x}$ associated with the L largest eigenvalues. We can define the transformation $\mathbf{y}_L = \mathbf{P}_L^T\mathbf{x} \in \mathbb{R}^L$, which gives the L principal components, characterized by the L largest variances $E[y_m^2] = \sigma_m^2 = \lambda_m, m \in \langle L \rangle$. This transformation allows us to reduce the dimension of the vector $\mathbf{y}$ by retaining only the L principal components $y_m, m \in \langle L \rangle$.

The choice of L is often based on the following criterion. For some fixed threshold δ, we can choose L as the smallest value such that:

$$\frac{\sum_{i=1}^{L} \lambda_i}{\sum_{i=1}^{M} \lambda_i} \geq \delta, \qquad [1.113]$$

which is equivalent to finding L such that the L first eigenvalues satisfy the variance percentage δ defined in [1.113], i.e. the percentage of information in the original data that we wish to be able to recover by PCA after reducing the dimensionality.

1.5.11.4. *PCA of data*

In practice, we have N samples of the random signal $\mathbf{x}$, assumed to be centered. Let $\mathbf{X}_N = [\mathbf{x}_1, \cdots, \mathbf{x}_N] \in \mathbb{R}^{M \times N}$ be the matrix whose columns are these samples. In a classification problem, the dimensions M and N might, for example, represent the number of data points to be classified and the number of variables characterizing each data point, respectively. The covariance $\mathbf{C}_{\mathbf{x}_N}$ is then estimated using the empirical covariance, which is defined as follows:

$$\widehat{\mathbf{C}}_{\mathbf{x}_N} = \frac{1}{N} \sum_{n=1}^{N} \mathbf{x}_n \mathbf{x}_n^T = \frac{1}{N} \mathbf{X}_N \mathbf{X}_N^T. \qquad [1.114]$$

The transformation matrix that allows us to reduce the dimensionality of $\mathbf{X}_N$ can be determined by computing the EVD of the estimated covariance matrix $\widehat{\mathbf{C}}_{\mathbf{x}_N}$. It can also be determined by computing the SVD of $\mathbf{Y} = \frac{1}{\sqrt{N}} \mathbf{X}_N$. Indeed, we have $\mathbf{Y}\mathbf{Y}^T = \frac{1}{N} \mathbf{X}_N \mathbf{X}_N^T = \widehat{\mathbf{C}}_{\mathbf{x}_N}$, and, as we saw earlier, the SVD $\mathbf{Y} = \mathbf{U}\boldsymbol{\Sigma}\mathbf{V}^T$ gives the matrix $\mathbf{U}$ of left singular vectors of $\mathbf{Y}$, which is the matrix of eigenvectors of $\mathbf{Y}\mathbf{Y}^T = \mathbf{U}\boldsymbol{\Sigma}\boldsymbol{\Sigma}^T\mathbf{U}^T$, and, therefore, of $\widehat{\mathbf{C}}_{\mathbf{x}_N}$, i.e. $\mathbf{P} = \mathbf{U}$.

This link between PCA and the SVD of $\mathbf{Y}$ allows us to determine the matrix $\mathbf{P}_L$ directly from the data matrix $\mathbf{X}_N$, without the intermediate step of estimating the covariance matrix $\widehat{\mathbf{C}}_{\mathbf{x}_N} = \frac{1}{N} \mathbf{X}_N \mathbf{X}_N^T$.

With the goal of dimensionality reduction, we can choose the column orthonormal matrix $\mathbf{P}_L = \mathbf{U}_L \in \mathbb{R}^{M \times L}$, formed by the L first left singular vectors of $\mathbf{Y}$ associated with the L largest singular values. The reduced-dimension data matrix is then defined as:

$$\mathbf{Y}_L = \mathbf{U}_L^T \mathbf{Y} = \frac{1}{\sqrt{N}} \mathbf{U}_L^T \mathbf{X}_N \in \mathbb{R}^{L \times N}, \qquad [1.115]$$

whose empirical covariance matrix is:

$$\widehat{\mathbf{C}}_{\mathbf{Y}_L} = \mathbf{Y}_L \mathbf{Y}_L^T = \mathbf{U}_L^T \mathbf{Y} \mathbf{Y}^T \mathbf{U}_L = \mathbf{U}_L^T (\mathbf{U} \mathbf{\Sigma} \mathbf{\Sigma}^T \mathbf{U}^T) \mathbf{U}_L$$

$$= \mathbf{U}_L^T [\mathbf{U}_L \ \mathbf{U}_{M-L}] \begin{bmatrix} \mathbf{\Sigma}_L^2 & \mathbf{0} \\ \mathbf{0} & \mathbf{\Sigma}_{M-L}^2 \end{bmatrix} [\mathbf{U}_L \ \mathbf{U}_{M-L}]^T \mathbf{U}_L$$

$$= [\mathbf{U}_L^T \mathbf{U}_L \ \mathbf{U}_L^T \mathbf{U}_{M-L}] \begin{bmatrix} \mathbf{\Sigma}_L^2 & \mathbf{0} \\ \mathbf{0} & \mathbf{\Sigma}_{M-L}^2 \end{bmatrix} [\mathbf{U}_L^T \mathbf{U}_L \ \mathbf{U}_L^T \mathbf{U}_{M-L}]^T.$$

Since the column vectors of $\mathbf{U}$ are orthonormal, we have $\mathbf{U}_L^T \mathbf{U}_L = \mathbf{I}_L$ and $\mathbf{U}_L^T \mathbf{U}_{M-L} = \mathbf{0}_{L \times M-L}$. Hence, the covariance $\widehat{\mathbf{C}}_{\mathbf{Y}_L}$ simplifies as follows:

$$\widehat{\mathbf{C}}_{\mathbf{Y}_L} = \mathbf{\Sigma}_L^2. \qquad \qquad [1.116]$$

The diagonal elements of the diagonal matrix $\mathbf{\Sigma}_L^2 \in \mathbb{R}^{L \times L}$ are the L largest eigenvalues $\lambda_l = \sigma_l^2, l \in \langle L \rangle$ of the covariance matrix $\widehat{\mathbf{C}}_{\mathbf{X}_N}$.

The reconstruction of $\mathbf{X}_N$ is then deduced from [1.115] as $\widehat{\mathbf{X}}_N = \sqrt{N} \mathbf{U}_L \mathbf{Y}_L = \mathbf{U}_L \mathbf{U}_L^T \mathbf{X}_N \in \mathbb{R}^{M \times N}$, i.e. by projecting $\mathbf{X}_N$ onto the L-dimensional subspace spanned by the columns $\{\mathbf{u}_1, \cdots, \mathbf{u}_L\}$ of $\mathbf{U}_L$, which are the L principal left singular vectors of $\mathbf{Y}$ and therefore the L principal eigenvectors of the empirical covariance matrix $\widehat{\mathbf{C}}_{\mathbf{X}_N}$.

REMARK 1.19.– Taking $L = M$, which implies that $\mathbf{U}_L = \mathbf{U}$, we have $\widehat{\mathbf{X}}_N = \sqrt{N} \mathbf{U} \mathbf{Y} = \mathbf{U} \mathbf{U}^T \mathbf{X}_N = \mathbf{X}_N$. Thus, we recover the original data matrix exactly.

In conclusion, the PCA method allows both the dimension of the data matrix to be reduced and the data to be decorrelated via the transformation $\mathbf{Y}_L = \mathbf{U}_L^T \mathbf{Y}$. This objective is achieved by projecting the original data (column vectors of $\mathbf{X}_N$) onto the subspace spanned by the L principal left singular vectors of $\mathbf{Y}$ associated with the L largest singular values. These projections, called the principal components, are characterized by having maximum variance, the maximized variances being equal to the L largest eigenvalues of $\widehat{\mathbf{C}}_{\mathbf{X}_N}$.

1.5.11.5. PCA algorithm

The various steps of the PCA algorithm are summarized as follows:

– Compute $\mathbf{Y} = \frac{1}{\sqrt{N}} \mathbf{X}_N$.

– Compute the matrix $\mathbf{U}$ of left singular vectors of $\mathbf{Y}$:

$$\mathbf{Y} = \mathbf{U} \mathbf{\Sigma} \mathbf{V}^T. \qquad \qquad [1.117]$$

– Determine L using the criterion [1.113].

– Compute the reduced-dimension data matrix:

$$\mathbf{Y}_L = \mathbf{U}_L^T \mathbf{Y} \in \mathbb{R}^{L \times N}. \tag{1.118}$$

– Reconstruct the original data matrix:

$$\widehat{\mathbf{X}}_N = \sqrt{N} \mathbf{U}_L \mathbf{Y}_L \in \mathbb{R}^{M \times N}. \tag{1.119}$$

– Reduce another data vector $\mathbf{x} \in \mathbb{R}^M$:

$$\mathbf{y}_L = \frac{1}{\sqrt{N}} \mathbf{U}_L^T \mathbf{x} \in \mathbb{R}^L. \tag{1.120}$$

– Reconstruct the data vector:

$$\widehat{\mathbf{x}} = \sqrt{N} \mathbf{U}_L \mathbf{y}_L \in \mathbb{R}^M. \tag{1.121}$$

REMARK 1.20.– We can make the following remarks:

– The reconstructed data vector can be expressed as follows:

$$\widehat{\mathbf{x}} = \mathbf{U}_L \mathbf{U}_L^T \mathbf{x} = [\mathbf{u}_1 \cdots \mathbf{u}_L] \begin{bmatrix} \langle \mathbf{x}, \mathbf{u}_1 \rangle \\ \vdots \\ \langle \mathbf{x}, \mathbf{u}_L \rangle \end{bmatrix} = \sum_{i=1}^{L} \langle \mathbf{x}, \mathbf{u}_i \rangle \mathbf{u}_i,$$

which amounts to projecting the data vector $\mathbf{x}$ onto the L principal components, i.e. onto the subspace $\mathbf{U}_L$ spanned by $\{\mathbf{u}_1, \cdots, \mathbf{u}_L\}$.

– If the goal is to reduce the dimension of the row space of $\mathbf{X}_N$ and decorrelate its row vectors, the same reasoning must be applied to $\mathbf{Y} = \frac{1}{\sqrt{M}} \mathbf{X}_N^T$, which amounts to diagonalizing the covariance matrix $\frac{1}{M} \mathbf{X}_N^T \mathbf{X}_N$.

1.5.12. *SVD and blind source separation*

In this section, we will show how the SVD can be used to solve the BSS problem in an instantaneous (i.e. memoryless) linear mixture, first for the noiseless case, and then for the noisy case.

1.5.12.1. *Noiseless case*

Consider the following noiseless instantaneous linear mixture of P sources:

$$\mathbf{x}_n = \mathbf{A}\mathbf{s}_n, \tag{1.122}$$

where $\mathbf{x}_n \in \mathbb{C}^M$ and $\mathbf{s}_n \in \mathbb{C}^P$ are the vectors containing the observations provided by M sensors and the P sources at time n, and $\mathbf{A} \in \mathbb{C}^{M \times P}$ is the mixture matrix. The sources are assumed to be mutually statistically independent, stationary, centered, with variance $\sigma_p^2 = E\big[|s_n(p)|^2\big]$, where $s_n(p)$ is the pth source at time n. The independence and centering hypotheses on the sources induce a diagonal spatial covariance matrix for the source vector:

$$\mathbf{C_s} \triangleq E[\mathbf{s}_n\mathbf{s}_n^H] = \mathrm{diag}(\sigma_1^2, \cdots, \sigma_P^2). \tag{1.123}$$

Furthermore, the sources are assumed to be time-uncorrelated. The spatiotemporal non-correlation means that:

$$E\big[s_n(p)s_t^*(q)\big] = \sigma_p^2\delta_{nt}\delta_{pq}. \tag{1.124}$$

We also assume that $M \geq P$, i.e. the system is over-dimensioned, with more sensors than sources, and that the matrix $\mathbf{A}$ has full column rank.

The covariance matrix of the observations is given by:

$$\mathbf{C_x} \triangleq E[\mathbf{x}_n\mathbf{x}_n^H] = \mathbf{A}E[\mathbf{s}_n\mathbf{s}_n^H]\mathbf{A}^H = \mathbf{A}\mathbf{C_s}\mathbf{A}^H. \tag{1.125}$$

Note that the assumption that the mixture matrix has full column rank implies that $\mathbf{C_x}$ has rank equal to P, which is the rank of $\mathbf{C_s}$.

The BSS problem involves estimating the sources and the mixture matrix from only signal measurements $\mathbf{x}_n$, for $n \in \langle N \rangle$. The next proposition presents a solution based on the SVD of a matrix of observations $\mathbf{Y}_N$. In Chapter 5, we will present two tensor-based solutions.

PROPOSITION 1.21.– Let $\mathbf{Y}_N$ be the observation matrix constructed from N samples of the signal $\mathbf{x}_n$, with $N \geq M$:

$$\mathbf{Y}_N \triangleq [\mathbf{x}_1 \cdots \mathbf{x}_N] = \mathbf{A}\mathbf{S}_N \in \mathbb{C}^{M \times N}, \tag{1.126}$$

where $\mathbf{S}_N \triangleq [\mathbf{s}_1 \cdots \mathbf{s}_N]$. Consider the reduced SVD of $\mathbf{Y}_N$ of order P:

$$\mathbf{Y}_N = \mathbf{U}_P\mathbf{\Sigma}_P\mathbf{V}_P^H, \tag{1.127}$$

where $\mathbf{U}_P \in \mathbb{C}^{M \times P}$ and $\mathbf{V}_P \in \mathbb{C}^{N \times P}$ are column orthonormal ($\mathbf{U}_P^H\mathbf{U}_P = \mathbf{V}_P^H\mathbf{V}_P = \mathbf{I}_P$), and $\mathbf{\Sigma}_P \in \mathbb{R}^{P \times P}$ is diagonal and non-singular, since $\mathbf{C_x}$ has rank P.

The source matrix $\mathbf{S}_N$ and the mixture matrix $\mathbf{A}$ can be estimated over a time window of length N using the following equations:

$$(\hat{\mathbf{S}}_{\mathrm{LS}})_N = \sqrt{N}\mathbf{C_s}^{1/2}\mathbf{\Sigma}_P^{-1}\mathbf{U}_P^H\mathbf{Y}_N = \sqrt{N}\mathbf{C_s}^{1/2}\mathbf{V}_P^H \tag{1.128}$$

$$\hat{\mathbf{A}} = \frac{1}{\sqrt{N}}\mathbf{U}_P\mathbf{\Sigma}_P\mathbf{C_s}^{-1/2}. \tag{1.129}$$

PROOF.– Let us first determine the covariance $\mathbf{C_S} \triangleq \frac{1}{N}E[\mathbf{S}_N\mathbf{S}_N^H] \in \mathbb{C}^{P \times P}$ of the source matrix $\mathbf{S}_N$. By the double spatiotemporal non-correlation [1.124] of the sources, we have:

$$(\mathbf{C_S})_{ij} = \frac{1}{N}E\Big[\sum_{n=1}^{N} s_n(i)s_n^*(j)\Big] = \frac{1}{N}\sum_{n=1}^{N} \sigma_i^2 \delta_{ij} \tag{1.130}$$

$$\mathbf{C_S} = \mathrm{diag}\big(\sigma_1^2, \cdots, \sigma_P^2\big) = \mathbf{C_s}. \tag{1.131}$$

This result can be recovered by writing: $\frac{1}{N}E[\mathbf{S}_N\mathbf{S}_N^H] = \frac{1}{N}\sum_{n=1}^{N} E[\mathbf{s}_n\mathbf{s}_n^H] = \mathbf{C_s}$. Using the definition [1.126], the covariance of the observation matrix $\mathbf{Y}_N$ is then given by:

$$\mathbf{C_Y} \triangleq \frac{1}{N}E[\mathbf{Y}_N\mathbf{Y}_N^H] = \frac{1}{N}\mathbf{A}E[\mathbf{S}_N\mathbf{S}_N^H]\mathbf{A}^H = \mathbf{AC_s}\mathbf{A}^H. \tag{1.132}$$

Moreover, from the SVD [1.127] of $\mathbf{Y}_N$, the empirical covariance of the observation matrix is given by:

$$\hat{\mathbf{C}}_\mathbf{Y} \triangleq \frac{1}{N}\mathbf{Y}_N\mathbf{Y}_N^H = \frac{1}{N}\mathbf{U}_P\mathbf{\Sigma}_P^2\mathbf{U}_P^H. \tag{1.133}$$

Identifying this empirical covariance with the exact covariance [1.132] allows us to deduce that $\mathbf{AC_s}^{1/2} = \frac{1}{\sqrt{N}}\mathbf{U}_P\mathbf{\Sigma}_P$, which gives the following estimate of the mixture matrix:

$$\hat{\mathbf{A}} = \frac{1}{\sqrt{N}}\mathbf{U}_P\mathbf{\Sigma}_P\mathbf{C_s}^{-1/2}, \tag{1.134}$$

which corresponds to the expression [1.129].

Applying the LS method to the observation equation $\mathbf{Y}_N = \mathbf{AS}_N$ with the hypothesis that the mixture matrix has full column rank yields the LS estimator of the source matrix:

$$(\hat{\mathbf{S}}_{\mathrm{LS}})_N = \mathbf{A}^\dagger\mathbf{Y}_N = (\mathbf{A}^H\mathbf{A})^{-1}\mathbf{A}^H\mathbf{Y}_N. \tag{1.135}$$

Replacing $\mathbf{A}$ by its estimate [1.129] gives:

$$(\hat{\mathbf{S}}_{\mathrm{LS}})_N = \sqrt{N}\mathbf{C_s}^{1/2}(\mathbf{\Sigma}_P\mathbf{U}_P^H\mathbf{U}_P\mathbf{\Sigma}_P)^{-1}\mathbf{C_s}^{1/2}\mathbf{C_s}^{-1/2}\mathbf{\Sigma}_P\mathbf{U}_P^H\mathbf{Y}_N,$$

or alternatively, after simplification:

$$(\hat{\mathbf{S}}_{\mathrm{LS}})_N = \sqrt{N}\mathbf{C_s}^{1/2}\mathbf{\Sigma}_P^{-1}\mathbf{U}_P^H\mathbf{Y}_N.$$

After replacing $\mathbf{Y}_N$ with its SVD [1.127], we also have:

$$(\hat{\mathbf{S}}_{\mathrm{LS}})_N = \sqrt{N}\mathbf{C_s}^{1/2}\mathbf{\Sigma}_P^{-1}\mathbf{U}_P^H(\mathbf{U}_P\mathbf{\Sigma}_P\mathbf{V}_P^H) = \sqrt{N}\mathbf{C_s}^{1/2}\mathbf{V}_P^H,$$

which proves the formula [1.128].   □

REMARK 1.22.–

– It can be checked that:

$$\hat{\mathbf{A}}(\hat{\mathbf{S}}_{\text{LS}})_N = \frac{1}{\sqrt{N}}\mathbf{U}_P\boldsymbol{\Sigma}_P\mathbf{C}_{\mathbf{s}}^{-1/2}(\sqrt{N}\mathbf{C}_{\mathbf{s}}^{1/2}\mathbf{V}_P^H) = \mathbf{U}_P\boldsymbol{\Sigma}_P\mathbf{V}_P^H = \mathbf{Y}_N.$$

The data are therefore recovered exactly by using the mixture and source matrices estimated using the SVD of the observation matrix $\mathbf{Y}_N$.

– In realistic situations, the variance of the sources is generally unknown and is therefore assumed to be one, which is equivalent to choosing the identity matrix $\mathbf{C_s} = \mathbf{I}_P$ as the spatial covariance matrix of the sources. The latter are then estimated up to a scaling factor, and equations [1.128] and [1.129] simplify as follows:

$$(\hat{\mathbf{S}}_{\text{LS}})_N = \sqrt{N}\boldsymbol{\Sigma}_P^{-1}\mathbf{U}_P^H\mathbf{Y}_N = \sqrt{N}\mathbf{V}_P^H \;\; ; \;\; \hat{\mathbf{A}} = \frac{1}{\sqrt{N}}\mathbf{U}_P\boldsymbol{\Sigma}_P. \qquad [1.136]$$

1.5.12.2. *Noisy case*

Let us now consider the noisy case, with:

$$\mathbf{y}_n = \mathbf{x}_n + \mathbf{b}_n = \mathbf{A}\mathbf{s}_n + \mathbf{b}_n, \qquad [1.137]$$

where the additive white noise $\mathbf{b}_n \in \mathbb{C}^M$ is assumed to be stationary, centered, with spatial covariance $\mathbf{C_b} = \sigma_b^2\mathbf{I}_M$, and independent of the source signals.

We define the noisy observation matrix $\mathbf{Y}_N$ constructed from N samples of the measured signal $\mathbf{y}_n$, with $N \geq M$, as follows:

$$\mathbf{Y}_N = [\mathbf{y}_1 \cdots \mathbf{y}_N] = \mathbf{A}\mathbf{S}_N + \mathbf{B}_N \in \mathbb{C}^{M \times N}. \qquad [1.138]$$

PROPOSITION 1.23.– In the noisy case defined above, the estimates of the source and mixture matrices are given by:

$$(\hat{\mathbf{S}}_{\text{LS}})_N = \sqrt{N}\mathbf{C}_{\mathbf{s}}^{1/2}\big(\boldsymbol{\Sigma}_P^2 - N\sigma_b^2\mathbf{I}_P\big)^{-1/2}\mathbf{U}_P^H\mathbf{Y}_N \qquad [1.139]$$

$$= \sqrt{N}\mathbf{C}_{\mathbf{s}}^{1/2}\big(\boldsymbol{\Sigma}_P^2 - N\sigma_b^2\mathbf{I}_P\big)^{-1/2}\boldsymbol{\Sigma}_P\mathbf{V}_P^H \qquad [1.140]$$

$$\hat{\mathbf{A}} = \frac{1}{\sqrt{N}}\mathbf{U}_P\big(\boldsymbol{\Sigma}_P^2 - N\sigma_b^2\mathbf{I}_P\big)^{1/2}\mathbf{C}_{\mathbf{s}}^{-1/2}. \qquad [1.141]$$

PROOF.– From [1.138], we can deduce the LS estimator of the source matrix, which is given by a formula[9] that is identical to [1.135]:

$$(\hat{\mathbf{S}}_{\text{LS}})_N = (\mathbf{A}^H\mathbf{A})^{-1}\mathbf{A}^H\mathbf{Y}_N. \qquad [1.142]$$

---

9 In the case of the observation equation $\mathbf{Y}_N = \mathbf{A}\mathbf{S}_N + \mathbf{B}_N$, with an additive white Gaussian noise (AWGN) with covariance matrix $\mathbf{C_B} = \sigma_b^2\mathbf{I}_M$, i.e. when the M components of $\mathbf{b}_n$ are spatially uncorrelated and have the same variance σ_b^2, the least squares estimator is optimal, in the sense of coinciding with the maximum likelihood (ML) estimator, whose expression is $(\hat{\mathbf{S}}_{\text{ML}})_N = (\mathbf{A}^H\mathbf{C}_{\mathbf{b}}^{-1}\mathbf{A})^{-1}\mathbf{A}^H\mathbf{C}_{\mathbf{b}}^{-1}\mathbf{Y}_N = (\mathbf{A}^H\mathbf{A})^{-1}\mathbf{A}^H\mathbf{Y}_N = (\hat{\mathbf{S}}_{\text{LS}})_N$.

As for the covariance $\mathbf{C_S}$ of the sources, defined in [1.131], the spatiotemporal non-correlation of the noise implies that $\mathbf{C_B} = \sigma_b^2 \mathbf{I}_M$. Hence, the covariance of $\mathbf{Y}_N$ deduced from the model [1.138] is given by:

$$\mathbf{C_Y} \triangleq E[\mathbf{Y}_N \mathbf{Y}_N^H] = \mathbf{A} \mathbf{C_s} \mathbf{A}^H + \sigma_b^2 \mathbf{I}_M. \qquad [1.143]$$

Taking into account the full SVD of $\mathbf{Y}_N = \mathbf{U} \mathbf{\Sigma} \mathbf{V}^H$, decomposed in the same way as [1.75], with $r = M$ and $k = P$, the empirical covariance matrix of the observations can be decomposed as follows:

$$\hat{\mathbf{C}}_\mathbf{Y} \triangleq \frac{1}{N} \mathbf{Y}_N \mathbf{Y}_N^H = \frac{1}{N} (\mathbf{U}_P \mathbf{\Sigma}_P^2 \mathbf{U}_P^H + \mathbf{U}_{M-P} \mathbf{\Sigma}_{M-P}^2 \mathbf{U}_{M-P}^H). \qquad [1.144]$$

This decomposition gives two sets of left singular vectors. The first P are associated with the signal subspace, and the $M - P$ others, associated with the $M - P$ smallest singular values, define the noise subspace. Since the eigenvectors form an orthonormal basis, these two subspaces are orthogonal.

By separating the signal space associated with the P largest singular values of the SVD from the noise space, we can write $\mathbf{I}_M = \mathbf{U} \mathbf{U}^H = [\mathbf{U}_P | \mathbf{U}_{M-P}]$ $[\mathbf{U}_P | \mathbf{U}_{M-P}]^H = \mathbf{U}_P \mathbf{U}_P^H + \mathbf{U}_{M-P} \mathbf{U}_{M-P}^H$. By identifying the signal components of the expressions [1.143] and [1.144] of $\mathbf{C_Y}$, we deduce:

$$\mathbf{A} \mathbf{C_s} \mathbf{A}^H = \frac{1}{N} \mathbf{U}_P (\mathbf{\Sigma}_P^2 - N\sigma_b^2 \mathbf{I}_P) \mathbf{U}_P^H, \qquad [1.145]$$

which gives the expression [1.141] of the estimated mixture matrix. Furthermore, by identifying the noise components of $\mathbf{C_Y}$ in [1.143] and [1.144], we have $\sigma_p^2 = \sigma_b^2$ for $p \in \{P+1, \cdots, M\}$, where σ_p is the pth singular value of $\mathbf{Y}_N$.

After replacing $\mathbf{A}$ with its estimate [1.141] in the expression [1.142] of the LS estimator, we obtain the following estimate for the source matrix:

$$(\hat{\mathbf{S}}_{\text{LS}})_N = \sqrt{N} \left(\mathbf{C_s}^{-1/2} (\mathbf{\Sigma}_P^2 - N\sigma_b^2 \mathbf{I}_P) \mathbf{C_s}^{-1/2} \right)^{-1} \mathbf{C_s}^{-1/2} (\mathbf{\Sigma}_P^2 - N\sigma_b^2 \mathbf{I}_P)^{1/2} \mathbf{U}_P^H \mathbf{Y}_N$$
$$= \sqrt{N} \mathbf{C_s}^{1/2} (\mathbf{\Sigma}_P^2 - N\sigma_b^2 \mathbf{I}_P)^{-1/2} \mathbf{U}_P^H \mathbf{Y}_N. \qquad [1.146]$$

Furthermore, from the SVD [1.127] of $\mathbf{Y}_N$, we deduce:

$$(\hat{\mathbf{S}}_{\text{LS}})_N = \sqrt{N} \mathbf{C_s}^{1/2} (\mathbf{\Sigma}_P^2 - N\sigma_b^2 \mathbf{I}_P)^{-1/2} \mathbf{\Sigma}_P \mathbf{V}_P^H. \qquad [1.147]$$

This proves the formula [1.140]. $\qquad\qquad\qquad\qquad\qquad\qquad\qquad\qquad\qquad$ □

REMARK 1.24.–

– Setting $\sigma_b = 0$ makes equations [1.139]–[1.141] identical to equations [1.128] and [1.129] for the noiseless case.

– If we assume that the sources have unit variance, equations [1.139]–[1.141] simplify as follows (Zarzoso and Nandi 1999):

$$(\hat{\mathbf{S}}_{\mathrm{LS}})_N = \sqrt{N}\left(\mathbf{\Sigma}_P^2 - N\sigma_b^2\mathbf{I}_P\right)^{-1/2}\mathbf{\Sigma}_P\mathbf{V}_P^H \ ; \ \hat{\mathbf{A}} = \frac{1}{\sqrt{N}}\mathbf{U}_P\left(\mathbf{\Sigma}_P^2 - N\sigma_b^2\mathbf{I}_P\right)^{1/2}.$$

1.6. CUR decomposition

As we saw above, the SVD gives the best rank-k approximation of a matrix $\mathbf{A} \in \mathbb{R}^{I \times J}$, denoted $\mathbf{A}_k$. Although very useful in many applications, the SVD has two drawbacks:

– the singular vectors are difficult to interpret in classification applications;

– the sparseness and non-negativity properties are not preserved.

These limitations led to the development of low-rank matrix decompositions that select certain columns, and possibly also rows, from the original matrix and are therefore expressed in terms of the data to analyze. This corresponds to the two following types of decomposition:

$$\mathbf{A} \approx \mathbf{CX} \text{ and } \mathbf{A} \approx \mathbf{CUR}, \qquad\qquad [1.148]$$

where the matrices $\mathbf{C} \in \mathbb{R}^{I \times c}$ and $\mathbf{R} \in \mathbb{R}^{l \times J}$ are formed by $c < J$ columns and $l < I$ rows of $\mathbf{A}$, respectively; in general, $l \ll I$ and $c \ll J$. The matrix $\mathbf{U} \in \mathbb{C}^{c \times l}$ is called the intersection matrix.

Column selection algorithms choose c columns from $\mathbf{A}$ in order to construct a matrix $\mathbf{C}$ that minimizes the following criterion (Boutsidis *et al.* 2010):

$$J_c = \|\mathbf{A} - \mathbf{CC}^\dagger\mathbf{A}\|_\xi, \qquad\qquad [1.149]$$

where $\mathbf{C}^\dagger$ is the Moore–Penrose pseudo-inverse of $\mathbf{C}$, the term $\mathbf{CC}^\dagger\mathbf{A}$ is the projection of $\mathbf{A}$ onto the subspace spanned by the columns of $\mathbf{C}$, and ξ denotes the norm being used: $\xi = 2$, or $\xi = F$, or $\xi = *$, depending on whether we are considering the spectral norm, the Frobenius norm, or the nuclear norm. Note that choosing $\mathbf{C} = \mathbf{A}$ makes the criterion zero, due to the properties of the pseudo-inverse ($\mathbf{AA}^\dagger\mathbf{A} = \mathbf{A}$). If the matrix $\mathbf{A}$ is large, making an optimal selection of c columns is a difficult combinatorial problem, since there are C_J^c possible combinations.

In the case of a sparse matrix $\mathbf{A}$, decompositions of the form $\mathbf{CX}$ allow a sparse structure to be preserved for $\mathbf{C}$, but not for the matrix $\mathbf{X} \in \mathbb{R}^{c \times J}$ of coefficients. This type of decomposition is only useful when $I \gg J$.

CUR decomposition methods aim to determine the triplet $(\mathbf{C}, \mathbf{U}, \mathbf{R})$ in such a way that the product $\mathbf{CUR}$ is as close as possible to $\mathbf{A}$, which amounts to minimizing an approximation error criterion of the form:

$$J_c = \|\mathbf{A} - \mathbf{CUR}\|_\xi, \qquad\qquad [1.150]$$

where the norm ξ is defined as above. Like the SVD, the CUR decomposition of a matrix $\mathbf{A} \in \mathbb{R}^{I \times J}$ provides a low-rank approximation (Mahoney and Drineas 2009). Of course, this decomposition is not unique. It is closely linked to the problem of selecting column and row subsets via random sampling of the data and assigning probabilities to the vectors of the original matrix or its truncated SVD as a function of the Euclidean norm of these vectors.

In the case of a large data matrix, the CUR decomposition can be used for purposes such as compression, representation, classification or even recovery and extrapolation of data. This type of decomposition has been used in various fields of applications, including the analysis of documents on the Internet and of medical data (Mahoney and Drineas 2009; Sorensen and Embree 2015), the extrapolation of urban traffic data (Mitrovic *et al.* 2013) and image compression (Voronin and Martinsson 2017).

Various algorithms have been proposed in the literature to compute a CUR decomposition (Drineas *et al.* 2006, 2008; Wang and Zhang 2013; Boutsidis and Woodruff 2014; Voronin and Martinsson 2017). Most algorithms are composed of two steps: first, a column and row selection method is applied to determine the matrices $\mathbf{C}$ and $\mathbf{R}$, then the matrix $\mathbf{U}$ is optimized to minimize the approximation error criterion [1.150].

For the Frobenius norm ($\xi = F$), the matrix $\mathbf{U}$ that minimizes the criterion [1.150] is given by:

$$\mathbf{U} = \mathbf{C}^\dagger \mathbf{A} \mathbf{R}^\dagger, \qquad\qquad [1.151]$$

where $\mathbf{C}^\dagger$ and $\mathbf{R}^\dagger$ denote the pseudo-inverses of $\mathbf{C}$ and $\mathbf{R}$, respectively. Determining $\mathbf{C}$ and $\mathbf{R}$ is equivalent to selecting a subset of columns of $\mathbf{A}$ and $\mathbf{A}^T$, respectively.

Drineas *et al.* (2008) presented polynomial-time random algorithms for the first time to compute CX and CUR decompositions of a matrix $\mathbf{A} \in \mathbb{R}^{I \times J}$. These algorithms require $l = O(k\epsilon^{-4}log^2 k)$ rows and $c = O(k\epsilon^{-2}log k)$ columns of $\mathbf{A}$ to be selected, achieving an approximation error $\mathbf{A} - \hat{\mathbf{A}}$, where $\hat{\mathbf{A}} = \mathbf{C}\mathbf{X}$ or $\hat{\mathbf{A}} = \mathbf{C}\mathbf{U}\mathbf{R}$, that is bounded as follows, with a probability of at least 0.7:

$$\|\mathbf{A} - \hat{\mathbf{A}}\|_F \leq (1 + \epsilon) \|\mathbf{A} - \mathbf{A}_k\|_F. \qquad\qquad [1.152]$$

The parameter ϵ (satisfying $0 < \epsilon < 1$) defines the precision of the approximation, k (satisfying $1 \leq k \ll \min(I, J)$) is a rank parameter and $\mathbf{A}_k$ is the best rank-k approximation of $\mathbf{A}$, obtained by truncating its SVD to order k, i.e. $\mathbf{A}_k = \mathbf{U}_k \mathbf{\Sigma}_k \mathbf{V}_k^T$. The approximation error is therefore equal to the approximation error provided by the SVD, truncated to order k up to a constant factor.

CUR algorithms that compute the k principal singular vectors of $\mathbf{A}$ use a subspace sampling technique based on sampling probabilities that are proportional to the Euclidean norm of the singular vectors[10]:

$$p_i = \frac{1}{k} \sum_{j=1}^{k} (\mathbf{U}_k)_{ij}^2, \quad q_j = \frac{1}{k} \sum_{i=1}^{k} (\mathbf{V}_k^T)_{ij}^2. \qquad [1.153]$$

The probabilities p_i and q_j, where $i \in \langle I \rangle$ and $j \in \langle J \rangle$, are used to select rows and columns, respectively. For example, one possible column selection algorithm is to retain the c columns of $\mathbf{A}$ with the largest probabilities q_j, for $j \in \langle J \rangle$, and to fix the corresponding columns of $\mathbf{C}$ as $\mathbf{C}_{\cdot j} = \frac{1}{\sqrt{cq_j}} \mathbf{A}_{\cdot j}$.

For a bound on the approximation error identical to [1.152], Wang and Zhang (2013) present a CUR algorithm that only requires $l = O(k\epsilon^{-2})$ rows and $c = O(k\epsilon^{-1})$ columns to be selected, while, for the algorithms proposed by Boutsidis and Woodruff (2014), these numbers are reduced to $l = c = O(k\epsilon^{-1})$. The latter reference compares several CUR algorithms in terms of the number of columns and rows to be selected, the bound on the approximation errors and the computation time.

The CUR decomposition was extended to tensors by Mahoney *et al.* (2008), who applied it to hyperspectral imagery for the purpose of classification and to recommendation systems in order to reconstruct missing data, as well as by Caiafa and Cichocki (2010).

---

10 Since the matrices $\mathbf{U}_k$ and $\mathbf{V}_k$ are column orthonormal, we have $\mathbf{U}_k^T \mathbf{U}_k = \mathbf{V}_k^T \mathbf{V}_k = \mathbf{I}_k$, and, hence, $\|\mathbf{U}_k\|_F^2 = \|\mathbf{V}_k\|_F^2 = k$, from which we deduce $\sum_{i=1}^{I} \sum_{j=1}^{k} (\mathbf{U}_k)_{ij}^2 = \sum_{i=1}^{k} \sum_{j=1}^{J} (\mathbf{V}_k^T)_{ij}^2 = k$, and so $\sum_{i=1}^{I} p_i = \sum_{j=1}^{J} q_j = 1$, which explains why the quantities p_i and q_j can be interpreted as probabilities.

2

Hadamard, Kronecker and Khatri–Rao Products

2.1. Introduction

In this chapter, we consider three matrix products that play a very important role in matrix computation: the Hadamard, Kronecker and Khatri–Rao products. The Kronecker product, also known as the tensor product, is widely used in many signal and image processing applications, such as compressed sampling using Kronecker dictionaries (Duarte and Baraniuk 2012), and for image restoration (Nagy *et al.* 2004). It is also widely used in systems theory (Brewer 1978) and in linear algebra to express and solve matrix equations such as Sylvester and Lyapunov equations (Bartels and Stewart 1972), as well as generalized Lyapunov equations (Lev Ari 2005). Additionally, it plays a key role in simplifying and implementing rapid transform algorithms, such as Fourier, Walsh-Hadamard, and Haar transforms (Regalia and Mitra 1989; Pitsianis 1997; Van Loan 2000). Recently, the identification of bilinear filters decomposed into Kronecker products was proposed in Paleologu *et al.* (2018), and so-called Kronecker receivers allowed various wireless communication systems based on different tensor models to be presented with a single, unified approach (da Costa *et al.* 2018). The Hadamard product has been applied in statistics (Styan 1973; Neudecker *et al.* 1995; Neudecker and Liu 2001). The Khatri–Rao product has been used to define space-time codes (Sidiropoulos and Budampati 2002) and space-time-frequency codes (de Almeida and Favier 2013) in the context of wireless communications. The Kronecker and Khatri–Rao products are also found in tensor calculus, since they naturally appear in the matrix unfoldings of basic tensor decompositions, such as PARAFAC (Harshman 1970) and Tucker (Tucker 1966) decompositions, and more generally constrained PARAFAC decompositions (Favier and de Almeida 2014).

For a block matrix, the block Kronecker product can also be defined (Tracy and Singh 1972; Hyland and Collins 1989; Koning *et al.* 1991).

Two other key notions are also considered in this chapter: the vectorization operator *vec* and the *vec*-permutation matrix (Tracy and Dwyer 1969), also called the commutation matrix (Magnus and Neudecker 1979 and 1988). The most common use of the *vec* operator is to vectorize a matrix by stacking its columns on top of each other. This operator was originally introduced by Sylvester in 1884 to solve linear matrix equations. However, there are several variants of the vectorization operator that have led to different notations, depending on whether the rows or the columns are being stacked, whether the operator is being applied to a symmetric square matrix, in which case the vectorization only includes the distinct elements of the matrix located on and below the diagonal (Henderson and Searle 1979), or alternatively whether the operator is being applied to a partitioned matrix. The *vec*-permutation matrix, introduced for the first time in Tracy and Dwyer (1969) to compute matrix derivatives, transforms the vectorized form $vec(\mathbf{A})$ of a matrix $\mathbf{A}$ into that of its transpose $vec(\mathbf{A}^T)$. It also transforms the Kronecker product of two vectors or two matrices ($\mathbf{u} \otimes \mathbf{v}; \mathbf{A} \otimes \mathbf{B}$) into the Kronecker product of the same vectors or matrices in reverse order ($\mathbf{v} \otimes \mathbf{u}; \mathbf{B} \otimes \mathbf{A}$).

A historical overview of vectorization operators and the Kronecker, Hadamard, and Khatri–Rao products can be found in previous studies (Magnus and Neudecker 1979; Henderson and Searle 1981; Henderson *et al.* 1983; Regalia and Mitra 1989; Horn 1990; Koning *et al.* 1991; Pitsianis 1997; Liu and Trenkler 2008; Van Loan 2009).

The following few paragraphs describe the content of this chapter.

The next section will define the notation used for the matrix products considered in this chapter. The Hadamard, Kronecker and Khatri–Rao products will then be presented in sections 2.3, 2.4 and 2.9, respectively, and the Kronecker sum will be defined in section 2.5. The key properties satisfied by these products will be presented.

This chapter will also provide an original contribution concerning the use of the index convention, which was introduced in Pollock (2011), in order to show some of the stated properties and relations. This convention, which generalizes Einstein's summation convention, allows us to make both the proofs more concise and the formulae more compact. In section 2.6, it will first be used to express matrix products, then to describe the vectorization of matrices and matrix products, as well as to establish formulae relating to the trace of matrix products. In Chapter 3, it will be employed to define matrix and vector unfoldings of a tensor, and in Chapter 5 it will be used for matricizing PARAFAC and Tucker decompositions.

In section 2.7, we will introduce the notion of the commutation matrix, and we will illustrate how to use it to permute the factors of simple Kronecker products, multiple Kronecker products and block Kronecker products, as well as to vectorize partitioned matrices. In sections 2.8 and 2.10, we will present several relations between the *diag* operator and the Kronecker product, as well as between the vectorization operator and the Kronecker and Khatri–Rao products. Further relations between the *diag* operator and the Hadamard product and among the three matrix products will also be presented in sections 2.3 and 2.11.

Finally, in section 2.12, we will describe various examples involving the use of Kronecker and Khatri–Rao products to:

– compute and arrange first-order partial derivatives of a function using the tensor formalism;

– solve matrix equations, in particular Lyapunov and Sylvester equations, which play an important role in the theory of linear systems.

Algorithms for estimating the factor matrices of a Khatri–Rao product and a Kronecker product will be described. These algorithms are based on the construction of rank-one matrices and tensors from the vectorization of the factor matrices. This reduces the estimation problem to that of rank-one matrix or tensor approximation using the SVD algorithm, or the HOSVD algorithm in the case of multiple Khatri–Rao and Kronecker products.

2.2. Notation

We will write $\mathbf{A}^*$, $\mathbf{A}^T$, $\mathbf{A}^H$, $\mathbf{A}^\dagger$, $\mathbf{A}_{i.}$, $\mathbf{A}_{.j}$, $\mathrm{r}(\mathbf{A})$ (or $\mathrm{r_A}$), and $\det(\mathbf{A})$, for the conjugate, the transpose, the transconjugate (also known as conjugate transpose or Hermitian transpose), the Moore–Penrose pseudo-inverse, the ith row, the jth column, the rank and the determinant of $\mathbf{A} \in \mathbb{K}^{I \times J}$, respectively.

Furthermore, a_i, $a_{ij} = (\mathbf{A})_{ij}$, $a_{ijk} = (\mathcal{A})_{ijk}$, $a_{i_1,\cdots,i_N} = (\mathcal{A})_{i_1,\cdots,i_N}$ are the elements of the vector $\mathbf{a} \in \mathbb{K}^I$, the matrix $\mathbf{A} = [a_{ij}] \in \mathbb{K}^{I \times J}$ and the tensors $\mathcal{A} \in \mathbb{K}^{I \times J \times K}$ and $\mathcal{A} \in \mathbb{K}^{I_1 \times \cdots \times I_N}$, respectively.

The symbol $\mathbf{1}_I$ denotes a column vector of size I whose elements are all equal to 1. The elements of the matrices $\mathbf{0}_{I \times J}$ and $\mathbf{1}_{I \times J}$ of size $(I \times J)$ are all equal to 0 and 1, respectively. The symbols $\mathbf{I}_N$ and $\mathbf{e}_n^{(N)}$ denote the identity matrix of order N and the nth vector of the canonical basis of the vector space $\mathbb{R}^N$, respectively.

The notation for the various matrix operations considered in this chapter is summarized in Table 2.1, with a reference to the section where it is introduced in each case[1]. Note that the symbols chosen to represent these operations are not universal. For example, in the literature, the Hadamard product is also denoted by $*$ or $\circ$. Similarly, the symbols $\odot$ and $\bowtie$ are also used for the Khatri–Rao and block Kronecker product, respectively.

Notation	Products	Section
$\odot$	Hadamard	2.3
$\otimes$	Kronecker	2.4
$\otimes_b$	block Kronecker (Tracy-Singh)	2.7.5
$\mid \otimes \mid$	strong Kronecker	2.7.6
$\diamond$	Khatri-Rao	2.9

Notation	Sum	Section
$\oplus$	Kronecker	2.5

Table 2.1. *Hadamard, Kronecker, and Khatri-Rao products and Kronecker sum*

2.3. Hadamard product

2.3.1. *Definition and identities*

Let $\mathbf{A}$ and $\mathbf{B} \in \mathbb{K}^{I \times J}$ be two matrices of the same size. The Hadamard product of $\mathbf{A}$ and $\mathbf{B}$ is the matrix $\mathbf{C} \in \mathbb{K}^{I \times J}$ defined as follows:

$$\mathbf{C} = \mathbf{A} \odot \mathbf{B} = \begin{bmatrix} a_{11}b_{11} & a_{12}b_{12} & \cdots & a_{1J}b_{1J} \\ a_{21}b_{21} & a_{22}b_{22} & \cdots & a_{2J}b_{2J} \\ \vdots & \vdots & \vdots & \\ a_{I1}b_{I1} & a_{I2}b_{I2} & \cdots & a_{IJ}b_{IJ} \end{bmatrix}, \qquad [2.1]$$

i.e. $c_{ij} = a_{ij}b_{ij}$, and therefore $\mathbf{C} = [a_{ij}b_{ij}]$, with $i \in \langle I \rangle, j \in \langle J \rangle$.

PROPOSITION 2.1.– For $\mathbf{A} \in \mathbb{K}^{I \times J}$ and $\mathbf{B} \in \mathbb{K}^{I \times I}$, the following identities follow directly from the definition of the Hadamard product:

$$\mathbf{A} \odot \mathbf{0}_{I \times J} = \mathbf{0}_{I \times J} \odot \mathbf{A} = \mathbf{0}_{I \times J} \qquad [2.2]$$

$$\mathbf{A} \odot \mathbf{1}_{I \times J} = \mathbf{1}_{I \times J} \odot \mathbf{A} = \mathbf{A} \qquad [2.3]$$

$$\mathbf{B} \odot \mathbf{I_I} = \mathbf{I_I} \odot \mathbf{B} = \text{diag}(b_{11}, b_{22}, \ldots, b_{II}). \qquad [2.4]$$

---

1 The Hadamard product is sometimes also called the Schur or Schur–Hadamard product. This is due to Schur's (product) theorem (1911), which establishes that the Hadamard product of two positive semi-definite matrices is also a positive semi-definite matrix. See Horn (1990) for a historical overview of the Hadamard product and its various names.

2.3.2. *Fundamental properties*

PROPOSITION 2.2.– The Hadamard product satisfies the properties of commutativity, associativity and distributivity with respect to addition:

$$\mathbf{A} \odot \mathbf{B} = \mathbf{B} \odot \mathbf{A} \qquad [2.5]$$

$$\mathbf{A} \odot (\mathbf{B} \odot \mathbf{C}) = (\mathbf{A} \odot \mathbf{B}) \odot \mathbf{C} = \mathbf{A} \odot \mathbf{B} \odot \mathbf{C} \qquad [2.6]$$

$$(\mathbf{A} + \mathbf{B}) \odot (\mathbf{C} + \mathbf{D}) = (\mathbf{A} \odot \mathbf{C}) + (\mathbf{A} \odot \mathbf{D}) + (\mathbf{B} \odot \mathbf{C}) + (\mathbf{B} \odot \mathbf{D}). \qquad [2.7]$$

The Hadamard product also satisfies the following properties:

$$(\mathbf{A} \odot \mathbf{B})^T = \mathbf{A}^T \odot \mathbf{B}^T \ , \ (\mathbf{A} \odot \mathbf{B})^H = \mathbf{A}^H \odot \mathbf{B}^H \qquad [2.8]$$

$$r_{\mathbf{A} \odot \mathbf{B}} \leq r_{\mathbf{A}} r_{\mathbf{B}}. \qquad [2.9]$$

The relations [2.8] continue to hold if the matrices are replaced by vectors.

The Hadamard product satisfies the following structural properties:

– Using the properties [2.8], it is easy to deduce that if $\mathbf{A}$ and $\mathbf{B} \in \mathbb{C}^{I \times I}$ are symmetric (resp. Hermitian), then their Hadamard product is itself a symmetric (respectively, Hermitian) matrix:

$$\left\{ \begin{array}{l} \mathbf{A}^T = \mathbf{A} \\ \mathbf{B}^T = \mathbf{B} \end{array} \right\} \Rightarrow (\mathbf{A} \odot \mathbf{B})^T = \mathbf{A}^T \odot \mathbf{B}^T = \mathbf{A} \odot \mathbf{B} \qquad [2.10]$$

$$\left\{ \begin{array}{l} \mathbf{A}^H = \mathbf{A} \\ \mathbf{B}^H = \mathbf{B} \end{array} \right\} \Rightarrow (\mathbf{A} \odot \mathbf{B})^H = \mathbf{A}^H \odot \mathbf{B}^H = \mathbf{A} \odot \mathbf{B}. \qquad [2.11]$$

– The Hadamard product of Hermitian positive definite or semi-definite matrices is itself a positive definite or semi-definite matrix, as summarized in Table 2.2.

Hypotheses	Properties
$\mathbf{A}, \mathbf{B} > 0$	$\mathbf{A} \odot \mathbf{B} > 0$
$\mathbf{A}, \mathbf{B} \geq 0$	$\mathbf{A} \odot \mathbf{B} \geq 0$

Table 2.2. *Hadamard product of Hermitian positive (semi-)definite matrices*

2.3.3. *Basic relations*

Table 2.3 states a few basic relations satisfied by the Hadamard product. In the last relation of this table, the matrices $\mathbf{A}, \mathbf{B}, \mathbf{C}$ can be permuted without changing the result.

$\mathbf{x}, \mathbf{y}, \mathbf{z} \in \mathbb{K}^I$ and $\mathbf{u}, \mathbf{v} \in \mathbb{K}^J$
$\mathbf{x}^T(\mathbf{y} \odot \mathbf{z}) = \mathbf{y}^T(\mathbf{x} \odot \mathbf{z}) = \mathbf{z}^T(\mathbf{x} \odot \mathbf{y}) = \sum_{i=1}^I x_i y_i z_i$
$(\mathbf{y} \odot \mathbf{z})\mathbf{u}^T = (\mathbf{y}\mathbf{u}^T) \odot (\mathbf{z}\mathbf{1}_J^T) = (\mathbf{y}\mathbf{1}_J^T) \odot (\mathbf{z}\mathbf{u}^T) = \left[y_i z_i u_j \right] \in \mathbb{K}^{I \times J}$
$(\mathbf{x}\mathbf{u}^T) \odot (\mathbf{y}\mathbf{v}^T) = (\mathbf{x} \odot \mathbf{y})(\mathbf{u} \odot \mathbf{v})^T = \left[x_i y_i u_j v_j \right] \in \mathbb{K}^{I \times J}$
$\mathbf{A}, \mathbf{B}, \mathbf{C}, \mathbf{D} \in \mathbb{K}^{I \times K}$
$(\mathbf{A} \odot \mathbf{B})(\mathbf{C} \odot \mathbf{D})^T = \left[\sum_{k=1}^K a_{ik} b_{ik} c_{jk} d_{jk} \right] \in \mathbb{K}^{I \times I}$
$\mathbf{I}_I \odot [(\mathbf{A} \odot \mathbf{B})\mathbf{C}^T] = \mathrm{diag}\left(\sum_{k=1}^K a_{ik} b_{ik} c_{ik} \right) \in \mathbb{K}^{I \times I}$

Table 2.3. *Basic relations satisfied by the Hadamard product*

2.3.4. *Relations between the* $diag$ *operator and Hadamard product*

Table 2.4 presents a few relations between the $diag$ operator and Hadamard product.

$\mathbf{u}, \mathbf{v} \in \mathbb{K}^I, \mathbf{w} \in \mathbb{K}^J, \mathbf{A}, \mathbf{B} \in \mathbb{K}^{I \times J}$
$\mathrm{diag}(\mathbf{u})\mathbf{v} = \mathbf{u} \odot \mathbf{v} = \mathrm{diag}(\mathbf{v})\mathbf{u} \in \mathbb{K}^I$
$\mathrm{diag}(\mathbf{u})\mathbf{A} = \mathbf{u}\mathbf{1}_J^T \odot \mathbf{A} = [u_i a_{ij}] \in \mathbb{K}^{I \times J}$
$\mathbf{A}\,\mathrm{diag}(\mathbf{w}) = \mathbf{A} \odot \mathbf{1}_I \mathbf{w}^T = [a_{ij} w_j] \in \mathbb{K}^{I \times J}$
$\begin{aligned} \left(\mathrm{diag}(\mathbf{u})\mathbf{A} \right) \odot \left(\mathbf{B}\,\mathrm{diag}(\mathbf{w}) \right) &= [u_i a_{ij} b_{ij} w_j] \in \mathbb{K}^{I \times J} \\ &= \mathrm{diag}(\mathbf{u})(\mathbf{A} \odot \mathbf{B})\mathrm{diag}(\mathbf{w}) \\ &= \left(\mathrm{diag}(\mathbf{u})\mathbf{A}\,\mathrm{diag}(\mathbf{w}) \right) \odot \mathbf{B} \\ &= \mathbf{A} \odot \left(\mathrm{diag}(\mathbf{u})\mathbf{B}\,\mathrm{diag}(\mathbf{w}) \right) \\ &= \left(\mathbf{A}\,\mathrm{diag}(\mathbf{w}) \right) \odot \left(\mathrm{diag}(\mathbf{u})\mathbf{B} \right) \end{aligned}$
$\mathrm{vec}(\mathbf{A} \odot \mathbf{B}) = \mathrm{diag}\big(\mathrm{vec}(\mathbf{A})\big)\mathrm{vec}(\mathbf{B}) = \mathrm{diag}\big(\mathrm{vec}(\mathbf{B})\big)\mathrm{vec}(\mathbf{A})$

Table 2.4. *The* $diag$ *operator and the Hadamard product*

PROOF.– Recall that the $diag$ operator transforms a vector argument into a diagonal matrix whose diagonal elements are the components of the vector:

$$\mathrm{diag}(\mathbf{u}) = \sum_{i=1}^I u_i \mathbf{e}_i^{(I)} (\mathbf{e}_i^{(I)})^T. \tag{2.12}$$

– For $\mathbf{u}, \mathbf{v} \in \mathbb{K}^I$, the orthonormality property of the basis vectors $\mathbf{e}_i^{(I)}$ and the commutativity of the Hadamard product give us:

$$\mathrm{diag}(\mathbf{u})\mathbf{v} = \sum_{i=1}^{I} u_i \mathbf{e}_i^{(I)} (\mathbf{e}_i^{(I)})^T \Big(\sum_{j=1}^{I} v_j \mathbf{e}_j^{(I)} \Big) = \sum_{i,j=1}^{I} u_i v_j \delta_{ij} \mathbf{e}_i^{(I)}$$

or alternatively:

$$\mathrm{diag}(\mathbf{u})\mathbf{v} = \sum_{i=1}^{I} u_i v_i \mathbf{e}_i^{(I)} = \mathbf{u} \odot \mathbf{v} = \mathbf{v} \odot \mathbf{u} = \mathrm{diag}(\mathbf{v})\mathbf{u} \in \mathbb{K}^I. \qquad [2.13]$$

– By decomposing the matrix $\mathbf{A}$ using its columns and the relation [2.13], with $\mathbf{v} = \mathbf{A}_{.j}$, together with the second relation in Table 2.3, we have:

$$\mathrm{diag}(\mathbf{u})\mathbf{A} = \sum_{j=1}^{J} \mathrm{diag}(\mathbf{u})\mathbf{A}_{.j} (\mathbf{e}_j^{(J)})^T = \sum_{j=1}^{J} \Big(\mathbf{u} \odot \mathbf{A}_{.j} \Big) (\mathbf{e}_j^{(J)})^T$$

$$= \mathbf{u}\,\mathbf{1}_J^T \odot \Big(\sum_{j=1}^{J} \mathbf{A}_{.j} (\mathbf{e}_j^{(J)})^T \Big) = \mathbf{u}\,\mathbf{1}_J^T \odot \mathbf{A} = [u_i a_{ij}] \in \mathbb{K}^{I \times J}.$$

$$[2.14]$$

– Using the relation [2.14] and the transposition and commutativity properties of the Hadamard product, we obtain:

$$\mathbf{A}\,\mathrm{diag}(\mathbf{w}) = \Big(\mathrm{diag}(\mathbf{w})\mathbf{A}^T \Big)^T = (\mathbf{w}\,\mathbf{1}_I^T \odot \mathbf{A}^T)^T$$

$$= \mathbf{A} \odot \mathbf{1}_I\,\mathbf{w}^T = [a_{ij} w_j] \in \mathbb{K}^{I \times J}. \qquad [2.15]$$

– Using the relations [2.14] and [2.15], as well as the commutativity and associativity properties of the Hadamard product, we obtain:

$$\Big(\mathrm{diag}(\mathbf{u})\mathbf{A} \Big) \odot \Big(\mathbf{B}\,\mathrm{diag}(\mathbf{w}) \Big) = [u_i a_{ij}] \odot [w_j b_{ij}] = [u_i w_j a_{ij} b_{ij}] \in \mathbb{K}^{I \times J}$$

$$= [u_i w_j a_{ij}] \odot [b_{ij}] = \Big(\mathrm{diag}(\mathbf{u})\mathbf{A}\,\mathrm{diag}(\mathbf{w}) \Big) \odot \mathbf{B}$$

$$= [a_{ij}] \odot [u_i w_j b_{ij}] = \mathbf{A} \odot \Big(\mathrm{diag}(\mathbf{u})\mathbf{B}\,\mathrm{diag}(\mathbf{w}) \Big)$$

$$= [w_j a_{ij}] \odot [u_i b_{ij}] = \Big(\mathbf{A}\,\mathrm{diag}(\mathbf{w}) \Big) \odot \Big(\mathrm{diag}(\mathbf{u})\mathbf{B} \Big),$$

and

$$\Big(\mathrm{diag}(\mathbf{u})\mathbf{A} \Big) \odot \Big(\mathbf{B}\,\mathrm{diag}(\mathbf{w}) \Big) = \Big(\mathbf{u}\,\mathbf{1}_J^T \odot \mathbf{A} \Big) \odot \Big(\mathbf{B} \odot \mathbf{1}_I\,\mathbf{w}^T \Big)$$

$$= \mathbf{u}\,\mathbf{1}_J^T \odot (\mathbf{A} \odot \mathbf{B}) \odot \mathbf{1}_I\,\mathbf{w}^T$$

$$= \mathrm{diag}(\mathbf{u})(\mathbf{A} \odot \mathbf{B})\,\mathrm{diag}(\mathbf{w}).$$

– Applying the formula [2.13] with $(\mathbf{u}, \mathbf{v}) = (\text{vec}(\mathbf{A}), \text{vec}(\mathbf{B}))$ gives:

$$\text{diag}\big(\text{vec}(\mathbf{A})\big)\text{vec}(\mathbf{B}) = \text{vec}(\mathbf{A}) \odot \text{vec}(\mathbf{B}) = \text{diag}\big(\text{vec}(\mathbf{B})\big)\text{vec}(\mathbf{A}).$$

The last relation in Table 2.4 follows from the property $\text{vec}(\mathbf{A}) \odot \text{vec}(\mathbf{B}) = \text{vec}(\mathbf{A} \odot \mathbf{B})$.

This concludes the proof of the relations in Table 2.4.   □

2.4. Kronecker product

2.4.1. *Kronecker product of vectors*

2.4.1.1. *Definition*

Below, the Kronecker product is defined for two vectors, then for three and N vectors.

– For $\mathbf{u} \in \mathbb{K}^I$ and $\mathbf{v} \in \mathbb{K}^J$, we have:

$$\mathbf{x} = \mathbf{u} \otimes \mathbf{v} \in \mathbb{K}^{IJ} \iff x_{j+(i-1)J} = u_i v_j$$

or equivalently:

$$\mathbf{u} \otimes \mathbf{v} = [u_1 v_1, u_1 v_2, \cdots, u_1 v_J, u_2 v_1, \cdots, u_I v_J]^T = \begin{bmatrix} u_1 \mathbf{v} \\ \vdots \\ u_I \mathbf{v} \end{bmatrix}.$$

– Similarly, for $\mathbf{u} \in \mathbb{K}^I$, $\mathbf{v} \in \mathbb{K}^J$, and $\mathbf{w} \in \mathbb{K}^K$, we have:

$$\mathbf{x} = \mathbf{u} \otimes \mathbf{v} \otimes \mathbf{w} \in \mathbb{K}^{IJK} \iff x_{k+(j-1)K+(i-1)JK} = u_i v_j w_k.$$

REMARK 2.3.– By convention, the order of the dimensions in a product IJK follows the order of variation of the corresponding indices (i, j, k). For example, $\mathbb{K}^{IJK}$ means that the index i varies more slowly than j, which itself varies more slowly than k.

– In general, for N vectors $\mathbf{u}^{(n)} \in \mathbb{K}^{I_n}$, the multiple Kronecker product $\overset{N}{\underset{n=1}{\otimes}} \mathbf{u}^{(n)} \triangleq \mathbf{u}^{(1)} \otimes \mathbf{u}^{(2)} \otimes \cdots \otimes \mathbf{u}^{(N)} \in \mathbb{K}^{I_1 I_2 \cdots I_N}$ is such that:

$$\left(\overset{N}{\underset{n=1}{\otimes}} \mathbf{u}^{(n)} \right)_i = \prod_{n=1}^{N} u_{i_n}^{(n)} \quad \text{with} \quad i = i_N + (i_{N-1})I_N + \sum_{n=1}^{N-2}(i_n - 1)\prod_{j=n+1}^{N} I_j.$$

$$[2.16]$$

Note that the Kronecker product of column vectors is identical to their Khatri–Rao product.

2.4.1.2. *Fundamental properties*

PROPOSITION 2.4.– The Kronecker product satisfies the properties of associativity and distributivity with respect to addition:

$$\mathbf{u} \otimes (\mathbf{v} \otimes \mathbf{w}) = (\mathbf{u} \otimes \mathbf{v}) \otimes \mathbf{w} = \mathbf{u} \otimes \mathbf{v} \otimes \mathbf{w} \qquad [2.17]$$

$$(\mathbf{u} + \mathbf{v}) \otimes (\mathbf{x} + \mathbf{y}) = (\mathbf{u} \otimes \mathbf{x}) + (\mathbf{u} \otimes \mathbf{y}) + (\mathbf{v} \otimes \mathbf{x}) + (\mathbf{v} \otimes \mathbf{y}). \qquad [2.18]$$

However, the Kronecker product is not commutative, i.e. if $\mathbf{u} \neq \mathbf{v}$, then, in general, $\mathbf{u} \otimes \mathbf{v} \neq \mathbf{v} \otimes \mathbf{u}$.

It also satisfies the following properties:

$$(\alpha \mathbf{u}) \otimes \mathbf{v} = \mathbf{u} \otimes (\alpha \mathbf{v}) = \alpha (\mathbf{u} \otimes \mathbf{v}) \, , \, \forall \alpha \in \mathbb{K}$$

$$(\mathbf{u} \otimes \mathbf{v})^T = \mathbf{u}^T \otimes \mathbf{v}^T \, , \, (\mathbf{u} \otimes \mathbf{v})^H = \mathbf{u}^H \otimes \mathbf{v}^H. \qquad [2.19]$$

2.4.1.3. *Basic relations*

In this section, we present various relations involving Kronecker products of vectors. Table 2.5 gives two basic relations satisfied by the Kronecker product of vectors.

$\mathbf{u} \in \mathbb{K}^I, \mathbf{v} \in \mathbb{K}^J$
$\mathbf{u}^H \otimes \mathbf{v} = \mathbf{v}\mathbf{u}^H = \mathbf{v} \otimes \mathbf{u}^H \in \mathbb{K}^{J \times I}$
$\mathbf{u} \otimes \mathbf{v}^H = \mathbf{u}\mathbf{v}^H = \mathbf{v}^H \otimes \mathbf{u} \in \mathbb{K}^{I \times J}$

Table 2.5. *Basic relations satisfied by the Kronecker product of vectors*

PROOF.– These Kronecker products can be developed as follows:

$$\mathbf{u}^H \otimes \mathbf{v} = [u_1^* \mathbf{v}, \cdots, u_I^* \mathbf{v}] = \mathbf{v}\mathbf{u}^H \qquad [2.20]$$

$$= \begin{bmatrix} v_1 u_1^* & \cdots & v_1 u_I^* \\ \vdots & & \vdots \\ v_J u_1^* & \cdots & v_J u_I^* \end{bmatrix} = \begin{bmatrix} v_1 \mathbf{u}^H \\ \vdots \\ v_J \mathbf{u}^H \end{bmatrix} = \mathbf{v} \otimes \mathbf{u}^H \qquad [2.21]$$

$$\mathbf{u} \otimes \mathbf{v}^H = \begin{bmatrix} u_1 \mathbf{v}^H \\ \vdots \\ u_I \mathbf{v}^H \end{bmatrix} = \mathbf{u}\mathbf{v}^H = \mathbf{v}^H \otimes \mathbf{u} \text{ by (2.20)}, \qquad [2.22]$$

which concludes the proof of the relations in Table 2.5.   □

These relations continue to hold if transconjugation is replaced by transposition, i.e.:

$$\mathbf{u}^T \otimes \mathbf{v} = \mathbf{v}\mathbf{u}^T = \mathbf{v} \otimes \mathbf{u}^T \qquad\qquad [2.23]$$

$$\mathbf{u} \otimes \mathbf{v}^T = \mathbf{u}\mathbf{v}^T = \mathbf{v}^T \otimes \mathbf{u}. \qquad\qquad [2.24]$$

PROPOSITION 2.5.– From the above, we can conclude that the Kronecker product of a column vector with a row vector is independent of the order of the vectors. Therefore, in a multiple Kronecker product of row vectors and column vectors, we can permute a row vector followed by a column vector (or vice versa) without changing the result, but we cannot permute two consecutive column or row vectors, since the Kronecker product is not commutative.

Thus, for $\mathbf{u} \in \mathbb{K}^I, \mathbf{v} \in \mathbb{K}^J$ and $\mathbf{w} \in \mathbb{K}^K$, we have:

$$\mathbf{u}^T \otimes \mathbf{v} \otimes \mathbf{w} = \mathbf{v} \otimes \mathbf{u}^T \otimes \mathbf{w} = \mathbf{v} \otimes \mathbf{w} \otimes \mathbf{u}^T \in \mathbb{K}^{JK \times I}$$

$$\neq \mathbf{w} \otimes \mathbf{v} \otimes \mathbf{u}^T \in \mathbb{K}^{KJ \times I}$$

$$\mathbf{u}^T \otimes \mathbf{v} \otimes \mathbf{w}^T = \mathbf{u}^T \otimes \mathbf{w}^T \otimes \mathbf{v} = \mathbf{v} \otimes \mathbf{u}^T \otimes \mathbf{w}^T \in \mathbb{K}^{J \times IK}$$

$$\neq \mathbf{v} \otimes \mathbf{w}^T \otimes \mathbf{u}^T \in \mathbb{K}^{J \times KI}.$$

EXAMPLE 2.6.– For $\mathbf{u} \in \mathbb{K}^3, \mathbf{v} \in \mathbb{K}^2$ and $\mathbf{w} \in \mathbb{K}^2$, we have:

$$\mathbf{u}^T \otimes \mathbf{v} \otimes \mathbf{w} = \begin{bmatrix} u_1v_1 & u_2v_1 & u_3v_1 \\ u_1v_2 & u_2v_2 & u_3v_2 \end{bmatrix} \otimes \mathbf{w}$$

$$= \begin{bmatrix} u_1v_1w_1 & u_2v_1w_1 & u_3v_1w_1 \\ u_1v_1w_2 & u_2v_1w_2 & u_3v_1w_2 \\ u_1v_2w_1 & u_2v_2w_1 & u_3v_2w_1 \\ u_1v_2w_2 & u_2v_2w_2 & u_3v_2w_2 \end{bmatrix}$$

and

$$\mathbf{v} \otimes \mathbf{w} \otimes \mathbf{u}^T = \begin{bmatrix} v_1w_1 \\ v_1w_2 \\ v_2w_1 \\ v_2w_2 \end{bmatrix} \otimes \mathbf{u}^T = \begin{bmatrix} v_1w_1u_1 & v_1w_1u_2 & v_1w_1u_3 \\ v_1w_2u_1 & v_1w_2u_2 & v_1w_2u_3 \\ v_2w_1u_1 & v_2w_1u_2 & v_2w_1u_3 \\ v_2w_2u_1 & v_2w_2u_2 & v_2w_2u_3 \end{bmatrix}$$

$$= \mathbf{u}^T \otimes \mathbf{v} \otimes \mathbf{w}.$$

By the property [2.24], we can deduce the following identity:

$$\mathbf{E}_{ij}^{(I \times J)} = \mathbf{e}_i^{(I)}(\mathbf{e}_j^{(J)})^T = \mathbf{e}_i^{(I)} \otimes (\mathbf{e}_j^{(J)})^T \in \mathbb{R}^{I \times J}. \qquad\qquad [2.25]$$

Omitting the dimensions of the basis vectors to alleviate the notation, $\mathbf{A} \in \mathbb{K}^{I \times J}$ and $\mathbf{A}^T$ can be expressed as follows in the canonical basis $\{\mathbf{E}_{ij}^{(I \times J)}\}$:

$$\mathbf{A} = \sum_{i=1}^{I} \sum_{j=1}^{J} a_{ij} \mathbf{E}_{ij}^{(I \times J)} = \sum_{i=1}^{I} \sum_{j=1}^{J} a_{ij} \mathbf{e}_i \mathbf{e}_j^T = \sum_{i=1}^{I} \sum_{j=1}^{J} a_{ij} \mathbf{e}_i \otimes \mathbf{e}_j^T \qquad [2.26]$$

$$\mathbf{A}^T = \sum_{i=1}^{I} \sum_{j=1}^{J} a_{ij} \mathbf{E}_{ji}^{(J \times I)} = \sum_{i=1}^{I} \sum_{j=1}^{J} a_{ij} \mathbf{e}_j \mathbf{e}_i^T = \sum_{i=1}^{I} \sum_{j=1}^{J} a_{ij} \mathbf{e}_j \otimes \mathbf{e}_i^T. \qquad [2.27]$$

2.4.1.4. *Rank-one matrices and Kronecker products of vectors*

Table 2.6 states a few relations satisfied by rank-one matrices expressed in terms of Kronecker products of vectors.

$\mathbf{u} \in \mathbb{K}^I, \mathbf{v} \in \mathbb{K}^J, \mathbf{x} \in \mathbb{K}^K, \mathbf{y} \in \mathbb{K}^L$
$(\mathbf{u} \otimes \mathbf{v})(\mathbf{x} \otimes \mathbf{y})^H = \mathbf{u}\mathbf{x}^H \otimes \mathbf{v}\mathbf{y}^H \in \mathbb{K}^{IJ \times KL}$
Generalization to P vectors
$\mathbf{u}_p \in \mathbb{K}^{I_p}, \mathbf{x}_p \in \mathbb{K}^{J_p}, \ I = \prod_{p=1}^{P} I_p, \ J = \prod_{p=1}^{P} J_p$
$\left(\overset{P}{\underset{p=1}{\otimes}} \mathbf{u}_p \right) \left(\overset{P}{\underset{p=1}{\otimes}} \mathbf{x}_p \right)^H = \overset{P}{\underset{p=1}{\otimes}} \left(\mathbf{u}_p \mathbf{x}_p^H \right) \in \mathbb{K}^{I \times J}$
Special cases
$(\mathbf{u} \otimes \mathbf{v})\mathbf{x}^H = (\mathbf{u} \otimes \mathbf{v})(\mathbf{x}^H \otimes 1) = \mathbf{u}\mathbf{x}^H \otimes \mathbf{v} \in \mathbb{K}^{IJ \times K}$
$(\mathbf{u} \otimes \mathbf{v})\mathbf{x}^H = (\mathbf{u} \otimes \mathbf{v})(1 \otimes \mathbf{x}^H) = \mathbf{u} \otimes \mathbf{v}\mathbf{x}^H \in \mathbb{K}^{IJ \times K}$
$\mathbf{x}(\mathbf{u} \otimes \mathbf{v})^H = \mathbf{x}\mathbf{u}^H \otimes \mathbf{v}^H = \mathbf{u}^H \otimes \mathbf{x}\mathbf{v}^H \in \mathbb{K}^{K \times IJ}$

Table 2.6. *Rank-one matrices and Kronecker products of vectors*

Using the associativity property [2.17] and the transconjugation property [2.19], as well as the properties stated in Tables 2.5 and 2.6, the product $(\mathbf{u} \otimes \mathbf{v})(\mathbf{x} \otimes \mathbf{y})^H$ can be decomposed in the various ways summarized in Table 2.7.

$(\mathbf{u} \otimes \mathbf{v})(\mathbf{x} \otimes \mathbf{y})^H = (\mathbf{u} \otimes \mathbf{v}) \otimes (\mathbf{x} \otimes \mathbf{y})^H = \mathbf{u} \otimes \mathbf{v} \otimes \mathbf{x}^H \otimes \mathbf{y}^H = \mathbf{u} \otimes \mathbf{v}\mathbf{x}^H \otimes \mathbf{y}^H$
$(\mathbf{u} \otimes \mathbf{v})(\mathbf{x} \otimes \mathbf{y})^H = (\mathbf{x} \otimes \mathbf{y})^H \otimes (\mathbf{u} \otimes \mathbf{v}) = \mathbf{x}^H \otimes \mathbf{y}^H \otimes \mathbf{u} \otimes \mathbf{v} = \mathbf{x}^H \otimes \mathbf{u}\mathbf{y}^H \otimes \mathbf{v}$

Table 2.7. *Other decompositions of* $(\mathbf{u} \otimes \mathbf{v})(\mathbf{x} \otimes \mathbf{y})^H$

2.4.1.5. *Vectorization of a rank-one matrix*

In this section, we present several relations between the vectorization operator applied to rank-one matrices and the Kronecker product of vectors. These relations can easily be shown using the definition of the vectorization operator and the associativity [2.17] and transposition/transconjugation [2.19] properties of the Kronecker product, together with the identity [2.22].

For $\mathbf{x} \in \mathbb{K}^I, \mathbf{y} \in \mathbb{K}^J, \mathbf{u} \in \mathbb{K}^K, \mathbf{v} \in \mathbb{K}^L$, we have:

$$\text{vec}(\mathbf{x} \otimes \mathbf{y}^H) = \text{vec}(\mathbf{x}\mathbf{y}^H) = \mathbf{y}^* \otimes \mathbf{x} \qquad [2.28]$$

$$\text{vec}(\mathbf{x} \otimes \mathbf{y} \otimes \mathbf{u}^H) = \mathbf{u}^* \otimes \mathbf{x} \otimes \mathbf{y} \qquad [2.29]$$

$$\text{vec}\left((\mathbf{x} \otimes \mathbf{y})(\mathbf{u} \otimes \mathbf{v})^H\right) = \mathbf{u}^* \otimes \mathbf{v}^* \otimes \mathbf{x} \otimes \mathbf{y}. \qquad [2.30]$$

These relations continue to hold if transconjugation is replaced by transposition, removing the conjugation of the vectors.

2.4.2. *Kronecker product of matrices*

2.4.2.1. *Definitions and identities*

Below, we define the Kronecker product of two matrices, followed by the multiple Kronecker product of more than two matrices.

– Given two matrices $\mathbf{A} \in \mathbb{K}^{I \times J}$ and $\mathbf{B} \in \mathbb{K}^{M \times N}$ of arbitrary size, the right Kronecker product of $\mathbf{A}$ by $\mathbf{B}$ is the matrix $\mathbf{C} \in \mathbb{K}^{IM \times JN}$ defined as follows:

$$\mathbf{C} = \mathbf{A} \otimes \mathbf{B} = \begin{bmatrix} a_{11}\mathbf{B} & a_{12}\mathbf{B} & \cdots & a_{1J}\mathbf{B} \\ a_{21}\mathbf{B} & a_{22}\mathbf{B} & \cdots & a_{2J}\mathbf{B} \\ \vdots & \vdots & & \vdots \\ a_{I1}\mathbf{B} & a_{I2}\mathbf{B} & \cdots & a_{IJ}\mathbf{B} \end{bmatrix} = [a_{ij}\mathbf{B}]. \qquad [2.31]$$

The Kronecker product is a matrix partitioned into (I, J) blocks, where the block (i, j) is given by the matrix $a_{ij}\mathbf{B} \in \mathbb{K}^{M \times N}$. The element $a_{ij}b_{mn}$ is located at the position $((i-1)M + m, (j-1)N + n)$ in $\mathbf{A} \otimes \mathbf{B}$.

– For $\mathbf{A}_p \in \mathbb{C}^{I_p \times J_p}, p \in \langle P \rangle$, we define the multiple Kronecker product as follows:

$$\overset{P}{\underset{p=1}{\otimes}} \mathbf{A}_p \triangleq \mathbf{A}_1 \otimes \mathbf{A}_2 \otimes \cdots \otimes \mathbf{A}_P \in \mathbb{C}^{I_1 \cdots I_P \times J_1 \cdots J_P}. \qquad [2.32]$$

EXAMPLE 2.7.– For $\mathbf{A} = \begin{pmatrix} a_{11} & a_{12} \\ a_{21} & a_{22} \end{pmatrix}$, $\mathbf{B} = \begin{pmatrix} b_{11} & b_{12} \\ b_{21} & b_{22} \end{pmatrix}$, we have:

$$\mathbf{A} \otimes \mathbf{B} = \begin{pmatrix} a_{11}\mathbf{B} & \vdots & a_{12}\mathbf{B} \\ \cdots & & \cdots \\ a_{21}\mathbf{B} & \vdots & a_{22}\mathbf{B} \end{pmatrix} = \begin{pmatrix} a_{11}b_{11} & a_{11}b_{12} & \vdots & a_{12}b_{11} & a_{12}b_{12} \\ a_{11}b_{21} & a_{11}b_{22} & \vdots & a_{12}b_{21} & a_{12}b_{22} \\ \cdots & \cdots & & \cdots & \cdots \\ a_{21}b_{11} & a_{21}b_{12} & \vdots & a_{22}b_{11} & a_{22}b_{12} \\ a_{21}b_{21} & a_{21}b_{22} & \vdots & a_{22}b_{21} & a_{22}b_{22} \end{pmatrix}.$$

The Kronecker product [2.31] can also be partitioned into column block or row block matrices. Indeed, the columns (respectively, rows) of $\mathbf{A} \otimes \mathbf{B}$ are formed by all the Kronecker products of one column (respectively, row) of $\mathbf{A}$ with one column (respectively, row) of $\mathbf{B}$, arranging the columns (respectively, rows) in lexicographical order in each case:

$$\mathbf{A} \otimes \mathbf{B} = \begin{bmatrix} \mathbf{A}_{.1} \otimes \mathbf{B}_{.1} & \cdots & \mathbf{A}_{.1} \otimes \mathbf{B}_{.N} & \cdots & \mathbf{A}_{.J} \otimes \mathbf{B}_{.1} & \cdots & \mathbf{A}_{.J} \otimes \mathbf{B}_{.N} \end{bmatrix} \tag{2.33}$$

and

$$\mathbf{A} \otimes \mathbf{B} = \begin{bmatrix} \mathbf{A}_{1.} \otimes \mathbf{B}_{1.} \\ \vdots \\ \mathbf{A}_{1.} \otimes \mathbf{B}_{M.} \\ \vdots \\ \mathbf{A}_{I.} \otimes \mathbf{B}_{1.} \\ \vdots \\ \mathbf{A}_{I.} \otimes \mathbf{B}_{M.} \end{bmatrix}.$$

PROPOSITION 2.8.– For $\mathbf{A} \in \mathbb{K}^{I \times J}$, it is easy to check the following identities:

$$\mathbf{I}_N \otimes \mathbf{A} = \mathrm{diag}\underbrace{\left(\mathbf{A}, \mathbf{A}, \cdots, \mathbf{A}\right)}_{N \text{ terms}} \in \mathbb{C}^{NI \times NJ} \tag{2.34}$$

$$\mathbf{1}_I \otimes \mathbf{1}_J^T = \mathbf{1}_J^T \otimes \mathbf{1}_I = \mathbf{1}_{I \times J} \tag{2.35}$$

$$\mathbf{I}_I \otimes \mathbf{I}_J = \mathbf{I}_{IJ} \tag{2.36}$$

$$\left(\alpha \mathbf{I}_M\right) \otimes \beta \mathbf{I}_N = \alpha\beta\,\mathbf{I}_{MN}, \tag{2.37}$$

where $\mathbf{1}_I$ denotes the vector of ones of size I.

2.4.2.2. Fundamental properties

For the rest of this chapter, we will assume that the matrices have suitable sizes to ensure that all operations are well defined.

PROPOSITION 2.9.– Like the Kronecker product of vectors, the Kronecker product of matrices satisfies the properties of associativity, distributivity with respect to addition, transposition/transconjugation and multiplication by a scalar, as summarized in the following:

$$\mathbf{A} \otimes (\mathbf{B} \otimes \mathbf{C}) = (\mathbf{A} \otimes \mathbf{B}) \otimes \mathbf{C} = \mathbf{A} \otimes \mathbf{B} \otimes \mathbf{C} \qquad [2.38]$$

$$(\mathbf{A} + \mathbf{B}) \otimes (\mathbf{C} + \mathbf{D}) = (\mathbf{A} \otimes \mathbf{C}) + (\mathbf{A} \otimes \mathbf{D}) + (\mathbf{B} \otimes \mathbf{C}) + (\mathbf{B} \otimes \mathbf{D})$$
$$[2.39]$$

$$(\mathbf{A} \otimes \mathbf{B})^T = \mathbf{A}^T \otimes \mathbf{B}^T \ , \ (\mathbf{A} \otimes \mathbf{B})^H = \mathbf{A}^H \otimes \mathbf{B}^H \qquad [2.40]$$

$$(\alpha\mathbf{A}) \otimes \mathbf{B} = \mathbf{A} \otimes (\alpha\mathbf{B}) = \alpha(\mathbf{A} \otimes \mathbf{B}) \ , \ \forall \alpha \in \mathbb{K}. \qquad [2.41]$$

If $\mathbf{A} \neq \mathbf{B}$, their Kronecker product is not commutative in general, i.e. $\mathbf{A} \otimes \mathbf{B} \neq \mathbf{B} \otimes \mathbf{A}$.

PROPOSITION 2.10.– If $\mathbf{A} \in \mathbb{C}^{I \times I}$ and $\mathbf{B} \in \mathbb{C}^{J \times J}$ are symmetric (respectively, Hermitian), then their Kronecker product is itself a symmetric (respectively, Hermitian) matrix.

PROOF.– Using the transposition/transconjugation property [2.40] of the Kronecker product, we can deduce that:

$$\left\{ \begin{array}{l} \mathbf{A}^T = \mathbf{A} \\ \mathbf{B}^T = \mathbf{B} \end{array} \right\} \Rightarrow (\mathbf{A} \otimes \mathbf{B})^T = \mathbf{A}^T \otimes \mathbf{B}^T = \mathbf{A} \otimes \mathbf{B} \qquad [2.42]$$

$$\left\{ \begin{array}{l} \mathbf{A}^H = \mathbf{A} \\ \mathbf{B}^H = \mathbf{B} \end{array} \right\} \Rightarrow (\mathbf{A} \otimes \mathbf{B})^H = \mathbf{A}^H \otimes \mathbf{B}^H = \mathbf{A} \otimes \mathbf{B}, \qquad [2.43]$$

which shows that $\mathbf{A} \otimes \mathbf{B}$ is symmetric (respectively, Hermitian). $\qquad \square$

PROPOSITION 2.11.– For $\mathbf{A} \in \mathbb{K}^{I \times J}, \mathbf{B} \in \mathbb{K}^{M \times N}, \mathbf{C} \in \mathbb{K}^{J \times K}$ and $\mathbf{D} \in \mathbb{K}^{N \times P}$, we have the following identities:

$$\mathbf{A} \otimes \mathbf{B} = (\mathbf{A} \otimes \mathbf{I}_M)(\mathbf{I}_J \otimes \mathbf{B}) = (\mathbf{I}_I \otimes \mathbf{B})(\mathbf{A} \otimes \mathbf{I}_N) \qquad [2.44]$$

$$\mathbf{A}\mathbf{C} \otimes \mathbf{I}_M = (\mathbf{A} \otimes \mathbf{I}_M)(\mathbf{C} \otimes \mathbf{I}_M) \qquad [2.45]$$

$$\mathbf{I}_N \otimes \mathbf{B}\mathbf{D} = (\mathbf{I}_N \otimes \mathbf{B})(\mathbf{I}_N \otimes \mathbf{D}) \qquad [2.46]$$

$$(\mathbf{I}_N \otimes \mathbf{A})^k = \mathbf{I}_N \otimes \mathbf{A}^k. \qquad [2.47]$$

PROOF.–

– The equation of the definition [2.31] can be rewritten as:

$$\mathbf{A} \otimes \mathbf{B} = \begin{bmatrix} a_{11}\mathbf{I}_M & a_{12}\mathbf{I}_M & \cdots & a_{1J}\mathbf{I}_M \\ a_{21}\mathbf{I}_M & a_{22}\mathbf{I}_M & \cdots & a_{2J}\mathbf{I}_M \\ \vdots & \vdots & & \vdots \\ a_{I1}\mathbf{I}_M & a_{I2}\mathbf{I}_M & \cdots & a_{IJ}\mathbf{I}_M \end{bmatrix} \begin{bmatrix} \mathbf{B} & 0 & \cdots & 0 \\ 0 & \mathbf{B} & \cdots & 0 \\ \vdots & \vdots & & \vdots \\ 0 & 0 & \cdots & \mathbf{B} \end{bmatrix}$$

$$= (\mathbf{A} \otimes \mathbf{I}_M)(\mathbf{I}_J \otimes \mathbf{B}). \tag{2.48}$$

Similarly, we have:

$$\mathbf{A} \otimes \mathbf{B} = \begin{bmatrix} \mathbf{B} & 0 & \cdots & 0 \\ 0 & \mathbf{B} & \cdots & 0 \\ \vdots & \vdots & & \vdots \\ 0 & 0 & \cdots & \mathbf{B} \end{bmatrix} \begin{bmatrix} a_{11}\mathbf{I}_N & a_{12}\mathbf{I}_N & \cdots & a_{1J}\mathbf{I}_N \\ a_{21}\mathbf{I}_N & a_{22}\mathbf{I}_N & \cdots & a_{2J}\mathbf{I}_N \\ \vdots & \vdots & & \vdots \\ a_{I1}\mathbf{I}_N & a_{I2}\mathbf{I}_N & \cdots & a_{IJ}\mathbf{I}_N \end{bmatrix}$$

$$= (\mathbf{I}_I \otimes \mathbf{B})(\mathbf{A} \otimes \mathbf{I}_N), \tag{2.49}$$

which proves the identities [2.44].

– Using the definition of the Kronecker product, we have:

$$(\mathbf{A} \otimes \mathbf{I}_M)(\mathbf{C} \otimes \mathbf{I}_M) = \begin{bmatrix} a_{11}\mathbf{I}_M & \cdots & a_{1J}\mathbf{I}_M \\ \vdots & & \vdots \\ a_{I1}\mathbf{I}_M & \cdots & a_{IJ}\mathbf{I}_M \end{bmatrix} \begin{bmatrix} c_{11}\mathbf{I}_M & \cdots & c_{1K}\mathbf{I}_M \\ \vdots & & \vdots \\ c_{J1}\mathbf{I}_M & \cdots & c_{JK}\mathbf{I}_M \end{bmatrix}$$

$$= \begin{bmatrix} \sum_{j=1}^{J} a_{1j}c_{j1}\mathbf{I}_M & \cdots & \sum_{j=1}^{J} a_{1j}c_{jK}\mathbf{I}_M \\ \vdots & & \vdots \\ \sum_{j=1}^{J} a_{Ij}c_{j1}\mathbf{I}_M & \cdots & \sum_{j=1}^{J} a_{Ij}c_{jK}\mathbf{I}_M \end{bmatrix},$$

with $\sum_{j=1}^{J} a_{ij}c_{jk} = (\mathbf{AC})_{ik}$, which implies:

$$(\mathbf{A} \otimes \mathbf{I}_M)(\mathbf{C} \otimes \mathbf{I}_M) = \mathbf{AC} \otimes \mathbf{I}_M.$$

This proves [2.45]. The identity [2.46] can be shown in the same way. The identity [2.47] follows directly from [2.46]. □

Using the properties [2.44]–[2.46], we can show the fundamental identity stated in the following proposition.

PROPOSITION 2.12.– For $\mathbf{A} \in \mathbb{K}^{I \times J}$, $\mathbf{B} \in \mathbb{K}^{M \times N}$, $\mathbf{C} \in \mathbb{K}^{J \times K}$ and $\mathbf{D} \in \mathbb{K}^{N \times P}$, we have the fundamental identity:

$$(\mathbf{A} \otimes \mathbf{B})(\mathbf{C} \otimes \mathbf{D}) = \mathbf{AC} \otimes \mathbf{BD}. \tag{2.50}$$

PROOF.– Using the results of the previous proposition, we obtain:

$$(\mathbf{A} \otimes \mathbf{B})(\mathbf{C} \otimes \mathbf{D}) = (\mathbf{A} \otimes \mathbf{I}_M)(\mathbf{I}_J \otimes \mathbf{B})(\mathbf{C} \otimes \mathbf{I}_N)(\mathbf{I}_K \otimes \mathbf{D})$$
$$= (\mathbf{A} \otimes \mathbf{I}_M)(\mathbf{C} \otimes \mathbf{B})(\mathbf{I}_K \otimes \mathbf{D})$$
$$= (\mathbf{A} \otimes \mathbf{I}_M)(\mathbf{C} \otimes \mathbf{I}_M)(\mathbf{I}_K \otimes \mathbf{B})(\mathbf{I}_K \otimes \mathbf{D})$$
$$= (\mathbf{AC} \otimes \mathbf{I}_M)(\mathbf{I}_K \otimes \mathbf{BD})$$
$$= \mathbf{AC} \otimes \mathbf{BD}.$$

This proves the identity [2.50]. □

The properties and identities presented above continue to hold if the matrices are replaced by vectors. For instance, from the relation [2.50], and using the fact that a scalar is equal to its transpose, we can deduce the relations stated in Table 2.8.

$\mathbf{x}, \mathbf{u} \in \mathbb{K}^I$, $\mathbf{y}, \mathbf{v} \in \mathbb{K}^J$
$(\mathbf{x} \otimes \mathbf{y})^T(\mathbf{u} \otimes \mathbf{v}) = \mathbf{x}^T\mathbf{u} \otimes \mathbf{y}^T\mathbf{v} = (\mathbf{x}^T\mathbf{u})(\mathbf{y}^T\mathbf{v})$
$\mathbf{x}, \mathbf{y} \in \mathbb{K}^I$
$(\mathbf{x} \otimes \mathbf{x}^*)^T(\mathbf{y}^* \otimes \mathbf{y}) = (\mathbf{x}^T\mathbf{y}^*)(\mathbf{x}^H\mathbf{y}) = (\mathbf{x}^T\mathbf{y}^*)^T(\mathbf{x}^H\mathbf{y}) = \mathbf{y}^H(\mathbf{x}\mathbf{x}^H)\mathbf{y}$
$\mathbf{x}, \mathbf{y}, \mathbf{u}, \mathbf{v} \in \mathbb{K}^I$
$(\mathbf{x} \otimes \mathbf{y}^T)(\mathbf{u}^T \otimes \mathbf{v}) = (\mathbf{x}\mathbf{u}^T) \otimes (\mathbf{y}^T\mathbf{v}) = (\mathbf{y}^T\mathbf{v})\mathbf{x}\mathbf{u}^T$

Table 2.8. *Relations involving Kronecker products of vectors*

2.4.2.3. *Multiple Kronecker product*

Table 2.9 states a generalization of the distributivity property [2.39] and the fundamental identity [2.50] to the case of the multiple Kronecker product.

2.4.2.4. *Interpretation as a tensor product*

Let $E_n = \mathbb{K}^{I_n}$ and $F_n = \mathbb{K}^{J_n}$, for $n \in \langle N \rangle$, be vector spaces with bases $\{\mathbf{b}_{i_n}^{(I_n)}\}$ and $\{\mathbf{c}_{j_n}^{(J_n)}\}$, respectively. Given N linear mappings $f_n : E_n \to F_n$ and the associated matrices $\mathbf{A}_n \in \mathbb{K}^{J_n \times I_n}$ with respect to these bases, consider the multilinear mapping:

$$\underset{n=1}{\overset{N}{\times}} E_n \ni (\mathbf{u}_1, \cdots, \mathbf{u}_N) \mapsto \underset{n=1}{\overset{N}{\otimes}} \left(f_n(\mathbf{u}_n)\right) \in \bigotimes_{n=1}^{N} F_n, \qquad [2.51]$$

where $\bigotimes_{n=1}^{N} F_n$ denotes the tensor space of the N vector spaces F_n. The multilinearity of this mapping follows from the linearity of the f_n, namely $f_n(\lambda\mathbf{u}_n + \mathbf{v}_n) = \lambda f_n(\mathbf{u}_n) + f_n(\mathbf{v}_n)$ for every $\mathbf{u}_n, \mathbf{v}_n \in E_n$ and all $\lambda \in \mathbb{K}$.

Distributivity of the multiple Kronecker product with respect to addition

$$\left(\sum_{p=1}^{P} \mathbf{A}_p\right) \otimes \left(\sum_{q=1}^{Q} \mathbf{B}_q\right) = \sum_{p=1}^{P} \sum_{q=1}^{Q} \left(\mathbf{A}_p \otimes \mathbf{B}_q\right)$$

Fundamental identities for the multiple Kronecker product

$$\prod_{p=1}^{P} \left(\mathbf{A}_p \otimes \mathbf{B}_p\right) = \left(\prod_{p=1}^{P} \mathbf{A}_p\right) \otimes \left(\prod_{p=1}^{P} \mathbf{B}_p\right)$$

$$\left(\overset{P}{\underset{p=1}{\otimes}} \mathbf{A}_p\right)\left(\overset{P}{\underset{p=1}{\otimes}} \mathbf{B}_p\right) = \mathbf{A}_1 \mathbf{B}_1 \otimes \cdots \otimes \mathbf{A}_P \mathbf{B}_P = \overset{P}{\underset{p=1}{\otimes}} \left(\mathbf{A}_p \mathbf{B}_p\right)$$

$$\mathbf{A} \in \mathbb{K}^{I \times I}, \mathbf{B} \in \mathbb{K}^{J \times J}$$

$$(\mathbf{A} \otimes \mathbf{B})^p = \mathbf{A}^p \otimes \mathbf{B}^p$$

Table 2.9. *Properties of the multiple Kronecker product*

By the universal property applied to this multilinear mapping, there exists a unique linear mapping $\overset{N}{\underset{n=1}{\otimes}} f_n$ from $\overset{N}{\underset{n=1}{\bigotimes}} E_n$ into $\overset{N}{\underset{n=1}{\bigotimes}} F_n$ such that (Favier 2019):

$$\overset{N}{\underset{n=1}{\otimes}} f_n \left(\overset{N}{\underset{n=1}{\otimes}} \mathbf{u}_n\right) \mapsto \overset{N}{\underset{n=1}{\otimes}} f_n(\mathbf{u}_n). \qquad [2.52]$$

This equation defines the tensor product $\overset{N}{\underset{n=1}{\otimes}} f_n$ of the linear mappings f_n. By the linearity of f_n, namely $f_n(\mathbf{u}_n) = \mathbf{A}_n \mathbf{u}_n$, and the third property of the Kronecker product in Table 2.9, we deduce that:

$$\overset{N}{\underset{n=1}{\bigotimes}} E_n \ni \overset{N}{\underset{n=1}{\otimes}} \mathbf{u}_n \mapsto \overset{N}{\underset{n=1}{\otimes}} f_n(\mathbf{u}_n) = \overset{N}{\underset{n=1}{\otimes}} (\mathbf{A}_n \mathbf{u}_n) \in \overset{N}{\underset{n=1}{\bigotimes}} F_n$$

$$\overset{N}{\underset{n=1}{\otimes}} \mathbf{u}_n \mapsto \left(\overset{N}{\underset{n=1}{\otimes}} \mathbf{A}_n\right)\left(\overset{N}{\underset{n=1}{\otimes}} \mathbf{u}_n\right). \qquad [2.53]$$

This equation shows that the multiple Kronecker product $\overset{N}{\underset{n=1}{\otimes}} \mathbf{A}_n$ is the matrix associated with the tensor product $\overset{N}{\underset{n=1}{\otimes}} f_n$ of the linear mappings f_n.

2.4.2.5. *Identities involving matrix-vector Kronecker products*

Using the fundamental identity [2.50] and the transposition property [2.40], it is easy to check the identities in Table 2.10.

Note that, for $\mathbf{x} \in \mathbb{K}^M$, we have:

$$\mathbf{x} \otimes \mathbf{I}_J = \begin{bmatrix} x_1 \mathbf{I}_J \\ \vdots \\ x_M \mathbf{I}_J \end{bmatrix} \in \mathbb{K}^{MJ \times J}, \ \mathbf{I}_J \otimes \mathbf{x} = \begin{bmatrix} \mathbf{x} & & \\ & \ddots & \\ & & \mathbf{x} \end{bmatrix} \in \mathbb{K}^{JM \times J}.$$

$\mathbf{x} \in \mathbb{K}^M, \mathbf{y} \in \mathbb{K}^J, \mathbf{A} \in \mathbb{K}^{I \times J}, \mathbf{B} \in \mathbb{K}^{J \times K}, \mathbf{C} \in \mathbb{K}^{K \times J}$
$(\mathbf{x} \otimes \mathbf{A})\mathbf{B} = (\mathbf{x} \otimes \mathbf{A})(1 \otimes \mathbf{B}) = \mathbf{x} \otimes (\mathbf{AB})$
$(\mathbf{A} \otimes \mathbf{x})\mathbf{B} = (\mathbf{A} \otimes \mathbf{x})(\mathbf{B} \otimes 1) = (\mathbf{AB}) \otimes \mathbf{x}$
$\mathbf{C}(\mathbf{x} \otimes \mathbf{A})^T = \mathbf{x}^T \otimes (\mathbf{CA}^T)$
$\mathbf{C}(\mathbf{A} \otimes \mathbf{x})^T = (\mathbf{CA}^T) \otimes \mathbf{x}^T$
Special case: $\mathbf{A} = \mathbf{I}_J$
$(\mathbf{x} \otimes \mathbf{I}_J)\mathbf{B} = \mathbf{x} \otimes \mathbf{B}$, $(\mathbf{I}_J \otimes \mathbf{x})\mathbf{B} = \mathbf{B} \otimes \mathbf{x}$
$\mathbf{C}(\mathbf{x} \otimes \mathbf{I}_J)^T = \mathbf{x}^T \otimes \mathbf{C}$, $\mathbf{C}(\mathbf{I}_J \otimes \mathbf{x})^T = \mathbf{C} \otimes \mathbf{x}^T$
Special case: $\mathbf{x} \in \mathbb{K}^M, \mathbf{y} \in \mathbb{K}^J$
$\mathbf{x} \otimes \mathbf{y} = (\mathbf{x} \otimes \mathbf{I}_J)\mathbf{y} = (\mathbf{I}_M \otimes \mathbf{y})\mathbf{x}$

Table 2.10. *Identities involving matrix-vector Kronecker products*

2.4.3. *Rank, trace, determinant and spectrum of a Kronecker product*

Table 2.11 presents a few properties of the rank, trace, determinant and spectrum of a Kronecker product.

$\mathbf{A} \in \mathbb{K}^{I \times J}, \mathbf{B} \in \mathbb{K}^{K \times L}$
$r_{\mathbf{A} \otimes \mathbf{B}} = r_{\mathbf{B} \otimes \mathbf{A}} = r_{\mathbf{A}} r_{\mathbf{B}}$
$\mathbf{A} \in \mathbb{K}^{I \times I}, \mathbf{B} \in \mathbb{K}^{J \times J}$
$\mathrm{sp}(\mathbf{A}) = \bigcup_{i=1}^{I}\{\lambda_i\}, \mathrm{sp}(\mathbf{B}) = \bigcup_{j=1}^{J}\{\mu_j\}$
$\mathrm{tr}(\mathbf{A} \otimes \mathbf{B}) = \mathrm{tr}(\mathbf{B} \otimes \mathbf{A}) = \mathrm{tr}(\mathbf{A})\mathrm{tr}(\mathbf{B})$
$\det(\mathbf{A} \otimes \mathbf{B}) = \det(\mathbf{B} \otimes \mathbf{A}) = [\det(\mathbf{A})]^J [\det(\mathbf{B})]^I$
$\mathrm{sp}(\mathbf{A} \otimes \mathbf{B}) = \mathrm{sp}(\mathbf{B} \otimes \mathbf{A}) = \bigcup_{i,j=1}^{I,J}\{\lambda_i \mu_j\}$
$\mathrm{eigenvectors}(\mathbf{A} \otimes \mathbf{B}) = \bigcup_{i,j=1}^{I,J}\{\mathbf{u}_i \otimes \mathbf{v}_j\}$

Table 2.11. *Rank, trace, determinant, and spectrum of a Kronecker product*

PROOF.–

– For a proof of the relations involving the rank and trace of $\mathbf{A} \otimes \mathbf{B}$, see sections 2.4.6 and 2.6.8.

– For $\mathbf{A} \in \mathbb{K}^{I \times I}$ and $\mathbf{B} \in \mathbb{K}^{J \times J}$, the identities [2.44] give:

$$\det(\mathbf{A} \otimes \mathbf{B}) = \det(\mathbf{A} \otimes \mathbf{I}_J)\det(\mathbf{I}_I \otimes \mathbf{B}) \qquad [2.54]$$

$$\det(\mathbf{B} \otimes \mathbf{A}) = \det(\mathbf{B} \otimes \mathbf{I}_I)\det(\mathbf{I}_J \otimes \mathbf{A}). \qquad [2.55]$$

Using the property [2.171], shown in section 2.7.3, which relates the Kronecker products $\mathbf{A} \otimes \mathbf{B}$ and $\mathbf{B} \otimes \mathbf{A}$, and noting that the determinant of a permutation matrix is equal to 1, we have:

$$\det(\mathbf{A} \otimes \mathbf{I}_J) = \det(\mathbf{I}_J \otimes \mathbf{A}) = [\det(\mathbf{A})]^J \qquad [2.56]$$

$$\det(\mathbf{B} \otimes \mathbf{I}_I) = \det(\mathbf{I}_I \otimes \mathbf{B}) = [\det(\mathbf{B})]^I. \qquad [2.57]$$

Therefore, by substituting these two expressions into [2.54] and [2.55], we obtain:

$$\det(\mathbf{A} \otimes \mathbf{B}) = \det(\mathbf{B} \otimes \mathbf{A}) = [\det(\mathbf{A})]^J [\det(\mathbf{B})]^I. \qquad [2.58]$$

From this relation, we can deduce that the product $\mathbf{A} \otimes \mathbf{B}$ is non-singular if and only if $\mathbf{A}$ and $\mathbf{B}$ are non-singular.

– Suppose that $\mathbf{A} \in \mathbb{C}^{I \times I}$ and $\mathbf{B} \in \mathbb{C}^{J \times J}$ admit the eigenpairs $(\lambda_i, \mathbf{u}_i)$, $i \in \langle I \rangle$, and $(\mu_j, \mathbf{v}_j), j \in \langle J \rangle$, respectively. The fundamental relation [2.50], with $\mathbf{C} = \mathbf{u}_i, \mathbf{D} = \mathbf{v}_j$, allows us to write:

$$(\mathbf{A} \otimes \mathbf{B})(\mathbf{u}_i \otimes \mathbf{v}_j) = \mathbf{A}\mathbf{u}_i \otimes \mathbf{B}\mathbf{v}_j = \lambda_i \mu_j (\mathbf{u}_i \otimes \mathbf{v}_j), \qquad [2.59]$$

from which we can conclude that $\mathbf{u}_i \otimes \mathbf{v}_j$ is an eigenvector of $\mathbf{A} \otimes \mathbf{B}$, associated with the eigenvalue $\lambda_i \mu_j$, for $i \in \langle I \rangle$ and $j \in \langle J \rangle$.

The spectrum of $\mathbf{A} \otimes \mathbf{B}$ therefore consists of the set of IJ products of eigenvalues of $\mathbf{A}$ and $\mathbf{B}$, namely:

$$\mathrm{sp}(\mathbf{A} \otimes \mathbf{B}) = \bigcup_{i,j=1}^{I,J} \{\lambda_i \mu_j\}, \qquad [2.60]$$

where $\mathrm{sp}(\mathbf{A})$ represents the spectrum of $\mathbf{A}$, i.e. the set of its eigenvalues. Furthermore, the eigenvectors of $\mathbf{A} \otimes \mathbf{B}$ are given by the IJ Kronecker products of an eigenvector of $\mathbf{A}$ with an eigenvector of $\mathbf{B}$, i.e. $\bigcup_{i,j=1}^{I,J} \{\mathbf{u}_i \otimes \mathbf{v}_j\}$.

Given that the determinant of a matrix is equal to the product of its eigenvalues, the formula [2.58] implies that $\mathrm{sp}(\mathbf{B} \otimes \mathbf{A}) = \mathrm{sp}(\mathbf{A} \otimes \mathbf{B})$. □

PROPOSITION 2.13.– From the relation $r_{\mathbf{A} \otimes \mathbf{B}} = r_{\mathbf{A}} r_{\mathbf{B}}$, we can conclude that, if $\mathbf{A}$ and $\mathbf{B}$ have full column rank, then the matrix $\mathbf{A} \otimes \mathbf{B}$ also has full column rank. Thus, in the case of square matrices, if $\mathbf{A}$ and $\mathbf{B}$ are non-singular, then $\mathbf{A} \otimes \mathbf{B}$ is also non-singular.

We can also deduce that:

$$r\left(\begin{bmatrix} \mathbf{A} & \mathbf{A} \\ \mathbf{A} & \mathbf{A} \end{bmatrix}\right) = r\left(\begin{bmatrix} 1 & 1 \\ 1 & 1 \end{bmatrix} \otimes \mathbf{A}\right) = r\left(\begin{bmatrix} 1 & 1 \\ 1 & 1 \end{bmatrix}\right) r_{\mathbf{A}} = r_{\mathbf{A}}$$

$$r\left(\begin{bmatrix} \mathbf{A} & -\mathbf{A} \\ \mathbf{A} & \mathbf{A} \end{bmatrix}\right) = r\left(\begin{bmatrix} 1 & -1 \\ 1 & 1 \end{bmatrix} \otimes \mathbf{A}\right) = r\left(\begin{bmatrix} 1 & -1 \\ 1 & 1 \end{bmatrix}\right) r_{\mathbf{A}} = 2 r_{\mathbf{A}}.$$

2.4.4. *Structural properties of a Kronecker product*

The Kronecker product satisfies the following structural properties:

– Using the relation [2.60], we can deduce the properties of the Kronecker product of positive/negative definite matrices, which are listed in Table 2.12. Similar properties can be deduced for positive/negative semi-definite matrices.

Hypotheses	Properties
$\mathbf{A}, \mathbf{B} > 0$	$\mathbf{A} \otimes \mathbf{B} > 0$
$\mathbf{A}, \mathbf{B} < 0$	$\mathbf{A} \otimes \mathbf{B} > 0$
$\mathbf{A} > 0, \mathbf{B} < 0$	$\mathbf{A} \otimes \mathbf{B} < 0$
$\mathbf{A} < 0, \mathbf{B} > 0$	$\mathbf{A} \otimes \mathbf{B} < 0$

Table 2.12. *Kronecker product of positive/negative definite matrices*

PROOF.– We have:

$$\left\{ \begin{array}{l} \mathbf{A}, \mathbf{B} > 0 \Leftrightarrow \lambda_i(\mathbf{A}) > 0 \text{ and } \mu_j(\mathbf{B}) > 0 \ \forall i, j \\ \mathbf{A}, \mathbf{B} < 0 \Leftrightarrow \lambda_i(\mathbf{A}) < 0 \text{ and } \mu_j(\mathbf{B}) < 0 \ \forall i, j \end{array} \right\} \Rightarrow \mathbf{A} \otimes \mathbf{B} > 0$$

$$\left\{ \begin{array}{l} \mathbf{A} > 0, \mathbf{B} < 0 \Leftrightarrow \lambda_i(\mathbf{A}) > 0 \text{ and } \mu_j(\mathbf{B}) < 0 \ \forall i, j \\ \mathbf{A} < 0, \mathbf{B} > 0 \Leftrightarrow \lambda_i(\mathbf{A}) < 0 \text{ and } \mu_j(\mathbf{B}) > 0 \ \forall i, j \end{array} \right\} \Rightarrow \mathbf{A} \otimes \mathbf{B} < 0,$$

from which we deduce that the Kronecker product of two positive/negative definite matrices is itself a positive/negative definite matrix.   □

– The Kronecker product of two diagonal matrices is a diagonal matrix. Similarly, the Kronecker product of two upper (respectively, lower) triangular matrices is an upper (respectively, lower) triangular matrix.

– The Kronecker product of two Toeplitz matrices[2] is a block Toeplitz matrix, i.e. a matrix consisting of identical Toeplitz blocks along the diagonals parallel to the main diagonal.

EXAMPLE 2.14.– Let $\mathbf{A}$ be a Toeplitz matrix of order 3, and let $\mathbf{B}$ be a Toeplitz matrix of order N. Then $\mathbf{A} \otimes \mathbf{B}$ is a block Toeplitz matrix of order $3N$:

$$\mathbf{A} = \begin{bmatrix} a_0 & a_{-1} & a_{-2} \\ a_1 & a_0 & a_{-1} \\ a_2 & a_1 & a_0 \end{bmatrix} \Rightarrow \mathbf{A} \otimes \mathbf{B} = \begin{bmatrix} a_0\mathbf{B} & a_{-1}\mathbf{B} & a_{-2}\mathbf{B} \\ a_1\mathbf{B} & a_0\mathbf{B} & a_{-1}\mathbf{B} \\ a_2\mathbf{B} & a_1\mathbf{B} & a_0\mathbf{B} \end{bmatrix}.$$

---

2 A square Toeplitz matrix $\mathbf{T} = [t_{ij}]$ of order I satisfies $t_{ij} = a_{i-j}$, $i, j \in \langle I \rangle$. It consists of identical coefficients along the diagonals parallel to the main diagonal, i.e. $t_{ij} = t_{kl}$ if $i - j = k - l$.

Similarly, the Kronecker product of two Hankel[3] matrices is a block Hankel matrix.

– Using the property [2.40] and the fundamental relation [2.50], we can conclude that the Kronecker product of two orthogonal (respectively, unitary) matrices $\mathbf{A} \in \mathbb{K}^{I \times I}$ and $\mathbf{B} \in \mathbb{K}^{J \times J}$ is itself an orthogonal (respectively, unitary) matrix:

$$\left\{ \begin{array}{c} \mathbf{A}^T \mathbf{A} = \mathbf{A} \mathbf{A}^T = \mathbf{I}_I \\ \mathbf{B}^T \mathbf{B} = \mathbf{B} \mathbf{B}^T = \mathbf{I}_J \end{array} \right\} \Rightarrow (\mathbf{A} \otimes \mathbf{B})^T (\mathbf{A} \otimes \mathbf{B}) = (\mathbf{A} \otimes \mathbf{B})(\mathbf{A} \otimes \mathbf{B})^T = \mathbf{I}_{IJ}$$

$$\left\{ \begin{array}{c} \mathbf{A}^H \mathbf{A} = \mathbf{A} \mathbf{A}^H = \mathbf{I}_I \\ \mathbf{B}^H \mathbf{B} = \mathbf{B} \mathbf{B}^H = \mathbf{I}_J \end{array} \right\} \Rightarrow (\mathbf{A} \otimes \mathbf{B})^H (\mathbf{A} \otimes \mathbf{B}) = (\mathbf{A} \otimes \mathbf{B})(\mathbf{A} \otimes \mathbf{B})^H = \mathbf{I}_{IJ}.$$

– Similarly, if $\mathbf{A} \in \mathbb{K}^{I \times J}$ and $\mathbf{B} \in \mathbb{K}^{K \times L}$ are column orthonormal matrices, then $\mathbf{A} \otimes \mathbf{B}$ is itself column orthonormal:

$$\left\{ \begin{array}{c} \mathbf{A}^H \mathbf{A} = \mathbf{I}_J \\ \mathbf{B}^H \mathbf{B} = \mathbf{I}_L \end{array} \right\} \Rightarrow (\mathbf{A} \otimes \mathbf{B})^H (\mathbf{A} \otimes \mathbf{B}) = \mathbf{A}^H \mathbf{A} \otimes \mathbf{B}^H \mathbf{B} = \mathbf{I}_{JL}.$$

The above results are summarized in Table 2.13.

Hypotheses $\mathbf{A}, \mathbf{B}$	Structural properties $\mathbf{A} \otimes \mathbf{B}$
symmetric/Hermitian	symmetric/Hermitian
diagonal	diagonal
upper/lower triangular	upper/lower triangular
column orthonormal	column orthonormal
orthogonal/unitary	orthogonal/unitary
Toeplitz (Hankel)	block Toeplitz (block Hankel)

Table 2.13. *Structural properties of the Kronecker product*

2.4.5. *Inverse and Moore–Penrose pseudo-inverse of a Kronecker product*

For non-singular square matrices $\mathbf{A}$ and $\mathbf{B}$, we have:

$$(\mathbf{A} \otimes \mathbf{B})^{-1} = \mathbf{A}^{-1} \otimes \mathbf{B}^{-1}. \tag{2.61}$$

This result can be deduced directly from the fundamental relation [2.50].

---

3 A square Hankel matrix $\mathbf{H} = [h_{ij}]$ of order I satisfies $h_{ij} = a_{i+j-1}$. It consists of identical coefficients along diagonals parallel to the antidiagonal.

The Moore–Penrose pseudo-inverse of the Kronecker product of two matrices $\mathbf{A} \in \mathbb{K}^{I \times J}$ and $\mathbf{B} \in \mathbb{K}^{M \times N}$ of full column rank is equal to the Kronecker product of the Moore–Penrose pseudo-inverses of these two matrices:

$$(\mathbf{A} \otimes \mathbf{B})^{\dagger} = \mathbf{A}^{\dagger} \otimes \mathbf{B}^{\dagger}. \tag{2.62}$$

PROOF.– Since the matrices $\mathbf{A}$ and $\mathbf{B}$ have full column rank, their Kronecker product also has full column rank.

The Moore–Penrose pseudo-inverse of $\mathbf{A} \otimes \mathbf{B}$ is then given by:

$$(\mathbf{A} \otimes \mathbf{B})^{\dagger} = \left[(\mathbf{A} \otimes \mathbf{B})^{H}(\mathbf{A} \otimes \mathbf{B}) \right]^{-1}(\mathbf{A} \otimes \mathbf{B})^{H}.$$

Using the properties [2.50] and [2.61], we obtain:

$$= (\mathbf{A}^{H}\mathbf{A} \otimes \mathbf{B}^{H}\mathbf{B})^{-1}(\mathbf{A}^{H} \otimes \mathbf{B}^{H})$$

$$= \left[(\mathbf{A}^{H}\mathbf{A})^{-1} \otimes (\mathbf{B}^{H}\mathbf{B})^{-1} \right](\mathbf{A}^{H} \otimes \mathbf{B}^{H})$$

$$= \left[(\mathbf{A}^{H}\mathbf{A})^{-1}\mathbf{A}^{H} \right] \otimes \left[(\mathbf{B}^{H}\mathbf{B})^{-1}\mathbf{B}^{H} \right] = \mathbf{A}^{\dagger} \otimes \mathbf{B}^{\dagger}.$$

This proves [2.62]. □

2.4.6. *Decompositions of a Kronecker product*

Using the relation [2.50], it is easy to deduce the Cholesky, UD, QR and SVD decompositions of $\mathbf{A} \otimes \mathbf{B}$ from the corresponding decompositions of $\mathbf{A}$ and $\mathbf{B}$.

– For symmetric positive definite matrices $\mathbf{A} \in \mathbb{R}^{I \times I}$ and $\mathbf{B} \in \mathbb{R}^{J \times J}$, we have:

$$\left\{ \begin{array}{l} \mathbf{A} = \mathbf{S}_A \mathbf{S}_A^T \\ \mathbf{B} = \mathbf{S}_B \mathbf{S}_B^T \end{array} \right\} \Rightarrow \mathbf{A} \otimes \mathbf{B} = (\mathbf{S}_A \otimes \mathbf{S}_B)(\mathbf{S}_A \otimes \mathbf{S}_B)^T$$

$$\left\{ \begin{array}{l} \mathbf{A} = \mathbf{U}_A \mathbf{D}_A \mathbf{U}_A^T \\ \mathbf{B} = \mathbf{U}_B \mathbf{D}_B \mathbf{U}_B^T \end{array} \right\} \Rightarrow \mathbf{A} \otimes \mathbf{B} = (\mathbf{U}_A \otimes \mathbf{U}_B)(\mathbf{D}_A \otimes \mathbf{D}_B)(\mathbf{U}_A \otimes \mathbf{U}_B)^T,$$

where $(\mathbf{S}_A, \mathbf{S}_B)$ are the upper or lower triangular matrices with positive diagonal elements that define the Cholesky decompositions of $\mathbf{A}$ and $\mathbf{B}$, respectively, $(\mathbf{U}_A, \mathbf{U}_B)$ are upper or lower triangular matrices whose diagonal elements are equal to 1, and $(\mathbf{D}_A, \mathbf{D}_B)$ are positive definite diagonal matrices from the UD decompositions of $\mathbf{A}$ and $\mathbf{B}$. Thus, we obtain the Cholesky and UD decompositions of $\mathbf{A} \otimes \mathbf{B}$ from the Cholesky and UD decompositions of $\mathbf{A}$ and $\mathbf{B}$.

– Regarding the QR decomposition, for matrices $\mathbf{A} \in \mathbb{R}^{I \times J}$ and $\mathbf{B} \in \mathbb{R}^{K \times L}$ of full column rank, we have:

$$\left\{ \begin{array}{l} \mathbf{A} = \mathbf{Q}_A \mathbf{R}_A \\ \mathbf{B} = \mathbf{Q}_B \mathbf{R}_B \end{array} \right\} \Rightarrow \mathbf{A} \otimes \mathbf{B} = (\mathbf{Q}_A \otimes \mathbf{Q}_B)(\mathbf{R}_A \otimes \mathbf{R}_B),$$

where $\mathbf{R}_A \in \mathbb{R}^{J \times J}$ and $\mathbf{R}_B \in \mathbb{R}^{L \times L}$ are upper triangular matrices, and $\mathbf{Q}_A \in \mathbb{R}^{I \times J}$ and $\mathbf{Q}_B \in \mathbb{R}^{K \times L}$ are column orthonormal matrices. Since the matrix $\mathbf{Q}_A \otimes \mathbf{Q}_B$ is itself column orthonormal, and $\mathbf{R}_A \otimes \mathbf{R}_B$ is upper triangular (see Table 2.13), we then obtain the QR decomposition of $\mathbf{A} \otimes \mathbf{B}$ from the QR decompositions of $\mathbf{A}$ and $\mathbf{B}$.

– Similarly, for the SVD decomposition, we have:

$$\left\{ \begin{array}{l} \mathbf{A} = \mathbf{U}_A \mathbf{\Sigma}_A \mathbf{V}_A^H \\ \mathbf{B} = \mathbf{U}_B \mathbf{\Sigma}_B \mathbf{V}_B^H \end{array} \right\} \Rightarrow \mathbf{A} \otimes \mathbf{B} = (\mathbf{U}_A \otimes \mathbf{U}_B)(\mathbf{\Sigma}_A \otimes \mathbf{\Sigma}_B)(\mathbf{V}_A \otimes \mathbf{V}_B)^H,$$

where $\mathbf{U}_A \in \mathbb{K}^{I \times I}$, $\mathbf{U}_B \in \mathbb{K}^{K \times K}$, $\mathbf{V}_A \in \mathbb{K}^{J \times J}$, $\mathbf{V}_B \in \mathbb{K}^{L \times L}$ are unitary matrices in the complex case ($\mathbb{K} = \mathbb{C}$) and orthogonal matrices in the real case ($\mathbb{K} = \mathbb{R}$), whereas $\mathbf{\Sigma}_A$ and $\mathbf{\Sigma}_B$ are pseudo-diagonal matrices.

From the last relation, we can deduce that the non-zero singular values of $\mathbf{A} \otimes \mathbf{B}$ are equal to $(\sigma_\mathbf{A})_i (\sigma_\mathbf{B})_j$, where the $(\sigma_\mathbf{A})_i, i \in \langle r_\mathbf{A} \rangle$, and the $(\sigma_\mathbf{B})_j, j \in \langle r_\mathbf{B} \rangle$, are the non-zero singular values of $\mathbf{A}$ and $\mathbf{B}$, and $r_\mathbf{A}$ and $r_\mathbf{B}$ denote the rank of $\mathbf{A}$ and $\mathbf{B}$, respectively. Hence, since the rank of a matrix is equal to the number of non-zero singular values, we can deduce that the rank of $\mathbf{A} \otimes \mathbf{B}$ is equal to $r_\mathbf{A} r_\mathbf{B}$, i.e. the product of the ranks of $\mathbf{A}$ and $\mathbf{B}$ (see Table 2.11).

2.5. Kronecker sum

In this section, we introduce the definition and a few properties of the Kronecker sum.

2.5.1. Definition

The Kronecker sum of $\mathbf{A} \in \mathbb{K}^{I \times I}$ and $\mathbf{B} \in \mathbb{K}^{J \times J}$ is defined as[4]:

$$\mathbf{A} \oplus \mathbf{B} = \mathbf{A} \otimes \mathbf{I}_J + \mathbf{I}_I \otimes \mathbf{B} \in \mathbb{K}^{IJ \times IJ}. \qquad [2.63]$$

EXAMPLE 2.15.– For $\mathbf{A} = \begin{pmatrix} a_{11} & a_{12} \\ a_{21} & a_{22} \end{pmatrix}, \mathbf{B} = \begin{pmatrix} b_{11} & b_{12} \\ b_{21} & b_{22} \end{pmatrix}$, we have:

$$\mathbf{A} \oplus \mathbf{B} = \begin{pmatrix} a_{11} + b_{11} & b_{12} & a_{12} & 0 \\ b_{21} & a_{11} + b_{22} & 0 & a_{12} \\ a_{21} & 0 & a_{22} + b_{11} & b_{12} \\ 0 & a_{21} & b_{21} & a_{22} + b_{22} \end{pmatrix}.$$

---

4 Some authors define the Kronecker sum as:

$$\mathbf{A} \oplus \mathbf{B} = \mathbf{I}_J \otimes \mathbf{A} + \mathbf{B} \otimes \mathbf{I}_I \in \mathbb{K}^{JI \times JI}.$$

2.5.2. *Properties*

The Kronecker sum $\mathbf{A} \oplus \mathbf{B}$ satisfies the property of associativity, and its spectrum has a simple expression in terms of the spectra of $\mathbf{A}$ and $\mathbf{B}$.

– *Associativity:* For $\mathbf{A} \in \mathbb{K}^{I \times I}, \mathbf{B} \in \mathbb{K}^{J \times J}, \mathbf{C} \in \mathbb{K}^{K \times K}$, we have:

$$\mathbf{A} \oplus (\mathbf{B} \oplus \mathbf{C}) = (\mathbf{A} \oplus \mathbf{B}) \oplus \mathbf{C} = \mathbf{A} \oplus \mathbf{B} \oplus \mathbf{C}.$$

– *Spectrum:* The spectrum of $\mathbf{A} \oplus \mathbf{B}$ is given by:

$$\mathrm{sp}(\mathbf{A} \oplus \mathbf{B}) = \mathrm{sp}(\mathbf{B} \oplus \mathbf{A}) = \bigcup_{i,j=1}^{I,J} \{\lambda_i + \mu_j\}, \qquad [2.64]$$

where $(\lambda_i, \mathbf{u}_i)$ and $(\mu_j, \mathbf{v}_j)$ are eigenpairs of $\mathbf{A}$ and $\mathbf{B}$, respectively, and $\mathbf{u}_i \otimes \mathbf{v}_j$ is the eigenvector associated with the eigenvalue $\lambda_i + \mu_j$.

PROOF.– Applying the fundamental relation [2.50] allows us to write:

$$(\mathbf{A} \oplus \mathbf{B})(\mathbf{u}_i \otimes \mathbf{v}_j) = (\mathbf{A} \otimes \mathbf{I}_J)(\mathbf{u}_i \otimes \mathbf{v}_j) + (\mathbf{I}_I \otimes \mathbf{B})(\mathbf{u}_i \otimes \mathbf{v}_j)$$
$$= \mathbf{A}\mathbf{u}_i \otimes \mathbf{v}_j + \mathbf{u}_i \otimes \mathbf{B}\mathbf{v}_j$$
$$= (\lambda_i \mathbf{u}_i \otimes \mathbf{v}_j) + (\mathbf{u}_i \otimes \mu_j \mathbf{v}_j)$$
$$= (\lambda_i + \mu_j)(\mathbf{u}_i \otimes \mathbf{v}_j), \qquad [2.65]$$

from which we deduce that $(\lambda_i + \mu_j, \mathbf{u}_i \otimes \mathbf{v}_j)$ is an eigenpair of $\mathbf{A} \oplus \mathbf{B}$. □

REMARK 2.16.– In general, the Kronecker sum is not distributive with respect to the Kronecker product, i.e.:

$$(\mathbf{A} \oplus \mathbf{B}) \otimes \mathbf{C} \neq (\mathbf{A} \otimes \mathbf{C}) \oplus (\mathbf{B} \otimes \mathbf{C})$$
$$\mathbf{C} \otimes (\mathbf{A} \oplus \mathbf{B}) \neq (\mathbf{C} \otimes \mathbf{A}) \oplus (\mathbf{C} \otimes \mathbf{B}).$$

This is easy to check, for example by choosing $\mathbf{A} = \mathbf{B} = 1$ and $\mathbf{C} = \mathbf{I}_J$:

$$(\mathbf{A} \oplus \mathbf{B}) \otimes \mathbf{C} = \mathbf{C} \otimes (\mathbf{A} \oplus \mathbf{B}) = 2\mathbf{C} = 2\mathbf{I}_J$$
$$(\mathbf{A} \otimes \mathbf{C}) \oplus (\mathbf{B} \otimes \mathbf{C}) = (\mathbf{C} \otimes \mathbf{A}) \oplus (\mathbf{C} \otimes \mathbf{B}) = \mathbf{C} \oplus \mathbf{C} = 2\mathbf{I}_{J^2}.$$

2.6. Index convention

Like Einstein's summation convention, the index convention enables us to eliminate the summation symbols to write certain formulae involving multi-index variables more compactly. The index convention also simplifies certain proofs, as illustrated later in this chapter.

If an index $i \in \langle I \rangle$ is repeated in an expression (or more generally in a term in an equation), the expression (or term) is interpreted as being summed over this index, from 1 to I. For example, $\sum_{i=1}^{I} a_i b_i$ is simply written as $a_i b_i$. However, there are two differences relative to Einstein's summation convention:

– each index can be repeated more than twice in an expression;

– ordered index sets are allowed.

The index notation can be interpreted in terms of two types of summation, the first associated with the row indices (superscripts) and the second associated with the column indices (subscripts), with the following rules, which follow from the relations [2.23] and [2.24]:

– the order of the column indices is independent of the order of the row indices;

– consecutive row and column indices (or index sets) can be permuted.

For the rest of this chapter, we will use the index convention to represent matrix products, in particular the Kronecker and Khatri–Rao products, to establish vectorization formulae for matrix products and partitioned matrices, and to express various relations involving the traces of matrix products, as well as relations between the Kronecker, Khatri–Rao and Hadamard products. In Chapter 3, the index convention will be used to define matrix and vector unfoldings, i.e. for the matricization and vectorization of a tensor. Finally, in Chapter 5, the index convention will allow us to write the Tucker and PARAFAC decompositions compactly, as well as to establish matricized forms of these decompositions.

2.6.1. *Writing vectors and matrices with the index convention*

Omitting the dimension of the vectors of the canonical basis to alleviate the notation, with $\mathbf{E}_{ij}^{(I \times J)}$ defined as in [2.25], the index convention allows us to write the vectors $\mathbf{u} \in \mathbb{K}^I$ and $\mathbf{v}^T \in \mathbb{K}^{1 \times J}$ and the matrices $\mathbf{A} \in \mathbb{K}^{I \times J}$ and $\mathbf{A}^T$ as:

$$\mathbf{u} = \sum_{i=1}^{I} u_i \mathbf{e}_i^{(I)} = u_i \mathbf{e}_i \qquad [2.66]$$

$$\mathbf{v}^T = \sum_{j=1}^{J} v_j (\mathbf{e}_j^{(J)})^T = v_j \mathbf{e}^j \;\; ; \;\; \mathbf{v}^H = v_j^* \mathbf{e}^j \qquad [2.67]$$

and

$$\mathbf{A} = \sum_{i=1}^{I} \sum_{j=1}^{j} a_{ij}(\mathbf{e}_i^{(I)} \otimes (\mathbf{e}_j^{(J)})^T) = a_{ij}\mathbf{e}_i^j = a_{ij}\mathbf{E}_{ij}^{(I \times J)} \qquad [2.68]$$

$$\mathbf{A}^T = a_{ij}\mathbf{e}_j^i \; ; \; \mathbf{A}^H = a_{ij}^*\mathbf{e}_j^i, \qquad [2.69]$$

Similarly, the identity matrix of order J, the matrix $\mathbf{1}_{I \times J}$ and the vector $\mathbf{1}_J$ of ones, of size $I \times J$ and J, respectively, and the integer I can be written as:

$$\mathbf{I}_J = \sum_{j=1}^{J} \mathbf{e}_j^{(J)} \otimes (\mathbf{e}_j^{(J)})^T = \mathbf{e}_j^j = \mathbf{E}_{jj}^{(J \times J)} \qquad [2.70]$$

$$\mathbf{1}_{I \times J} = u_{ij}\mathbf{e}_i^j \; , \; \text{with } u_{ij} = 1 \; \forall i \in \langle I \rangle, \forall j \in \langle J \rangle \qquad [2.71]$$

$$\mathbf{1}_J = \sum_{j=1}^{J} \mathbf{e}_j^{(J)} = \delta_{jj}\mathbf{e}_j \qquad [2.72]$$

$$I = \mathbf{e}^i\mathbf{e}_i. \qquad [2.73]$$

2.6.2. *Basic rules and identities with the index convention*

Consider the ordered set $\mathbb{S} = \{n_1, \ldots, n_N\}$ obtained by permuting the elements of $\langle N \rangle \triangleq \{1, \ldots, N\}$, and define the index combination $\mathbb{I} = i_{n_1} \cdots i_{n_N}$ associated with $\mathbb{S}$. Let $\mathbf{e}_{\mathbb{I}}$ and $\mathbf{e}^{\mathbb{I}}$ denote the following Kronecker products:

$$\mathbf{e}_{\mathbb{I}} \triangleq \underset{n \in \mathbb{S}}{\otimes} \mathbf{e}_{i_n}^{(I_n)} \triangleq \mathbf{e}_{i_{n_1}}^{(I_{n_1})} \otimes \mathbf{e}_{i_{n_2}}^{(I_{n_2})} \otimes \cdots \otimes \mathbf{e}_{i_{n_N}}^{(I_{n_N})} \in \mathbb{R}^{I_{n_1} I_{n_2} \cdots I_{n_N}}$$

$$\mathbf{e}^{\mathbb{I}} \triangleq \underset{n \in \mathbb{S}}{\otimes} (\mathbf{e}_{i_n}^{(I_n)})^T \triangleq (\mathbf{e}_{i_{n_1}}^{(I_{n_1})})^T \otimes \cdots \otimes (\mathbf{e}_{i_{n_N}}^{(I_{n_N})})^T \in \mathbb{R}^{1 \times I_{n_1} I_{n_2} \cdots I_{n_N}}.$$

Using the above definition, together with the expression $\mathbf{u}^{(n)} = u_{i_n}^{(n)}\mathbf{e}_{i_n}^{(I_n)} \in \mathbb{K}^{I_n}$, the multiple Kronecker product $\underset{n \in \mathbb{S}}{\otimes} \mathbf{u}^{(n)}$ can be written as:

$$\underset{n \in \mathbb{S}}{\otimes} \mathbf{u}^{(n)} = u_{i_{n_1}}^{(n_1)}\mathbf{e}_{i_{n_1}}^{(I_{n_1})} \otimes \cdots \otimes u_{i_{n_N}}^{(n_N)}\mathbf{e}_{i_{n_N}}^{(I_{n_N})}$$

$$= \Big(\prod_{n \in \mathbb{S}} u_{i_n}^{(n)}\Big) \underset{n \in \mathbb{S}}{\otimes} \mathbf{e}_{i_n}^{(I_n)} = \Big(\prod_{n \in \mathbb{S}} u_{i_n}^{(n)}\Big)\mathbf{e}_{\mathbb{I}}. \qquad [2.74]$$

PROPOSITION 2.17.– Let $\mathbb{I}$ and $\mathbb{J}$ be two ordered index sets, partitioned into two subsets $(\mathbb{I}_1, \mathbb{I}_2)$ and $(\mathbb{J}_1, \mathbb{J}_2)$, respectively. The rules stated above imply the following identities:

$$\mathbf{e}_{\mathbb{I}}^{\mathbb{J}} = \mathbf{e}_{\mathbb{I}} \otimes \mathbf{e}^{\mathbb{J}} = \mathbf{e}^{\mathbb{J}} \otimes \mathbf{e}_{\mathbb{I}} = \mathbf{e}_{\mathbb{I}_1 \mathbb{I}_2}^{\mathbb{J}_1 \mathbb{J}_2} \qquad [2.75]$$

$$= \mathbf{e}_{\mathbb{I}_1} \otimes \mathbf{e}_{\mathbb{I}_2} \otimes \mathbf{e}^{\mathbb{J}_1} \otimes \mathbf{e}^{\mathbb{J}_2} = \mathbf{e}_{\mathbb{I}_1} \otimes \mathbf{e}^{\mathbb{J}_1} \otimes \mathbf{e}_{\mathbb{I}_2} \otimes \mathbf{e}^{\mathbb{J}_2}$$

$$= \mathbf{e}_{\mathbb{I}_1} \otimes \mathbf{e}^{\mathbb{J}_1} \otimes \mathbf{e}^{\mathbb{J}_2} \otimes \mathbf{e}_{\mathbb{I}_2} = \mathbf{e}^{\mathbb{J}_1} \otimes \mathbf{e}^{\mathbb{J}_2} \otimes \mathbf{e}_{\mathbb{I}_1} \otimes \mathbf{e}_{\mathbb{I}_2}$$

$$= \mathbf{e}^{\mathbb{J}_1} \otimes \mathbf{e}_{\mathbb{I}_1} \otimes \mathbf{e}^{\mathbb{J}_2} \otimes \mathbf{e}_{\mathbb{I}_2} = \mathbf{e}^{\mathbb{J}_1} \otimes \mathbf{e}_{\mathbb{I}_1} \otimes \mathbf{e}_{\mathbb{I}_2} \otimes \mathbf{e}^{\mathbb{J}_2}.$$

These identities generalize the identities stated by Pollock (2011) for a term of the form $\mathbf{e}_{ij}^{kl}$ to two index sets partitioned into two subsets. They follow from the properties [2.23] and [2.24] of the Kronecker product of a column vector with a row vector, which namely state that this product is independent of the order of the vectors $(\mathbf{u}^T \otimes \mathbf{v} = \mathbf{v} \otimes \mathbf{u}^T)$. This implies that, in a sequence of Kronecker products of column vectors and row vectors, a column vector can be permuted with a row vector without changing the end result, provided that the order of the column vectors and the order of the row vectors are not changed in the sequence. In particular, we have:

$$\mathbf{e}_{ij}^{ji}\mathbf{e}_{ji}^{ij} = (\mathbf{e}_{ij} \otimes \mathbf{e}^{ji})(\mathbf{e}^{ij} \otimes \mathbf{e}_{ji}) = (\mathbf{e}_{ij}\mathbf{e}^{ij}) \otimes (\mathbf{e}^{ji}\mathbf{e}_{ji}) \quad \text{(by [2.50])}$$

$$= \mathbf{e}_{ij}^{ij} = \mathbf{I}_{IJ} \quad \text{(because } \mathbf{e}^{ji}\mathbf{e}_{ji} = 1 \text{ for fixed } i \text{ and } j, \text{ and by [2.70])} \qquad [2.76]$$

from which we deduce:

$$(\mathbf{e}_{ij}^{ji})^{-1} = \mathbf{e}_{ji}^{ij}. \qquad [2.77]$$

EXAMPLE 2.18.– For $I = 2, J = 3$, using the relations [2.75] and [2.24], we have:

$$\mathbf{e}_{ij}^{ji} = \mathbf{e}_{ij} \otimes \mathbf{e}^{ji} = \mathbf{e}_{ij}\mathbf{e}^{ji}$$

$$= \sum_{i=1}^{2}\sum_{j=1}^{3}\left(\mathbf{e}_i^{(2)} \otimes \mathbf{e}_j^{(3)}\right)\left(\mathbf{e}_j^{(3)} \otimes \mathbf{e}_i^{(2)}\right)^T$$

$$= \left(\begin{pmatrix}1\\0\end{pmatrix} \otimes \begin{pmatrix}1\\0\\0\end{pmatrix}\right)\left(\begin{pmatrix}1\\0\\0\end{pmatrix} \otimes \begin{pmatrix}1\\0\end{pmatrix}\right)^T + \left(\begin{pmatrix}1\\0\end{pmatrix} \otimes \begin{pmatrix}0\\1\\0\end{pmatrix}\right)\left(\begin{pmatrix}0\\1\\0\end{pmatrix} \otimes \begin{pmatrix}1\\0\end{pmatrix}\right)^T$$

$$+ \left(\begin{pmatrix}1\\0\end{pmatrix} \otimes \begin{pmatrix}0\\0\\1\end{pmatrix}\right)\left(\begin{pmatrix}0\\0\\1\end{pmatrix} \otimes \begin{pmatrix}1\\0\end{pmatrix}\right)^T + \left(\begin{pmatrix}0\\1\end{pmatrix} \otimes \begin{pmatrix}1\\0\\0\end{pmatrix}\right)\left(\begin{pmatrix}1\\0\\0\end{pmatrix} \otimes \begin{pmatrix}0\\1\end{pmatrix}\right)^T$$

$$+ \left(\begin{pmatrix}0\\1\end{pmatrix} \otimes \begin{pmatrix}0\\1\\0\end{pmatrix}\right)\left(\begin{pmatrix}0\\1\\0\end{pmatrix} \otimes \begin{pmatrix}0\\1\end{pmatrix}\right)^T + \left(\begin{pmatrix}0\\1\end{pmatrix} \otimes \begin{pmatrix}0\\0\\1\end{pmatrix}\right)\left(\begin{pmatrix}0\\0\\1\end{pmatrix} \otimes \begin{pmatrix}0\\1\end{pmatrix}\right)^T,$$

which gives:

$$
\mathbf{e}_{ij}^{ji} =
\begin{bmatrix}
1 & 0 & 0 & 0 & 0 & 0 \\
0 & 0 & 1 & 0 & 0 & 0 \\
0 & 0 & 0 & 0 & 1 & 0 \\
0 & 1 & 0 & 0 & 0 & 0 \\
0 & 0 & 0 & 1 & 0 & 0 \\
0 & 0 & 0 & 0 & 0 & 1
\end{bmatrix}
, \
\mathbf{e}_{ji}^{ij} =
\begin{bmatrix}
1 & 0 & 0 & 0 & 0 & 0 \\
0 & 0 & 0 & 1 & 0 & 0 \\
0 & 1 & 0 & 0 & 0 & 0 \\
0 & 0 & 0 & 0 & 1 & 0 \\
0 & 0 & 1 & 0 & 0 & 0 \\
0 & 0 & 0 & 0 & 0 & 1
\end{bmatrix}
= (\mathbf{e}_{ij}^{ji})^{-1}.
$$

REMARK 2.19.– Since the matrix $\mathbf{e}_{ij}^{ji}$ is a permutation matrix, we have $(\mathbf{e}_{ij}^{ji})^{-1} = (\mathbf{e}_{ij}^{ji})^T$ (see [2.168]).

2.6.3. *Matrix products and index convention*

Using the index convention, the products $\mathbf{v}^T\mathbf{u}$ and $\mathbf{A}\mathbf{u}$ can be rewritten as follows, with $\mathbf{u}, \mathbf{v} \in \mathbb{K}^J$, and $\mathbf{A} \in \mathbb{K}^{I \times J}$:

$$
\mathbf{v}^T\mathbf{u} = \sum_{i=1}^{I} u_i v_i = u_i v_i \tag{2.78}
$$

$$
\mathbf{A}\mathbf{u} = \sum_{i=1}^{I} (\mathbf{A}_{i.}\mathbf{u})\mathbf{e}_i^{(I)} = (\mathbf{A}_{i.}\mathbf{u})\mathbf{e}_i \tag{2.79}
$$

$$
= \sum_{i=1}^{I} (\sum_{j=1}^{J} a_{ij} u_j)\mathbf{e}_i = a_{ij} u_j \mathbf{e}_i \in \mathbb{K}^I. \tag{2.80}
$$

Similarly, for $\mathbf{u} \in \mathbb{K}^J, \mathbf{A} \in \mathbb{K}^{I \times J}, \mathbf{B} \in \mathbb{K}^{J \times K}$ and $\mathbf{C} \in \mathbb{K}^{K \times J}$, we have:

$$
\mathbf{A}\mathbf{B} = \sum_{i=1}^{I} \sum_{k=1}^{K} (\sum_{j=1}^{J} a_{ij} b_{jk})\mathbf{e}_i^k = a_{ij} b_{jk} \mathbf{e}_i^k \in \mathbb{K}^{I \times K} \tag{2.81}
$$

$$
= (a_{ij}\mathbf{e}_i)(b_{jk}\mathbf{e}^k) = \mathbf{A}_{.j}\mathbf{B}_{j.} \tag{2.82}
$$

$$
\mathbf{A}\mathbf{C}^T = a_{ij} c_{kj} \mathbf{e}_i^k \in \mathbb{K}^{I \times K}. \tag{2.83}
$$

We also have:

$$
\mathbf{A}\mathrm{diag}(u_1, \cdots, u_J)\mathbf{B} = u_j \mathbf{A}_{.j}\mathbf{B}_{j.}, \tag{2.84}
$$

and the Hadamard product of $\mathbf{A} \in \mathbb{K}^{I \times J}$ with $\mathbf{B} \in \mathbb{K}^{I \times J}$ can be written as:

$$
\mathbf{A} \odot \mathbf{B} = a_{ij} b_{ij} \mathbf{e}_i^j. \tag{2.85}
$$

2.6.4. *Kronecker products and index convention*

The Kronecker product of vectors can be written compactly using the index convention, as illustrated in Table 2.14.

$\mathbf{u} \in \mathbb{K}^I, \mathbf{v} \in \mathbb{K}^J, \mathbf{w} \in \mathbb{K}^K$
$\mathbf{x}^{(n)} \in \mathbb{K}^{I_n}$, with $n \in \langle N \rangle$
$\mathbf{u} \otimes \mathbf{v} = u_i v_j \mathbf{e}_{ij}$
$\mathbf{u}^T \otimes \mathbf{v}^T = u_i v_j \mathbf{e}^{ij}$
$\mathbf{u} \otimes \mathbf{v}^T = u_i v_j \mathbf{e}_i^j$
$\mathbf{u} \otimes \mathbf{v} \otimes \mathbf{w} = u_i v_j w_k \mathbf{e}_{ijk}$
$\mathbf{u} \otimes \mathbf{v}^T \otimes \mathbf{w} = u_i v_j w_k \mathbf{e}_{ik}^j$
$\overset{N}{\underset{n=1}{\otimes}} \mathbf{x}^{(n)} = \left(\overset{N}{\underset{n=1}{\prod}} x_{i_n}^{(n)} \right) \mathbf{e}_{i_1 \cdots i_N}$

Table 2.14. *Kronecker products of vectors and index convention*

PROOF.– First, note that:

$$\mathbf{e}_i^{(I)} \otimes \mathbf{e}_j^{(J)} = \mathbf{e}_{(i-1)J+j}^{(IJ)} = \mathbf{e}_{ij} \qquad [2.86]$$

$$\mathbf{e}_i^{(I)} \otimes \mathbf{e}_j^{(J)} \otimes \mathbf{e}_k^{(K)} = \mathbf{e}_{(i-1)JK+(j-1)K+k}^{(IJK)} = \mathbf{e}_{ijk} \qquad [2.87]$$

$$\mathbf{e}_{i_1}^{(I_1)} \otimes \mathbf{e}_{i_2}^{(I_2)} \otimes \cdots \otimes \mathbf{e}_{i_N}^{(I_N)} = \mathbf{e}_{(i_1-1)I_2\cdots I_N+(i_2-1)I_3\cdots I_N+\cdots+i_N}^{(I_1 I_2 \cdots I_N)} = \mathbf{e}_{i_1 \cdots i_N}$$
$$[2.88]$$

The expressions in Table 2.14 can be proven using the properties [2.75]. Thus, for $\mathbf{u} \in \mathbb{K}^I$, $\mathbf{v} \in \mathbb{K}^J$, the index convention allows us to write:

$$\mathbf{u} \otimes \mathbf{v} = \left(\sum_{i=1}^I u_i \mathbf{e}_i^{(I)} \right) \otimes \left(\sum_{j=1}^J v_j \mathbf{e}_j^{(J)} \right)$$

$$= \sum_{i=1}^I \sum_{j=1}^J u_i v_j (\mathbf{e}_i^{(I)} \otimes \mathbf{e}_j^{(J)}) = u_i v_j \, \mathbf{e}_{ij}, \qquad [2.89]$$

which shows the first relation in Table 2.14. This relation implies that:

$$u_i v_j = [\mathbf{u} \otimes \mathbf{v}]_{(i-1)J+j} = (\mathbf{e}_{(i-1)J+j}^{(IJ)})^T (\mathbf{u} \otimes \mathbf{v}) \qquad [2.90]$$

$$= (\mathbf{e}_i^{(I)} \otimes \mathbf{e}_j^{(J)})^T (\mathbf{u} \otimes \mathbf{v}) = (\mathbf{u} \otimes \mathbf{v})^T (\mathbf{e}_i^{(I)} \otimes \mathbf{e}_j^{(J)})$$

$$= \mathbf{e}^{ij} (\mathbf{u} \otimes \mathbf{v}) = (\mathbf{u} \otimes \mathbf{v})^T \mathbf{e}_{ij}. \qquad [2.91]$$

Similarly, we have:

$$\mathbf{u}^T \otimes \mathbf{v}^T = (u_i \mathbf{e}^i) \otimes (v_j \mathbf{e}^j) = u_i v_j \mathbf{e}^{ij} \qquad [2.92]$$

$$\mathbf{u} \otimes \mathbf{v}^T = (u_i \mathbf{e}_i) \otimes (v_j \mathbf{e}^j) = u_i v_j \mathbf{e}_i^j. \qquad [2.93]$$

For the Kronecker product of three vectors, we have:

$$\mathbf{u} \otimes \mathbf{v} \otimes \mathbf{w} = \sum_{i=1}^{I} \sum_{j=1}^{J} \sum_{k=1}^{K} u_i v_j w_k \; \mathbf{e}_{(i-1)JK+(j-1)K+k}^{(IJK)}$$

$$= u_i v_j w_k \; \mathbf{e}_{(i-1)JK+(j-1)K+k}^{(IJK)} = u_i v_j w_k \; \mathbf{e}_{ijk} \qquad [2.94]$$

$$\mathbf{u} \otimes \mathbf{v}^T \otimes \mathbf{w} = (u_i \mathbf{e}_i) \otimes (v_j \mathbf{e}^j) \otimes (w_k \mathbf{e}_k) = u_i v_j w_k \mathbf{e}_{ik}^j. \qquad [2.95]$$

In general, for $\mathbf{x}^{(n)} \in \mathbb{K}^{I_n}$, with $n \in \langle N \rangle$, we have:

$$\overset{N}{\underset{n=1}{\otimes}} \mathbf{x}^{(n)} = \sum_{i_1=1}^{I_1} \sum_{i_2=1}^{I_2} \cdots \sum_{i_N=1}^{I_N} \Big(\prod_{n=1}^{N} x_{i_n}^{(n)} \Big) \; \mathbf{e}_{(i_1-1)I_2 \cdots I_N+(i_2-1)I_3 \cdots I_N+\cdots+i_N}^{(I_1 I_2 \cdots I_N)}$$

$$= \Big(\prod_{n=1}^{N} x_{i_n}^{(n)} \Big) \; \mathbf{e}_{i_1 \cdots i_N}. \qquad [2.96]$$

This concludes the proof of the expressions in Table 2.14.   □

For the Kronecker product of matrices, we have the equations listed in Table 2.15.

$\mathbf{A} \in \mathbb{K}^{I \times J}, \mathbf{B} \in \mathbb{K}^{K \times L}$
$\mathbf{A} \otimes \mathbf{B} = a_{ij} b_{kl} \mathbf{e}_{ik}^{jl}$
$\mathbf{A}^T \otimes \mathbf{B}^T = a_{ij} b_{kl} \mathbf{e}_{jl}^{ik}$
$\mathbf{A} \otimes \mathbf{B}^T = a_{ij} b_{kl} \mathbf{e}_{il}^{jk}$
$\mathbf{A}^T \otimes \mathbf{B} = a_{ij} b_{kl} \mathbf{e}_{jk}^{il}$

Table 2.15. *Kronecker products of matrices and index convention*

PROOF.– Using the expressions [2.68] and [2.69] of a matrix and its transpose using the index convention, we have:

$$\mathbf{A} \otimes \mathbf{B} = (a_{ij} \mathbf{e}_i^j) \otimes (b_{kl} \mathbf{e}_k^l) = a_{ij} b_{kl} \mathbf{e}_{ik}^{jl} \qquad [2.97]$$

$$\mathbf{A}^T \otimes \mathbf{B}^T = (a_{ij} \mathbf{e}_j^i) \otimes (b_{kl} \mathbf{e}_l^k) = a_{ij} b_{kl} \mathbf{e}_{jl}^{ik} \qquad [2.98]$$

and

$$\mathbf{A} \otimes \mathbf{B}^T = (a_{ij}\mathbf{e}_i^j) \otimes (b_{kl}\mathbf{e}_l^k) = a_{ij}b_{kl}\mathbf{e}_{il}^{jk} \qquad [2.99]$$

$$\mathbf{A}^T \otimes \mathbf{B} = (a_{ij}\mathbf{e}_j^i) \otimes (b_{kl}\mathbf{e}_k^l) = a_{ij}b_{kl}\mathbf{e}_{jk}^{il}. \qquad [2.100]$$

This proves the formulae in Table 2.15.   □

Note that, from [2.97], we can deduce that:

$$a_{ij}b_{kl} = \mathbf{e}^{ik}(\mathbf{A} \otimes \mathbf{B})\mathbf{e}_{jl} = (\mathbf{e}_i^{(I)} \otimes \mathbf{e}_k^{(K)})^T(\mathbf{A} \otimes \mathbf{B})(\mathbf{e}_j^{(J)} \otimes \mathbf{e}_l^{(L)}). \qquad [2.101]$$

The expression [2.97] can be generalized to the case of a multiple Kronecker product, as formulated in the following proposition.

PROPOSITION 2.20.– Using the index convention, the multiple Kronecker product $\underset{n\in\langle N\rangle}{\otimes} \mathbf{A}^{(n)}$ can alternatively be written as:

$$\underset{n\in\langle N\rangle}{\otimes} \mathbf{A}^{(n)} = \Big(\prod_{n=1}^{N} a_{i_n,r_n}^{(n)}\Big)\mathbf{e}_{i_1\cdots i_N}^{r_1\cdots r_N} = \Big(\prod_{n=1}^{N} a_{i_n,r_n}^{(n)}\Big)\mathbf{e}_{\mathbb{I}}^{\mathbb{R}}, \qquad [2.102]$$

where $a_{i_n,r_n}^{(n)}$ is the current element of $\mathbf{A}^{(n)} \in \mathbb{K}^{I_n \times R_n}$, $\mathbb{I} = i_1 \cdots i_N$, and $\mathbb{R} = r_1 \cdots r_N$.

2.6.5. *Vectorization and index convention*

Noting that:

$$\mathrm{vec}(\mathbf{ab}^T) = \mathbf{b} \otimes \mathbf{a} \qquad [2.103]$$

and therefore:

$$\mathrm{vec}\Big(\mathbf{e}_i^{(I)}\mathbf{e}_j^{(J)T}\Big) = \mathbf{e}_j^{(J)} \otimes \mathbf{e}_i^{(I)} = \mathbf{e}_{ji}, \qquad [2.104]$$

the index convention allows us to write the vectorization operator compactly, as illustrated in the following proposition.

PROPOSITION 2.21.– For $\mathbf{A} \in \mathbb{K}^{I \times J}$ and $\mathbf{B} \in \mathbb{K}^{K \times L}$, we have:

$$\mathrm{vec}(\mathbf{A}) = \mathrm{vec}\Big(\sum_{i=1}^{I}\sum_{j=1}^{J} a_{ij}\mathbf{e}_i^{(I)}\mathbf{e}_j^{(J)T}\Big) = \sum_{i=1}^{I}\sum_{j=1}^{J} a_{ij}\Big(\mathbf{e}_j^{(J)} \otimes \mathbf{e}_i^{(I)}\Big)$$

$$= a_{ij}\mathbf{e}_{ji} \in \mathbb{K}^{JI} \qquad [2.105]$$

$$\mathrm{vec}(\mathbf{A}^T) = a_{ij}\mathbf{e}_{ij} \in \mathbb{K}^{IJ} \qquad [2.106]$$

$$\mathrm{vec}(\mathbf{A} \otimes \mathbf{B}) = a_{ij}b_{kl}\mathbf{e}_{jlik} \in \mathbb{K}^{JLIK}, \qquad [2.107]$$

where, as a special case of the last relation, for $\mathbf{x} \in \mathbb{K}^I$ and $\mathbf{y} \in \mathbb{K}^L$:

$$\text{vec}(\mathbf{x} \otimes \mathbf{y}^T) = x_i \, y_l \, \mathbf{e}_{li} = \mathbf{y} \otimes \mathbf{x} \in \mathbb{K}^{LI}. \qquad [2.108]$$

Thus, we recover the relation [2.28] with transposition instead of transconjugation.

It should be noted that $\text{vec}(\mathbf{A}^T)$ is equivalent to forming a column vector by stacking the row vectors of $\mathbf{A}$, rearranged as column vectors.

The element a_{ij} of $\mathbf{A}$ is located at the position $i + (j - 1)I$ in $\text{vec}(\mathbf{A})$ and at $j + (i - 1)J$ in $\text{vec}(\mathbf{A}^T)$.

The element $a_{ij}b_{kl}$ of $\mathbf{A} \otimes \mathbf{B}$ is located at the position $k + (i - 1)K + (l - 1)IK + (j - 1)LIK$ in the vector $\text{vec}(\mathbf{A} \otimes \mathbf{B})$.

EXAMPLE 2.22.– From the vectorization [2.105] of the matrix $\mathbf{A}$, with $a_{ij} = \delta_{ij}$, we can deduce the vectorization of the identity matrix $\mathbf{I}_I$:

$$\text{vec}(\mathbf{I}_I) = \mathbf{e}_{ii}. \qquad [2.109]$$

In particular, for $I = 2$, we have:

$$\text{vec}(\mathbf{I}_2) = \text{vec}\left(\begin{bmatrix} 1 & 0 \\ 0 & 1 \end{bmatrix} \right) = \mathbf{e}_1^{(2)} \otimes \mathbf{e}_1^{(2)} + \mathbf{e}_2^{(2)} \otimes \mathbf{e}_2^{(2)}$$

$$= \begin{bmatrix} 1 \\ 0 \end{bmatrix} \otimes \begin{bmatrix} 1 \\ 0 \end{bmatrix} + \begin{bmatrix} 0 \\ 1 \end{bmatrix} \otimes \begin{bmatrix} 0 \\ 1 \end{bmatrix}$$

$$= \begin{bmatrix} 1 \\ 0 \\ 0 \\ 0 \end{bmatrix} + \begin{bmatrix} 0 \\ 0 \\ 0 \\ 1 \end{bmatrix} = \begin{bmatrix} 1 \\ 0 \\ 0 \\ 1 \end{bmatrix}.$$

EXAMPLE 2.23.– For $\mathbf{A} = [a_{11}, a_{12}]$ and $\mathbf{B} = \begin{bmatrix} b_{11} & b_{12} \\ b_{21} & b_{22} \end{bmatrix}$, we have:

$$\mathbf{A} \otimes \mathbf{B} = [a_{11}\mathbf{B}, a_{12}\mathbf{B}] = \begin{bmatrix} a_{11}b_{11} & a_{11}b_{12} & a_{12}b_{11} & a_{12}b_{12} \\ a_{11}b_{21} & a_{11}b_{22} & a_{12}b_{21} & a_{12}b_{22} \end{bmatrix}$$

$$\text{vec}(\mathbf{A} \otimes \mathbf{B}) = [\underbrace{a_{11}b_{11}, a_{11}b_{21}}_{k=1,2}, a_{11}b_{12}, a_{11}b_{22}, a_{12}b_{11}, a_{12}b_{21}, a_{12}b_{12}, a_{12}b_{22}]^T$$

$$\underbrace{\qquad\qquad\qquad\qquad\qquad}_{l=1,2}$$

$$\underbrace{\qquad\qquad\qquad\qquad\qquad\qquad\qquad\qquad\qquad}_{j=1,2}$$

which corresponds to a variation of the indices of $\mathbf{e}_{jlik}$ in [2.107] such that k varies faster than l and j, and l varies faster than j, with $i = 1$.

2.6.6. *Vectorization formulae*

Here, we present several vectorization formulae for the Kronecker product of two matrices, and then for other matrix products. These formulae can be proven using the index convention.

PROPOSITION 2.24.– For $\mathbf{A} \in \mathbb{K}^{I \times J}$ and $\mathbf{B} \in \mathbb{K}^{K \times L}$, we have:

$$\text{vec}(\mathbf{A} \otimes \mathbf{B}) = (\mathbf{I}_J \otimes \mathbf{e}_{li}^{il} \otimes \mathbf{I}_K)\Big(\text{vec}(\mathbf{A}) \otimes \text{vec}(\mathbf{B})\Big), \qquad [2.110]$$

where $\mathbf{e}_{li}^{il}$ represents the commutation matrix defined in [2.162].

PROOF.– Using the relations [2.105], [2.107] and [2.75], as well as the second property in Table 2.9, and noting that:

$$\mathbf{e}_{li}^{il}\mathbf{e}_{il} = \mathbf{e}_{li}(\mathbf{e}_i^T \otimes \mathbf{e}_l^T)(\mathbf{e}_i \otimes \mathbf{e}_l) = \mathbf{e}_{li}(\mathbf{e}_i^T \mathbf{e}_i \otimes \mathbf{e}_l^T \mathbf{e}_l) = \mathbf{e}_{li}, \qquad [2.111]$$

we deduce:

$$\text{vec}(\mathbf{A}) \otimes \text{vec}(\mathbf{B}) = a_{ij}\mathbf{e}_{ji} \otimes b_{kl}\mathbf{e}_{lk} = a_{ij}b_{kl}\mathbf{e}_{jilk} = a_{ij}b_{kl}(\mathbf{e}_j \otimes \mathbf{e}_{il} \otimes \mathbf{e}_k),$$

$$\text{vec}(\mathbf{A} \otimes \mathbf{B}) = a_{ij}b_{kl}\mathbf{e}_{jlik} = a_{ij}b_{kl}(\mathbf{e}_j \otimes \mathbf{e}_{li} \otimes \mathbf{e}_k)$$

$$= a_{ij}b_{kl}(\mathbf{I}_J \otimes \mathbf{e}_{li}^{il} \otimes \mathbf{I}_K)(\mathbf{e}_j \otimes \mathbf{e}_{il} \otimes \mathbf{e}_k)$$

$$= (\mathbf{I}_J \otimes \mathbf{e}_{li}^{il} \otimes \mathbf{I}_K)\Big(\text{vec}(\mathbf{A}) \otimes \text{vec}(\mathbf{B})\Big),$$

which proves the relation [2.110].   □

PROPOSITION 2.25.– From the relation [2.110], with $\mathbf{A} = \mathbf{u} \in \mathbb{K}^I$, $\mathbf{B} = \mathbf{v}^T \in \mathbb{K}^{1 \times L}$, and $J = K = 1$, we deduce that:

$$\text{vec}(\mathbf{u} \otimes \mathbf{v}^T) = \mathbf{v} \otimes \mathbf{u} = \mathbf{e}_{li}^{il}(\mathbf{u} \otimes \mathbf{v}) \in \mathbb{K}^{LI}. \qquad [2.112]$$

EXAMPLE 2.26.– For $\mathbf{A} = \mathbf{u} \in \mathbb{K}^2$, $\mathbf{B} = \mathbf{v}^T \in \mathbb{K}^{1 \times 2}$, i.e. $I = L = 2, J = K = 1$, the formula [2.110] becomes:

$$\text{vec}(\mathbf{u} \otimes \mathbf{v}^T) = \text{vec}\left(\begin{bmatrix} u_1 v_1 & u_1 v_2 \\ u_2 v_1 & u_2 v_2 \end{bmatrix} \right) = \begin{bmatrix} u_1 v_1 \\ u_2 v_1 \\ u_1 v_2 \\ u_2 v_2 \end{bmatrix} = \mathbf{v} \otimes \mathbf{u}$$

$$\mathbf{e}_{li}^{il}\Big(\mathbf{u} \otimes \mathbf{v}\Big) = \begin{bmatrix} 1 & 0 & 0 & 0 \\ 0 & 0 & 1 & 0 \\ 0 & 1 & 0 & 0 \\ 0 & 0 & 0 & 1 \end{bmatrix} \begin{bmatrix} u_1 v_1 \\ u_1 v_2 \\ u_2 v_1 \\ u_2 v_2 \end{bmatrix} = \begin{bmatrix} u_1 v_1 \\ u_2 v_1 \\ u_1 v_2 \\ u_2 v_2 \end{bmatrix} = \text{vec}(\mathbf{u} \otimes \mathbf{v}^T).$$

PROPOSITION 2.27.– For $\mathbf{A} \in \mathbb{K}^{I \times J}, \mathbf{B} \in \mathbb{K}^{J \times M}, \mathbf{C} \in \mathbb{K}^{M \times N}$, we have:

$$\text{vec}(\mathbf{ABC}) = (\mathbf{C}^T \otimes \mathbf{A})\text{vec}(\mathbf{B}). \tag{2.113}$$

$$= (\mathbf{I}_N \otimes \mathbf{AB})\text{vec}(\mathbf{C}) \tag{2.114}$$

$$= (\mathbf{C}^T\mathbf{B}^T \otimes \mathbf{I}_I)\text{vec}(\mathbf{A}) \tag{2.115}$$

$$\text{vec}(\mathbf{AB}) = (\mathbf{I}_M \otimes \mathbf{A})\text{vec}(\mathbf{B}) \tag{2.116}$$

$$= (\mathbf{B}^T \otimes \mathbf{I}_I)\text{vec}(\mathbf{A}) \tag{2.117}$$

$$= (\mathbf{B}^T \otimes \mathbf{A})\text{vec}(\mathbf{I}_J). \tag{2.118}$$

PROOF.– Let us first show the relation [2.113] using the definition of the vectorization operation. The product $\mathbf{ABC} \in \mathbb{K}^{I \times N}$ expands into:

$$\mathbf{ABC} = (a_{ij}\mathbf{e}_i^j)(b_{jm}\mathbf{e}_j^m)(c_{mn}\mathbf{e}_m^n)$$

$$= a_{ij}b_{jm}c_{mn}\mathbf{e}_i^j\mathbf{e}_j^m\mathbf{e}_m^n = a_{ij}b_{jm}c_{mn}\mathbf{e}_i^n. \tag{2.119}$$

Applying the vectorization formula [2.105] then gives us:

$$\text{vec}(\mathbf{ABC}) = (\mathbf{ABC})_{in}\mathbf{e}_{ni} = a_{ij}b_{jm}c_{mn}\mathbf{e}_{ni}. \tag{2.120}$$

Noting that:

$$\mathbf{e}^{mj}\mathbf{e}_{m'j'} = (\mathbf{e}^m \otimes \mathbf{e}^j)(\mathbf{e}_{m'} \otimes \mathbf{e}_{j'}) = \mathbf{e}^m\mathbf{e}_{m'} \otimes \mathbf{e}^j\mathbf{e}_{j'} = \delta_{mm'}\delta_{jj'} \tag{2.121}$$

$$\mathbf{e}_{ni}^{mj}\mathbf{e}_{m'j'} = (\mathbf{e}_{ni} \otimes \mathbf{e}^{mj})\mathbf{e}_{m'j'} = \mathbf{e}_{ni} \otimes \mathbf{e}^{mj}\mathbf{e}_{m'j'} = \delta_{mm'}\delta_{jj'}\mathbf{e}_{ni}, \tag{2.122}$$

applying the formulae [2.105] and [2.100] allows us to rewrite $\text{vec}(\mathbf{ABC})$ as:

$$\text{vec}(\mathbf{ABC}) = c_{mn}a_{ij}b_{j'm'}\mathbf{e}_{ni}^{mj}\mathbf{e}_{m'j'}$$

$$= (c_{mn}a_{ij}\mathbf{e}_{ni}^{mj})(b_{j'm'}\mathbf{e}_{m'j'}) = (\mathbf{C}^T \otimes \mathbf{A})\text{vec}(\mathbf{B}), \tag{2.123}$$

which proves [2.113]. According to Horn and Jonhson (1991), this formula appears to have been proposed for the first time by Roth (1934).

It is important to note that introducing the double Kronecker product $\mathbf{e}^{mj}\mathbf{e}_{m'j'}$ is what led to the formation of the factors $\mathbf{C}^T \otimes \mathbf{A}$ and $\text{vec}(\mathbf{B})$. This double Kronecker product allows us to replace the double sum over the indices m and j associated with the product $\mathbf{ABC}$ followed by the vectorization of a matrix of size $I \times N$, by the product of the matrix $\mathbf{C}^T \otimes \mathbf{A} \in \mathbb{K}^{NI \times MJ}$ with the vector $\text{vec}(\mathbf{B}) \in \mathbb{K}^{MJ}$.

The relations [2.116]–[2.118] can be deduced directly from [2.113] by replacing the triplet $(\mathbf{A}, \mathbf{B}, \mathbf{C})$ with $(\mathbf{A}, \mathbf{B}, \mathbf{I}_M)$, $(\mathbf{I}_I, \mathbf{A}, \mathbf{B})$ and $(\mathbf{A}, \mathbf{I}_J, \mathbf{B})$, respectively. Similarly, the relations [2.114] and [2.115] can be deduced from [2.113] by replacing $(\mathbf{A}, \mathbf{B}, \mathbf{C})$ with $(\mathbf{AB}, \mathbf{C}, \mathbf{I}_N)$ and $(\mathbf{I}_I, \mathbf{A}, \mathbf{BC})$, respectively. $\square$

REMARK 2.28.– The reasoning used to prove [2.113] will be reused later in this chapter. To illustrate this reasoning, let us apply it to prove the identities [2.116] and [2.117] that allow us to compute vec($\mathbf{AB}$).

A classical method to compute vec($\mathbf{AB}$) is to compute $\mathbf{C} = \mathbf{AB} \in \mathbb{K}^{I \times M}$ then vectorize $\mathbf{C}$. By replacing K with M and using the expressions [2.81] and [2.105], $\mathbf{AB}$ and vec($\mathbf{AB}$) can be written as:

$$\mathbf{AB} = \mathbf{C} = c_{im}\mathbf{e}_i^m = a_{ij}b_{jm}\mathbf{e}_i^m \qquad [2.124]$$

$$\mathrm{vec}(\mathbf{AB}) = \mathrm{vec}(\mathbf{C}) = c_{im}\mathbf{e}_{mi} = a_{ij}b_{jm}\mathbf{e}_{mi}. \qquad [2.125]$$

Another method is to replace $\mathbf{e}_{mi}$ with $\mathbf{e}_{mi}^{mj}\mathbf{e}_{m'j'}$. Equation [2.125] can then be written as:

$$\mathrm{vec}(\mathbf{AB}) = (a_{ij}\mathbf{e}_{mi}^{mj})(b_{j'm'}\mathbf{e}_{m'j'}) \qquad [2.126]$$

$$= (\mathbf{e}_m^m \otimes a_{ij}\mathbf{e}_i^j)\mathrm{vec}(\mathbf{B}) \qquad [2.127]$$

$$= (\mathbf{I}_M \otimes \mathbf{A})\mathrm{vec}(\mathbf{B}), \qquad [2.128]$$

which corresponds to the identity [2.116].

Similarly, by replacing $\mathbf{e}_{mi}$ with $\mathbf{e}_{mi}^{ji}\mathbf{e}_{j'i'}$, equation [2.125] becomes:

$$\mathrm{vec}(\mathbf{AB}) = (b_{jm}\mathbf{e}_{mi}^{ji})(a_{i'j'}\mathbf{e}_{j'i'}) \qquad [2.129]$$

$$= (b_{jm}\mathbf{e}_m^j \otimes \mathbf{e}_i^i)\mathrm{vec}(\mathbf{A}) \qquad [2.130]$$

$$= (\mathbf{B}^T \otimes \mathbf{I}_I)\mathrm{vec}(\mathbf{A}), \qquad [2.131]$$

which is the identity [2.117].

Thus, the decompositions of $\mathbf{e}_{mi}$ used above allowed us to replace the computation of the matrix $\mathbf{C} = \mathbf{AB}$ followed by its vectorization with two matrix products involving the vectorization of $\mathbf{A}$ or $\mathbf{B}$.

The identities [2.113]–[2.118] play a very important role in solving systems of matrix equations, as will be illustrated in section 2.12.

PROPOSITION 2.29.– Using the relation [2.113], it is easy to deduce the following vectorization formulae for $\mathbf{A} \in \mathbb{K}^{I \times J}, \mathbf{B} \in \mathbb{K}^{J \times M}, \mathbf{C} \in \mathbb{K}^{M \times N}, \mathbf{D} \in \mathbb{K}^{N \times P}$:

$$\mathrm{vec}(\mathbf{ABCD}) = (\mathbf{D}^T \otimes \mathbf{AB})\mathrm{vec}(\mathbf{C}) \qquad [2.132]$$

$$= (\mathbf{D}^T\mathbf{C}^T \otimes \mathbf{A})\mathrm{vec}(\mathbf{B}) \qquad [2.133]$$

$$= (\mathbf{D}^T\mathbf{C}^T\mathbf{B}^T \otimes \mathbf{I}_I)\mathrm{vec}(\mathbf{A}) \qquad [2.134]$$

$$= (\mathbf{I}_P \otimes \mathbf{ABC})\mathrm{vec}(\mathbf{D}). \qquad [2.135]$$

2.6.7. *Vectorization of partitioned matrices*

In this section, we will use the index convention to compute the block vectorization of a partitioned matrix.

PROPOSITION 2.30.– Let $\mathbf{A} \in \mathbb{K}^{I \times J}$ be a matrix partitioned into blocks $\mathbf{A}_{rs} \in \mathbb{K}^{I_r \times J_s}$, with $r \in \langle R \rangle$, $s \in \langle S \rangle$, and $\sum_{r=1}^{R} I_r = I$, $\sum_{s=1}^{S} J_s = J$. The block vectorization of $\mathbf{A}$ is given as:

$$\text{vec}_b(\mathbf{A}) = a_{i_r j_s}^{(rs)} \mathbf{e}_{s,r,j_s,i_r}^{(SRJ_sI_r)} = (\mathbf{e}_{s,r,j_s,i_r}^{s,j_s,r,i_r}) \text{vec}(\mathbf{A}) \qquad [2.136]$$

$$= (\mathbf{I}_S \otimes \mathbf{e}_{r,j_s}^{j_s,r} \otimes \mathbf{I}_{I_r}) \text{vec}(\mathbf{A}). \qquad [2.137]$$

PROOF.– Since the matrix $\mathbf{A}$ is partitioned in the form:

$$\mathbf{A} = \begin{bmatrix} \mathbf{A}_{11} & \cdots & \mathbf{A}_{1S} \\ \vdots & & \vdots \\ \mathbf{A}_{R1} & \cdots & \mathbf{A}_{RS} \end{bmatrix}, \qquad [2.138]$$

it can be written as follows with the index convention:

$$\mathbf{A} = \sum_{r=1}^{R} \sum_{s=1}^{S} (\mathbf{E}_{rs}^{(R \times S)} \otimes \mathbf{A}_{rs}) = \mathbf{e}_r^s \otimes \mathbf{A}_{rs}, \qquad [2.139]$$

where $\mathbf{E}_{rs}^{(R \times S)} = \mathbf{e}_r^{(R)} (\mathbf{e}_s^{(S)})^T = \mathbf{e}_r^s$ is the matrix of size $R \times S$ with 1 at the position (r, s) and 0 everywhere else, defined as in [2.25]. Using the formulae [2.110] and [2.105], we have:

$$\text{vec}(\mathbf{A}) = \text{vec}(\mathbf{e}_r^s \otimes \mathbf{A}_{rs})$$

$$= a_{i_r,j_s}^{(rs)} \text{vec}(\mathbf{e}_r^s \otimes \mathbf{e}_{i_r}^{j_s}) = a_{i_r,j_s}^{(rs)} \text{vec}(\mathbf{e}_{r,i_r}^{s,j_s})$$

$$= a_{i_r,j_s}^{(rs)} \mathbf{e}_{s,j_s,r,i_r}^{(SJ_sRI_r)}. \qquad [2.140]$$

The difference between block vectorization and standard vectorization is the order of variation of the indices r and j_s, with:

$$\text{vec}_b(\mathbf{A}) = a_{i_r,j_s}^{(rs)} \mathbf{e}_{s,r,j_s,i_r}^{(SRJ_sI_r)}, \qquad [2.141]$$

with the index r of the block varying more slowly than the index j_s of the element $a_{i_r,j_s}^{(rs)}$ of $\mathbf{A}$ in the block vectorization, whereas the reverse holds for $\text{vec}(\mathbf{A})$.

Therefore, we deduce that:

$$\text{vec}_b(\mathbf{A}) = \mathbf{e}_{s,r,j_s,i_r}^{s,j_s,r,i_r} \text{vec}(\mathbf{A}) = (\mathbf{e}_s^s \otimes \mathbf{e}_{r,j_s}^{j_s,r} \otimes \mathbf{e}_{i_r}^{i_r}) \text{vec}(\mathbf{A}) \qquad [2.142]$$

$$= (\mathbf{I}_S \otimes \mathbf{e}_{r,j_s}^{j_s,r} \otimes \mathbf{I}_{I_r}) \text{vec}(\mathbf{A}), \qquad [2.143]$$

which proves [2.137]. □

REMARK 2.31.– This formula means that the position of the element $a_{i_r,j_s}^{(rs)}$ in $\text{vec}(\mathbf{A})$ is given by the canonical basis vector $\mathbf{e}_{s,j_s,r,i_r}^{(SJ_sRI_r)} = \mathbf{e}_s^{(S)} \otimes \mathbf{e}_{j_s}^{(J_s)} \otimes \mathbf{e}_r^{(R)} \otimes \mathbf{e}_{i_r}^{(I_r)}$, whereas in $\text{vec}_b(\mathbf{A})$ it is given by $\mathbf{e}_{s,r,j_s,i_r}^{(SJ_sRI_r)} = \mathbf{e}_s^{(S)} \otimes \mathbf{e}_r^{(R)} \otimes \mathbf{e}_{j_s}^{(J_s)} \otimes \mathbf{e}_{i_r}^{(I_r)}$.

EXAMPLE 2.32.– Consider the case where $R = S = 2$, with $I_r = 1, J_s = 2, r, s \in \{1, 2\}$:

$$
\mathbf{A} = \begin{bmatrix} \mathbf{A}_{11} & \mathbf{A}_{12} \\ \mathbf{A}_{21} & \mathbf{A}_{22} \end{bmatrix} = \begin{bmatrix} a_{11}^{(11)} & a_{12}^{(11)} & \vdots & a_{11}^{(12)} & a_{12}^{(12)} \\ \cdots & \cdots & & \cdots & \cdots \\ a_{11}^{(21)} & a_{12}^{(21)} & \vdots & a_{11}^{(22)} & a_{12}^{(22)} \end{bmatrix}.
\qquad [2.144]
$$

We have:

$$
\text{vec}(\mathbf{A}) = [a_{11}^{(11)}, a_{11}^{(21)}, a_{12}^{(11)}, a_{12}^{(21)}, a_{11}^{(12)}, a_{11}^{(22)}, a_{12}^{(12)}, a_{12}^{(22)}]^T.
$$

By definition, $\text{vec}_b(\mathbf{A})$ is given by:

$$
\text{vec}_b(\mathbf{A}) = \begin{bmatrix} \text{vec}(\mathbf{A}_{11}) \\ \text{vec}(\mathbf{A}_{21}) \\ \text{vec}(\mathbf{A}_{12}) \\ \text{vec}(\mathbf{A}_{22}) \end{bmatrix}
$$

$$
= \big[\underbrace{\underbrace{a_{11}^{(11)}, a_{12}^{(11)}}_{i_r=1}, a_{11}^{(21)}, a_{12}^{(21)}}_{\substack{j_s=1,2 \\ r=1,2\,;\,s=1}}, \underbrace{a_{11}^{(12)}, a_{12}^{(12)}, a_{11}^{(22)}, a_{12}^{(22)}}_{r=1,2\,;\,s=2}\big]^T.
$$

Applying the formula [2.137], we obtain:

$$
\text{vec}_b(\mathbf{A}) = [\mathbf{I}_2 \otimes \mathbf{e}_{rj_s}^{j_s r}]\text{vec}(\mathbf{A}) = \left(\mathbf{I}_2 \otimes \begin{bmatrix} 1 & 0 & 0 & 0 \\ 0 & 0 & 1 & 0 \\ 0 & 1 & 0 & 0 \\ 0 & 0 & 0 & 1 \end{bmatrix}\right)\text{vec}(\mathbf{A}),
$$

which gives:

$$
\text{vec}_b(\mathbf{A}) =
\begin{bmatrix}
1 & 0 & 0 & 0 & \vdots & 0 & 0 & 0 & 0 \\
0 & 0 & 1 & 0 & \vdots & 0 & 0 & 0 & 0 \\
0 & 1 & 0 & 0 & \vdots & 0 & 0 & 0 & 0 \\
0 & 0 & 0 & 1 & \vdots & 0 & 0 & 0 & 0 \\
\cdots & \cdots & \cdots & \cdots & & \cdots & \cdots & \cdots & \cdots \\
0 & 0 & 0 & 0 & \vdots & 1 & 0 & 0 & 0 \\
0 & 0 & 0 & 0 & \vdots & 0 & 0 & 1 & 0 \\
0 & 0 & 0 & 0 & \vdots & 0 & 1 & 0 & 0 \\
0 & 0 & 0 & 0 & \vdots & 0 & 0 & 0 & 1
\end{bmatrix}
\begin{bmatrix}
a_{11}^{(11)} \\
a_{11}^{(21)} \\
a_{12}^{(11)} \\
a_{12}^{(21)} \\
a_{11}^{(12)} \\
a_{11}^{(22)} \\
a_{12}^{(12)} \\
a_{12}^{(22)}
\end{bmatrix}.
$$

To illustrate remark 2.31, consider the element $a_{12}^{(12)}$ corresponding to $(r = i_r = 1; s = j_s = 2)$. Its position in $\text{vec}(\mathbf{A})$ is given by $\mathbf{e}_{s,j_s,r,i_r}^{(SJ_sRI_r)} = \mathbf{e}_2^{(2)} \otimes \mathbf{e}_2^{(2)} \otimes \mathbf{e}_1^{(2)} \otimes \mathbf{e}_1^{(1)} = \mathbf{e}_7^{(8)}$, which is the seventh row of $\text{vec}(\mathbf{A})$, whereas in $\text{vec}_b(\mathbf{A})$ its position is given by $\mathbf{e}_{s,r,j_s,i_r}^{(SRJ_sI_r)} = \mathbf{e}_2^{(2)} \otimes \mathbf{e}_1^{(2)} \otimes \mathbf{e}_2^{(2)} \otimes \mathbf{e}_1^{(1)} = \mathbf{e}_6^{(8)}$, which is the sixth row.

2.6.8. *Traces of matrix products and index convention*

Using the index convention, we can prove the following relations for traces of matrices:

– For $\mathbf{A}, \mathbf{B}, \mathbf{C}, \mathbf{D} \in \mathbb{K}^{I \times J}$, we have:

$$
\text{tr}\left[(\mathbf{A} \odot \mathbf{B})(\mathbf{C} \odot \mathbf{D})^T\right] = \sum_{i=1}^{I} \sum_{j=1}^{J} (a_{ij}b_{ij})(c_{ij}d_{ij}) = a_{ij}b_{ij}c_{ij}d_{ij} \qquad [2.145]
$$

$$
\text{tr}\left[(\mathbf{A} \odot \mathbf{B})\mathbf{C}^T\right] = \sum_{i=1}^{I} \sum_{j=1}^{J} (a_{ij}b_{ij})c_{ij} = a_{ij}b_{ij}c_{ij} \qquad [2.146]
$$

$$
= (a_{ij}c_{ij})b_{ij} = \text{tr}\left[(\mathbf{A} \odot \mathbf{C})\mathbf{B}^T\right]. \qquad [2.147]
$$

– For $\mathbf{A} \in \mathbb{K}^{I \times J}, \mathbf{B} \in \mathbb{K}^{J \times I}$, we have:

$$
\text{tr}(\mathbf{AB}) = a_{ij}b_{ji} \qquad [2.148]
$$

$$
= \text{vec}^T(\mathbf{A}^T)\text{vec}(\mathbf{B}) = \mathbf{1}_I^T(\mathbf{A} \odot \mathbf{B}^T)\mathbf{1}_J. \qquad [2.149]
$$

PROOF.– The index convention allows us to write:

$$
\text{tr}(\mathbf{AB}) = (\mathbf{AB})_{ii} = \mathbf{A}_{i.}\mathbf{B}_{.i} = a_{ij}b_{ji}. \qquad [2.150]
$$

Furthermore, like in the proof of the formula [2.123], introducing the scalar product $e^{ij}e_{i'j'} = \delta_{ii'}\delta_{jj'}$ allows us to replace the summation of scalars $a_{ij}b_{ji}$ over the indices i and j with the scalar product of the vectors $\text{vec}(\mathbf{A}^T)$ and $\text{vec}(\mathbf{B})$ of size IJ. Using the relations [2.105] and [2.106], we obtain:

$$\text{tr}(\mathbf{AB}) = (a_{ij}e^{ij})(b_{j'i'}e_{i'j'}) = \text{vec}^T(\mathbf{A}^T)\text{vec}(\mathbf{B}).$$

We can also write $\text{tr}(\mathbf{AB})$ as:

$$\text{tr}(\mathbf{AB}) = a_{ij}b_{ji} = \sum_{i=1}^{I}\sum_{j=1}^{I}(\mathbf{A} \odot \mathbf{B}^T)_{ij} = \mathbf{1}_I^T(\mathbf{A} \odot \mathbf{B}^T)\mathbf{1}_J,$$

which proves [2.149]. $\square$

We also have:

– For $\mathbf{A} \in \mathbb{K}^{I \times I}$ and $\mathbf{B} \in \mathbb{K}^{J \times J}$:

$$\text{tr}(\mathbf{A} \otimes \mathbf{B}) = \text{tr}(\mathbf{A})\text{tr}(\mathbf{B}). \tag{2.151}$$

– For $\mathbf{A} \in \mathbb{K}^{I \times J}, \mathbf{B} \in \mathbb{K}^{J \times M}, \mathbf{C} \in \mathbb{K}^{M \times N}, \mathbf{D} \in \mathbb{K}^{N \times I}$:

$$\text{tr}(\mathbf{ABCD}) = \text{vec}^T(\mathbf{D}^T)(\mathbf{C}^T \otimes \mathbf{A})\text{vec}(\mathbf{B}) \tag{2.152}$$

$$= \text{vec}^T(\mathbf{B})(\mathbf{C} \otimes \mathbf{A}^T)\text{vec}(\mathbf{D}^T) \tag{2.153}$$

$$= \text{vec}^T(\mathbf{D})(\mathbf{A} \otimes \mathbf{C}^T)\text{vec}(\mathbf{B}^T) \tag{2.154}$$

$$= \text{vec}^T(\mathbf{A})(\mathbf{B} \otimes \mathbf{D}^T)\text{vec}(\mathbf{C}^T). \tag{2.155}$$

– For $\mathbf{A} \in \mathbb{K}^{I \times J}, \mathbf{B} \in \mathbb{K}^{J \times M}, \mathbf{C} \in \mathbb{K}^{M \times I}$:

$$\text{tr}(\mathbf{ABC}) = \text{vec}^T(\mathbf{I}_I)(\mathbf{C}^T \otimes \mathbf{A})\text{vec}(\mathbf{B}) \tag{2.156}$$

$$= \text{vec}^T(\mathbf{A})(\mathbf{B} \otimes \mathbf{I}_I)\text{vec}(\mathbf{C}^T). \tag{2.157}$$

PROOF.– First, note that the trace of $\mathbf{A}$ can be written $\text{tr}(\mathbf{A}) = \sum_{i=1}^{I} a_{ii} = \mathbf{e}_i^T\mathbf{A}\mathbf{e}_i$.

– By the expression [2.101], the trace of the Kronecker product can be written as:

$$\text{tr}(\mathbf{A} \otimes \mathbf{B}) = \sum_{i=1}^{I}\sum_{j=1}^{J} a_{ii}b_{jj}$$

$$= (\mathbf{e}_i \otimes \mathbf{e}_j)^T(\mathbf{A} \otimes \mathbf{B})(\mathbf{e}_i \otimes \mathbf{e}_j)$$

$$= (\mathbf{e}_i^T\mathbf{A}\mathbf{e}_i) \otimes (\mathbf{e}_j^T\mathbf{B}\mathbf{e}_j) = (\mathbf{e}_i^T\mathbf{A}\mathbf{e}_i)(\mathbf{e}_j^T\mathbf{B}\mathbf{e}_j)$$

$$= \text{tr}(\mathbf{A})\text{tr}(\mathbf{B}).$$

The fourth equality follows from the fact that the quantities in parentheses are scalars. This proves the relation [2.151].

– Following the same approach as before, by the relations $e^{ni}e_{n'i'}^{m'j'}e_{mj} = \delta_{ii'}\delta_{jj'}\delta_{mm'}\delta_{nn'}$, [2.100], [2.99], [2.105] and [2.106], we can write:

$$\text{tr}(\mathbf{ABCD}) = a_{ij}b_{jm}c_{mn}d_{ni}$$

$$= d_{ni}e^{ni}\left(c_{m'n'}a_{i'j'}e_{n'i'}^{m'j'}\right)b_{jm}e_{mj}$$

$$= \text{vec}^{T}(\mathbf{D}^{T})(\mathbf{C}^{T}\otimes\mathbf{A})\text{vec}(\mathbf{B}),$$

which proves [2.152]. Similarly, by introducing the term $e^{ji}e_{j'i'}^{m'n'}e_{mn}$, we obtain:

$$\text{tr}(\mathbf{ABCD}) = a_{ij}e^{ji}(b_{j'm'}d_{n'i'}e_{j'i'}^{m'n'})c_{mn}e_{mn}$$

$$= \text{vec}^{T}(\mathbf{A})(\mathbf{B}\otimes\mathbf{D}^{T})\text{vec}(\mathbf{C}^{T}),$$

i.e. the relation [2.155].

The formula [2.152] can also be shown using the relations [2.149] and [2.113], together with the relation $\text{tr}(\mathbf{AB}) = \text{tr}(\mathbf{BA})$:

$$\text{tr}(\mathbf{ABCD}) = \text{tr}\Big(\mathbf{D}(\mathbf{ABC})\Big) = \text{vec}^{T}(\mathbf{D}^{T})\text{vec}(\mathbf{ABC})$$

$$= \text{vec}^{T}(\mathbf{D}^{T})(\mathbf{C}^{T}\otimes\mathbf{A})\text{vec}(\mathbf{B}).$$

The relation [2.153] is obtained by transposing the right-hand side of [2.152], which is a scalar and therefore equal to its transpose. Similarly, since $\text{tr}(\mathbf{AB}) = \text{tr}(\mathbf{A}^{T}\mathbf{B}^{T})$, we have:

$$\text{tr}(\mathbf{ABCD}) = \text{tr}\Big(\mathbf{D}^{T}(\mathbf{ABC})^{T}\Big) = \text{tr}\Big(\mathbf{D}^{T}(\mathbf{C}^{T}\mathbf{B}^{T}\mathbf{A}^{T})\Big),$$

and, using the relations [2.149] and [2.113], we obtain:

$$\text{tr}(\mathbf{ABCD}) = \text{vec}^{T}(\mathbf{D})\text{vec}(\mathbf{C}^{T}\mathbf{B}^{T}\mathbf{A}^{T})$$

$$= \text{vec}^{T}(\mathbf{D})(\mathbf{A}\otimes\mathbf{C}^{T})\text{vec}(\mathbf{B}^{T}),$$

which is the relation [2.154].

The formulae [2.156] and [2.157] can be deduced from [2.152] and [2.155], respectively, by taking $N = I$ and $\mathbf{D} = \mathbf{I}_{I}$. $\qquad\square$

2.7. Commutation matrices

In this section, we introduce the notion of commutation matrix via the transformation that transforms $\text{vec}(\mathbf{A})$ to $\text{vec}(\mathbf{A}^{T})$. Then, after presenting a few properties of these matrices, we will describe several of their uses to permute the factors of a simple or multiple Kronecker product, as well as to define the block Kronecker product.

2.7.1. *Definition*

PROPOSITION 2.33.– For $\mathbf{A} \in \mathbb{K}^{I \times J}$, we can transform $\text{vec}(\mathbf{A})$ to $\text{vec}(\mathbf{A}^T)$ using the following formula:

$$\text{vec}(\mathbf{A}^T) = \mathbf{K}_{IJ}\text{vec}(\mathbf{A}), \qquad\qquad\qquad [2.158]$$

$$\mathbf{K}_{IJ} = \mathbf{e}_{ij}^{ji} \in \mathbb{R}^{IJ \times JI}, \qquad\qquad\qquad [2.159]$$

where $\mathbf{K}_{IJ}$ is called the *vec*-permutation matrix, or commutation matrix (Magnus and Neudecker 1979).

PROOF.– We have:

$$\mathbf{e}_{ij} = \mathbf{e}_{ij} \otimes \mathbf{e}^{j'i'}\mathbf{e}_{ji} = (\delta_{ii'}\delta_{jj'}\mathbf{e}_{ij}^{j'i'})\mathbf{e}_{ji}. \qquad\qquad [2.160]$$

The term in parentheses means that there is no summation over the indices i and j, which enables us to separate this term from the factor $\mathbf{e}_{ji}$. Hence, from the relations [2.105] and [2.106], we deduce:

$$\text{vec}(\mathbf{A}^T) = a_{ij}\mathbf{e}_{ij} = (\delta_{ii'}\delta_{jj'}\mathbf{e}_{ij}^{j'i'})(a_{ij}\mathbf{e}_{ji}) = \mathbf{e}_{ij}^{ji}\text{vec}(\mathbf{A}), \qquad [2.161]$$

which proves [2.158]. $\qquad\qquad\qquad\qquad\qquad\qquad\qquad\qquad\qquad\qquad\Box$

The commutation matrix $\mathbf{K}_{IJ}$ can be expanded as follows:

$$\mathbf{K}_{IJ} = \mathbf{e}_{ij}^{ji} = (\mathbf{e}_i^{(I)} \otimes \mathbf{e}_j^{(J)})(\mathbf{e}_j^{(J)} \otimes \mathbf{e}_i^{(I)})^T \qquad\qquad [2.162]$$

$$= \mathbf{e}_{(i-1)J+j}^{(IJ)}(\mathbf{e}_{(j-1)I+i}^{(JI)})^T \qquad\qquad\qquad [2.163]$$

$$= \mathbf{e}_i^j \otimes \mathbf{e}_j^i = \mathbf{E}_{ij}^{(I \times J)} \otimes \mathbf{E}_{ji}^{(J \times I)}, \qquad\qquad [2.164]$$

where $\mathbf{E}_{ij}^{(I \times J)}$ is defined in [2.25] as:

$$\mathbf{E}_{ij}^{(I \times J)} = \mathbf{e}_i^j = \mathbf{e}_i^{(I)} \otimes (\mathbf{e}_j^{(J)})^T = \mathbf{e}_i^{(I)}(\mathbf{e}_j^{(J)})^T. \qquad\qquad [2.165]$$

REMARK 2.34.– By the definition [2.159], the vectorization formula [2.110] can also be written as:

$$\text{vec}(\mathbf{A} \otimes \mathbf{B}) = (\mathbf{I}_J \otimes \mathbf{K}_{LI} \otimes \mathbf{I}_K)(\text{vec}(\mathbf{A}) \otimes \text{vec}(\mathbf{B})). \qquad [2.166]$$

PROPOSITION 2.35.– The commutation matrix only depends on the size of $\mathbf{A}$ and not on its elements. The 1 in each column n of $\mathbf{K}_{IJ}$ is obtained by computing the matrix $\mathbf{e}_{ij}^{ji}$ for the index pair (i, j) from the element a_{ij} located on the nth row of $\text{vec}(\mathbf{A})$, with $n = (j-1)I + i$.

EXAMPLE 2.36.– For $\mathbf{A} = \begin{bmatrix} a_{11} & a_{12} \\ a_{21} & a_{22} \end{bmatrix}$, with $I = J = 2$, we have:

$$\text{vec}(\mathbf{A}) = [a_{11}, a_{21}, a_{12}, a_{22}]^T \quad , \quad \text{vec}(\mathbf{A}^T) = [a_{11}, a_{12}, a_{21}, a_{22}]^T$$

and:

$$\mathbf{K}_{22} = \mathbf{e}_{ij}^{ji} = \mathbf{e}_{11}\mathbf{e}_{11}^T + \mathbf{e}_{12}\mathbf{e}_{21}^T + \mathbf{e}_{21}\mathbf{e}_{12}^T + \mathbf{e}_{22}\mathbf{e}_{22}^T$$

$$= \left(\begin{pmatrix} 1 \\ 0 \end{pmatrix} \otimes \begin{pmatrix} 1 \\ 0 \end{pmatrix}\right)\left(\begin{pmatrix} 1 \\ 0 \end{pmatrix} \otimes \begin{pmatrix} 1 \\ 0 \end{pmatrix}\right)^T + \left(\begin{pmatrix} 1 \\ 0 \end{pmatrix} \otimes \begin{pmatrix} 0 \\ 1 \end{pmatrix}\right)\left(\begin{pmatrix} 0 \\ 1 \end{pmatrix} \otimes \begin{pmatrix} 1 \\ 0 \end{pmatrix}\right)^T +$$

$$\left(\begin{pmatrix} 0 \\ 1 \end{pmatrix} \otimes \begin{pmatrix} 1 \\ 0 \end{pmatrix}\right)\left(\begin{pmatrix} 1 \\ 0 \end{pmatrix} \otimes \begin{pmatrix} 0 \\ 1 \end{pmatrix}\right)^T + \left(\begin{pmatrix} 0 \\ 1 \end{pmatrix} \otimes \begin{pmatrix} 0 \\ 1 \end{pmatrix}\right)\left(\begin{pmatrix} 0 \\ 1 \end{pmatrix} \otimes \begin{pmatrix} 0 \\ 1 \end{pmatrix}\right)^T$$

$$= \begin{bmatrix} 1 & 0 & 0 & 0 \\ 0 & 0 & 1 & 0 \\ 0 & 1 & 0 & 0 \\ 0 & 0 & 0 & 1 \end{bmatrix}. \tag{2.167}$$

2.7.2. Properties

PROPOSITION 2.37.– Since the commutation matrix is a permutation matrix, it satisfies the following properties (Magnus and Neudecker 1979):

$$\mathbf{K}_{JI} = \mathbf{K}_{IJ}^T = \mathbf{K}_{IJ}^{-1}. \tag{2.168}$$

PROOF.– These relations can be shown with the index convention:

$$\mathbf{K}_{JI} = \mathbf{e}_{ji}^{ij} = (\mathbf{e}_{ij}^{ji})^T = \mathbf{K}_{IJ}^T.$$

Furthermore, by the relation [2.76], we deduce that:

$$\mathbf{K}_{IJ}\mathbf{K}_{JI} = \mathbf{e}_{ij}^{ji}\mathbf{e}_{ji}^{ij} = \mathbf{I}_{IJ} \quad \Rightarrow \quad \mathbf{K}_{IJ}^{-1} = \mathbf{K}_{JI}, \tag{2.169}$$

which concludes the proof of the relations [2.168].   □

2.7.3. Kronecker product and permutation of factors

Given the vectors $\mathbf{u} \in \mathbb{K}^I$ and $\mathbf{v} \in \mathbb{K}^K$, the Kronecker products $\mathbf{u} \otimes \mathbf{v} \in \mathbb{K}^{IK}$ and $\mathbf{v} \otimes \mathbf{u} \in \mathbb{K}^{KI}$ both contain the same components, consisting of all products of a component of $\mathbf{u}$ with a component of $\mathbf{v}$. These two Kronecker products are related by a row permutation matrix.

Similarly, for $\mathbf{A} \in \mathbb{K}^{I \times J}$ and $\mathbf{B} \in \mathbb{K}^{K \times L}$, the Kronecker products $\mathbf{A} \otimes \mathbf{B} \in \mathbb{K}^{IK \times JL}$ and $\mathbf{B} \otimes \mathbf{A} \in \mathbb{K}^{KI \times LJ}$ are related by row and column permutation matrices, as we will show in the next proposition using an original approach based on the index convention.

PROPOSITION 2.38.– Given the vectors $\mathbf{u} \in \mathbb{K}^I, \mathbf{v} \in \mathbb{K}^K$ and the matrices $\mathbf{A} \in \mathbb{K}^{I \times J}, \mathbf{B} \in \mathbb{K}^{K \times L}$, we have:

$$\mathbf{v} \otimes \mathbf{u} = \mathbf{K}_{KI}(\mathbf{u} \otimes \mathbf{v}) \tag{2.170}$$

$$\mathbf{B} \otimes \mathbf{A} = \mathbf{K}_{KI}(\mathbf{A} \otimes \mathbf{B})\mathbf{K}_{JL}, \tag{2.171}$$

where $\mathbf{K}_{KI}$ and $\mathbf{K}_{JL}$ are row and column permutation matrices, i.e. commutation matrices defined as in [2.162]:

$$\mathbf{K}_{KI} \triangleq \mathbf{e}_{ki}^{ik} = (\mathbf{e}_k^{(K)} \otimes \mathbf{e}_i^{(I)})(\mathbf{e}_i^{(I)} \otimes \mathbf{e}_k^{(K)})^T \tag{2.172}$$

$$= \sum_{k=1}^{K}\sum_{i=1}^{I} \mathbf{e}_k^{(K)}\mathbf{e}_i^{(I)^T} \otimes \mathbf{e}_i^{(I)}\mathbf{e}_k^{(K)^T} \in \mathbb{R}^{KI \times IK}, \tag{2.173}$$

and

$$\mathbf{K}_{JL} \triangleq \mathbf{e}_{jl}^{lj} = (\mathbf{e}_j^{(J)} \otimes \mathbf{e}_l^{(L)})(\mathbf{e}_l^{(L)} \otimes \mathbf{e}_j^{(J)})^T \tag{2.174}$$

$$= \sum_{j=1}^{J}\sum_{l=1}^{L} \mathbf{e}_j^{(J)}\mathbf{e}_l^{(L)^T} \otimes \mathbf{e}_l^{(L)}\mathbf{e}_j^{(J)^T} \in \mathbb{R}^{JL \times LJ}. \tag{2.175}$$

PROOF.– Using the expression of $\mathbf{u} \otimes \mathbf{v}$ given in Table 2.14, we have:

$$\mathbf{u} \otimes \mathbf{v} = u_i v_k \mathbf{e}_{ik} \;\; ; \;\; \mathbf{v} \otimes \mathbf{u} = u_i v_k \mathbf{e}_{ki}. \tag{2.176}$$

Noting that $\mathbf{e}_{ki}^{i'k'}\mathbf{e}_{ik} = \mathbf{e}_{ki}$, we deduce that:

$$\mathbf{v} \otimes \mathbf{u} = (\delta_{ii'}\delta_{kk'}\mathbf{e}_{ki}^{i'k'})(u_i v_k \mathbf{e}_{ik}) = \mathbf{K}_{KI}(\mathbf{u} \otimes \mathbf{v}), \tag{2.177}$$

which proves the relation [2.170].

Similarly, in the matrix case, by [2.97], we have:

$$\mathbf{A} \otimes \mathbf{B} = a_{ij}b_{kl}\mathbf{e}_{ik}^{jl} \;\; ; \;\; \mathbf{B} \otimes \mathbf{A} = a_{ij}b_{kl}\mathbf{e}_{ki}^{lj} \tag{2.178}$$

and, noting that:

$$\mathbf{e}_{ki}^{i'k'}\mathbf{e}_{ik}^{jl}\mathbf{e}_{j'l'}^{lj} = \mathbf{e}_{ki}^{lj}, \tag{2.179}$$

we can rewrite $\mathbf{B} \otimes \mathbf{A}$ as:

$$\mathbf{B} \otimes \mathbf{A} = a_{ij}b_{kl}\mathbf{e}_{ki}^{i'k'}\mathbf{e}_{ik}^{jl}\mathbf{e}_{j'l'}^{lj}$$

$$= (\delta_{ii'}\delta_{kk'}\mathbf{e}_{ki}^{i'k'})(a_{ij}b_{kl}\mathbf{e}_{ik}^{jl})(\delta_{jj'}\delta_{ll'}\mathbf{e}_{j'l'}^{lj})$$

$$= \mathbf{e}_{ki}^{ik}(\mathbf{A} \otimes \mathbf{B})\mathbf{e}_{jl}^{lj}$$

$$= \mathbf{K}_{KI}(\mathbf{A} \otimes \mathbf{B})\mathbf{K}_{JL},$$

which proves [2.171], with $\mathbf{K}_{KI}$ and $\mathbf{K}_{JL}$ defined as in [2.173] and [2.175]. □

PROPOSITION 2.39.– Using the property [2.168] of the commutation matrix, the relation [2.171] can also be written as:

$$(\mathbf{B} \otimes \mathbf{A})\mathbf{K}_{LJ} = \mathbf{K}_{KI}(\mathbf{A} \otimes \mathbf{B}).$$ [2.180]

In the case where $\mathbf{A}$ and $\mathbf{B}$ are square matrices $(I = J, K = L)$, the relations [2.171] and [2.180] become:

$$\mathbf{B} \otimes \mathbf{A} = \mathbf{K}_{KI}(\mathbf{A} \otimes \mathbf{B})\mathbf{K}_{IK}$$ [2.181]

$$(\mathbf{B} \otimes \mathbf{A})\mathbf{K}_{KI} = \mathbf{K}_{KI}(\mathbf{A} \otimes \mathbf{B}).$$ [2.182]

Table 2.16 summarizes the formulae established for the vectorization of a transposed matrix and a Kronecker product, as well as for the permutation of factors of Kronecker products of vectors and matrices.

2.7.4. *Multiple Kronecker product and commutation matrices*

PROPOSITION 2.40.– The relations [2.170] and [2.171] can be generalized to the Kronecker product of three vectors $\mathbf{u} \in \mathbb{K}^I, \mathbf{v} \in \mathbb{K}^J, \mathbf{w} \in \mathbb{K}^K$ and of three matrices $\mathbf{A} \in \mathbb{K}^{I \times J}, \mathbf{B} \in \mathbb{K}^{K \times L}, \mathbf{C} \in \mathbb{K}^{M \times N}$. For example:

$$\mathbf{u} \otimes \mathbf{w} \otimes \mathbf{v} = \mathbf{e}_{ikj}^{ijk}(\mathbf{u} \otimes \mathbf{v} \otimes \mathbf{w}) = (\mathbf{I}_I \otimes \mathbf{e}_{kj}^{jk})(\mathbf{u} \otimes \mathbf{v} \otimes \mathbf{w})$$ [2.183]

$$\mathbf{w} \otimes \mathbf{v} \otimes \mathbf{u} = \mathbf{e}_{kji}^{ijk}(\mathbf{u} \otimes \mathbf{v} \otimes \mathbf{w}) = (\mathbf{e}_k^i \otimes \mathbf{I}_J \otimes \mathbf{e}_i^k)(\mathbf{u} \otimes \mathbf{v} \otimes \mathbf{w})$$ [2.184]

and

$$\mathbf{C} \otimes \mathbf{A} \otimes \mathbf{B} = \mathbf{e}_{mik}^{ikm}(\mathbf{A} \otimes \mathbf{B} \otimes \mathbf{C})\mathbf{e}_{jln}^{njl}$$ [2.185]

$$\mathbf{B} \otimes \mathbf{C} \otimes \mathbf{A} = \mathbf{e}_{kmi}^{ikm}(\mathbf{A} \otimes \mathbf{B} \otimes \mathbf{C})\mathbf{e}_{jln}^{lnj}$$ [2.186]

$$\mathbf{C} \otimes \mathbf{B} \otimes \mathbf{A} = (\mathbf{e}_m^i \otimes \mathbf{I}_K \otimes \mathbf{e}_i^m)(\mathbf{A} \otimes \mathbf{B} \otimes \mathbf{C})(\mathbf{e}_j^n \otimes \mathbf{I}_L \otimes \mathbf{e}_n^j).$$ [2.187]

EXAMPLE 2.41.– For $I = J = K = 2$, we have:

$$\mathbf{u} \otimes \mathbf{v} \otimes \mathbf{w} = [u_1 v_1 w_1, u_1 v_1 w_2, u_1 v_2 w_1, u_1 v_2 w_2,$$
$$u_2 v_1 w_1, u_2 v_1 w_2, u_2 v_2 w_1, u_2 v_2 w_2]^T$$

$$\mathbf{w} \otimes \mathbf{v} \otimes \mathbf{u} = [u_1 v_1 w_1, u_2 v_1 w_1, u_1 v_2 w_1, u_2 v_2 w_1,$$
$$u_1 v_1 w_2, u_2 v_1 w_2, u_1 v_2 w_2, u_2 v_2 w_2]^T$$
$$= \mathbf{e}_{kji}^{ijk}(\mathbf{u} \otimes \mathbf{v} \otimes \mathbf{w})$$ [2.188]

with the following row permutation matrix:

$$\mathbf{e}^{ijk}_{kji} = \mathbf{e}^i_k \otimes \mathbf{e}^j_j \otimes \mathbf{e}^k_i$$

$$= \begin{bmatrix} 1 & 0 & 0 & 0 & 0 & 0 & 0 & 0 \\ 0 & 0 & 0 & 0 & 1 & 0 & 0 & 0 \\ 0 & 0 & 1 & 0 & 0 & 0 & 0 & 0 \\ 0 & 0 & 0 & 0 & 0 & 0 & 1 & 0 \\ 0 & 1 & 0 & 0 & 0 & 0 & 0 & 0 \\ 0 & 0 & 0 & 0 & 0 & 1 & 0 & 0 \\ 0 & 0 & 0 & 1 & 0 & 0 & 0 & 0 \\ 0 & 0 & 0 & 0 & 0 & 0 & 0 & 1 \end{bmatrix}.$$

[2.189]

$$\mathbf{K}_{IJ} = \mathbf{e}^{ji}_{ij} = \mathbf{E}^{(I \times J)}_{ij} \otimes \mathbf{E}^{(J \times I)}_{ji}$$

$$\mathbf{K}_{JI} = \mathbf{K}^T_{IJ} = \mathbf{K}^{-1}_{IJ}$$

$$\mathbf{u} \in \mathbb{K}^I, \mathbf{v} \in \mathbb{K}^K$$

$$\mathbf{v} \otimes \mathbf{u} = \mathbf{K}_{KI}(\mathbf{u} \otimes \mathbf{v})$$

$$\mathbf{A} \in \mathbb{K}^{I \times J}, \mathbf{B} \in \mathbb{K}^{K \times L}$$

$$\text{vec}(\mathbf{A}^T) = \mathbf{K}_{IJ}\text{vec}(\mathbf{A})$$

$$\text{vec}(\mathbf{A} \otimes \mathbf{B}) = (\mathbf{I}_J \otimes \mathbf{K}_{LI} \otimes \mathbf{I}_K)(\text{vec}(\mathbf{A}) \otimes \text{vec}(\mathbf{B}))$$

$$\mathbf{B} \otimes \mathbf{A} = \mathbf{K}_{KI}(\mathbf{A} \otimes \mathbf{B})\mathbf{K}_{JL}$$

or

$$(\mathbf{B} \otimes \mathbf{A})\mathbf{K}_{LJ} = \mathbf{K}_{KI}(\mathbf{A} \otimes \mathbf{B})$$

Table 2.16. *Commutation matrices, vectorization,*
and Kronecker product

For example, the 1 in the fifth and seventh columns of the permutation matrix are obtained for the triplets $(i, j, k) = (2, 1, 1)$ and $(2, 2, 1)$, respectively, i.e. by computing:

$$\mathbf{e}^{211}_{112} = \mathbf{e}^2_1 \otimes \mathbf{e}^1_1 \otimes \mathbf{e}^1_2 = \begin{bmatrix} 0 & 1 \\ 0 & 0 \end{bmatrix} \otimes \begin{bmatrix} 1 & 0 \\ 0 & 0 \end{bmatrix} \otimes \begin{bmatrix} 0 & 0 \\ 1 & 0 \end{bmatrix}$$

$$\mathbf{e}^{221}_{122} = \mathbf{e}^2_1 \otimes \mathbf{e}^2_2 \otimes \mathbf{e}^1_2 = \begin{bmatrix} 0 & 1 \\ 0 & 0 \end{bmatrix} \otimes \begin{bmatrix} 0 & 0 \\ 0 & 1 \end{bmatrix} \otimes \begin{bmatrix} 0 & 0 \\ 1 & 0 \end{bmatrix}.$$

The formulae [2.183]–[2.187] can easily be generalized to arbitrary numbers of Kronecker products.

2.7.5. *Block Kronecker product*

Below, we illustrate another use of the index convention for the block Kronecker product of two partitioned vectors and of two partitioned matrices.

PROPOSITION 2.42.– For $\mathbf{u} \in \mathbb{K}^I$ and $\mathbf{v} \in \mathbb{K}^K$, partitioned into R blocks $\mathbf{u}^{(r)} \in \mathbb{K}^{I_r}$ and M blocks $\mathbf{v}^{(m)} \in \mathbb{K}^{K_m}$, respectively, with $r \in \langle R \rangle$, $m \in \langle M \rangle$ and $\sum_{r=1}^{R} I_r = I, \sum_{m=1}^{M} K_m = K$, the block Kronecker product $\mathbf{u} \otimes_b \mathbf{v}$ is defined as:

$$
\mathbf{u} \otimes_b \mathbf{v} = \begin{bmatrix} \mathbf{u}^{(1)} \otimes \mathbf{v}^{(1)} \\ \vdots \\ \mathbf{u}^{(1)} \otimes \mathbf{v}^{(M)} \\ \vdots \\ \mathbf{u}^{(R)} \otimes \mathbf{v}^{(1)} \\ \vdots \\ \mathbf{u}^{(R)} \otimes \mathbf{v}^{(M)} \end{bmatrix}.
$$

It can be written in terms of $\mathbf{u} \otimes \mathbf{v}$ using the following formula:

$$
\mathbf{u} \otimes_b \mathbf{v} = u_{i_r}^{(r)} v_{k_m}^{(m)} \mathbf{e}_{r,m,i_r,k_m}^{(RMI_rK_m)} \tag{2.190}
$$

$$
= \mathbf{e}_{r,m,i_r,k_m}^{r,i_r,m,k_m} (\mathbf{u} \otimes \mathbf{v}), \tag{2.191}
$$

where the commutation matrix is defined as in section 2.7.1.

PROOF.– For $\mathbf{u} = \begin{bmatrix} \mathbf{u}^{(1)} \\ \vdots \\ \mathbf{u}^{(R)} \end{bmatrix}$ and $\mathbf{v} = \begin{bmatrix} \mathbf{v}^{(1)} \\ \vdots \\ \mathbf{v}^{(M)} \end{bmatrix}$, we can write:

$$
\mathbf{u} = \mathbf{e}_r^{(R)} \otimes \mathbf{u}^{(r)} \ , \ \ \mathbf{v} = \mathbf{e}_m^{(M)} \otimes \mathbf{v}^{(m)}. \tag{2.192}
$$

Hence, the Kronecker product $\mathbf{u} \otimes \mathbf{v}$ can be expanded as follows:

$$
\begin{aligned}
\mathbf{u} \otimes \mathbf{v} &= (\mathbf{e}_r^{(R)} \otimes \mathbf{u}^{(r)}) \otimes (\mathbf{e}_m^{(M)} \otimes \mathbf{v}^{(m)}) \\
&= (\mathbf{e}_r^{(R)} \otimes u_{i_r}^{(r)} \mathbf{e}_{i_r}^{(I_r)}) \otimes (\mathbf{e}_m^{(M)} \otimes v_{k_m}^{(m)} \mathbf{e}_{k_m}^{(K_m)}) \\
&= u_{i_r}^{(r)} v_{k_m}^{(m)} \mathbf{e}_{r,i_r,m,k_m}^{(RI_rMK_m)}.
\end{aligned}
$$

With the index convention, we can expand the block Kronecker product into the Kronecker products of each pair of vectors $\mathbf{u}^{(r)} \otimes \mathbf{v}^{(m)}$, which gives:

$$
\begin{aligned}
\mathbf{u} \otimes_b \mathbf{v} &= \mathbf{e}_r^{(R)} \otimes \mathbf{e}_m^{(M)} \otimes \mathbf{u}^{(r)} \otimes \mathbf{v}^{(m)} \\
&= u_{i_r}^{(r)} v_{k_m}^{(m)} \mathbf{e}_r^{(R)} \otimes \mathbf{e}_m^{(M)} \otimes \mathbf{e}_{i_r}^{(I_r)} \otimes \mathbf{e}_{k_m}^{(K_m)} \\
&= u_{i_r}^{(r)} v_{k_m}^{(m)} \mathbf{e}_{r,m,i_r,k_m}^{(RMI_rK_m)} \\
&= \mathbf{e}_{r,m,i_r,k_m}^{r,i_r,m,k_m}(\mathbf{u} \otimes \mathbf{v}).
\end{aligned}
\qquad [2.193]
$$

This proves the relation [2.191]. $\square$

PROPOSITION 2.43.– For $\mathbf{A} \in \mathbb{K}^{I \times J}$ and $\mathbf{B} \in \mathbb{K}^{K \times L}$ partitioned into blocks $\mathbf{A}_{rs} \in \mathbb{K}^{I_r \times J_s}$, and $\mathbf{B}_{mn} \in \mathbb{K}^{K_m \times L_n}$, respectively, with $r \in \langle R \rangle$, $s \in \langle S \rangle$, $m \in \langle M \rangle$, $n \in \langle N \rangle$, and $\sum_{r=1}^R I_r = I, \sum_{s=1}^S J_s = J, \sum_{m=1}^M K_m = K, \sum_{n=1}^N L_n = L$, the block Kronecker product of $\mathbf{A}$ with $\mathbf{B}$, called the Tracy–Singh product (1972) and denoted $\mathbf{A} \otimes_b \mathbf{B}$, is defined as:

$$
\mathbf{A} \otimes_b \mathbf{B} = \begin{bmatrix}
\mathbf{A}_{11} \otimes \mathbf{B}_{11} & \cdots & \mathbf{A}_{11} \otimes \mathbf{B}_{1N} & \cdots & \mathbf{A}_{1S} \otimes \mathbf{B}_{11} & \cdots & \mathbf{A}_{1S} \otimes \mathbf{B}_{1N} \\
\vdots & \vdots & \vdots & \vdots & \vdots & \vdots & \vdots \\
\mathbf{A}_{11} \otimes \mathbf{B}_{M1} & \cdots & \mathbf{A}_{11} \otimes \mathbf{B}_{MN} & \cdots & \mathbf{A}_{1S} \otimes \mathbf{B}_{M1} & \cdots & \mathbf{A}_{1S} \otimes \mathbf{B}_{MN} \\
\vdots & \vdots & \vdots & \vdots & \vdots & \vdots & \vdots \\
\mathbf{A}_{R1} \otimes \mathbf{B}_{11} & \cdots & \mathbf{A}_{R1} \otimes \mathbf{B}_{1N} & \cdots & \mathbf{A}_{RS} \otimes \mathbf{B}_{11} & \cdots & \mathbf{A}_{RS} \otimes \mathbf{B}_{1N} \\
\vdots & \vdots & \vdots & \vdots & \vdots & \vdots & \vdots \\
\mathbf{A}_{R1} \otimes \mathbf{B}_{M1} & \cdots & \mathbf{A}_{R1} \otimes \mathbf{B}_{MN} & \cdots & \mathbf{A}_{RS} \otimes \mathbf{B}_{M1} & \cdots & \mathbf{A}_{RS} \otimes \mathbf{B}_{MN}
\end{bmatrix},
$$

of size $(\sum_{r=1}^R \sum_{m=1}^M I_r K_m, \sum_{s=1}^S \sum_{n=1}^N J_s L_n)$.

The block Kronecker product can be written concisely as:

$$
\mathbf{A} \otimes_b \mathbf{B} = [\mathbf{A}_{rs} \otimes \mathbf{B}]_{rs} = \left[[\mathbf{A}_{rs} \otimes \mathbf{B}_{mn}]_{mn} \right]_{rs},
\qquad [2.194]
$$

where $\mathbf{A}_{rs} \otimes \mathbf{B}$ is the (r,s)th sub-block of $\mathbf{A} \otimes_b \mathbf{B}$, of size $I_r K \times J_s L$, which itself admits the matrix $\mathbf{A}_{rs} \otimes \mathbf{B}_{mn}$, of size $I_r K_m \times J_s L_n$, as its (m,n)th sub-block. The block Kronecker product therefore corresponds to the Kronecker product of all pairs of blocks of the partitioned matrices $\mathbf{A}$ and $\mathbf{B}$.

It can be written compactly in terms of the Kronecker product $\mathbf{A} \otimes \mathbf{B}$ using the commutation matrix:

$$
\mathbf{A} \otimes_b \mathbf{B} = a_{i_r,j_s}^{(r,s)} b_{k_m,l_n}^{(m,n)} \mathbf{e}_{r,m,i_r,k_m}^{s,n,j_s,l_n}
\qquad [2.195]
$$

$$
= \mathbf{e}_{r,m,i_r,k_m}^{r,i_r,m,k_m}(\mathbf{A} \otimes \mathbf{B})\mathbf{e}_{s,j_s,n,l_n}^{s,n,j_s,l_n}.
\qquad [2.196]
$$

The relations [2.195] and [2.196] can be proven with the same approach as for the block Kronecker product of two vectors, with the following correspondences between rows and columns: $(r, m, i_r, k_m) \leftrightarrow (s, n, j_s, l_n)$.

We also define the block Kronecker product of two matrices $\mathbf{A} = [\mathbf{A}_1, \cdots, \mathbf{A}_K]$ and $\mathbf{B} = [\mathbf{B}_1, \cdots, \mathbf{B}_K]$ partitioned into K column blocks $\mathbf{A}_k \in \mathbb{K}^{I \times M}$ and $\mathbf{B}_k \in \mathbb{K}^{J \times N}$, with $k \in \langle K \rangle$, denoted $\mathbf{A} \otimes_b \mathbf{B}$, as follows:

$$\mathbf{A} \otimes_b \mathbf{B} = [\mathbf{A}_1 \otimes \mathbf{B}_1, \cdots, \mathbf{A}_K \otimes \mathbf{B}_K] \in \mathbb{K}^{IJ \times KMN}. \tag{2.197}$$

2.7.6. *Strong Kronecker product*

Another Kronecker product of partitioned matrices, called the strong Kronecker product and denoted $| \otimes |$, was introduced by de Launey and Seberry (1994) to generate orthogonal matrices from Hadamard matrices. This Kronecker product is also used to represent tensor train decompositions in the case of large-scale tensors (Lee and Cichocki 2017). Given the matrices $\mathbf{A}$ and $\mathbf{B}$ partitioned into (R, S) blocks $\mathbf{A}_{rs} \in \mathbb{C}^{I \times J}$, with $(r \in \langle R \rangle, s \in \langle S \rangle)$, and (S, N) blocks $\mathbf{B}_{sn} \in \mathbb{C}^{K \times L}$, with $n \in \langle N \rangle$, respectively, we define the strong Kronecker product $\mathbf{A}| \otimes |\mathbf{B}$ as the matrix $\mathbf{C}$ partitioned into (R, N) blocks $\mathbf{C}_{rn} \in \mathbb{C}^{IK \times JL}$ such that:

$$\mathbf{C}_{rn} = \sum_{s=1}^{S} \mathbf{A}_{rs} \otimes \mathbf{B}_{sn} , \ r \in \langle R \rangle, \ n \in \langle N \rangle.$$

This operation is fully determined by the parameters (R, S, N).

2.8. Relations between the $diag$ operator and the Kronecker product

PROPOSITION 2.44.– Given the vectors $\mathbf{u} \in \mathbb{C}^I$, $\mathbf{w} \in \mathbb{C}^J$, $\mathbf{x} \in \mathbb{C}^{IJ}$, we have:

$$\mathrm{diag}(\mathbf{u}) \otimes \mathrm{diag}(\mathbf{w}) = \mathrm{diag}(\mathbf{u} \otimes \mathbf{w}) \tag{2.198}$$

$$\left(\mathrm{diag}(\mathbf{u} \otimes \mathbf{w})\right)\mathbf{x} = \mathrm{diag}(\mathbf{x})(\mathbf{u} \otimes \mathbf{w}). \tag{2.199}$$

PROOF.–

– by the expression [2.12] of *diag* and using index notation, we have:
$$\mathrm{diag}(\mathbf{u}) \otimes \mathrm{diag}(\mathbf{w}) = u_i \mathbf{e}_i^i \otimes w_j \mathbf{e}_j^j = u_i w_j \mathbf{e}_{ij}^{ij} = \mathrm{diag}(\mathbf{u} \otimes \mathbf{w}).$$

– by the relations [2.198] and [2.13], we deduce that:
$$\left(\mathrm{diag}(\mathbf{u}) \otimes \mathrm{diag}(\mathbf{w})\right)\mathbf{x} = \left(\mathrm{diag}(\mathbf{u} \otimes \mathbf{w})\right)\mathbf{x} = \mathrm{diag}(\mathbf{x})(\mathbf{u} \otimes \mathbf{w}), \tag{2.200}$$

which completes the proof. $\square$

EXAMPLE 2.45.– For $I = J = 2$, we have:

$$\Big(\text{diag}(\mathbf{u}) \otimes \text{diag}(\mathbf{w})\Big)\mathbf{x} = \begin{bmatrix} u_1 w_1 & & & 0 \\ & u_1 w_2 & & \\ & & u_2 w_1 & \\ 0 & & & u_2 w_2 \end{bmatrix} \begin{bmatrix} x_1 \\ x_2 \\ x_3 \\ x_4 \end{bmatrix}$$

$$= \begin{bmatrix} u_1 w_1 x_1 & & & 0 \\ & u_1 w_2 x_2 & & \\ & & u_2 w_1 x_3 & \\ 0 & & & u_2 w_2 x_4 \end{bmatrix}$$

$$= \text{diag}(\mathbf{x})(\mathbf{u} \otimes \mathbf{w}).$$

2.9. Khatri–Rao product

The Khatri–Rao product of two matrices with the same number of columns is identical to the column-wise Kronecker product of these matrices. As we will see in Chapter 5, this product plays an important role in the matricization of the PARAFAC decomposition of a tensor. After introducing the definition of the simple and multiple Khatri–Rao products of matrices, we will present a few identities that it satisfies, then we will show that it can be used to express the trace of a product of matrices.

2.9.1. *Definition*

Let $\mathbf{A} \in \mathbb{K}^{I \times J}$ and $\mathbf{B} \in \mathbb{K}^{K \times J}$ be two matrices with the same number of columns. The Khatri–Rao product of $\mathbf{A}$ with $\mathbf{B}$ is the matrix denoted $\mathbf{A} \diamond \mathbf{B} \in \mathbb{K}^{IK \times J}$ and defined as (Khatri and Rao 1968, 1972):

$$\mathbf{A} \diamond \mathbf{B} = \begin{bmatrix} \mathbf{A}_{.1} \otimes \mathbf{B}_{.1} & , & \mathbf{A}_{.2} \otimes \mathbf{B}_{.2} & , & \cdots & , & \mathbf{A}_{.J} \otimes \mathbf{B}_{.J} \end{bmatrix}. \quad [2.201]$$

We say that $\mathbf{A} \diamond \mathbf{B}$ is the column-wise Kronecker product of $\mathbf{A}$ and $\mathbf{B}$. It is a matrix partitioned into J column blocks, where the jth block is equal to the Kronecker product of the jth column vector of $\mathbf{A}$ with the jth column vector of $\mathbf{B}$.

PROPOSITION 2.46.– The Khatri–Rao product can also be written as a matrix partitioned into I row blocks:

$$\mathbf{A} \diamond \mathbf{B} = \begin{bmatrix} \mathbf{BD}_1(\mathbf{A}) \\ \mathbf{BD}_2(\mathbf{A}) \\ \vdots \\ \mathbf{BD}_I(\mathbf{A}) \end{bmatrix}, \quad [2.202]$$

where $\mathbf{D}_i(\mathbf{A}) = \text{diag}(a_{i1}, a_{i2}, \cdots, a_{iJ})$ denotes the diagonal matrix with the elements of the ith row of $\mathbf{A}$ along the diagonal.

PROOF.– By the definition of the Khatri–Rao product, the ith row block can be written as:

$$\left[a_{i1}\mathbf{B}_{.1},\cdots,a_{iJ}\mathbf{B}_{.J}\right] = \left[\mathbf{B}_{.1},\cdots,\mathbf{B}_{.J}\right]\mathrm{diag}(a_{i1},\cdots,a_{iJ})$$

$$= \mathbf{B}D_i(\mathbf{A}) \in \mathbb{C}^{K\times J}. \qquad [2.203]$$

By stacking the row blocks $i \in \langle I\rangle$, we deduce the expression [2.202]. $\qquad \square$

EXAMPLE 2.47.– For $\mathbf{A} = \begin{pmatrix} a_{11} & a_{12} \\ a_{21} & a_{22} \end{pmatrix}, \mathbf{B} = \begin{pmatrix} b_{11} & b_{12} \\ b_{21} & b_{22} \end{pmatrix}$, we have:

$$\mathbf{A}\diamond\mathbf{B} = \begin{pmatrix} \mathbf{A}_{.1}\otimes\mathbf{B}_{.1} & \mathbf{A}_{.2}\otimes\mathbf{B}_{.2} \end{pmatrix} = \begin{pmatrix} a_{11}b_{11} & a_{12}b_{12} \\ a_{11}b_{21} & a_{12}b_{22} \\ \cdots\cdots & \cdots\cdots \\ a_{21}b_{11} & a_{22}b_{12} \\ a_{21}b_{21} & a_{22}b_{22} \end{pmatrix}$$

or alternatively:

$$\mathbf{A}\diamond\mathbf{B} = \begin{pmatrix} \mathbf{B}\begin{pmatrix} a_{11} & 0 \\ 0 & a_{12} \end{pmatrix} \\ \cdots\cdots\cdots\cdots \\ \mathbf{B}\begin{pmatrix} a_{21} & 0 \\ 0 & a_{22} \end{pmatrix} \end{pmatrix} = \begin{pmatrix} \mathbf{B}D_1(\mathbf{A}) \\ \cdots \\ \mathbf{B}D_2(\mathbf{A}) \end{pmatrix}.$$

2.9.2. Khatri–Rao product and index convention

Table 2.17 presents three examples of the Khatri–Rao product expressed using the index convention. These formulae can be viewed as simplified versions of the formulae involving the Kronecker product listed in Table 2.15.

$\mathbf{A} \in \mathbb{K}^{I\times R}, \mathbf{B} \in \mathbb{K}^{K\times R}$, and $\mathbf{C} \in \mathbb{K}^{R\times N}$
$\mathbf{A}\diamond\mathbf{B} = a_{ir}b_{kr}\mathbf{e}_{ik}^{r}$
$(\mathbf{A}\diamond\mathbf{B})^T = a_{ir}b_{kr}\mathbf{e}_{r}^{ik}$
$\mathbf{A}\diamond\mathbf{C}^T = a_{ir}\mathbf{e}_i^r \diamond c_{rn}\mathbf{e}_n^r = a_{ir}c_{rn}\mathbf{e}_{in}^r$

Table 2.17. Khatri-Rao product and index convention

2.9.3. *Multiple Khatri–Rao product*

Given the vectors $\mathbf{u}^{(n)} \in \mathbb{K}^{I_n}$ and the matrices $\mathbf{A}^{(n)} \in \mathbb{K}^{I_n \times R}$, with $n \in \langle N \rangle$, we define the multiple Khatri–Rao product as:

$$\underset{n=1}{\overset{N}{\diamond}} \mathbf{u}^{(n)} \triangleq \mathbf{u}^{(1)} \diamond \mathbf{u}^{(2)} \diamond \cdots \diamond \mathbf{u}^{(N)} = \underset{n=1}{\overset{N}{\otimes}} \mathbf{u}^{(n)} \in \mathbb{K}^{I_1 \cdots I_N} \qquad [2.204]$$

$$\underset{n=1}{\overset{N}{\diamond}} \mathbf{A}^{(n)} \triangleq \mathbf{A}^{(1)} \diamond \mathbf{A}^{(2)} \diamond \cdots \diamond \mathbf{A}^{(N)} \in \mathbb{K}^{I_1 \cdots I_N \times R}. \qquad [2.205]$$

Using the index convention, and writing $a_{i_n,r}^{(n)}$ for the current element of $\mathbf{A}^{(n)}$, the above equations can be rewritten compactly as:

$$\underset{n=1}{\overset{N}{\diamond}} \mathbf{u}^{(n)} = \Big(\prod_{n=1}^{N} u_{i_n}^{(n)} \Big) \mathbf{e}_{\mathbb{I}} \qquad [2.206]$$

$$\underset{n=1}{\overset{N}{\diamond}} \mathbf{A}^{(n)} = \Big(\prod_{n=1}^{N} a_{i_n,r}^{(n)} \Big) \mathbf{e}_{\mathbb{I}}^{r}, \qquad [2.207]$$

where $\mathbb{I} = i_1 \cdots i_N$, $\mathbf{e}_{\mathbb{I}} = \underset{n=1}{\overset{N}{\otimes}} \mathbf{e}_{i_n}^{(I_n)}$ and $\mathbf{e}_{\mathbb{I}}^{r} = \big(\underset{n=1}{\overset{N}{\otimes}} \mathbf{e}_{i_n}^{(I_n)} \big) \otimes (\mathbf{e}_r^{(R)})^T$.

2.9.4. *Properties*

The Khatri–Rao product satisfies the following properties:

– For $\mathbf{A} \in \mathbb{K}^{I \times J}, \mathbf{B} \in \mathbb{K}^{M \times J}$, and $\forall \alpha \in \mathbb{K}$:

$$(\alpha \mathbf{A}) \diamond \mathbf{B} = \mathbf{A} \diamond (\alpha \mathbf{B}) = \alpha(\mathbf{A} \diamond \mathbf{B}). \qquad [2.208]$$

– Associativity: For $\mathbf{A} \in \mathbb{K}^{I \times J}, \mathbf{B} \in \mathbb{K}^{K \times J}, \mathbf{C} \in \mathbb{K}^{M \times J}$:

$$\mathbf{A} \diamond (\mathbf{B} \diamond \mathbf{C}) = (\mathbf{A} \diamond \mathbf{B}) \diamond \mathbf{C} = \mathbf{A} \diamond \mathbf{B} \diamond \mathbf{C}. \qquad [2.209]$$

– Distributivity with respect to addition: For $\mathbf{A}, \mathbf{B} \in \mathbb{K}^{I \times J}, \mathbf{C} \in \mathbb{K}^{K \times J}$:

$$(\mathbf{A} + \mathbf{B}) \diamond \mathbf{C} = \mathbf{A} \diamond \mathbf{C} + \mathbf{B} \diamond \mathbf{C} \qquad [2.210]$$

$$\mathbf{C} \diamond (\mathbf{A} + \mathbf{B}) = \mathbf{C} \diamond \mathbf{A} + \mathbf{C} \diamond \mathbf{B}. \qquad [2.211]$$

– Transconjugation: For $\mathbf{A} \in \mathbb{K}^{I \times J}, \mathbf{B} \in \mathbb{K}^{K \times J}$:

$$(\mathbf{A} \diamond \mathbf{B})^H = \big[\, \mathbf{D}_1^*(\mathbf{A})\mathbf{B}^H, \quad \cdots, \quad \mathbf{D}_I^*(\mathbf{A})\mathbf{B}^H \, \big] \in \mathbb{K}^{J \times IK} \qquad [2.212]$$

$$\mathbf{D}_i^*(\mathbf{A}) = \mathrm{diag}(a_{i1}^*, \cdots, a_{iJ}^*). \qquad [2.213]$$

PROOF.– The properties [2.208]–[2.210] are obvious, whereas [2.212] can be deduced by transconjugating the right-hand side of equation [2.202]. □

– For $\mathbf{A} \in \mathbb{K}^{I \times J}, \mathbf{B} \in \mathbb{K}^{K \times J}$, we have:

$$\mathrm{r}(\mathbf{A} \diamond \mathbf{B}) \geq \max\big(\mathrm{r}(\mathbf{A}), \mathrm{r}(\mathbf{B})\big). \qquad [2.214]$$

In the case where $\mathbf{A}$ and $\mathbf{B}$ have full column rank, $\mathbf{A} \diamond \mathbf{B}$ also has full column rank, as shown in section 2.11.

Like the Kronecker product, the Khatri–Rao product is not commutative in general. However, for $\mathbf{a} \in \mathbb{K}^J$ and $\mathbf{B} \in \mathbb{K}^{I \times J}$, we have:

$$\mathbf{a}^T \diamond \mathbf{B} = \mathbf{B} \diamond \mathbf{a}^T = \mathbf{B} \operatorname{diag}(\mathbf{a}) \in \mathbb{K}^{I \times J}. \qquad [2.215]$$

2.9.5. *Identities*

PROPOSITION 2.48.– The Khatri–Rao product $\mathbf{A} \diamond \mathbf{B}$, where $\mathbf{A} \in \mathbb{K}^{I \times R}, \mathbf{B} \in \mathbb{K}^{K \times R}$, satisfies the following identity:

$$\mathbf{A} \diamond \mathbf{B} = (\mathbf{A} \otimes \mathbf{1}_K) \odot (\mathbf{1}_I \otimes \mathbf{B}). \qquad [2.216]$$

PROOF.– Noting that $\mathbf{1}_K = \delta_{kk}\mathbf{e}_k$, and using the first relation in Table 2.17, we have:

$$(\mathbf{A} \otimes \mathbf{1}_K) \odot (\mathbf{1}_I \otimes \mathbf{B}) = a_{ir}\delta_{kk}\mathbf{e}_{ik}^r \odot b_{kr}\delta_{ii}\mathbf{e}_{ik}^r = a_{ir}b_{kr}\mathbf{e}_{ik}^r$$
$$= \mathbf{A} \diamond \mathbf{B},$$

which proves [2.216]. □

By choosing $\mathbf{A} = \mathbf{I}_I \otimes \mathbf{1}_K^T$ and $\mathbf{B} = \mathbf{1}_I^T \otimes \mathbf{I}_K$ with $R = IK$, in the relation [2.216], we obtain the following identity:

$$(\mathbf{I}_I \otimes \mathbf{1}_K^T) \diamond (\mathbf{1}_I^T \otimes \mathbf{I}_K) = (\mathbf{I}_I \otimes \mathbf{1}_K^T \otimes \mathbf{1}_K) \odot (\mathbf{1}_I \otimes \mathbf{1}_I^T \otimes \mathbf{I}_K)$$
$$= (\mathbf{I}_I \otimes \mathbf{1}_{K \times K}) \odot (\mathbf{1}_{I \times I} \otimes \mathbf{I}_K) = \mathbf{I}_{IK}. \qquad [2.217]$$

EXAMPLE 2.49.– For $I = K = R = 2$, we have:

$$(\mathbf{A} \otimes \mathbf{1}_2) \odot (\mathbf{1}_2 \otimes \mathbf{B}) = \begin{bmatrix} a_{11} & a_{12} \\ a_{11} & a_{12} \\ a_{21} & a_{22} \\ a_{21} & a_{22} \end{bmatrix} \odot \begin{bmatrix} b_{11} & b_{12} \\ b_{21} & b_{22} \\ b_{11} & b_{12} \\ b_{21} & b_{22} \end{bmatrix}$$

$$= \begin{bmatrix} a_{11}b_{11} & a_{12}b_{12} \\ a_{11}b_{21} & a_{12}b_{22} \\ a_{21}b_{11} & a_{22}b_{12} \\ a_{21}b_{21} & a_{22}b_{22} \end{bmatrix} = \mathbf{A} \diamond \mathbf{B}.$$

2.9.6. *Khatri–Rao product and permutation of factors*

From [2.170] and [2.171], and with $\mathbf{K}_{KI}$ as defined in [2.173], we can deduce that, for $\mathbf{u} \in \mathbb{K}^I, \mathbf{v} \in \mathbb{K}^K, \mathbf{A} \in \mathbb{K}^{I \times J}$, and $\mathbf{B} \in \mathbb{K}^{K \times J}$:

$$\mathbf{v} \diamond \mathbf{u} = \mathbf{K}_{KI}(\mathbf{u} \diamond \mathbf{v}) \tag{2.218}$$

$$\mathbf{B} \diamond \mathbf{A} = \mathbf{K}_{KI}(\mathbf{A} \diamond \mathbf{B}). \tag{2.219}$$

Similarly, for $\mathbf{A} \in \mathbb{K}^{I \times J}, \mathbf{B} \in \mathbb{K}^{K \times J}, \mathbf{C} \in \mathbb{K}^{M \times J}$, the formulae [2.185]–[2.187] become:

$$\mathbf{C} \diamond \mathbf{A} \diamond \mathbf{B} = \mathbf{e}_{mik}^{ikm}(\mathbf{A} \diamond \mathbf{B} \diamond \mathbf{C}) \tag{2.220}$$

$$\mathbf{B} \diamond \mathbf{C} \diamond \mathbf{A} = \mathbf{e}_{kmi}^{ikm}(\mathbf{A} \diamond \mathbf{B} \diamond \mathbf{C}) \tag{2.221}$$

$$\mathbf{C} \diamond \mathbf{B} \diamond \mathbf{A} = (\mathbf{e}_m^i \otimes \mathbf{I}_K \otimes \mathbf{e}_i^m)(\mathbf{A} \diamond \mathbf{B} \diamond \mathbf{C}). \tag{2.222}$$

In the case of a multiple Khatri–Rao product $\overset{N}{\underset{n=1}{\diamond}} \mathbf{u}^{(n)}$ of N vectors $\mathbf{u}^{(n)} \in \mathbb{K}^{(I_n)}$, we can permute $\mathbf{u}^{(p)}$ into the first position of the Khatri–Rao product, with $p \in \{2, \cdots, N\}$, using the following formula:

$$\mathbf{u}^{(p)} \diamond \left(\overset{p-1}{\underset{n=1}{\diamond}} \mathbf{u}^{(n)} \right) \diamond \left(\overset{N}{\underset{n=p+1}{\diamond}} \mathbf{u}^{(n)} \right) = \mathbf{\Pi} \left(\overset{N}{\underset{n=1}{\diamond}} \mathbf{u}^{(n)} \right) \tag{2.223}$$

with the permutation matrix:

$$\mathbf{\Pi} = \mathbf{e}_{i_p i_1 \cdots i_{p-1}}^{i_1 \cdots i_p} \otimes \mathbf{I}_{I_{p+1} \cdots I_N} \tag{2.224}$$

$$= \mathbf{e}_{i_p}^{(I_p)} \otimes \left(\overset{p-1}{\underset{n=1}{\otimes}} \mathbf{e}_{i_n}^{(I_n)} \right) \left(\overset{p}{\underset{n=1}{\otimes}} \mathbf{e}_{i_n}^{(I_n)} \right)^T \otimes \mathbf{I}_{I_{p+1} \cdots I_N}, \tag{2.225}$$

where $\mathbf{I}_{I_{p+1} \cdots I_N}$ is the identity matrix of order $I_{p+1} \cdots I_N$.

EXAMPLE 2.50.– For $\mathbf{u} \in \mathbb{K}^I, \mathbf{v} \in \mathbb{K}^J$ and $\mathbf{w} \in \mathbb{K}^K$, with $I = J = K = 2$, we have:

$$\mathbf{v} \diamond \mathbf{u} \diamond \mathbf{w} = \mathbf{e}_{jik}^{ijk}(\mathbf{u} \diamond \mathbf{v} \diamond \mathbf{w}) = (\mathbf{e}_{ji}^{ij} \otimes \mathbf{I}_2)(\mathbf{u} \diamond \mathbf{v} \diamond \mathbf{w}),$$

with:

$$\mathbf{e}_{ji}^{ij} \otimes \mathbf{I}_2 = \begin{bmatrix} 1 & 0 & 0 & 0 \\ 0 & 0 & 1 & 0 \\ 0 & 1 & 0 & 0 \\ 0 & 0 & 0 & 1 \end{bmatrix} \otimes \begin{bmatrix} 1 & 0 \\ 0 & 1 \end{bmatrix}$$

$$= \begin{bmatrix} \mathbf{I}_2 & \mathbf{0}_2 & \mathbf{0}_2 & \mathbf{0}_2 \\ \mathbf{0}_2 & \mathbf{0}_2 & \mathbf{I}_2 & \mathbf{0}_2 \\ \mathbf{0}_2 & \mathbf{I}_2 & \mathbf{0}_2 & \mathbf{0}_2 \\ \mathbf{0}_2 & \mathbf{0}_2 & \mathbf{0}_2 & \mathbf{I}_2 \end{bmatrix},$$

which gives:

$$\mathbf{e}_{ji}^{ij} \otimes \mathbf{I}_2 = \begin{bmatrix} 1 & 0 & 0 & 0 & 0 & 0 & 0 & 0 \\ 0 & 1 & 0 & 0 & 0 & 0 & 0 & 0 \\ 0 & 0 & 0 & 0 & 1 & 0 & 0 & 0 \\ 0 & 0 & 0 & 0 & 0 & 1 & 0 & 0 \\ 0 & 0 & 1 & 0 & 0 & 0 & 0 & 0 \\ 0 & 0 & 0 & 1 & 0 & 0 & 0 & 0 \\ 0 & 0 & 0 & 0 & 0 & 0 & 1 & 0 \\ 0 & 0 & 0 & 0 & 0 & 0 & 0 & 1 \end{bmatrix}.$$

It is now easy to check that:

$$\begin{aligned} \mathbf{v} \diamond \mathbf{u} \diamond \mathbf{w} &= [u_1 v_1 w_1, u_1 v_1 w_2, u_2 v_1 w_1, u_2 v_1 w_2, \\ &\quad u_1 v_2 w_1, u_1 v_2 w_2, u_2 v_2 w_1, u_2 v_2 w_2]^T \\ &= (\mathbf{e}_{ji}^{ij} \otimes \mathbf{I}_2)(\mathbf{u} \otimes \mathbf{v} \otimes \mathbf{w}) \\ &= (\mathbf{e}_{ji}^{ij} \otimes \mathbf{I}_2)[u_1 v_1 w_1, u_1 v_1 w_2, u_1 v_2 w_1, u_1 v_2 w_2, \\ &\quad u_2 v_1 w_1, u_2 v_1 w_2, u_2 v_2 w_1, u_2 v_2 w_2]^T. \end{aligned}$$

2.9.7. *Trace of a product of matrices and Khatri–Rao product*

PROPOSITION 2.51.– For $\mathbf{A} \in \mathbb{K}^{I \times J}, \mathbf{B} \in \mathbb{K}^{J \times M}, \mathbf{C} \in \mathbb{K}^{M \times I}$, we have:

$$\text{tr}(\mathbf{ABC}) = \text{vec}^T(\mathbf{A})(\mathbf{B} \diamond \mathbf{C}^T)\mathbf{1}_M. \qquad [2.226]$$

PROOF.– Analogously to [2.150], we have:

$$\text{tr}(\mathbf{ABC}) = a_{ij} b_{jm} c_{mi}. \qquad [2.227]$$

By the relations [2.72] and [2.105], the expression of $\mathbf{B} \diamond \mathbf{C}^T$ deduced from Table 2.17, and after introducing the term $\mathbf{e}^{ji} \mathbf{e}_{j'i'}^{m'} \mathbf{e}_m = \delta_{ii'} \delta_{jj'} \delta_{mm'}$, we obtain:

$$\text{tr}(\mathbf{ABC}) = (a_{ij} \mathbf{e}^{ji})(b_{j'm'} c_{m'i'} \mathbf{e}_{j'i'}^{m'})(\delta_{mm} \mathbf{e}_m) = \text{vec}^T(\mathbf{A})(\mathbf{B} \diamond \mathbf{C}^T)\mathbf{1}_M,$$

which corresponds to the relation [2.226]. □

Comparing the formulae [2.226] and [2.157], which express $\text{tr}(\mathbf{ABC})$ in terms of the Khatri–Rao and Kronecker products, respectively, we deduce:

$$(\mathbf{B} \diamond \mathbf{C}^T)\mathbf{1}_M = (\mathbf{B} \otimes \mathbf{I}_I)\text{vec}(\mathbf{C}^T). \qquad [2.228]$$

2.10. Relations between vectorization and Kronecker and Khatri–Rao products

The following proposition presents two expressions for the vectorized form of the matrix product $\mathbf{U}\mathbf{V}^T$ in terms of the Kronecker and Khatri–Rao products of the factor matrices $\mathbf{U}$ and $\mathbf{V}$.

PROPOSITION 2.52.– Consider the product $\mathbf{U}\mathbf{V}^T \in \mathbb{K}^{I \times J}$ with $\mathbf{U} = [\mathbf{u}_1 \cdots \mathbf{u}_R] \in \mathbb{K}^{I \times R}$ and $\mathbf{V} = [\mathbf{v}_1 \cdots \mathbf{v}_R] \in \mathbb{K}^{J \times R}$. We have:

$$\text{vec}(\mathbf{U}\mathbf{V}^T) = (\mathbf{V} \otimes \mathbf{U})\text{vec}(\mathbf{I}_R) \qquad [2.229]$$

$$= (\mathbf{V} \diamond \mathbf{U})\mathbf{1}_R, \qquad [2.230]$$

where $\mathbf{I}_R$ and $\mathbf{1}_R$, respectively, denote the identity matrix of order R and the vector of size R composed of ones.

PROOF.– The relation [2.229] can be deduced directly from [2.113] with $(\mathbf{A}, \mathbf{B}, \mathbf{C}) = (\mathbf{U}, \mathbf{I}_R, \mathbf{V}^T)$. Moreover, using the relation [2.28], we have:

$$\text{vec}(\mathbf{U}\mathbf{V}^T) = \text{vec}\left(\sum_{r=1}^{R} \mathbf{u}_r \mathbf{v}_r^T\right) = \sum_{r=1}^{R} \text{vec}(\mathbf{u}_r \mathbf{v}_r^T) = \sum_{r=1}^{R} \mathbf{v}_r \diamond \mathbf{u}_r$$

$$= (\mathbf{V} \diamond \mathbf{U})\mathbf{1}_R,$$

which proves [2.230]. $\qquad\qquad\qquad\qquad\qquad\qquad\qquad\qquad\qquad\qquad\qquad\quad \square$

2.11. Relations between the Kronecker, Khatri–Rao and Hadamard products

There is a certain hierarchy of complexity between the Hadamard, Khatri–Rao, Kronecker and block Kronecker products.

– The Hadamard product of $\mathbf{A} \in \mathbb{K}^{I \times J}$ with $\mathbf{B} \in \mathbb{K}^{I \times J}$ consists of a subset of rows of their Khatri–Rao product:

$$\mathbf{A} \odot \mathbf{B} = \mathbf{S}_I^T(\mathbf{A} \diamond \mathbf{B}), \qquad [2.231]$$

where $\mathbf{S}_I^T$ is a row selection matrix defined as:

$$\mathbf{S}_I^T = \sum_{i=1}^{I} \mathbf{e}_i^{(I)}\left(\mathbf{e}_i^{(I)} \otimes \mathbf{e}_i^{(I)}\right)^T$$

$$= \mathbf{e}_i^{ii} \text{ (with the index convention)} \tag{2.232}$$

$$= \left[\mathbf{e}_1^{(I^2)}, \mathbf{e}_{I+2}^{(I^2)}, \mathbf{e}_{2I+3}^{(I^2)}, \cdots, \mathbf{e}_{I^2}^{(I^2)} \right]^T \in \mathbb{R}^{I \times I^2}. \tag{2.233}$$

PROOF.– Noting that the ith row of $\mathbf{A} \odot \mathbf{B}$, for $i \in \langle I \rangle$, is given by:

$$(\mathbf{A} \odot \mathbf{B})_{i.} = [a_{i1}b_{i1}, a_{i2}b_{i2}, \cdots, a_{iJ}b_{iJ}]$$

$$= \left(\mathbf{e}_i^{(I)} \otimes \mathbf{e}_i^{(I)} \right)^T (\mathbf{A} \diamond \mathbf{B}) = \left(\mathbf{e}_{ii}^{(I^2)} \right)^T (\mathbf{A} \diamond \mathbf{B}),$$

the matrix $\mathbf{A} \odot \mathbf{B}$ is obtained by varying i from 1 to I:

$$\mathbf{A} \odot \mathbf{B} = \sum_{i=1}^{I} \mathbf{e}_i^{(I)} (\mathbf{A} \odot \mathbf{B})_{i.} = \sum_{i=1}^{I} \mathbf{e}_i^{(I)} \left(\mathbf{e}_{ii}^{(I^2)} \right)^T (\mathbf{A} \diamond \mathbf{B}) = \mathbf{e}_i^{ii} (\mathbf{A} \diamond \mathbf{B})$$

$$= \left[\mathbf{e}_1^{(I)} \left(\mathbf{e}_1^{(I^2)} \right)^T + \mathbf{e}_2^{(I)} \left(\mathbf{e}_{I+2}^{(I^2)} \right)^T + \cdots + \left(\mathbf{e}_{I^2}^{(I^2)} \right)^T \right] (\mathbf{A} \diamond \mathbf{B}) \tag{2.234}$$

or alternatively:

$$\mathbf{A} \odot \mathbf{B} = \begin{bmatrix} (\mathbf{e}_1^{(I^2)})^T \\ (\mathbf{e}_{I+2}^{(I^2)})^T \\ \vdots \\ (\mathbf{e}_{I^2}^{(I^2)})^T \end{bmatrix} (\mathbf{A} \diamond \mathbf{B}) = \mathbf{S}_I^T (\mathbf{A} \diamond \mathbf{B}), \tag{2.235}$$

where the row selection matrix $\mathbf{S}_I^T$ is defined as in [2.233]. This proves [2.231].   □

– The Khatri–Rao product of $\mathbf{A} \in \mathbb{K}^{I \times J}$ with $\mathbf{B} \in \mathbb{K}^{K \times J}$ consists of a subset of columns of their Kronecker product:

$$\mathbf{A} \diamond \mathbf{B} = (\mathbf{A} \otimes \mathbf{B}) \mathbf{S}_J, \tag{2.236}$$

where $\mathbf{S}_J$ is a column selection matrix defined as:

$$\mathbf{S}_J = \left[\mathbf{e}_1^{(J^2)}, \mathbf{e}_{J+2}^{(J^2)}, \mathbf{e}_{2J+3}^{(J^2)}, \cdots, \mathbf{e}_{J^2}^{(J^2)} \right] \in \mathbb{R}^{J^2 \times J} \tag{2.237}$$

$$= \mathbf{e}_{jj}^j \text{ (with the index convention).} \tag{2.238}$$

This relation can be shown by applying the same reasoning to the columns of $\mathbf{A} \otimes \mathbf{B}$ as used earlier for the rows of $\mathbf{A} \diamond \mathbf{B}$ to establish the relation [2.231].

REMARK 2.53.– If $\mathbf{A}$ and $\mathbf{B}$ have full column rank, which implies $J \leq \min(I, K)$, then $\mathbf{A} \otimes \mathbf{B}$ is itself a matrix with full column rank (see the rank property of the

Kronecker product in Table 2.11). Hence, by [2.236], we can deduce that $\mathbf{A} \diamond \mathbf{B}$ has the same rank as $\mathbf{S}_J$, which is therefore equal to J.

This proves that, if $\mathbf{A}$ and $\mathbf{B}$ have full column rank, then their Khatri–Rao product also has full column rank, like their Kronecker product.

– By combining the relations [2.231] and [2.236], we deduce the following relation between the Kronecker and Hadamard products of two matrices $\mathbf{A}, \mathbf{B} \in \mathbb{K}^{I \times J}$, of same size (Marcus and Khan 1959; Lev Ari 2005):

$$\mathbf{A} \odot \mathbf{B} = \mathbf{S}_I^T (\mathbf{A} \otimes \mathbf{B}) \mathbf{S}_J \qquad [2.239]$$

with the row and column selection matrices defined in [2.233] and [2.237].

Table 2.18 summarizes the relations between the three products considered in this chapter.

$\mathbf{A}, \mathbf{B} \in \mathbb{K}^{I \times J}$
$\mathbf{A} \odot \mathbf{B} = \mathbf{S}_I^T (\mathbf{A} \diamond \mathbf{B})$ $\mathbf{S}_I = \left[\mathbf{e}_1^{(I^2)}, \mathbf{e}_{I+2}^{(I^2)}, \mathbf{e}_{2I+3}^{(I^2)}, \cdots, \mathbf{e}_{I^2}^{(I^2)} \right]$
$\mathbf{A} \in \mathbb{K}^{I \times J} , \ \mathbf{B} \in \mathbb{K}^{K \times J}$
$\mathbf{A} \diamond \mathbf{B} = (\mathbf{A} \otimes \mathbf{B}) \mathbf{S}_J$ $\mathbf{S}_J = \left[\mathbf{e}_1^{(J^2)}, \mathbf{e}_{J+2}^{(J^2)}, \mathbf{e}_{2J+3}^{(J^2)}, \cdots, \mathbf{e}_{J^2}^{(J^2)} \right]$
$\mathbf{A}, \mathbf{B} \in \mathbb{K}^{I \times J}$
$\mathbf{A} \odot \mathbf{B} = \mathbf{S}_I^T (\mathbf{A} \otimes \mathbf{B}) \mathbf{S}_J$

Table 2.18. *Relations between the Hadamard, Khatri-Rao, and Kronecker products*

EXAMPLE 2.54.– For $\mathbf{A}, \mathbf{B} \in \mathbb{K}^{2 \times 2}$, after transposition, the definition [2.232] becomes:

$$\mathbf{S}_2 = \sum_{j=1}^{2} \left(\mathbf{e}_j^{(2)} \otimes \mathbf{e}_j^{(2)} \right) (\mathbf{e}_j^{(2)})^T = \begin{bmatrix} 1 & 0 \\ 0 & 0 \\ 0 & 0 \\ 0 & 1 \end{bmatrix}$$

and

$$\mathbf{A} \otimes \mathbf{B} = \begin{bmatrix} a_{11}b_{11} & a_{11}b_{12} & a_{12}b_{11} & a_{12}b_{12} \\ a_{11}b_{21} & a_{11}b_{22} & a_{12}b_{21} & a_{12}b_{22} \\ a_{21}b_{11} & a_{21}b_{12} & a_{22}b_{11} & a_{22}b_{12} \\ a_{21}b_{21} & a_{21}b_{22} & a_{22}b_{21} & a_{22}b_{22} \end{bmatrix}$$

$$\mathbf{A} \diamond \mathbf{B} = \begin{bmatrix} a_{11}b_{11} & a_{12}b_{12} \\ a_{11}b_{21} & a_{12}b_{22} \\ a_{21}b_{11} & a_{22}b_{12} \\ a_{21}b_{21} & a_{22}b_{22} \end{bmatrix} = (\mathbf{A} \otimes \mathbf{B})\mathbf{S}_2$$

$$\mathbf{A} \odot \mathbf{B} = \begin{bmatrix} a_{11}b_{11} & a_{12}b_{12} \\ a_{21}b_{21} & a_{22}b_{22} \end{bmatrix} = \mathbf{S}_2^T(\mathbf{A} \diamond \mathbf{B}) = \mathbf{S}_2^T(\mathbf{A} \otimes \mathbf{B})\mathbf{S}_2.$$

Below, we present a few identities relating the Hadamard and Kronecker products of matrices, and then of vectors.

– Let $\mathbf{A}, \mathbf{C} \in \mathbb{K}^{I \times J}$, and $\mathbf{B}, \mathbf{D} \in \mathbb{K}^{K \times L}$. We have the identity:

$$(\mathbf{A} \otimes \mathbf{B}) \odot (\mathbf{C} \otimes \mathbf{D}) = (\mathbf{A} \odot \mathbf{C}) \otimes (\mathbf{B} \odot \mathbf{D}). \qquad [2.240]$$

This identity is easy to prove using the index convention:

$$\begin{aligned}(\mathbf{A} \otimes \mathbf{B}) \odot (\mathbf{C} \otimes \mathbf{D}) &= (a_{ij}b_{kl}\mathbf{e}_{ik}^{jl}) \odot (c_{ij}d_{kl}\mathbf{e}_{ik}^{jl}) = a_{ij}c_{ij}b_{kl}d_{kl}\mathbf{e}_{ik}^{jl} \\ &= a_{ij}c_{ij}b_{kl}d_{kl}(\mathbf{e}_i^j \otimes \mathbf{e}_k^l) = (a_{ij}c_{ij}\mathbf{e}_i^j) \otimes (b_{kl}d_{kl}\mathbf{e}_k^l) \\ &= (\mathbf{A} \odot \mathbf{C}) \otimes (\mathbf{B} \odot \mathbf{D}). \qquad [2.241]\end{aligned}$$

– Similarly, for $\mathbf{u}, \mathbf{x} \in \mathbb{K}^I$, and $\mathbf{v}, \mathbf{y} \in \mathbb{K}^J$, we have:

$$(\mathbf{u} \otimes \mathbf{v}) \odot (\mathbf{x} \otimes \mathbf{y}) = (\mathbf{u} \odot \mathbf{x}) \otimes (\mathbf{v} \odot \mathbf{y}). \qquad [2.242]$$

– Given the vectors $\mathbf{u}, \mathbf{v}, \mathbf{x}, \mathbf{y}$, of respective sizes I, J, K, L, we have the following identities:

$$(\mathbf{u} \otimes \mathbf{v})(\mathbf{x} \otimes \mathbf{y})^T = (\mathbf{u} \otimes \mathbf{1}_J)(\mathbf{1}_K \otimes \mathbf{y})^T \odot (\mathbf{1}_I \otimes \mathbf{v})(\mathbf{x} \otimes \mathbf{1}_L)^T \qquad [2.243]$$

$$= (\mathbf{1}_I \otimes \mathbf{v})(\mathbf{1}_K \otimes \mathbf{y})^T \odot (\mathbf{u} \otimes \mathbf{1}_J)(\mathbf{x} \otimes \mathbf{1}_L)^T \qquad [2.244]$$

$$\mathbf{u} \otimes \mathbf{v} \otimes \mathbf{x} \otimes \mathbf{y} = (\mathbf{u} \otimes \mathbf{v} \otimes \mathbf{1}_{KL}) \odot (\mathbf{1}_{IJ} \otimes \mathbf{x} \otimes \mathbf{y}) \qquad [2.245]$$

$$= (\mathbf{u} \otimes \mathbf{1}_J \otimes \mathbf{x} \otimes \mathbf{1}_L) \odot (\mathbf{1}_I \otimes \mathbf{v} \otimes \mathbf{1}_K \otimes \mathbf{y}) \qquad [2.246]$$

$$= (\mathbf{u} \otimes \mathbf{1}_{JK} \otimes \mathbf{y}) \odot (\mathbf{1}_I \otimes \mathbf{v} \otimes \mathbf{x} \otimes \mathbf{1}_L). \qquad [2.247]$$

PROOF.– As an example, we will check the relation [2.243]. The other relations can be checked similarly. Using [2.72] and [2.89], we have:

$$(\mathbf{u} \otimes \mathbf{1}_J)(\mathbf{1}_K \otimes \mathbf{y})^T \odot (\mathbf{1}_I \otimes \mathbf{v})(\mathbf{x} \otimes \mathbf{1}_L)^T = (u_i \delta_{jj} \mathbf{e}_{ij})(\delta_{kk} y_l \mathbf{e}^{kl}) \odot (\delta_{ii} v_j \mathbf{e}_{ij})$$

$$(x_k \delta_{ll} \mathbf{e}^{kl})$$

$$= (u_i y_l \delta_{jj} \delta_{kk} \mathbf{e}_{ij}^{kl}) \odot (v_j x_k \delta_{ii} \delta_{ll} \mathbf{e}_{ij}^{kl})$$

$$= u_i y_l v_j x_k \mathbf{e}_{ij}^{kl} = (u_i v_j \mathbf{e}_{ij})(x_k y_l \mathbf{e}^{kl})$$

$$= (\mathbf{u} \otimes \mathbf{v})(\mathbf{x} \otimes \mathbf{y})^T.$$

$$\square$$

– From [2.243]–[2.247], it is easy to deduce the other relations below, choosing $(\mathbf{y} = L = 1)$ and $(\mathbf{v} = J = 1)$:

$$(\mathbf{u} \otimes \mathbf{v})\mathbf{x}^T = (\mathbf{u} \otimes \mathbf{1}_J)\mathbf{1}_K^T \odot (\mathbf{1}_I \otimes \mathbf{v})\mathbf{x}^T \qquad [2.248]$$

$$= (\mathbf{1}_I \otimes \mathbf{v})\mathbf{1}_K^T \odot (\mathbf{u} \otimes \mathbf{1}_J)\mathbf{x}^T \qquad [2.249]$$

$$\mathbf{u}(\mathbf{x} \otimes \mathbf{y})^T = \mathbf{u}(\mathbf{1}_K \otimes \mathbf{y})^T \odot \mathbf{1}_I(\mathbf{x} \otimes \mathbf{1}_L)^T \qquad [2.250]$$

$$= \mathbf{1}_I(\mathbf{1}_K \otimes \mathbf{y})^T \odot \mathbf{u}(\mathbf{x} \otimes \mathbf{1}_L)^T \qquad [2.251]$$

$$\mathbf{u} \otimes \mathbf{v} \otimes \mathbf{x} = (\mathbf{u} \otimes \mathbf{v} \otimes \mathbf{1}_K) \odot (\mathbf{1}_{IJ} \otimes \mathbf{x}) \qquad [2.252]$$

$$= (\mathbf{u} \otimes \mathbf{1}_J \otimes \mathbf{x}) \odot (\mathbf{1}_I \otimes \mathbf{v} \otimes \mathbf{1}_K) \qquad [2.253]$$

$$= (\mathbf{u} \otimes \mathbf{1}_{JK}) \odot (\mathbf{1}_I \otimes \mathbf{v} \otimes \mathbf{x}). \qquad [2.254]$$

The three products also satisfy the relations in Table 2.19, which are proven below.

– Using the relation [2.236] between the Khatri–Rao and Kronecker products, then the properties [2.40] and [2.50] of the Kronecker product, and finally the relation [2.239] between the Hadamard and Kronecker products, we have:

$$(\mathbf{A} \diamond \mathbf{B})^H (\mathbf{C} \diamond \mathbf{D}) = \mathbf{S}_J^T (\mathbf{A} \otimes \mathbf{B})^H (\mathbf{C} \otimes \mathbf{D}) \mathbf{S}_N$$

$$= \mathbf{S}_J^T [(\mathbf{A}^H \mathbf{C}) \otimes (\mathbf{B}^H \mathbf{D})] \mathbf{S}_N$$

$$= (\mathbf{A}^H \mathbf{C}) \odot (\mathbf{B}^H \mathbf{D}) \in \mathbb{K}^{J \times N}. \qquad [2.255]$$

– Similarly, using the relation [2.236] between the Khatri–Rao and Kronecker products, then the property [2.50] of the Kronecker product, and finally the relation [2.236] once again, we obtain:

$$(\mathbf{A} \otimes \mathbf{B})(\mathbf{C} \diamond \mathbf{D}) = (\mathbf{A} \otimes \mathbf{B})(\mathbf{C} \otimes \mathbf{D}) \mathbf{S}_N$$

$$= (\mathbf{AC} \otimes \mathbf{BD}) \mathbf{S}_N = (\mathbf{AC} \diamond \mathbf{BD}) \in \mathbb{K}^{IM \times N}. \qquad [2.256]$$

$\mathbf{A} \in \mathbb{K}^{I \times J}, \mathbf{B} \in \mathbb{K}^{K \times J}, \mathbf{C} \in \mathbb{K}^{I \times N}, \mathbf{D} \in \mathbb{K}^{K \times N}$
$(\mathbf{A} \diamond \mathbf{B})^H (\mathbf{C} \diamond \mathbf{D}) = (\mathbf{A}^H \mathbf{C}) \odot (\mathbf{B}^H \mathbf{D}) \in \mathbb{K}^{J \times N}$
$\mathbf{A} \in \mathbb{K}^{I \times J}, \mathbf{B} \in \mathbb{K}^{M \times K}, \mathbf{C} \in \mathbb{K}^{J \times N}, \mathbf{D} \in \mathbb{K}^{K \times N}$
$(\mathbf{A} \otimes \mathbf{B})(\mathbf{C} \diamond \mathbf{D}) = (\mathbf{A}\mathbf{C}) \diamond (\mathbf{B}\mathbf{D}) \in \mathbb{K}^{IM \times N}$
$\mathbf{A} \in \mathbb{K}^{I \times J}, \mathbf{C} \in \mathbb{K}^{J \times K}$
$\mathrm{vec}\Big(\mathbf{A}\mathrm{diag}(\lambda_1, \cdots, \lambda_J)\mathbf{C}\Big) = (\mathbf{C}^T \diamond \mathbf{A}) \begin{bmatrix} \lambda_1 \\ \vdots \\ \lambda_J \end{bmatrix} \in \mathbb{K}^{KI}$
$\mathbf{A} \in \mathbb{K}^{I \times J}, \mathbf{B} \in \mathbb{K}^{K \times J},\ \text{where}\ \mathbf{A} \diamond \mathbf{B}\ \text{has full column rank}$
$(\mathbf{A} \diamond \mathbf{B})^{\dagger} = \Big[(\mathbf{A}^H \mathbf{A} \odot \mathbf{B}^H \mathbf{B})\Big]^{-1} (\mathbf{A} \diamond \mathbf{B})^H$
$\mathbf{A}_n \in \mathbb{K}^{I_n \times J}, \mathbf{C}_n \in \mathbb{K}^{I_n \times K},\ n \in \langle N \rangle$
$(\overset{N}{\underset{n=1}{\diamond}} \mathbf{A}_n)^H (\overset{N}{\underset{n=1}{\diamond}} \mathbf{C}_n) = \overset{N}{\underset{n=1}{\odot}} (\mathbf{A}_n^H \mathbf{C}_n) \in \mathbb{K}^{J \times K}$

Table 2.19. *Relations between the Kronecker, Khatri-Rao, and Hadamard products of matrices*

– Noting that:

$$\mathbf{A}\mathrm{diag}(\lambda_1, \cdots, \lambda_J)\mathbf{C} = \sum_{j=1}^{J} \lambda_j \mathbf{A}_{.j} \mathbf{C}_{j.}$$

and using the property [2.28] with $\mathbf{x} = \mathbf{A}_{.j}, \mathbf{y}^H = \mathbf{C}_{j.}$, we deduce:

$$\mathrm{vec}\Big(\mathbf{A}\mathrm{diag}(\lambda_1, \cdots, \lambda_J)\mathbf{C}\Big) = \sum_{j=1}^{J} \lambda_j \mathrm{vec}\Big(\mathbf{A}_{.j}\mathbf{C}_{j.}\Big) = \sum_{j=1}^{J} \lambda_j (\mathbf{C}_{j.}^T \diamond \mathbf{A}_{.j})$$

$$= (\mathbf{C}^T \diamond \mathbf{A}) \begin{bmatrix} \lambda_1 \\ \vdots \\ \lambda_J \end{bmatrix}. \qquad [2.257]$$

This relation can also be deduced from [2.113] by dropping the zeroes from $\mathrm{vec}\big(\mathrm{diag}(\lambda_1, \cdots, \lambda_J)\big)$.

– Using the property [2.255], the Moore–Penrose pseudo-inverse of $\mathbf{A} \diamond \mathbf{B}$ can be written as:

$$(\mathbf{A} \diamond \mathbf{B})^\dagger = \left[(\mathbf{A} \diamond \mathbf{B})^H (\mathbf{A} \diamond \mathbf{B}) \right]^{-1} (\mathbf{A} \diamond \mathbf{B})^H$$

$$= \left[(\mathbf{A}^H \mathbf{A} \odot \mathbf{B}^H \mathbf{B}) \right]^{-1} (\mathbf{A} \diamond \mathbf{B})^H, \qquad [2.258]$$

which proves the formula of the Moore–Penrose pseudo-inverse of the Khatri–Rao product of two full column rank matrices.

– The last relation in Table 2.19 is a generalization of the first relation to the case of multiple Khatri–Rao products.

We also have the relations listed in Table 2.20.

$\mathbf{u}, \mathbf{a} \in \mathbb{K}^I, \mathbf{v}, \mathbf{b} \in \mathbb{K}^J, \mathbf{x}, \mathbf{y} \in \mathbb{K}^K$
$(\mathbf{u} \otimes \mathbf{v})\left(\mathbf{x}^T \diamond \mathbf{y}^T \right) = \mathbf{u}\mathbf{x}^T \diamond \mathbf{v}\mathbf{y}^T$
$(\mathbf{u} \otimes \mathbf{v})\mathbf{x}^T = \mathbf{u}\mathbf{x}^T \diamond \mathbf{v}\mathbf{1}_K^T$
$(\mathbf{u} \otimes \mathbf{v}) \odot (\mathbf{a} \otimes \mathbf{b}) = (\mathbf{u} \odot \mathbf{a}) \diamond (\mathbf{v} \odot \mathbf{b})$
$\mathbf{U}, \mathbf{A} \in \mathbb{K}^{I \times K}$, $\mathbf{V}, \mathbf{B} \in \mathbb{K}^{J \times K}$
$(\mathbf{U} \diamond \mathbf{V}) \odot (\mathbf{A} \diamond \mathbf{B}) = (\mathbf{U} \odot \mathbf{A}) \diamond (\mathbf{V} \odot \mathbf{B})$

Table 2.20. *Other relations between the Kronecker, Khatri-Rao, and Hadamard products of vectors and matrices*

PROOF.– The first relation can be obtained from the property [2.256] by setting $\mathbf{A} = \mathbf{u}, \mathbf{B} = \mathbf{v}, \mathbf{C} = \mathbf{x}^T$ and $\mathbf{D} = \mathbf{y}^T$, whereas the second relation follows from the first, with $\mathbf{y} = \mathbf{1}_K$, and from the identity $\mathbf{x}^T = \mathbf{x}^T \diamond \mathbf{1}_K^T$. The third relation can be shown using the index convention, noting that $\mathbf{u} \odot \mathbf{a} = u_i \mathbf{e}_i \odot a_i \mathbf{e}_i = u_i a_i \mathbf{e}_i$ and $\mathbf{u} \otimes \mathbf{v} = \mathbf{u} \diamond \mathbf{v} = u_i v_j \mathbf{e}_{ij}$ (see Table 2.14):

$$(\mathbf{u} \otimes \mathbf{v}) \odot (\mathbf{a} \otimes \mathbf{b}) = (u_i v_j \mathbf{e}_{ij}) \odot (a_i b_j \mathbf{e}_{ij}) = u_i v_j a_i b_j \mathbf{e}_{ij} \qquad [2.259]$$

$$= (u_i a_i \mathbf{e}_i) \diamond (v_j b_j \mathbf{e}_j) = (\mathbf{u} \odot \mathbf{a}) \diamond (\mathbf{v} \odot \mathbf{b}). \qquad [2.260]$$

Finally, the last relation can be deduced from the previous one by decomposing the matrix Khatri–Rao products column by column. □

Tables 2.21 and 2.22 summarize the key relations presented in this chapter.

$\mathbf{A} \in \mathbb{K}^{I \times J}, \mathbf{B} \in \mathbb{K}^{M \times N}, \mathbf{C} \in \mathbb{K}^{J \times K}, \mathbf{D} \in \mathbb{K}^{N \times P}$
$(\mathbf{A} \otimes \mathbf{B})(\mathbf{C} \otimes \mathbf{D}) = \mathbf{AC} \otimes \mathbf{BD}$
$\mathbf{A} \in \mathbb{K}^{I \times J}, \mathbf{B} \in \mathbb{K}^{M \times N}, \mathbf{C} \in \mathbb{K}^{J \times K}, \mathbf{D} \in \mathbb{K}^{N \times K}$
$(\mathbf{A} \otimes \mathbf{B})(\mathbf{C} \diamond \mathbf{D}) = (\mathbf{AC}) \diamond (\mathbf{BD})$
$\mathbf{A} \in \mathbb{K}^{I \times J}, \mathbf{B} \in \mathbb{K}^{K \times J}, \mathbf{C} \in \mathbb{K}^{I \times N}, \mathbf{D} \in \mathbb{K}^{K \times N}$
$(\mathbf{A} \diamond \mathbf{B})^{H}(\mathbf{C} \diamond \mathbf{D}) = (\mathbf{A}^{H}\mathbf{C}) \odot (\mathbf{B}^{H}\mathbf{D})$
$\mathbf{A}, \mathbf{C} \in \mathbb{K}^{I \times J}, \ \mathbf{B}, \mathbf{D} \in \mathbb{K}^{K \times L}$
$(\mathbf{A} \otimes \mathbf{B}) \odot (\mathbf{C} \otimes \mathbf{D}) = (\mathbf{A} \odot \mathbf{C}) \otimes (\mathbf{B} \odot \mathbf{D})$

Table 2.21. *Basic relations*

$\mathbf{A} \in \mathbb{K}^{I \times J}, \ \mathbf{B} \in \mathbb{K}^{J \times M}, \ \mathbf{C} \in \mathbb{K}^{M \times N}$
$\mathrm{vec}(\mathbf{ABC}) = (\mathbf{C}^{T} \otimes \mathbf{A})\mathrm{vec}(\mathbf{B})$
$\mathbf{A} \in \mathbb{K}^{I \times J}, \ \mathbf{C} \in \mathbb{K}^{J \times K}$
$\mathrm{vec}\Big(\mathbf{A}\mathrm{diag}(\lambda_1, \cdots, \lambda_J)\mathbf{C}\Big) = (\mathbf{C}^{T} \diamond \mathbf{A}) \begin{bmatrix} \lambda_1 \\ \vdots \\ \lambda_J \end{bmatrix}$

Table 2.22. *Vectorization formulae*

2.12. Applications

In this section, we present two applications of the Kronecker product as examples. The first concerns the computation and arrangement of first-order partial derivatives. Multiple definitions of these derivatives are provided using the index convention. The second application relates to solving matrix equations such as Sylvester and Lyapunov equations, both continuous-time and discrete-time, as well as other equations that often appear when estimating the parameters of tensor models. The problem of estimating the factors of a Khatri–Rao product and a Kronecker product is also solved at the end of the chapter.

2.12.1. *Partial derivatives and index convention*

Matrix differential calculus, i.e. computing the derivative of a matrix function with respect to a matrix variable, was developed by Magnus and Neudecker (1985, 1988) in the case with real functions and real variables (see also Magnus 2010).

In the following, we will consider real scalar, vector and matrix functions of a real variable that can itself be a scalar, vector or matrix. The nine cases of partial derivatives that we will consider are summarized in Table 2.23. All functions are assumed to be differentiable.

Our objective here is to introduce different ways of arranging the first-order partial derivatives[5] as a vector or a matrix using the index convention. We will also illustrate these results with the Jacobian matrix of certain functions.

Function \ Variable	$f \in \mathbb{R}$	$\mathbf{f} \in \mathbb{R}^I$	$\mathbf{F} \in \mathbb{R}^{I \times J}$
$x \in \mathbb{R}$	$\frac{\partial f(x)}{\partial x}$	$\frac{\partial \mathbf{f}(x)}{\partial x}$	$\frac{\partial \mathbf{F}(x)}{\partial x}$
$\mathbf{x} \in \mathbb{R}^N$	$\frac{\partial f(\mathbf{x})}{\partial \mathbf{x}}$	$\frac{\partial \mathbf{f}(\mathbf{x})}{\partial \mathbf{x}}$	$\frac{\partial \mathbf{F}(\mathbf{x})}{\partial \mathbf{x}}$
$\mathbf{X} \in \mathbb{R}^{M \times N}$	$\frac{\partial f(\mathbf{X})}{\partial \mathbf{X}}$	$\frac{\partial \mathbf{f}(\mathbf{X})}{\partial \mathbf{X}}$	$\frac{\partial \mathbf{F}(\mathbf{X})}{\partial \mathbf{X}}$

Table 2.23. *Partial derivatives of various functions*

Table 2.24 gives various definitions of the first-order partial derivative of a real function with respect to a real variable, stating the size of the derivatives in each case, according to whether the function and the variable are scalars or vectors, and using the index convention. These definitions characterize how the partial derivatives are arranged within a vector or a matrix.

For example, for $(\mathbf{f}, \mathbf{x}) \in \mathbb{R}^I \times \mathbb{R}^N$, the partial derivatives $\frac{\partial f_i(\mathbf{x})}{\partial x_n}$, with $i \in \langle I \rangle$ and $n \in \langle N \rangle$, can be arranged in a matrix in two different ways:

$$\frac{\partial f(\mathbf{x})}{\partial \mathbf{x}^T} = \frac{\partial f_i(\mathbf{x})}{\partial x_n} \mathbf{e}_i^n = \begin{bmatrix} \frac{\partial f_1(\mathbf{x})}{\partial x_1} & \cdots & \frac{\partial f_1(\mathbf{x})}{\partial x_N} \\ \vdots & \vdots & \vdots \\ \frac{\partial f_I(\mathbf{x})}{\partial x_1} & \cdots & \frac{\partial f_I(\mathbf{x})}{\partial x_N} \end{bmatrix} \in \mathbb{R}^{I \times N} \qquad [2.261]$$

and

$$\frac{\partial f^T(\mathbf{x})}{\partial \mathbf{x}} = \frac{\partial f_i(\mathbf{x})}{\partial x_n} \mathbf{e}_n^i = \begin{bmatrix} \frac{\partial f_1(\mathbf{x})}{\partial x_1} & \cdots & \frac{\partial f_I(\mathbf{x})}{\partial x_1} \\ \vdots & \vdots & \vdots \\ \frac{\partial f_1(\mathbf{x})}{\partial x_N} & \cdots & \frac{\partial f_I(\mathbf{x})}{\partial x_N} \end{bmatrix} \in \mathbb{R}^{N \times I}. \qquad [2.262]$$

---

[5] A partial derivative of a function $f(x_1, ..., x_N)$ of several independent variables is the derivative with respect to one of the variables, say x_n, while assuming that the other variables are constant. This partial derivative is denoted $\frac{\partial f(\mathbf{x})}{\partial x_n}$, where $\mathbf{x} = (x_1, ..., x_N)$.

(Function, variable)	Derivatives	Size
$(f, \mathbf{x}) \in \mathbb{R} \times \mathbb{R}^N$	$\frac{\partial f(\mathbf{x})}{\partial \mathbf{x}} = \frac{\partial f(\mathbf{x})}{\partial x_n} \mathbf{e}_n$	$\mathbb{R}^N$
$(f, \mathbf{x}) \in \mathbb{R} \times \mathbb{R}^N$	$\frac{\partial f(\mathbf{x})}{\partial \mathbf{x}^T} = \frac{\partial f(\mathbf{x})}{\partial x_n} \mathbf{e}^n$	$\mathbb{R}^{1 \times N}$
$(\mathbf{f}, x) \in \mathbb{R}^I \times \mathbb{R}$	$\frac{\partial \mathbf{f}(x)}{\partial x} = \frac{\partial f_i(x)}{\partial x} \mathbf{e}_i$	$\mathbb{R}^I$
$(\mathbf{f}, x) \in \mathbb{R}^I \times \mathbb{R}$	$\frac{\partial \mathbf{f}^T(x)}{\partial x} = \frac{\partial f_i(x)}{\partial x} \mathbf{e}^i$	$\mathbb{R}^{1 \times I}$
$(\mathbf{f}, \mathbf{x}) \in \mathbb{R}^I \times \mathbb{R}^N$	$\frac{\partial \mathbf{f}(\mathbf{x})}{\partial \mathbf{x}^T} = \frac{\partial f_i(\mathbf{x})}{\partial x_n} \mathbf{e}_i^n$	$\mathbb{R}^{I \times N}$
$(\mathbf{f}, \mathbf{x}) \in \mathbb{R}^I \times \mathbb{R}^N$	$\frac{\partial \mathbf{f}^T(\mathbf{x})}{\partial \mathbf{x}} = \frac{\partial f_i(\mathbf{x})}{\partial x_n} \mathbf{e}_n^i$	$\mathbb{R}^{N \times I}$
$(\mathbf{f}, \mathbf{x}) \in \mathbb{R}^I \times \mathbb{R}^N$	$\frac{\partial \mathbf{f}(\mathbf{x})}{\partial \mathbf{x}} = \frac{\partial f_i(\mathbf{x})}{\partial x_n} \mathbf{e}_{in}$	$\mathbb{R}^{IN}$

Table 2.24. *Derivatives of a scalar function and a vector function*

The form [2.261] is called the Jacobian matrix of $\mathbf{f}$. The ith row vector of this matrix is given by: $\frac{\partial f_i(\mathbf{x})}{\partial \mathbf{x}^T} = [\frac{\partial f_i(\mathbf{x})}{\partial x_1} \cdots \frac{\partial f_i(\mathbf{x})}{\partial x_N}]$. This is the transpose of the gradient vector of the function $f_i(\mathbf{x})$, with $i \in \langle I \rangle$. When $I = N$, the determinant of the Jacobian matrix is called the Jacobian of $\mathbf{f}$.

EXAMPLE 2.55.– For $\mathbf{x} \in \mathbb{R}^I$, we have:
$$\frac{\partial \mathbf{x}}{\partial \mathbf{x}^T} = \frac{\partial \mathbf{x}^T}{\partial \mathbf{x}} = \mathbf{I}_I.$$

EXAMPLE 2.56.– For $\mathbb{R}^2 \ni \mathbf{x} = \begin{bmatrix} r \\ \theta \end{bmatrix} \mapsto \begin{bmatrix} r \cos \theta \\ r \sin \theta \end{bmatrix} \in \mathbb{R}^2$, we have:
$$\frac{\partial \mathbf{f}(\mathbf{x})}{\partial \mathbf{x}^T} = \begin{bmatrix} \cos \theta & -r \sin \theta \\ \sin \theta & r \cos \theta \end{bmatrix}, \quad \frac{\partial \mathbf{f}^T(\mathbf{x})}{\partial \mathbf{x}} = \begin{bmatrix} \cos \theta & \sin \theta \\ -r \sin \theta & r \cos \theta \end{bmatrix},$$
and the Jacobian is equal to $r(\cos^2 \theta + \sin^2 \theta) = r$.

EXAMPLE 2.57.– For $\mathbf{f}(\mathbf{x}) = \mathbf{A}\mathbf{x} \in \mathbb{R}^I$, where $\mathbf{x} \in \mathbb{R}^N$ and $\mathbf{A} \in \mathbb{R}^{I \times N}$ is a constant matrix, the matrix of first-order partial derivatives can be defined as:

$$\frac{\partial (\mathbf{A}\mathbf{x})}{\partial \mathbf{x}^T} = \mathbf{A} \in \mathbb{R}^{I \times N}$$

$$\frac{\partial (\mathbf{A}\mathbf{x})^T}{\partial \mathbf{x}} = \frac{\partial (\mathbf{x}^T \mathbf{A}^T)}{\partial \mathbf{x}} = \mathbf{A}^T \in \mathbb{R}^{N \times I}.$$

EXAMPLE 2.58.– For $f(\mathbf{x}) = \mathbf{x}^T \mathbf{A}\mathbf{x} = x_m a_{mn} x_n = x_n a_{nm} x_m$, where $\mathbf{x} \in \mathbb{R}^M$ and $\mathbf{A} \in \mathbb{R}^{M \times M}$ is a constant matrix, we have:

$$\frac{\partial (\mathbf{x}^T \mathbf{A}\mathbf{x})}{\partial \mathbf{x}} = \frac{\partial f(\mathbf{x})}{\partial x_m} \mathbf{e}_m = (a_{mn} x_n + x_n a_{nm}) \mathbf{e}_m$$

$$= (\mathbf{A} + \mathbf{A}^T) \mathbf{x}.$$

If $\mathbf{A}$ is a symmetric matrix, then the above expression becomes:

$$\frac{\partial (\mathbf{x}^T \mathbf{A}\mathbf{x})}{\partial \mathbf{x}} = 2\mathbf{A}\mathbf{x}. \qquad [2.263]$$

In the case of a real matrix function $\mathbf{F}(\mathbf{X}) \in \mathbb{R}^{I \times J}$ and a real matrix variable $\mathbf{X} \in \mathbb{R}^{M \times N}$, the partial derivatives $\frac{\partial f_{ij}}{\partial x_{mn}}$ form a fourth-order tensor of size $I \times J \times M \times N$. There are therefore several ways to arrange these partial derivatives in a vector or matrix, depending on which combinations of the modes (i, j, m, n) are considered. To reduce the problem to that of the partial derivatives of a vector function with respect to a vector variable, one solution is to vectorize $\mathbf{F}$ or $\mathbf{F}^T$ and $\mathbf{X}$ or $\mathbf{X}^T$. Table 2.25 presents four definitions that depend on these vectorizations of $\mathbf{F}$, $\mathbf{F}^T$, $\mathbf{X}$ and $\mathbf{X}^T$.

Derivatives
$\dfrac{\partial \,\text{vec}\left[\mathbf{F}(\mathbf{X})\right]}{\partial \,(\text{vec}(\mathbf{X}))^T} = \dfrac{\partial f_{ij}(\mathbf{X})}{\partial x_{mn}} \mathbf{e}_{ji}^{nm} \in \mathbb{R}^{JI \times NM}$
$\dfrac{\partial \,\text{vec}\left[\mathbf{F}^T(\mathbf{X})\right]}{\partial \,(\text{vec}[\mathbf{X}^T])^T} = \dfrac{\partial f_{ij}(\mathbf{X})}{\partial x_{mn}} \mathbf{e}_{ij}^{mn} \in \mathbb{R}^{IJ \times MN}$
$\dfrac{\partial \,\text{vec}\left[\mathbf{F}(\mathbf{X})\right]}{\partial \,(\text{vec}[\mathbf{X}^T])^T} = \dfrac{\partial f_{ij}(\mathbf{X})}{\partial x_{mn}} \mathbf{e}_{ji}^{mn} \in \mathbb{R}^{JI \times MN}$
$\dfrac{\partial \,\text{vec}\left[\mathbf{F}^T(\mathbf{X})\right]}{\partial \,(\text{vec}(\mathbf{X}))^T} = \dfrac{\partial f_{ij}(\mathbf{X})}{\partial x_{mn}} \mathbf{e}_{ij}^{nm} \in \mathbb{R}^{IJ \times NM}$

Table 2.25. *Derivatives of a matrix function with respect to a matrix variable*

For example, for $\frac{\partial \text{vec}[\mathbf{F}(\mathbf{X})]}{\partial (\text{vec}(\mathbf{X}))^T} \in \mathbb{R}^{JI \times NM}$, the matrices $\mathbf{F}$ and $\mathbf{X}$ are arranged as two column vectors $\mathbf{g} = \text{vec}(\mathbf{F}) \in \mathbb{R}^{JI}$ and $\mathbf{y} = \text{vec}(\mathbf{X}) \in \mathbb{R}^{NM}$ such that[6]: $g_{(j-1)I+i} = f_{ij}$ and $y_{(n-1)M+m} = x_{mn}$. The derivative can then be rewritten as $\frac{\partial \mathbf{g}(\mathbf{y})}{\partial \mathbf{y}^T}$. The pth row vector of this matrix, with $p = (j-1)I + i$, contains the partial derivatives of the function $f_{ij}(\mathbf{X})$ with respect to the various components of the matrix variable $\mathbf{X}$ arranged into the vector $\text{vec}(\mathbf{X})$.

We can also define $\frac{\partial \mathbf{g}^T(\mathbf{y})}{\partial \mathbf{y}} = \frac{\partial \text{vec}^T[\mathbf{F}(\mathbf{X})]}{\partial \text{vec}(\mathbf{X})} \in \mathbb{R}^{NM \times JI}$. The derivatives $\frac{\partial \mathbf{g}(\mathbf{y})}{\partial \mathbf{y}^T}$ and $\frac{\partial \mathbf{g}^T(\mathbf{y})}{\partial \mathbf{y}}$ correspond to the definitions $\frac{\partial \mathbf{f}(\mathbf{x})}{\partial \mathbf{x}^T}$ and $\frac{\partial \mathbf{f}^T(\mathbf{x})}{\partial \mathbf{x}}$ in Table 2.24, respectively.

To illustrate the results in Table 2.25, let us consider several examples of functions $\mathbf{F}(\mathbf{X})$. Let $\mathbf{A} \in \mathbb{R}^{I \times M}$ and $\mathbf{B} \in \mathbb{R}^{J \times N}$ be constant matrices, and let $\mathbf{X} \in \mathbb{R}^{M \times N}$ be a matrix variable. Recalling that $\text{vec}(\mathbf{A}\mathbf{X}\mathbf{B}^T) = (\mathbf{B} \otimes \mathbf{A})\text{vec}(\mathbf{X}) \in \mathbb{R}^{JI}$, and using the definition $\frac{\partial \text{vec}[\mathbf{F}(\mathbf{X})]}{\partial (\text{vec}(\mathbf{X}))^T}$ of the derivative, we obtain:

$$\mathbf{F}(\mathbf{X}) = \mathbf{A}\mathbf{X}\mathbf{B}^T \;\Rightarrow\; \frac{\partial \text{vec}[\mathbf{F}(\mathbf{X})]}{\partial (\text{vec}(\mathbf{X}))^T} = \mathbf{B} \otimes \mathbf{A} \in \mathbb{R}^{JI \times NM} \qquad [2.264]$$

$$\mathbf{F}(\mathbf{X}) = \mathbf{A}\mathbf{X} \;\Rightarrow\; \frac{\partial \text{vec}[\mathbf{F}(\mathbf{X})]}{\partial (\text{vec}(\mathbf{X}))^T} = \mathbf{I}_N \otimes \mathbf{A} \in \mathbb{R}^{NI \times NM} \qquad [2.265]$$

$$\mathbf{F}(\mathbf{X}) = \mathbf{X}\mathbf{B}^T \;\Rightarrow\; \frac{\partial \text{vec}[\mathbf{F}(\mathbf{X})]}{\partial (\text{vec}(\mathbf{X}))^T} = \mathbf{B} \otimes \mathbf{I}_M \in \mathbb{R}^{JM \times NM}. \qquad [2.266]$$

For $\mathbf{F}(\mathbf{X}) = \mathbf{X} \in \mathbb{R}^{M \times N}$, we have:

$$\frac{\partial \text{vec}[\mathbf{F}(\mathbf{X})]}{\partial (\text{vec}(\mathbf{X}))^T} = \mathbf{e}_{nm}^{nm} = \mathbf{I}_{NM} \qquad [2.267]$$

$$\frac{\partial \text{vec}[\mathbf{F}(\mathbf{X})]}{\partial (\text{vec}(\mathbf{X}^T))^T} = \mathbf{e}_{nm}^{mn} \neq \mathbf{I}_{NM}. \qquad [2.268]$$

This last example shows that the definition $\frac{\partial \text{vec}[\mathbf{F}(\mathbf{X})]}{\partial (\text{vec}(\mathbf{X}))^T}$ gives a matrix of partial derivatives equal to the identity matrix. This definition can be viewed as a generalization of the Jacobian matrix of a vector function of a vector variable (example [2.55]), which justifies it being called the Jacobian matrix of the matrix function $\mathbf{F}(\mathbf{X})$ (Magnus and Neudecker, 1985).

---

6 Recall that, by convention, the order of the dimensions in a product JI is related to the order of variation of the indices, with j varying more slowly than i.

Let us use the same definition of the derivative to compute the derivative of the product of two matrix functions $\mathbf{F}(\mathbf{X}) \in \mathbb{R}^{I \times J}$ and $\mathbf{G}(\mathbf{X}) \in \mathbb{R}^{J \times K}$ of a matrix variable $\mathbf{X} \in \mathbb{R}^{M \times N}$. Using the index convention, we obtain:

$$\frac{\partial \operatorname{vec}[\mathbf{FG}(\mathbf{X})]}{\partial (\operatorname{vec}(\mathbf{X}))^T} = \frac{\partial (\mathbf{FG})_{ik}(\mathbf{X})}{\partial x_{mn}} \mathbf{e}_{ki}^{nm} = \frac{\partial (f_{ij} g_{jk})}{\partial x_{mn}} \mathbf{e}_{ki}^{nm}$$

$$= [g_{jk} \frac{\partial f_{ij}}{\partial x_{mn}} + f_{ij} \frac{\partial g_{jk}}{\partial x_{mn}}] \mathbf{e}_{ki}^{nm}. \qquad [2.269]$$

Noting that $\mathbf{e}_{ki}^{nm} = \mathbf{e}_{ki}^{ji} \mathbf{e}_{j'i'}^{nm} = \mathbf{e}_{ki}^{kj} \mathbf{e}_{k'j'}^{nm}$, the expansion [2.269] can be rewritten as:

$$\frac{\partial \operatorname{vec}[\mathbf{FG}(\mathbf{X})]}{\partial (\operatorname{vec}(\mathbf{X}))^T} = \left(g_{jk} \mathbf{e}_{ki}^{ji}\right) \left(\frac{\partial f_{i'j'}}{\partial x_{mn}} \mathbf{e}_{j'i'}^{nm}\right) + \left(f_{ij} \mathbf{e}_{ki}^{kj}\right) \left(\frac{\partial g_{j'k'}}{\partial x_{mn}} \mathbf{e}_{k'j'}^{nm}\right),$$

$$[2.270]$$

with:

$$g_{jk} \mathbf{e}_{ki}^{ji} = (g_{jk} \mathbf{e}_k^j) \otimes \mathbf{e}_i^i = \mathbf{G}^T \otimes \mathbf{I}_I$$

$$f_{ij} \mathbf{e}_{ki}^{kj} = \mathbf{e}_k^k \otimes (f_{ij} \mathbf{e}_i^j) = \mathbf{I}_K \otimes \mathbf{F}.$$

The formula [2.270] can therefore be written as:

$$\frac{\partial \operatorname{vec}[\mathbf{FG}(\mathbf{X})]}{\partial (\operatorname{vec}(\mathbf{X}))^T} = (\mathbf{G}^T \otimes \mathbf{I}_I) \frac{\partial \operatorname{vec}(\mathbf{F}(\mathbf{X}))}{\partial (\operatorname{vec}(\mathbf{X}))^T} + (\mathbf{I}_K \otimes \mathbf{F}) \frac{\partial \operatorname{vec}(\mathbf{G}(\mathbf{X}))}{\partial (\operatorname{vec}(\mathbf{X}))^T} \in \mathbb{R}^{KI \times NM}.$$

$$[2.271]$$

This formula was proven by Magnus and Neudecker (1985) with a different proof that does not use the index convention.

Consider now the Hadamard product of two matrix functions $\mathbf{F}(\mathbf{X})$, $\mathbf{G}(\mathbf{X}) \in \mathbb{R}^{I \times J}$ of the matrix variable $\mathbf{X} \in \mathbb{R}^{M \times N}$. Using the vectorization formula of a Hadamard product from Table 2.4, we can write:

$$\operatorname{vec}(\mathbf{F} \odot \mathbf{G}) = \operatorname{diag}\left(\operatorname{vec}(\mathbf{F})\right) \operatorname{vec}(\mathbf{G}) = \operatorname{diag}\left(\operatorname{vec}(\mathbf{G})\right) \operatorname{vec}(\mathbf{F}). \qquad [2.272]$$

We then deduce the following formula for the derivative of the Hadamard product $\mathbf{F} \odot \mathbf{G}$ (Magnus and Neudecker 1985):

$$\frac{\partial \operatorname{vec}(\mathbf{F} \odot \mathbf{G})}{\partial \operatorname{vec}(\mathbf{X})} = \operatorname{diag}\left(\operatorname{vec}(\mathbf{F})\right) \frac{\partial \operatorname{vec}(\mathbf{G})}{\partial \operatorname{vec}(\mathbf{X})} + \operatorname{diag}\left(\operatorname{vec}(\mathbf{G})\right) \frac{\partial \operatorname{vec}(\mathbf{F})}{\partial \operatorname{vec}(\mathbf{X})}.$$

For $\mathbf{F}(\mathbf{X}) \in \mathbb{R}^{I \times J}, \mathbf{X} \in \mathbb{R}^{M \times N}$, we can also define the matrix of partial derivatives as a partitioned matrix, divided either into $I \times J$ blocks of size $M \times N$, where each block (i, j) is the matrix $\frac{\partial f_{ij}(\mathbf{X})}{\partial \mathbf{X}}$, with $i \in \langle I \rangle$ and $j \in \langle J \rangle$, or into

$M \times N$ blocks of size $I \times J$, where each block (m, n) is the matrix $\frac{\partial F(X)}{\partial x_{mn}}$, with $m \in \langle M \rangle$ and $n \in \langle N \rangle$. These two partitioned forms are presented below using the index convention:

$$\mathbf{e}_i^j \otimes \frac{\partial f_{ij}(\mathbf{X})}{\partial \mathbf{X}} = \begin{bmatrix} \frac{\partial f_{11}(\mathbf{X})}{\partial \mathbf{X}} & \cdots & \frac{\partial f_{1J}(\mathbf{X})}{\partial \mathbf{X}} \\ \vdots & \vdots & \vdots \\ \frac{\partial f_{I1}(\mathbf{X})}{\partial \mathbf{X}} & \cdots & \frac{\partial f_{IJ}(\mathbf{X})}{\partial \mathbf{X}} \end{bmatrix} \in \mathbb{R}^{IM \times JN} \qquad [2.273]$$

$$\mathbf{e}_m^n \otimes \frac{\partial \mathbf{F}(\mathbf{X})}{\partial x_{mn}} = \begin{bmatrix} \frac{\partial \mathbf{F}(\mathbf{X})}{\partial x_{11}} & \cdots & \frac{\partial \mathbf{F}(\mathbf{X})}{\partial x_{1N}} \\ \vdots & \vdots & \vdots \\ \frac{\partial \mathbf{F}(\mathbf{X})}{\partial x_{M1}} & \cdots & \frac{\partial \mathbf{F}(\mathbf{X})}{\partial x_{MN}} \end{bmatrix} \in \mathbb{R}^{MI \times NJ} \qquad [2.274]$$

with:

$$\frac{\partial f_{ij}(\mathbf{X})}{\partial \mathbf{X}} = \frac{\partial f_{ij}(\mathbf{X})}{\partial x_{mn}} \mathbf{e}_m^n \in \mathbb{R}^{M \times N} \qquad [2.275]$$

$$\frac{\partial \mathbf{F}(\mathbf{X})}{\partial x_{mn}} = \frac{\partial f_{ij}(\mathbf{X})}{\partial x_{mn}} \mathbf{e}_i^j \in \mathbb{R}^{I \times J}. \qquad [2.276]$$

This gives us the differentiation formulae stated in Table 2.26.

$\mathbf{F} \in \mathbb{R}^{I \times J}, \mathbf{X} \in \mathbb{R}^{M \times N}$
$\sum_{i=1}^{I} \sum_{j=1}^{J} \mathbf{E}_{ij}^{(I \times J)} \otimes \frac{\partial f_{ij}}{\partial \mathbf{X}} = \mathbf{e}_i^j \otimes \frac{\partial f_{ij}}{\partial \mathbf{X}} = \frac{\partial f_{ij}}{\partial x_{mn}} \mathbf{e}_{im}^{jn} \in \mathbb{R}^{IM \times JN}$
$\sum_{m=1}^{M} \sum_{n=1}^{N} \mathbf{E}_{mn}^{(M \times N)} \otimes \frac{\partial \mathbf{F}}{\partial x_{mn}} = \mathbf{e}_m^n \otimes \frac{\partial \mathbf{F}}{\partial x_{mn}} = \frac{\partial f_{ij}}{\partial x_{mn}} \mathbf{e}_{mi}^{nj} \in \mathbb{R}^{MI \times NJ}$

Table 2.26. *Definitions of $\frac{\partial \mathbf{F}(\mathbf{X})}{\partial \mathbf{X}}$ in partitioned form*

– For $\mathbf{F}(\mathbf{X}) = \mathbf{X} \in \mathbb{R}^{M \times N}$, the two differentiation formulae in Table 2.26 give the same partial derivative:

$$\frac{\partial x_{ij}}{\partial x_{mn}} \mathbf{e}_{mi}^{nj} = \delta_{im} \delta_{jn} \mathbf{e}_{mi}^{nj} = \mathbf{e}_{mm}^{nn} = \text{vec}(\mathbf{I}_M) \text{vec}^T(\mathbf{I}_N) \in \mathbb{R}^{M^2 \times N^2},$$

since, from [2.109], we have $\mathbf{e}_{mm}^{nn} = \text{vec}(\mathbf{I}_M)\text{vec}^T(\mathbf{I}_N)$, which is a rank-one matrix. This differentiation formula should be compared against [2.267] and [2.268]. Since the Jacobian matrix of the identity function $(\mathbf{X} \mapsto \mathbf{F}(\mathbf{X}) = \mathbf{X})$ is the identity matrix, we can conclude that the definition $\frac{\partial \text{vec}[\mathbf{F}(\mathbf{X})]}{\partial (\text{vec}(\mathbf{X}))^T}$ is preferable to the other definitions (Magnus and Neudecker, 1985).

– For $\mathbf{F}(\mathbf{X}) \in \mathbb{R}^{I \times J}$ and $\mathbf{G}(\mathbf{X}) \in \mathbb{R}^{J \times K}$, with $\mathbf{X} \in \mathbb{R}^{M \times N}$, the second differentiation formula in Table 2.26 gives:

$$\frac{\partial(\mathbf{FG}(\mathbf{X}))}{\partial \mathbf{X}} = \sum_{m=1}^{M} \sum_{n=1}^{N} \mathbf{E}_{mn}^{(M \times N)} \otimes \frac{\partial(\mathbf{FG}(\mathbf{X}))}{\partial x_{mn}}. \qquad [2.277]$$

Noting that $\mathbf{e}_i^k = \mathbf{e}_i^j \mathbf{e}_{j'}^k$, $\mathbf{F} = f_{ij} \mathbf{e}_i^j$, $\mathbf{G} = g_{jk} \mathbf{e}_j^k$, and $\mathbf{FG} = f_{ij} g_{jk} \mathbf{e}_i^k$, the property [2.50] of the Kronecker product allows us to expand the above equation as follows using the index convention:

$$\frac{\partial(\mathbf{FG}(\mathbf{X}))}{\partial \mathbf{X}} = \mathbf{e}_m^n \otimes \frac{\partial(f_{ij} g_{jk})}{\partial x_{mn}} \mathbf{e}_i^k = \mathbf{e}_m^n \otimes \left[g_{jk} \frac{\partial f_{ij}}{\partial x_{mn}} + f_{ij} \frac{\partial g_{jk}}{\partial x_{mn}} \right] \mathbf{e}_i^k$$

$$= (\mathbf{e}_m^n \otimes \frac{\partial f_{ij}}{\partial x_{mn}} \mathbf{e}_i^j)(\mathbf{I}_N \otimes g_{j'k} \mathbf{e}_{j'}^k) + (\mathbf{I}_M \otimes f_{ij} \mathbf{e}_i^j)(\mathbf{e}_m^n \otimes \frac{\partial g_{j'k}}{\partial x_{mn}} \mathbf{e}_{j'}^k)$$

$$= \frac{\partial \mathbf{F}(\mathbf{X})}{\partial \mathbf{X}}(\mathbf{I}_N \otimes \mathbf{G}) + (\mathbf{I}_M \otimes \mathbf{F}) \frac{\partial \mathbf{G}(\mathbf{X})}{\partial \mathbf{X}} \in \mathbb{R}^{MI \times NK}, \qquad [2.278]$$

where $\frac{\partial \mathbf{F}(\mathbf{X})}{\partial \mathbf{X}} \in \mathbb{R}^{MI \times NJ}$ and $\frac{\partial \mathbf{G}(\mathbf{X})}{\partial \mathbf{X}} \in \mathbb{R}^{MJ \times NK}$ are in partitioned form, as defined in Table 2.26.

Table 2.27 presents several definitions of the partial derivatives of a scalar function of a matrix variable, as well as of a matrix function of a scalar variable.

Below, we prove several formulae for computing the derivatives of traces of matrix products using the index convention.

EXAMPLE 2.59.– For $\mathbf{X} \in \mathbb{R}^{M \times M}$ and $f(\mathbf{X}) = \text{tr}(\mathbf{X}) = \sum_{j=1}^{M} x_{jj} = x_{jj}$, we have:

$$\frac{\partial \text{tr}(\mathbf{X})}{\partial \text{vec}(\mathbf{X}^T)} = \frac{\partial \text{tr}(\mathbf{X})}{\partial x_{mn}} \mathbf{e}_{mn} = \frac{\partial x_{jj}}{\partial x_{mn}} \mathbf{e}_{mn} = \delta_{jm} \delta_{jn} \mathbf{e}_{mn} = \mathbf{e}_{mm} = \text{vec}(\mathbf{I}_M) \in \mathbb{R}^{M^2}$$

$$\frac{\partial \text{tr}(\mathbf{X})}{\partial \mathbf{X}} = \frac{\partial \text{tr}(\mathbf{X})}{\partial x_{mn}} \mathbf{e}_m^n = \frac{\partial x_{jj}}{\partial x_{mn}} \mathbf{e}_m^n = \delta_{jm} \delta_{jn} \mathbf{e}_m^n = \mathbf{e}_m^m = \mathbf{I}_M \in \mathbb{R}^{M \times M}.$$

EXAMPLE 2.60.– Let $f(\mathbf{X}) = \text{tr}(\mathbf{AX}) = a_{ji} x_{ij}, i \in \langle M \rangle, j \in \langle N \rangle$, where $\mathbf{A} \in \mathbb{R}^{N \times M}$ is a constant matrix, and $\mathbf{X} \in \mathbb{R}^{M \times N}$. We have:

$$\frac{\partial \text{tr}(\mathbf{AX})}{\partial \mathbf{X}} = \frac{\partial \text{tr}(\mathbf{AX})}{\partial x_{mn}} \mathbf{e}_m^n = a_{nm} \mathbf{e}_m^n = \mathbf{A}^T. \qquad [2.279]$$

(Function, variable)	Derivatives	Size
$(f, \mathbf{X}) \in \mathbb{R} \times \mathbb{R}^{M \times N}$	$\dfrac{\partial f(\mathbf{X})}{\partial \left(\text{vec}(\mathbf{X})\right)^T} = \dfrac{\partial f(\mathbf{X})}{\partial x_{mn}} \mathbf{e}^{nm}$	$\mathbb{R}^{1 \times NM}$
$(f, \mathbf{X}) \in \mathbb{R} \times \mathbb{R}^{M \times N}$	$\dfrac{\partial f(\mathbf{X})}{\partial \text{vec}(\mathbf{X}^T)} = \dfrac{\partial f(\mathbf{X})}{\partial x_{mn}} \mathbf{e}_{mn}$	$\mathbb{R}^{MN}$
$(f, \mathbf{X}) \in \mathbb{R} \times \mathbb{R}^{M \times N}$	$\dfrac{\partial f(\mathbf{X})}{\partial \mathbf{X}} = \dfrac{\partial f(\mathbf{X})}{\partial x_{mn}} \mathbf{e}_m^n$	$\mathbb{R}^{M \times N}$
$(f, \mathbf{X}) \in \mathbb{R} \times \mathbb{R}^{M \times N}$	$\dfrac{\partial f(\mathbf{X})}{\partial \mathbf{X}^T} = \dfrac{\partial f(\mathbf{X})}{\partial x_{mn}} \mathbf{e}_n^m$	$\mathbb{R}^{N \times M}$
$(\mathbf{F}, x) \in \mathbb{R}^{I \times J} \times \mathbb{R}$	$\dfrac{\partial \text{vec}\left(\mathbf{F}^T(x)\right)}{\partial x} = \dfrac{\partial f_{ij}(x)}{\partial x} \mathbf{e}_{ij}$	$\mathbb{R}^{IJ}$
$(\mathbf{F}, x) \in \mathbb{R}^{I \times J} \times \mathbb{R}$	$\dfrac{\partial \mathbf{F}(x)}{\partial x} = \dfrac{\partial f_{ij}(x)}{\partial x} \mathbf{e}_i^j$	$\mathbb{R}^{I \times J}$

Table 2.27. *Other derivatives*

EXAMPLE 2.61.– Let $f(\mathbf{X}) = \text{tr}(\mathbf{AXB}) = a_{im} x_{mn} b_{ni}$, where $\mathbf{A} \in \mathbb{R}^{I \times M}$ and $\mathbf{B} \in \mathbb{R}^{N \times I}$ are constant matrices, and $\mathbf{X} \in \mathbb{R}^{M \times N}$. We have:

$$\frac{\partial \, \text{tr}(\mathbf{AXB})}{\partial \mathbf{X}} = \frac{\partial \, \text{tr}(\mathbf{AXB})}{\partial x_{mn}} \mathbf{e}_m^n = a_{im} b_{ni} \, \mathbf{e}_m^n$$

$$= (\mathbf{BA})_{nm} \, \mathbf{e}_m^n = (\mathbf{BA})^T = \mathbf{A}^T \mathbf{B}^T. \qquad [2.280]$$

In Table 2.28, we present several derivatives of traces of matrix products that can be proven from the above examples. Thus, noting that $\text{tr}(\mathbf{X}^T \mathbf{A}) = \text{tr}(\mathbf{A}^T \mathbf{X})$, the formula for the derivative $\frac{\partial \, \text{tr}(\mathbf{X}^T \mathbf{A})}{\partial \mathbf{X}}$ can be deduced from [2.279] by replacing $\mathbf{A}$ with $\mathbf{A}^T$. Similarly, since $\text{tr}(\mathbf{AX}^T \mathbf{B}) = \text{tr}(\mathbf{B}^T \mathbf{XA}^T)$, the formula for the derivative $\frac{\partial \, \text{tr}(\mathbf{AX}^T \mathbf{B})}{\partial \mathbf{X}}$ can be obtained from [2.280] by replacing $(\mathbf{A}, \mathbf{B})$ with $(\mathbf{B}^T, \mathbf{A}^T)$. Other differentiation formulae involving Kronecker products can be found in Tracy and Dwyer (1969) and Brewer (1978). From the above, we can conclude that there are several possible definitions for arranging partial derivatives. Consequently, it is important to always state the definitions used in any given computation.

2.12.2. *Solving matrix equations*

The Kronecker product can be used to represent matrix systems of linear equations in such a way that the unknowns initially contained in a matrix are arranged into a

Size of the matrices	Derivatives
$\mathbf{X} \in \mathbb{R}^{M \times N}$	$\frac{\partial \operatorname{tr}(\mathbf{X})}{\partial \mathbf{X}} = \mathbf{I}_M$
$(\mathbf{A}, \mathbf{X}) \in \mathbb{R}^{N \times M} \times \mathbb{R}^{M \times N}$	$\frac{\partial \operatorname{tr}(\mathbf{AX})}{\partial \mathbf{X}} = \mathbf{A}^T$
$(\mathbf{A}, \mathbf{X}) \in \mathbb{R}^{M \times N} \times \mathbb{R}^{M \times N}$	$\frac{\partial \operatorname{tr}(\mathbf{X}^T \mathbf{A})}{\partial \mathbf{X}} = \frac{\partial \operatorname{tr}(\mathbf{A}^T \mathbf{X})}{\partial \mathbf{X}} = \mathbf{A}$
$(\mathbf{A}, \mathbf{X}, \mathbf{B}) \in \mathbb{R}^{I \times M} \times \mathbb{R}^{M \times N} \times \mathbb{R}^{N \times I}$	$\frac{\partial \operatorname{tr}(\mathbf{AXB})}{\partial \mathbf{X}} = \mathbf{A}^T \mathbf{B}^T$
$(\mathbf{A}, \mathbf{X}, \mathbf{B}) \in \mathbb{R}^{I \times N} \times \mathbb{R}^{M \times N} \times \mathbb{R}^{M \times I}$	$\frac{\partial \operatorname{tr}(\mathbf{AX}^T \mathbf{B})}{\partial \mathbf{X}} = \mathbf{BA}$

Table 2.28. *Derivatives of matrix traces*

vector. A new expression for the equations is obtained by vectorizing both sides of each equation. The system to solve is then expressed in the standard form $\mathbf{Ax} = \mathbf{b}$.

PROPOSITION 2.62.– For $\mathbf{T} \in \mathbb{C}^{M \times I}$, $\mathbf{S} \in \mathbb{C}^{N \times J}$, and $\mathbf{X} \in \mathbb{C}^{I \times J}$, using the relations [2.116], [2.117] and [2.113], we can deduce the following equivalences:

$$\mathbf{TX} = \mathbf{C} \iff (\mathbf{I}_J \otimes \mathbf{T})\operatorname{vec}(\mathbf{X}) = \operatorname{vec}(\mathbf{C}) \ , \quad \mathbf{C} \in \mathbb{C}^{M \times J} \tag{2.281}$$

$$\mathbf{XS}^T = \mathbf{C} \iff (\mathbf{S} \otimes \mathbf{I}_I)\operatorname{vec}(\mathbf{X}) = \operatorname{vec}(\mathbf{C}) \ , \quad \mathbf{C} \in \mathbb{C}^{I \times N} \tag{2.282}$$

$$\mathbf{TXS}^T = \mathbf{C} \iff (\mathbf{S} \otimes \mathbf{T})\operatorname{vec}(\mathbf{X}) = \operatorname{vec}(\mathbf{C}) \ , \quad \mathbf{C} \in \mathbb{C}^{M \times N}. \tag{2.283}$$

More generally, for $\mathbf{T}_p \in \mathbb{C}^{M \times I}$, $\mathbf{S}_p \in \mathbb{C}^{N \times J}$, $\mathbf{C} \in \mathbb{C}^{M \times N}$, and $\mathbf{X} \in \mathbb{C}^{I \times J}$, we have:

$$\sum_{p=1}^{P} \mathbf{T}_p \mathbf{XS}_p^T = \mathbf{C} \iff \sum_{p=1}^{P} (\mathbf{S}_p \otimes \mathbf{T}_p)\operatorname{vec}(\mathbf{X}) = \operatorname{vec}(\mathbf{C}). \tag{2.284}$$

2.12.2.1. *Continuous-time Sylvester and Lyapunov equations*

Using the relations [2.281] and [2.282], the continuous-time Sylvester equation:

$$\mathbf{TX} + \mathbf{XS} = \mathbf{C} \tag{2.285}$$

with $\mathbf{T} \in \mathbb{C}^{I \times I}, \mathbf{S} \in \mathbb{C}^{J \times J}$, and $\mathbf{C}, \mathbf{X} \in \mathbb{C}^{I \times J}$, can be written as a system of IJ linear equations with IJ unknowns in the following form:

$$\Big((\mathbf{I}_J \otimes \mathbf{T}) + (\mathbf{S}^T \otimes \mathbf{I}_I)\Big)\operatorname{vec}(\mathbf{X}) = \operatorname{vec}(\mathbf{C}), \tag{2.286}$$

or alternatively:

$$(\mathbf{S}^T \oplus \mathbf{T})\operatorname{vec}(\mathbf{X}) = \operatorname{vec}(\mathbf{C}), \tag{2.287}$$

where $\oplus$ denotes the Kronecker sum defined in [2.63]. If the matrix $\mathbf{S}^T \oplus \mathbf{T}$ is non-singular, we obtain the following LS solution:

$$\text{vec}(\hat{\mathbf{X}}) = (\mathbf{S}^T \oplus \mathbf{T})^{-1}\text{vec}(\mathbf{C}). \qquad [2.288]$$

Given the property [2.64], the eigenvalues of $\mathbf{S}^T \oplus \mathbf{T}$ are $\lambda_i + \mu_j$, where λ_i and μ_j are eigenvalues of $\mathbf{T}$ and $\mathbf{S}$, and therefore of $\mathbf{S}^T$, respectively. Hence, the uniqueness condition of the LS solution [2.288] is $\lambda_i + \mu_j \neq 0, \forall i \in \langle I \rangle, \forall j \in \langle J \rangle$, which means that the eigenvalues of $\mathbf{S}$ must not be equal to the eigenvalues of $\mathbf{T}$ with opposite sign.

In the case where $\mathbf{S} = \mathbf{T}^T$, with $I = J$, Sylvester's equation [2.285] corresponds to a continuous-time Lyapunov equation that is encountered in filtering and optimal control problems. The uniqueness condition stated above then becomes $\lambda_i + \lambda_j \neq 0, \forall i, j \in \langle I \rangle$.

2.12.2.2. *Discrete-time Sylvester and Lyapunov equations*

In the discrete-time case, the Sylvester and Lyapunov equations are written as follows, respectively:

$$\mathbf{X} - \mathbf{TXS} = \mathbf{C} \qquad [2.289]$$

and

$$\mathbf{X} - \mathbf{TXT}^T = \mathbf{C}. \qquad [2.290]$$

In the case where $I = J$, these equations can be rewritten as:

$$\left(\mathbf{I}_{I^2} - (\mathbf{S}^T \otimes \mathbf{T})\right)\text{vec}(\mathbf{X}) = \text{vec}(\mathbf{C}) \qquad [2.291]$$

$$\left(\mathbf{I}_{I^2} - (\mathbf{T} \otimes \mathbf{T})\right)\text{vec}(\mathbf{X}) = \text{vec}(\mathbf{C}), \qquad [2.292]$$

which leads us to the following LS solutions:

$$\text{vec}(\hat{\mathbf{X}}) = \left(\mathbf{I}_{I^2} - (\mathbf{S}^T \otimes \mathbf{T})\right)^{-1}\text{vec}(\mathbf{C}) \qquad [2.293]$$

$$\text{vec}(\hat{\mathbf{X}}) = \left(\mathbf{I}_{I^2} - (\mathbf{T} \otimes \mathbf{T})\right)^{-1}\text{vec}(\mathbf{C}) \qquad [2.294]$$

if the matrices $\mathbf{I}_{I^2} - (\mathbf{S}^T \otimes \mathbf{T})$ and $\mathbf{I}_{I^2} - (\mathbf{T} \otimes \mathbf{T})$ are non-singular, or equivalently if $\det(\mathbf{I}_{I^2} - (\mathbf{S}^T \otimes \mathbf{T})) \neq 0$ and $\det(\mathbf{I}_{I^2} - (\mathbf{T} \otimes \mathbf{T})) \neq 0$, i.e. if all the eigenvalues of $\mathbf{S}^T \otimes \mathbf{T}$ and $\mathbf{T} \otimes \mathbf{T}$ are not 1. Noting the property [2.60] concerning the spectrum of a Kronecker product, this can be translated into the following conditions: $\lambda_i \mu_j \neq 1, \forall i \in \langle I \rangle, \forall j \in \langle J \rangle$, and $\lambda_i \neq \pm 1, \forall i \in \langle I \rangle$, respectively, where λ_i and μ_j represent eigenvalues of $\mathbf{T}$ and $\mathbf{S}$, respectively.

2.12.2.3. *Equations of the form* $\mathbf{Y} = \mathbf{AXB}^T$

Consider the equation $\mathbf{AXB}^T = \mathbf{Y}$, where $\mathbf{A}$ and $\mathbf{B}$ have full column rank, which implies that $\mathbf{B} \otimes \mathbf{A}$ has full column rank. Using the relation [2.283] and the property [2.62], the optimal LS solution is given by:

$$\min_{\mathbf{X}} \|\mathbf{Y} - \mathbf{AXB}^T\|_F^2 \;\Leftrightarrow\; \min_{\mathbf{X}} \|\text{vec}(\mathbf{Y}) - (\mathbf{B} \otimes \mathbf{A})\text{vec}(\mathbf{X})\|_2^2$$

$$\Downarrow$$

$$\begin{aligned}
\text{vec}(\hat{\mathbf{X}}) &= (\mathbf{B} \otimes \mathbf{A})^\dagger \text{vec}(\mathbf{Y}) = [(\mathbf{B} \otimes \mathbf{A})^H (\mathbf{B} \otimes \mathbf{A})]^{-1} (\mathbf{B} \otimes \mathbf{A})^H \text{vec}(\mathbf{Y}) \\
&= (\mathbf{B}^H \mathbf{B} \otimes \mathbf{A}^H \mathbf{A})^{-1} (\mathbf{B}^H \otimes \mathbf{A}^H) \text{vec}(\mathbf{Y}) \\
&= ((\mathbf{B}^H \mathbf{B})^{-1} \mathbf{B}^H \otimes (\mathbf{A}^H \mathbf{A})^{-1} \mathbf{A}^H) \text{vec}(\mathbf{Y}) \\
&= (\mathbf{B}^\dagger \otimes \mathbf{A}^\dagger) \text{vec}(\mathbf{Y}). \qquad\qquad\qquad [2.295]
\end{aligned}$$

This solution can also be written as:

$$\hat{\mathbf{X}} = \mathbf{A}^\dagger \mathbf{Y} (\mathbf{B}^T)^\dagger. \qquad\qquad [2.296]$$

PROOF.– Applying the property [2.113] of the Kronecker product to the right-hand side of equation [2.296], we obtain $\text{vec}(\hat{\mathbf{X}}) = \left[(\mathbf{B}^T)^\dagger \right]^T \otimes \mathbf{A}^\dagger \right] \text{vec}(\mathbf{Y})$, where $[(\mathbf{B}^T)^\dagger]^T = [\mathbf{B}^*(\mathbf{B}^T \mathbf{B}^*)^{-1}]^T = (\mathbf{B}^H \mathbf{B})^{-1} \mathbf{B}^H = \mathbf{B}^\dagger$, and so we recover [2.295]. □

2.12.2.4. *Equations of the form* $\mathbf{Y} = (\mathbf{B} \diamond \mathbf{A})\mathbf{X}$

Consider the equation $(\mathbf{B} \diamond \mathbf{A})\mathbf{X} = \mathbf{Y}$, where $\mathbf{B} \in \mathbb{C}^{I \times J}$, $\mathbf{A} \in \mathbb{C}^{K \times J}$, $\mathbf{X} \in \mathbb{C}^{J \times N}$, $\mathbf{Y} \in \mathbb{C}^{IK \times N}$. This type of equation will be encountered in Chapter 5 when estimating the parameters of a third-order PARAFAC model. Assuming that $\mathbf{B} \diamond \mathbf{A}$ has full column rank, which implies the necessary but not sufficient condition $IK \geq J$, and using the relation [2.255], the LS estimate of $\mathbf{X}$ is given by:

$$\begin{aligned}
\hat{\mathbf{X}} &= (\mathbf{B} \diamond \mathbf{A})^\dagger \mathbf{Y} = [(\mathbf{B} \diamond \mathbf{A})^H (\mathbf{B} \diamond \mathbf{A})]^{-1} (\mathbf{B} \diamond \mathbf{A})^H \mathbf{Y} \\
&= \left(\mathbf{B}^H \mathbf{B} \odot \mathbf{A}^H \mathbf{A} \right)^{-1} (\mathbf{B} \diamond \mathbf{A})^H \mathbf{Y}. \qquad\qquad [2.297]
\end{aligned}$$

If $\mathbf{B}$ is column orthonormal, then using the orthonormality property of $\mathbf{B}$ allows us to simplify this LS solution as:

$$\begin{aligned}
\hat{\mathbf{X}} &= \left(\mathbf{I}_J \odot \mathbf{A}^H \mathbf{A} \right)^{-1} (\mathbf{B} \diamond \mathbf{A})^H \mathbf{Y} \\
&= \left(\text{diag}(\|\mathbf{A}_{.1}\|_2^2, \cdots, \|\mathbf{A}_{.J}\|_2^2) \right)^{-1} (\mathbf{B} \diamond \mathbf{A})^H \mathbf{Y}. \qquad [2.298]
\end{aligned}$$

This type of simplified solution is used in Freitas *et al.* (2018) to compute a semi-blind receiver in a communication system with relays using column orthonormal coding matrices.

2.12.2.5. *Estimation of the factors of a Khatri–Rao product*

In many applications based on tensor approaches using PARAFAC or Tucker models, parametric estimation of these models plays a fundamental role. In some cases, it is desirable to estimate the factors of Khatri–Rao or Kronecker products, respectively. Some of these factors, denoted $\mathbf{A}$, $\mathbf{B}$, $\mathbf{C}$ and $\mathbf{D}$, are estimated by minimizing the following LS criteria:

$$\min_{\mathbf{A},\mathbf{B}}\|\mathbf{Y} - \mathbf{A} \diamond \mathbf{B}\|_F^2 \text{ and } \min_{\mathbf{C},\mathbf{D}}\|\mathbf{Z} - \mathbf{C} \otimes \mathbf{D}\|_F^2, \qquad [2.299]$$

where $\mathbf{A} = [\mathbf{a}_1, \cdots, \mathbf{a}_R] \in \mathbb{K}^{I \times R}$, $\mathbf{B} = [\mathbf{b}_1, \cdots, \mathbf{b}_R] \in \mathbb{K}^{J \times R}$, $\mathbf{C} \in \mathbb{K}^{I \times J}$, and $\mathbf{D} \in \mathbb{K}^{K \times L}$, and $\mathbf{Y} = [\mathbf{y}_1, \cdots, \mathbf{y}_R] \in \mathbb{K}^{IJ \times R}$ and $\mathbf{Z} \in \mathbb{K}^{IK \times JL}$ are given noisy matrices, i.e. $\mathbf{Y} = \mathbf{A} \diamond \mathbf{B} + \mathbf{E}$ and $\mathbf{Z} = \mathbf{C} \otimes \mathbf{D} + \mathbf{F}$, where $\mathbf{E}$ and $\mathbf{F}$ represent additive noise matrices caused by estimation and/or modeling errors.

First, consider the case of a Khatri–Rao product. By vectorizing $\mathbf{Y}$, the LS criterion can be rewritten as:

$$\min_{\mathbf{A},\mathbf{B}}\|\text{vec}(\mathbf{Y}) - \text{vec}(\mathbf{A} \diamond \mathbf{B})\|_2^2 = \min_{\substack{\mathbf{a}_r,\mathbf{b}_r \\ r\in\langle R\rangle}} \sum_{r=1}^{R} \|\mathbf{y}_r - \mathbf{a}_r \diamond \mathbf{b}_r\|_2^2. \qquad [2.300]$$

Since each term in this sum can be minimized separately, the columns $\mathbf{a}_r \in \mathbb{K}^I$ and $\mathbf{b}_r \in \mathbb{K}^J$ are estimated by minimizing the criterion $\min\limits_{\mathbf{a}_r,\mathbf{b}_r} \|\mathbf{y}_r - \mathbf{a}_r \diamond \mathbf{b}_r\|_2^2$.

Let us define $\mathbf{Y}_r \triangleq \text{unvec}(\mathbf{y}_r) \in \mathbb{K}^{J \times I}$ as the matrix obtained by inverting the vectorization operation. Applying the identity [2.103] then gives us:

$$\min_{\mathbf{a}_r,\mathbf{b}_r} \|\mathbf{Y}_r - \mathbf{b}_r \mathbf{a}_r^T\|_F^2 = \min_{\mathbf{a}_r,\mathbf{b}_r} \|\mathbf{Y}_r - \mathbf{b}_r \circ \mathbf{a}_r\|_F^2 , \quad r \in \langle R\rangle, \qquad [2.301]$$

where $\circ$ denotes the outer product defined in [3.107]. Now that the criterion has been rewritten in terms of a rank-one matrix, the vectors $\mathbf{a}_r$ and $\mathbf{b}_r$ can be estimated by computing the rank-one reduced SVD of $\mathbf{Y}_r$:

$$\mathbf{Y}_r = \sigma_r^{(1)}\mathbf{u}_r^{(1)}(\mathbf{v}_r^{(1)})^H, \qquad [2.302]$$

where $\sigma_r^{(1)}$ denotes the largest singular value, and $\mathbf{u}_r^{(1)}$ and $\mathbf{v}_r^{(1)}$ are the left and right singular vectors associated with $\sigma_r^{(1)}$, respectively. Identifying the reduced SVD [2.302] with $\mathbf{b}_r\mathbf{a}_r^T$ leads us to the following estimation formulae (Kibangou and Favier 2009b):

$$\hat{\mathbf{a}}_r = \sqrt{\sigma_r^{(1)}}\mathbf{v}_r^{(1)*} , \quad \hat{\mathbf{b}}_r = \sqrt{\sigma_r^{(1)}}\mathbf{u}_r^{(1)}. \qquad [2.303]$$

REMARK 2.63.–

– The matrix $\mathbf{Y}_r$ can be formed from the vector $\mathbf{y}_r$ using the following transformation:

$$(\mathbf{Y}_r)_{j,i} = (\mathbf{y}_r)_{j+(i-1)J} \ , \ i \in \langle I \rangle \, , \, j \in \langle J \rangle. \qquad [2.304]$$

– Note that the R columns of $\mathbf{A}$ and $\mathbf{B}$ can be estimated in parallel.

– Each vector $\mathbf{a}_r$ and $\mathbf{b}_r$, $r \in \langle R \rangle$, is only estimated up to a scaling factor, since $(\lambda_r \mathbf{a}_r) \diamond (\frac{1}{\lambda_r} \mathbf{b}_r) = \mathbf{a}_r \diamond \mathbf{b}_r$ for every $\lambda_r \in \mathbb{K}$. To eliminate this scaling ambiguity, we simply need to know one component of one of these two vectors. For example, if we assume that the first component $(\mathbf{a}_r)_1$ is equal to 1, then $\lambda_r = \frac{1}{(\hat{\mathbf{a}}_r)_1}$, and the estimated vectors are as follows after removing the ambiguity:

$$\hat{\hat{\mathbf{a}}}_r = \frac{1}{(\hat{\mathbf{a}}_r)_1} \hat{\mathbf{a}}_r \ , \ \hat{\hat{\mathbf{b}}}_r = (\hat{\mathbf{a}}_r)_1 \hat{\mathbf{b}}_r, \qquad [2.305]$$

where $\hat{\mathbf{a}}_r$ and $\hat{\mathbf{b}}_r$ are defined in [2.303]. From the above, we can conclude that the factors $\mathbf{A}$ and $\mathbf{B}$ of a Khatri–Rao product can be estimated without scaling ambiguity if we have *a priori* knowledge of one row of $\mathbf{A}$ or $\mathbf{B}$. This follows from the fact that $\mathbf{A}\boldsymbol{\Lambda} \diamond \mathbf{B}\boldsymbol{\Lambda}^{-1} = \mathbf{A} \diamond \mathbf{B}$ for any non-singular matrix $\boldsymbol{\Lambda} = \operatorname{diag}(\lambda_1, \cdots, \lambda_R) \in \mathbb{K}^{R \times R}$.

– In the special case of a Khatri–Rao product of two vectors of which one is known, the other can be estimated using the LS method. Let $\mathbf{y} = \mathbf{a} \diamond \mathbf{b}$, where $\mathbf{a} \in \mathbb{K}^I$, $\mathbf{b} \in \mathbb{K}^J$ and $\mathbf{y} \in \mathbb{K}^{IJ}$. The LS estimates of the components a_i, $i \in \langle I \rangle$ and b_j, $j \in \langle J \rangle$ of $\mathbf{a}$ and $\mathbf{b}$ are given by:

- If $\mathbf{b}$ is known:

$$\hat{a}_i = \frac{\mathbf{b}^T \mathbf{y}(i)}{\|\mathbf{b}\|_2^2} \ , \ i \in \langle I \rangle \qquad [2.306]$$

$$\mathbf{y}(i) \triangleq [y_{1+(i-1)J}, \cdots, y_{iJ-1}, y_{iJ}]^T = a_i \mathbf{b} \in \mathbb{K}^J. \qquad [2.307]$$

- If $\mathbf{a}$ is known:

$$\hat{b}_j = \frac{\mathbf{a}^T \mathbf{y}(j)}{\|\mathbf{a}\|_2^2} \ , \ j \in \langle J \rangle \qquad [2.308]$$

$$\mathbf{y}(j) \triangleq [y_j, y_{j+J}, \cdots y_{j+(I-1)J}]^T = b_j \mathbf{a} \in \mathbb{K}^I. \qquad [2.309]$$

Note that, from [2.306] and [2.308], we can construct an iterative alternating least squares (ALS) algorithm for estimating $\mathbf{a}$ and $\mathbf{b}$ by applying each formula in turn. See section 5.2.5.7 for a description of the ALS algorithm in the context of estimating a PARAFAC model. The advantage of this algorithm compared to the previous method based on computing the SVD is its high numerical simplicity, since it does not require any matrix computations. However, like any iterative algorithm, its speed of convergence depends on the initialization.

The algorithm that estimates the factors of a Khatri–Rao product by finding rank-one approximations of R matrices can be generalized to the case of a multiple Khatri–Rao product: $\mathbf{Y} = \overset{N}{\underset{n=1}{\diamond}} \mathbf{A}^{(n)}$, with $\mathbf{A}^{(n)} = [\mathbf{a}_1^{(n)}, \cdots, \mathbf{a}_R^{(n)}] \in \mathbb{K}^{I_n \times R}$, $n \in \langle N \rangle$. This generalization is based on the rank-one approximation of R tensors of order N constructed from the columns of the factors $\mathbf{A}^{(n)}$. Consider the column vector $\mathbf{y}_r = \overset{N}{\underset{n=1}{\diamond}} \mathbf{a}_r^{(n)} \in \mathbb{K}^{I_1 \cdots I_N}$, for $r \in \langle R \rangle$, and the associated rank-one tensor $\mathcal{X}_r = \overset{N}{\underset{n=1}{\circ}} \mathbf{a}_r^{(n)} \in \mathbb{K}^{I_1 \times \cdots \times I_N}$ defined using the following transformation:

$$(\mathcal{X}_r)_{i_1, \cdots, i_N} = (\mathbf{y}_r)_{\overline{i_1 \cdots i_N}}, \tag{2.310}$$

where $\overline{i_1 \cdots i_N} \triangleq i_N + \sum_{n=1}^{N-1}(i_n - 1)\prod_{k=n+1}^{N} I_k$, as defined in [3.118] for the vectorization of a rank-one tensor. We can now determine the columns $\mathbf{a}_r^{(n)}$ from a rank-one approximation of the tensor $\mathcal{X}_r$ using the THOSVD algorithm, which will be presented in section 5.2.1.8. Like for a Khatri–Rao product of two vectors, the R columns of the factors $\mathbf{A}^{(n)}$ can be estimated in parallel.

2.12.2.6. *Estimation of the factors of a Kronecker product*

Like for the Khatri–Rao product, an LS estimate of the factors of $\mathbf{Z} = \mathbf{C} \otimes \mathbf{D}$ can be found by solving a rank-one matrix approximation problem using the SVD (Van Loan and Pitsianis 1993). By the identity [2.33] and the property $\|\mathbf{u} \otimes \mathbf{v}\|_2^2 = \|\mathbf{u}\mathbf{v}^T\|_F^2$, the Frobenius norm of the Kronecker product can be written as follows:

$$\|\mathbf{C} \otimes \mathbf{D}\|_F^2 = \sum_{j=1}^{J}\sum_{l=1}^{L} \|\mathbf{C}_{.j} \otimes \mathbf{D}_{.l}\|_2^2 = \sum_{j=1}^{J}\sum_{l=1}^{L} \|\mathbf{C}_{.j}\mathbf{D}_{.l}^T\|_F^2. \tag{2.311}$$

By concatenating the columns of $\mathbf{C}$ and $\mathbf{D}$, i.e. by vectorizing $\mathbf{C}$ and $\mathbf{D}$, we also have:

$$\|\mathbf{C} \otimes \mathbf{D}\|_F^2 = \|\text{vec}(\mathbf{C})\text{vec}^T(\mathbf{D})\|_F^2. \tag{2.312}$$

This transformation of $\mathbf{Z} = \mathbf{C} \otimes \mathbf{D}$ into $\mathbf{X} = \text{vec}(\mathbf{C})\text{vec}^T(\mathbf{D})$ amounts to placing the element $c_{ij}d_{kl} = z_{k+(i-1)K, l+(j-1)L}$ in $\mathbf{X}$ at the position defined by the matrix $\mathbf{e}_{ji}^{lk} = \mathbf{e}_j^{(J)} \otimes \mathbf{e}_i^{(I)} \otimes (\mathbf{e}_l^{(L)})^T \otimes (\mathbf{e}_k^{(K)})^T$, which is equivalent to performing the following transformation:

$$x_{i+(j-1)I, k+(l-1)K} = z_{k+(i-1)K, l+(j-1)L}. \tag{2.313}$$

The LS criterion can be rewritten in terms of the matrix $\mathbf{X}$ as:

$$\|\mathbf{Z} - \mathbf{C} \otimes \mathbf{D}\|_F^2 = \|\mathbf{X} - \text{vec}(\mathbf{C})\text{vec}^T(\mathbf{D})\|_F^2 = \|\mathbf{X} - \text{vec}(\mathbf{C}) \circ \text{vec}(\mathbf{D})\|_F^2. \tag{2.314}$$

Since the matrix $\mathbf{X}$ is of rank one, its SVD can be written as $\mathbf{X} = \mathbf{U}\boldsymbol{\Sigma}\mathbf{V}^H = \sigma^{(1)}\mathbf{u}^{(1)}\mathbf{v}^{(1)H}$, where $\sigma^{(1)}$ is the largest singular value, and $\mathbf{u}^{(1)}$ and $\mathbf{v}^{(1)}$ are the left and right singular vectors associated with $\sigma^{(1)}$. The factor matrices $\mathbf{C}$ and $\mathbf{D}$ can therefore be estimated in vectorized form as follows:

$$\mathrm{vec}(\hat{\mathbf{C}}) = \sqrt{\sigma^{(1)}}\mathbf{u}^{(1)} \ , \ \ \mathrm{vec}(\hat{\mathbf{D}}) = \sqrt{\sigma^{(1)}}\mathbf{v}^{(1)*}. \qquad [2.315]$$

EXAMPLE 2.64.– For $I = J = K = L = 2$, we have:

$$\mathbf{Z} = \mathbf{C} \otimes \mathbf{D} = \begin{bmatrix} c_{11}d_{11} & c_{11}d_{12} & c_{12}d_{11} & c_{12}d_{12} \\ c_{11}d_{21} & c_{11}d_{22} & c_{12}d_{21} & c_{12}d_{22} \\ c_{21}d_{11} & c_{21}d_{12} & c_{22}d_{11} & c_{22}d_{12} \\ c_{21}d_{21} & c_{21}d_{22} & c_{22}d_{21} & c_{22}d_{22} \end{bmatrix}$$

$$\mathbf{X} = \mathrm{vec}(\mathbf{C})\mathrm{vec}^T(\mathbf{D}) = \begin{bmatrix} c_{11}d_{11} & c_{11}d_{21} & c_{11}d_{12} & c_{11}d_{22} \\ c_{21}d_{11} & c_{21}d_{21} & c_{21}d_{12} & c_{21}d_{22} \\ c_{12}d_{11} & c_{12}d_{21} & c_{12}d_{12} & c_{12}d_{22} \\ c_{22}d_{11} & c_{22}d_{21} & c_{22}d_{12} & c_{22}d_{22} \end{bmatrix}$$

$$= \begin{bmatrix} z_{11} & z_{21} & z_{12} & z_{22} \\ z_{31} & z_{41} & z_{32} & z_{42} \\ z_{13} & z_{23} & z_{14} & z_{24} \\ z_{33} & z_{43} & z_{34} & z_{44} \end{bmatrix}.$$

REMARK 2.65.– Unlike a Khatri–Rao product of two matrices, which admits a diagonal ambiguity matrix, a Kronecker product is ambiguous up to a scalar scaling factor, since $(\lambda\mathbf{C}) \otimes (\frac{1}{\lambda}\mathbf{D}) = \mathbf{C} \otimes \mathbf{D}$. The factors $\mathbf{C}$ and $\mathbf{D}$ can therefore be estimated without ambiguity, provided that one element of $\mathbf{C}$ or $\mathbf{D}$ is known.

In the case of a Kronecker product $\mathbf{Z} = \mathbf{C} \otimes \mathbf{D}$ where one of the two factors is known, it is possible to estimate the other factor using the LS method. Thus, for example, if $\mathbf{D}$ is known, then each element c_{ij} of $\mathbf{C}$ can be estimated from the block $\mathbf{Z}_{ij} = c_{ij}\mathbf{D}$ of $\mathbf{Z}$ associated with the coefficient c_{ij}. By vectorizing both sides of this equation, we obtain the following LS estimate:

$$\hat{c}_{ij} = \frac{\left(\mathrm{vec}(\mathbf{D})\right)^H \mathrm{vec}(\mathbf{Z}_{ij})}{\|\mathbf{D}\|_F^2} \ , \ i \in \langle I \rangle \, , j \in \langle J \rangle. \qquad [2.316]$$

Similarly, in the case where $\mathbf{C}$ is known, after permuting the factors $\mathbf{C}$ and $\mathbf{D}$ using the relation [2.171]:

$$\mathbf{Y} = \mathbf{D} \otimes \mathbf{C} = \mathbf{K}_{KI}(\mathbf{C} \otimes \mathbf{D})\mathbf{K}_{JL}, \qquad [2.317]$$

with the commutation matrices $\mathbf{K}_{KI}$ and $\mathbf{K}_{JL}$ defined in [2.173] and [2.175], it is possible to estimate the factor $\mathbf{D}$ by applying the LS method to the equation $\mathbf{Y}_{kl} = d_{kl}\mathbf{C}$, which gives:

$$\hat{d}_{kl} = \frac{(\text{vec}(\mathbf{C}))^H \text{vec}(\mathbf{Y}_{kl})}{\|\mathbf{C}\|_F^2} \ , \ k \in \langle K \rangle, \ l \in \langle L \rangle. \qquad [2.318]$$

Like for the Khatri–Rao product, we can construct an iterative estimation algorithm for $\mathbf{C}$ and $\mathbf{D}$ based on the ALS method using equations [2.316] and [2.318] in turn.

The algorithm for estimating the factors of a Kronecker product presented above can be generalized to the case of a multiple Kronecker product: $\mathbf{Z} = \overset{N}{\underset{n=1}{\otimes}} \mathbf{A}^{(n)}$, with $\mathbf{A}^{(n)} \in \mathbb{K}^{I_n \times J_n}, n \in \langle N \rangle$.

The idea is to construct a rank-one tensor from $\mathbf{Z}$, defined as the outer product of the vectorized forms $\text{vec}(\mathbf{A}^{(n)}) \in \mathbb{K}^{J_n I_n}$ of the matrices $\mathbf{A}^{(n)}$:

$$\mathcal{X} = \text{vec}(\mathbf{A}^{(1)}) \circ \text{vec}(\mathbf{A}^{(2)}) \circ \cdots \circ \text{vec}(\mathbf{A}^{(N)}) \in \mathbb{K}^{J_1 I_1 \times J_2 I_2 \times \cdots \times J_N I_N}. \qquad [2.319]$$

This is equivalent to performing the following transformation:

$$(\mathcal{X})_{k_1, \cdots, k_N} = z_{\overline{i_1 \cdots i_N}, \overline{j_1 \cdots j_N}} \ , \ k_n \in \langle K_n \rangle, \ n \in \langle N \rangle, \qquad [2.320]$$

where $k_n = (j_n - 1)I_n + i_n$, $K_n = J_n I_n$, and $\overline{i_1 \cdots i_N}$ and $\overline{j_1 \cdots j_N}$ are defined as in [3.118].

The problem of estimating the factors $\mathbf{A}^{(n)}$ can be solved in the sense of minimizing the following LS criterion, which generalizes the criterion [2.314] to the case of a Kronecker product of N matrix factors:

$$\min_{\mathbf{A}^{(1)}, \cdots, \mathbf{A}^{(N)}} \|\mathbf{Z} - \overset{N}{\underset{n=1}{\otimes}} \mathbf{A}^{(n)}\|_F^2 = \min_{\mathbf{A}^{(1)}, \cdots, \mathbf{A}^{(N)}} \|\mathcal{X} - \text{vec}(\mathbf{A}^{(1)}) \circ \cdots \circ \text{vec}(\mathbf{A}^{(N)})\|_F^2.$$

This minimization amounts to finding the best rank-one approximation for the tensor $\mathcal{X}$ using the THOSVD method. This approximation gives an estimate of the vectorized forms $\text{vec}(\mathbf{A}^{(n)})$, from which it is then easy to deduce an estimate of the factors $\mathbf{A}^{(n)}$. This estimate is subject to a scalar scaling ambiguity λ_n for each factor $\mathbf{A}^{(n)}$, where $\prod_{n=1}^{N} \lambda_n = 1$.

Tensor Operations

3.1. Introduction

Tensor operations play an important role in tensor calculus. They allow us to rearrange the elements of a tensor into a vector or a matrix, and more generally into a reduced order tensor, to define tensor models and decompositions, and to compute the eigenvalues of a tensor. In this chapter, we will study two tensor multiplication operations, called the Tucker and Einstein products, more thoroughly. We will use this last product to introduce the notions of inverse and pseudo-inverse tensors, as well as tensor decompositions in the form of factorizations. The Hadamard and Kronecker tensor products will also be defined. Finally, we will describe a few examples of tensor systems that can be solved using the least squares (LS) method.

As we saw in Volume 1 (Favier 2019), there is a very close link between tensors, polynomials and multilinear forms. We will now use this connection to present different classes of tensors. Like in the matrix case, where the notions of symmetric bilinear form and positive definite quadratic form lead us to symmetric and positive definite matrices. The notion of symmetric positive definite multilinear form will lead us to symmetric positive definite tensors, as presented in Chapter 4. Several types of symmetry are presented in detail in this chapter, whose key objectives are to:

– present some links between multilinear forms, homogeneous polynomials and different sets of tensors;

– describe various operations with tensors (matricization, vectorization, transposition, multiplication, inversion, pseudo-inversion, extension, tensorization and Hankelization);

– present certain tensor decompositions, e.g. singular value decomposition (SVD) and full-rank factorization;

– introduce the notion of tensor systems and illustrate the solution of some of them using the LS method.

After introducing some notations relating to tensors and various sets of tensors, we will define the notions of slice (fibers, matrix slices and tensor slices) and mode combination. We will then present the most common ways of arranging the elements of a tensor as matrices and vectors, corresponding to the operations of matricization (also known as matrix unfolding) and vectorization. Next, we will describe the operation of transposition/transconjugation, introducing the notions of orthogonal/unitary tensors, as well as idempotence for square tensors. Various tensor multiplication operations and their properties will be presented, with a particular focus on the mode-p product, also known as the Tucker product, and the Einstein product. We will show how this last product can be used to define the notion of the generalized inverse and therefore of the Moore–Penrose pseudo-inverse of a tensor, to develop tensor factorizations, such as the eigendecomposition and SVD of a tensor, and to solve tensor systems. The Hadamard and Kronecker products of tensors will also be defined, as well as the operations of tensor extension, tensorization and Hankelization.

The organization and content of this chapter is as follows:

– in section 3.2, we define the notations and the various sets of tensors considered throughout the chapter;

– in section 3.3, we present the notions of matrix and tensor slices;

– the modal contraction operation resulting from a mode combination is presented in section 3.4;

– the notion of the partitioned tensor is introduced in section 3.5;

– (partially) diagonal tensors are defined in section 3.6;

– the operations of matricization and vectorization based on mode combinations are presented in sections 3.7 and 3.9. Mode-n matricized forms, with $n \in \langle N \rangle$, allow us to define the modal subspaces and the multilinear rank of an Nth-order tensor in section 3.8;

– section 3.10 focuses on the operation of transposition;

– a few structured tensors, including (partially) symmetric tensors and triangular tensors are presented in sections 3.11 and 3.12;

– various products with tensors that generalize matrix–matrix multiplication and their properties are described, distinguishing between the outer product (section 3.13.1), tensor–matrix (also called mode-p) multiplication (section 3.13.2), tensor–vector multiplication (section 3.13.3), mode-(p, n) product (section 3.13.4) and the Einstein product (section 3.13.5). The TT model will be briefly introduced using the mode-(p, n) product;

– the notions of the inverse, generalized inverse and Moore–Penrose pseudo-inverse of a tensor are introduced in section 3.14;

– tensor decompositions expressed as factorizations, such as the eigendecomposition of a square tensor and the SVD and full-rank decompositions of a rectangular tensor, are presented in section 3.15;

– the notions of the inner product, Frobenius norm and trace of a tensor are defined in section 3.16;

– some links between tensor products and (linear and multilinear) tensor systems, as well as the LS solution of certain systems, are established in section 3.17;

– the Hadamard and Kronecker tensor products are defined in section 3.18;

– the operations of tensor extension and tensorization, which consists of constructing a data tensor, are described in sections 3.19 and 3.20, respectively;

– the Hankelization operation, which can be viewed as a particular tensorization method allowing us to build a Hankel matrix or a higher order Hankel tensor from a long data vector, will also be briefly discussed in section 3.21.

3.2. Notation and particular sets of tensors

Let $\mathcal{X} \in \mathbb{K}^{I_1 \times \cdots \times I_N}$ be a tensor of order N and size $I_1 \times \cdots \times I_N$. The order corresponds to the number of indices that characterize its elements $x_{i_1,\cdots,i_N} \in \mathbb{K}$, also denoted $x_{i_1 \cdots i_N}$ or $(\mathcal{X})_{i_1,\cdots,i_N}$. Each index $i_n \in \langle I_n \rangle$, for $n \in \langle N \rangle$, is associated with a mode, also called a way, and I_n denotes the dimension of the nth mode. The number of elements in $\mathcal{X}$ is equal to $\prod_{n=1}^{N} I_n$. The tensor $\mathcal{X}$ is said to be real (respectively, complex) if its elements are real numbers (respectively, complex numbers), which corresponds to $\mathbb{K} = \mathbb{R}$ (respectively, $\mathbb{K} = \mathbb{C}$). It is said to have even order (respectively, odd order) if N is even (respectively, odd). The special cases $N = 2$ and $N = 1$ correspond to the sets of matrices $\mathbf{X} \in \mathbb{K}^{I \times J}$ and column vectors $\mathbf{x} \in \mathbb{K}^{I}$, respectively.

A real tensor $\mathcal{X} \in \mathbb{R}^{I_1 \times \cdots \times I_N}$ is said to be non-negative (respectively, positive) if all of its elements are positive or zero (respectively, positive), i.e. if $x_{i_1,\cdots,i_N} \geq 0$ (respectively, $x_{i_1,\cdots,i_N} > 0$) for all $i_n \in \langle I_n \rangle$ and all $n \in \langle N \rangle$. The tensor $\mathcal{X}$ is said to be the zero tensor if all of its elements are zero, regardless of its size. The zero tensor is denoted as $\mathcal{O}$.

In applications, a tensor is typically viewed as an array of numbers $[x_{i_1,\cdots,i_N}]$. This explains some of the other names given to tensors, such as N-way or multiway array. As we saw in Volume 1, the notion of a tensor of order N can be introduced by defining the tensor product (denoted $\otimes$) of N vector spaces $\mathcal{U}_n$ of dimension I_n defined over the same field $\mathbb{K}$ (Lim 2013). The resulting tensor space, denoted $\mathcal{U}_1 \otimes \cdots \otimes \mathcal{U}_N$, is the set of tensors of order N and size $I_1 \times \cdots \times I_N$, where $I_n = \dim(\mathcal{U}_n)$. By defining a basis $\mathcal{B}^{(n)} = \{\mathbf{b}_1^{(n)}, \cdots, \mathbf{b}_{I_n}^{(n)}\}$ for each vector space $\mathcal{U}_n$, any tensor $\mathcal{X}$ can be represented with respect to the bases $\{\mathcal{B}^{(n)}, n \in \langle N \rangle\}$ as a linear combination of tensor products of the basis vectors, i.e.:

$$\mathcal{X} = \sum_{i_1=1}^{I_1} \cdots \sum_{i_N=1}^{I_N} c_{i_1,\cdots,i_N} \mathbf{b}_{i_1}^{(1)} \otimes \mathbf{b}_{i_2}^{(2)} \otimes \cdots \otimes \mathbf{b}_{i_N}^{(N)}. \qquad [3.1]$$

The coordinates $c_{i_1,\cdots,i_N}$, for $i_n \in \langle I_n \rangle$, define a hypermatrix $\mathcal{C} = \left[c_{i_1,\cdots,i_N} \right] \in \mathbb{K}^{I_1 \times \cdots \times I_N}$ and characterize the tensor $\mathcal{X}$ with respect to the bases $\{\mathcal{B}^{(n)}, n \in \langle N \rangle\}$.

Note that the operator $\otimes$ applied to any general vectors of the vector spaces $\mathcal{U}_n$ is called the tensor product. But when the vector spaces $\mathcal{U}_n$ are defined as $\mathbb{K}^{I_n}$, the tensor space $\mathcal{U}_1 \otimes \cdots \otimes \mathcal{U}_N$ is more specifically the space $\mathbb{K}^{I_1 \times \cdots \times I_N}$ of dimension $I_1 \times \cdots \times I_N$, and the tensor product of the basis vectors $\overset{N}{\underset{n=1}{\otimes}} \mathbf{b}_{i_n}^{(n)} \triangleq \mathbf{b}_{i_1}^{(1)} \otimes \cdots \otimes \mathbf{b}_{i_N}^{(N)}$, with $\mathbf{b}_{i_n}^{(n)} \in \mathbb{K}^{I_n}$, can be replaced by the outer product of these vectors, denoted $\overset{N}{\underset{n=1}{\circ}} \mathbf{b}_{i_n}^{(n)} \triangleq \mathbf{b}_{i_1}^{(1)} \circ \cdots \circ \mathbf{b}_{i_N}^{(N)}$, which defines a rank-one tensor. In particular, in the canonical basis, we have:

$$\mathcal{X} = \sum_{i_1=1}^{I_1} \cdots \sum_{i_N=1}^{I_N} x_{i_1,\cdots,i_N} \overset{N}{\underset{n=1}{\circ}} \mathbf{e}_{i_n}^{(I_n)} \qquad [3.2]$$

$$= x_{i_1,\cdots,i_N} \mathcal{E}_{i_1 i_2 \cdots i_N}^{(I_1 \times I_2 \times \cdots \times I_N)},$$

where $\mathbf{e}_{i_n}^{(I_n)}$ is the i_nth vector of the canonical basis of $\mathbb{R}^{I_n}$, and $\mathcal{E}_{i_1 i_2 \cdots i_N}^{(I_1 \times I_2 \times \cdots \times I_N)}$ is the tensor of $\mathbb{R}^{I_1 \times I_2 \times \cdots \times I_N}$, with 1 at the position $(i_1, i_2, \cdots, i_N)$ and zeros everywhere else. The last equality uses the index convention. The coordinate hypermatrix is then written as $\mathcal{X} = \left[x_{i_1,\cdots,i_N} \right]$, i.e. as the tensor itself, with $i_n \in \langle I_n \rangle$ for $n \in \langle N \rangle$.

In the following, tensors will be identified with their coordinate hypermatrices, which means that the bases $\{\mathcal{B}^{(n)}, n \in \langle N \rangle\}$ are implicitly fixed.

If $I_1 = \cdots = I_N = I$, the Nth-order tensor $\mathcal{X} = \left[x_{i_1,\cdots,i_N} \right] \in \mathbb{K}^{I \times I \times \cdots \times I}$ is said to be hypercubic, of dimensions I, with $i_n \in \langle I \rangle$, for $n \in \langle N \rangle$. The number of elements in $\mathcal{X}$ is then equal to I^N. The set of (real or complex) hypercubic tensors of order N and dimensions I will be denoted $\mathbb{K}^{[N;I]}$. The special cases $N = 3$ and $N = 2$ correspond to the cubic tensors of size $I \times I \times I$ and the square matrices of size $I \times I$, respectively.

Hypercubic tensors play an important role in statistical signal processing approaches to solving blind source separation (BSS), and system identification problems using tensors of cumulants of order greater than two, in which each mode has the same dimension.

The identity tensor of order N and dimensions I is denoted $\mathcal{I}_{N,I} = [\delta_{i_1,\cdots,i_N}]$, with $i_n \in \langle I \rangle$ for $n \in \langle N \rangle$, or simply $\mathcal{I}$. It is a hypercubic tensor whose elements are defined using the generalized Kronecker delta:

$$\delta_{i_1,\cdots,i_N} = \begin{cases} 1 & \text{if} \quad i_1 = \cdots = i_N \\ 0 & \text{otherwise} \end{cases}.$$

The identity tensor $\mathcal{I}_{N,I} \in \mathbb{R}^{[N;I]}$ is a diagonal tensor whose diagonal elements are equal to 1 and whose other elements are all zero.

For the rest of this chapter, we will write $\mathbb{K}^{I_1 \times \cdots \times I_P \times J_1 \times \cdots \times J_N} \triangleq \mathbb{K}^{\underline{I}_P \times \underline{J}_N}$, with $\underline{I}_P = I_1 \times \cdots \times I_P$ and $\underline{J}_N = J_1 \times \cdots \times J_N$ for the set of tensors of order $P + N$ and of size $I_1 \times \cdots \times I_P \times J_1 \times \cdots \times J_N$. In the case where $I_p = I, \forall p \in \langle P \rangle$, and $J_n = J, \forall n \in \langle N \rangle$, the corresponding set of tensors will be denoted $\mathbb{K}^{[P+N;I,J]}$. If we also have $N = P$ and $I = J$, the set of tensors of order $2P$ and of dimensions I will be denoted $\mathbb{K}^{[2P;I]}$. Some authors write $\mathbb{K}^{I^{2P}}$ for this set instead.

Table 3.1 summarizes the notation used for sets of indices and dimensions.

$\underline{i}_P \triangleq \{i_1, \cdots, i_P\} \; ; \; \underline{j}_N \triangleq \{j_1, \cdots, j_N\}$
$\underline{I}_P \triangleq \{I_1, \cdots, I_P\} \; ; \; \underline{J}_N \triangleq \{J_1, \cdots, J_N\}$
$\underline{I}_P \triangleq I_1 \times \cdots \times I_P \; ; \; \underline{J}_N \triangleq J_1 \times \cdots \times J_N$
$\underline{I}_P \times \underline{J}_N = I_1 \times \cdots \times I_P \times J_1 \times \cdots \times J_N$
$\underline{I}_P \times \underline{I}_P = I_1 \times \cdots \times I_P \times I_1 \times \cdots \times I_P$
$\Pi_{I_P} \triangleq I_1 \cdots I_P = \prod_{p=1}^{P} I_p$

Table 3.1. *Notation for sets of indices and dimensions*

Using this notation and the index convention, the multiple sum over the indices of $x_{i_1,\cdots,i_P} y_{i_1,\cdots,i_P}$ will be abbreviated to:

$$\sum_{i_1=1}^{I_1} \cdots \sum_{i_P=1}^{I_P} x_{i_1,\cdots,i_P} y_{i_1,\cdots,i_P} = \sum_{\underline{i}_P=\underline{1}}^{\underline{I}_P} x_{\underline{i}_P} y_{\underline{i}_P} = x_{\underline{i}_P} y_{\underline{i}_P}, \qquad [3.3]$$

where $\underline{1}$ denotes a set of ones, whose number is fixed by the index P of the set $\underline{I}_P$. The notations $\underline{i}_P$ and $\underline{I}_P$ allow us to simplify the expression of the multiple sum into a single sum over an index set, which is further simplified by using the index convention.

Table 3.2 summarizes the various sets of tensors that will be considered in this chapter.

Order	Size	Sets of tensors
P	$\underline{I}_P = I_1 \times \cdots \times I_P$	$\mathbb{K}^{I_1 \times \cdots \times I_P} \triangleq \mathbb{K}^{\underline{I}_P}$
P	$\underline{I}_P = I_1 \times \cdots \times I_P$ with $I_p = I, \forall p \in \langle P \rangle$	$\mathbb{K}^{[P;I]}$
$P+N$	$\underline{I}_P \times \underline{J}_N = I_1 \times \cdots \times I_P \times J_1 \times \cdots \times J_N$	$\mathbb{K}^{\underline{I}_P \times \underline{J}_N}$
$P+N$	$\underline{I}_P \times \underline{J}_N = I \times \cdots \times I \times J \times \cdots \times J$ with $I_p = I, \forall p \in \langle P \rangle$ and $J_n = J, \forall n \in \langle N \rangle$	$\mathbb{K}^{[P+N;I,J]}$
$2P$	$\underline{I}_P \times \underline{I}_P$ with $I_p = I, \forall p \in \langle P \rangle$	$\mathbb{K}^{[2P;I]}$
$2P$	$I_1 \times J_1 \times \cdots \times I_P \times J_P$	$\mathbb{K}_p^{[2P\,;\,I_p \times J_p]}$

Table 3.2. *Various sets of tensors*

REMARK 3.1.– We can make the following remarks about the sets of tensors defined in Table 3.2:

– for $P = N = 1$, the set $\mathbb{K}^{[2;I,J]}$ is the set $\mathbb{K}^{I \times J}$ of (real or complex) matrices of size $I \times J$;

– the set $\mathbb{K}^{[P;I]}$ is also denoted $\mathbb{K}^{I^P}$ or $T^P(\mathbb{K}^I)$ by some authors;

– the set $\mathbb{K}^{\underline{I}_P \times \underline{I}_P}$ is called the set of even-order (or square) tensors of order $2P$ and size $\underline{I}_P \times \underline{I}_P$. The name square tensor comes from the fact that the index set is divided into two identical subsets of dimension $\underline{I}_P$;

– analogously to matrices, tensors in the sets $\mathbb{K}^{\underline{I}_P \times \underline{J}_P}$ with $J_p \neq I_p$ and $\mathbb{K}^{\underline{I}_P \times \underline{J}_N}$ are said to be rectangular. The set $\mathbb{K}^{\underline{I}_P \times \underline{J}_N}$ is called the set of rectangular tensors with index blocks of dimensions $\underline{I}_P$ and $\underline{J}_N$;

– the set $\mathbb{K}_p^{[2P\,;\,I_p \times J_p]} \triangleq \mathbb{K}^{I_1 \times J_1 \times \cdots \times I_P \times J_P}$, introduced by Huang and Qi (2018) and called the set of even-order paired tensors, corresponds to the case where the indices are divided into adjacent pairs $\{i_p, j_p\}$ associated with the dimensions $I_p \times J_p$, for $p \in \langle P \rangle$. The index p in $\mathbb{K}_p$ refers to this adjacent index pairing. These tensors play an important role in elasticity theory.

The various tensor classes introduced above can be associated with scalar multilinear forms in vector variables and homogeneous polynomials. Analogously to matrices, which can be associated with bilinear and quadratic forms in the real case and sesquilinear and Hermitian forms in the complex case, we will distinguish between real-valued multilinear forms and complex-valued multilinear forms. Like in the matrix case, we will distinguish between homogeneous polynomials of degree P that depend on the components of P vector variables and those that depend on just one vector variable. Another difference between the tensor case (for orders greater than two) and the matrix case is the number of conjugated variables compared to the number of non-conjugated variables.

These various multilinear forms are summarized in Table 3.3, which also states the transformations corresponding to each of them, as well as the associated tensors.

Multilin. forms	Transformations	Tensors
real-valued in P vectors	$\underset{p=1}{\overset{P}{\times}} \mathbb{R}^{I_P} \ni (\mathbf{x}^{(1)}, \cdots \mathbf{x}^{(P)}) \longmapsto f(\mathbf{x}^{(1)}, \cdots \mathbf{x}^{(P)}) \in \mathbb{R}$	$\mathcal{A} \in \mathbb{R}^{\underline{I}_P}$
real-valued in one vector	$\mathbb{R}^I \ni \mathbf{x} \longmapsto f(\underbrace{\mathbf{x}, \cdots, \mathbf{x}}_{P \text{ terms}}) \in \mathbb{R}$	$\mathcal{A} \in \mathbb{R}^{[P;I]}$
complex-valued in $P + N$ vectors	$\underset{p=1}{\overset{P}{\times}} \mathbb{C}^{I_P} \times \underset{n=1}{\overset{N}{\times}} \mathbb{C}^{J_n} \ni (\mathbf{y}^{(1)}, \cdots, y^{(P)}, \mathbf{x}^{(1)}, \cdots, \mathbf{x}^{(N)}) \longmapsto$ $f\big(\underbrace{(\mathbf{y}^{(1)})^*, \cdots, (\mathbf{y}^{(P)})^*}_{P \text{ terms}}, \underbrace{\mathbf{x}^{(1)}, \cdots, \mathbf{x}^{(N)}}_{N \text{ terms}}\big) \in \mathbb{C}$	$\mathcal{A} \in \mathbb{C}^{\underline{I}_P \times \underline{J}_N}$
complex-valued in one vector	$\mathbb{C}^I \ni \mathbf{x} \longmapsto f(\underbrace{\mathbf{x}^*, \cdots, \mathbf{x}^*}_{P \text{ terms}}, \underbrace{\mathbf{x}, \cdots, \mathbf{x}}_{P \text{ terms}}) \in \mathbb{C}$	$\mathcal{A} \in \mathbb{C}^{[2P;I]}$

Table 3.3. *Multilinear forms and associated tensors*

Table 3.4 recalls the definitions of bilinear/quadratic and sesquilinear/Hermitian forms using the index convention, and then presents the multilinear forms defined in Table 3.3, as well as the associated tensors from Table 3.2 and the corresponding homogeneous polynomials.

For instance, the real-valued multilinear form in P vectors $\mathbf{x}^{(p)} \in \mathbb{R}^{I_p}, p \in \langle P \rangle$ can be written as:

$$f\big(\mathbf{x}^{(1)}, \cdots, \mathbf{x}^{(P)}\big) = \sum_{i_1=1}^{I_1} \cdots \sum_{i_P=1}^{I_P} a_{i_1, \cdots, i_P} x_{i_1}^{(1)} \cdots x_{i_P}^{(P)} = a_{\underline{\mathbf{i}}_P} \prod_{p=1}^{P} x_{i_p}^{(p)}.$$

[3.4]

Similarly, the complex multilinear form in P conjugated vectors and N non-conjugated vectors can be written as:

$$f\big((\mathbf{y}^{(1)})^*, \cdots, (\mathbf{y}^{(P)})^*, \mathbf{x}^{(1)}, \cdots, \mathbf{x}^{(N)}\big) = \sum_{i_1=1}^{I_1} \cdots \sum_{i_P=1}^{I_P} \sum_{j_1=1}^{J_1} \cdots$$

$$\sum_{j_N=1}^{J_N} a_{i_1, \cdots, i_P, j_1, \cdots, j_N}$$

$$(\mathbf{y}_{i_1}^{(1)})^* \cdots (\mathbf{y}_{i_P}^{(P)})^* x_{j_1}^{(1)} \cdots x_{j_N}^{(N)} = a_{\underline{\mathbf{i}}_P, \underline{\mathbf{j}}_N} \prod_{p=1}^{P} (y_{i_p}^{(p)})^* \prod_{n=1}^{N} x_{j_n}^{(n)}. \qquad [3.5]$$

Forms	Matrices/Tensors	Homogeneous polynomials
Bilinear	$\mathbf{A} \in \mathbb{R}^{I \times J}; \mathbf{y} \in \mathbb{R}^I, \mathbf{x} \in \mathbb{R}^J$	$f(\mathbf{x}, \mathbf{y}) = \mathbf{y}^T \mathbf{A} \mathbf{x} = a_{ij} y_i x_j$
Quadratic	$\mathbf{A} \in \mathbb{R}^{I \times I}; \mathbf{x} \in \mathbb{R}^I$	$f(\mathbf{x}) = \mathbf{x}^T \mathbf{A} \mathbf{x} = a_{ij} x_i x_j$
Sesquilinear	$\mathbf{A} \in \mathbb{C}^{I \times J}; \mathbf{y} \in \mathbb{C}^I, \mathbf{x} \in \mathbb{C}^J$	$f(\mathbf{x}, \mathbf{y}) = \mathbf{y}^H \mathbf{A} \mathbf{x} = a_{ij} y_i^* x_j$
Hermitian	$\mathbf{A} \in \mathbb{C}^{I \times I}; \mathbf{x} \in \mathbb{C}^I$	$f(\mathbf{x}) = \mathbf{x}^H \mathbf{A} \mathbf{x} = a_{ij} x_i^* x_j$
Real multilinear in P vectors	$\mathcal{A} \in \mathbb{R}^{\underline{I}_P}; \mathbf{x}^{(p)} \in \mathbb{R}^{I_p}$	$f(\mathbf{x}^{(1)}, \cdots \mathbf{x}^{(P)}) = a_{\underline{\mathbf{i}}_P} \prod_{p=1}^{P} x_{i_p}^{(p)}$
Real multilinear in one vector	$\mathcal{A} \in \mathbb{R}^{[P;I]}; \mathbf{x} \in \mathbb{R}^I$	$f(\underbrace{\mathbf{x}, \cdots, \mathbf{x}}_{P \text{ terms}}) = a_{\underline{\mathbf{i}}_P} \prod_{p=1}^{P} x_{i_p}$
Complex multilinear in one vector	$\mathcal{A} \in \mathbb{C}^{[2P;I]}; \mathbf{x} \in \mathbb{C}^I$	$f(\underbrace{\mathbf{x}^*, \cdots, \mathbf{x}^*}_{P \text{ terms}}, \underbrace{\mathbf{x}, \cdots, \mathbf{x}}_{P \text{ terms}}) =$ $a_{\underline{\mathbf{i}}_P, \underline{\mathbf{j}}_P} \prod_{p=1}^{P} x_{i_p}^* \prod_{n=1}^{P} x_{j_n}$
Complex multilinear in $P + N$ vectors	$\mathcal{A} \in \mathbb{C}^{\underline{I}_P \times \underline{J}_N};$ $\mathbf{y}^{(p)} \in \mathbb{C}^{I_p}, p \in \langle P \rangle;$ $\mathbf{x}^{(n)} \in \mathbb{C}^{J_n}, n \in \langle N \rangle$	$f(\underbrace{(\mathbf{y}^{(1)})^*, \cdots, (\mathbf{y}^{(P)})^*}_{P \text{ terms}}, \underbrace{\mathbf{x}^{(1)}, \cdots, \mathbf{x}^{(N)}}_{N \text{ terms}}) =$ $a_{\underline{\mathbf{i}}_P, \underline{\mathbf{j}}_N} \prod_{p=1}^{P} (y_{i_p}^{(p)})^* \prod_{n=1}^{N} x_{j_n}^{(n)}$

Table 3.4. *Multilinear forms and associated homogeneous polynomials*

REMARK 3.2.– We can make the following remarks:

– In the same way that bilinear/sesquilinear forms depend on two variables that do not belong to the same vector space, general real and complex multilinear forms depend on variables that belong to different vector spaces: P variables $\mathbf{x}^{(p)} \in \mathbb{R}^{I_p}$ in the real case; P conjugated variables $\mathbf{y}^{(p)} \in \mathbb{C}^{I_p}$ and N non-conjugated variables $\mathbf{x}^{(n)} \in \mathbb{C}^{J_n}$ in the complex case, with $p \in \langle P \rangle$ and $n \in \langle N \rangle$.

– Analogously to the quadratic and Hermitian forms obtained from bilinear and sesquilinear forms by replacing the pair $(\mathbf{x}, \mathbf{y})$ with the vector $\mathbf{x}$, real and complex multilinear forms expressed using just one vector $\mathbf{x} \in \mathbb{K}^I$ will be called multiquadratic forms ($\mathbb{K} = \mathbb{R}$) and multi-Hermitian forms ($\mathbb{K} = \mathbb{C}$), respectively. In section 3.11, we will see that, in the same way the symmetric quadratic/Hermitian forms lead to the notion of a symmetric/Hermitian matrix, the symmetry of multiquadratic/multi-Hermitian forms is directly linked to the symmetry of their associated tensors.

3.3. Notion of slice

A slice is a sub-tensor obtained by fixing one or more indices. In the case of a tensor of order N, if we fix $N-1$ indices, we obtain a vector called a fiber, which generalizes the column and row vectors of a matrix to tensors of order greater than two.

3.3.1. *Fibers*

In the case of a third-order tensor $\mathcal{X} \in \mathbb{K}^{I \times J \times K}$, we obtain the following three types of fiber, called columns, rows, and tubes, or mode-1, -2 and -3 fibers:

– Columns: j and k fixed $\Rightarrow$ JK columns $\mathbf{x}_{\bullet jk} \in \mathbb{K}^{I}$.

– Rows: i and k fixed $\Rightarrow$ IK rows $\mathbf{x}_{i \bullet k} \in \mathbb{K}^{J}$.

– Tubes: i and j fixed $\Rightarrow$ IJ tubes $\mathbf{x}_{ij \bullet} \in \mathbb{K}^{K}$.

The fibers defined above will also be denoted $\mathbf{x}_{\bullet,j,k}$, $\mathbf{x}_{i,\bullet,k}$ and $\mathbf{x}_{i,j,\bullet}$, where the dot indicates which index varies. More generally, in the case of a tensor $\mathcal{X} \in \mathbb{K}^{I_N}$ of order N, the mode-n fiber will be denoted $\mathbf{x}_{i_1,\cdots,i_{n-1},\bullet,i_{n+1},\cdots,i_N}$, where the bold dot ($\bullet$) makes it easier to see which index varies. These fibers are illustrated in Figure 3.1 for a third-order tensor $\mathcal{X} \in \mathbb{K}^{I \times J \times K}$, with $I = J = K = 3$.

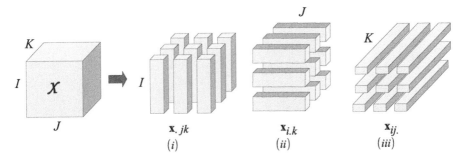

Figure 3.1. *Fibers of a third-order tensor*

3.3.2. *Matrix and tensor slices*

If $N-2$ indices are fixed, we obtain a matrix slice. Thus, for a third-order tensor, there are three possible types of matrix slice, called horizontal, lateral and frontal when the indices i, j and k are fixed, respectively:

$$\mathbf{X}_{i..} \in \mathbb{K}^{J \times K}, \mathbf{X}_{.j.} \in \mathbb{K}^{K \times I}, \mathbf{X}_{..k} \in \mathbb{K}^{I \times J}, \qquad [3.6]$$

defined as:

$$\mathbf{X}_{i..} = \begin{pmatrix} x_{i11} & x_{i12} & \cdots & x_{i1K} \\ x_{i21} & x_{i22} & \cdots & x_{i2K} \\ \vdots & \vdots & \ddots & \vdots \\ x_{iJ1} & x_{iJ2} & \cdots & x_{iJK} \end{pmatrix} \in \mathbb{K}^{J \times K}, \qquad [3.7]$$

$$\mathbf{X}_{.j.} = \begin{pmatrix} x_{1j1} & x_{2j1} & \cdots & x_{Ij1} \\ x_{1j2} & x_{2j2} & \cdots & x_{Ij2} \\ \vdots & \vdots & \ddots & \vdots \\ x_{1jK} & x_{2jK} & \cdots & x_{IjK} \end{pmatrix} \in \mathbb{K}^{K \times I}, \qquad [3.8]$$

$$\mathbf{X}_{..k} = \begin{pmatrix} x_{11k} & x_{12k} & \cdots & x_{1Jk} \\ x_{21k} & x_{22k} & \cdots & x_{2Jk} \\ \vdots & \vdots & \ddots & \vdots \\ x_{I1k} & x_{I2k} & \cdots & x_{IJk} \end{pmatrix} \in \mathbb{K}^{I \times J}. \qquad [3.9]$$

These matrix slices are illustrated in Figure 3.2.

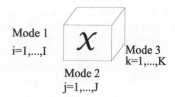

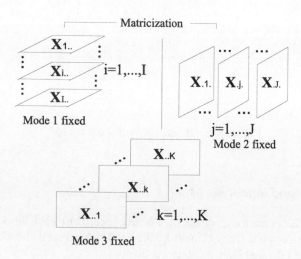

Figure 3.2. *Matrix slices of a third-order tensor*

Using the index convention introduced in section 2.6, the horizontal, lateral and frontal slices can be written as follows:

$$\mathbf{X}_{i..} = \sum_{j=1}^{J} \sum_{k=1}^{K} x_{ijk}\, \mathbf{e}_j^{(J)} \otimes (\mathbf{e}_k^{(K)})^T = x_{ijk}\mathbf{e}_j^k \qquad [3.10]$$

$$\mathbf{X}_{.j.} = \sum_{i=1}^{I} \sum_{k=1}^{K} x_{ijk}\, \mathbf{e}_k^{(K)} \otimes (\mathbf{e}_i^{(I)})^T = x_{ijk}\mathbf{e}_k^i \qquad [3.11]$$

$$\mathbf{X}_{..k} = \sum_{i=1}^{I} \sum_{j=1}^{J} x_{ijk}\, \mathbf{e}_i^{(I)} \otimes (\mathbf{e}_j^{(J)})^T = x_{ijk}\mathbf{e}_i^j. \qquad [3.12]$$

In general, for a tensor $\mathcal{X} \in \mathbb{K}^{I_1 \times \cdots \times I_N}$ of order N, we obtain tensors with a reduced order $N - p$ by fixing p indices. Thus, by fixing the mode n, we obtain a tensor slice of order $N - 1$, called a mode-n slice, denoted $\mathcal{X}_{...i_n...}$, of size $I_{n+1} \times \cdots \times I_N \times I_1 \times \cdots \times I_{n-1}$.

REMARK 3.3.– Some authors define this mode-n slice as having the size $I_1 \times \cdots \times I_{n-1} \times I_{n+1} \times \cdots \times I_N$.

EXAMPLE 3.4.– For $\mathcal{X} \in \mathbb{K}^{I \times J \times K}$, with $I = J = K = 2$, we have:

$$\mathbf{X}_{1..} = \begin{pmatrix} x_{111} & x_{112} \\ x_{121} & x_{122} \end{pmatrix} \;,\; \mathbf{X}_{2..} = \begin{pmatrix} x_{211} & x_{212} \\ x_{221} & x_{222} \end{pmatrix}$$

$$\mathbf{X}_{.1.} = \begin{pmatrix} x_{111} & x_{211} \\ x_{112} & x_{212} \end{pmatrix} \;,\; \mathbf{X}_{.2.} = \begin{pmatrix} x_{121} & x_{221} \\ x_{122} & x_{222} \end{pmatrix}$$

$$\mathbf{X}_{..1} = \begin{pmatrix} x_{111} & x_{121} \\ x_{211} & x_{221} \end{pmatrix} \;,\; \mathbf{X}_{..2} = \begin{pmatrix} x_{112} & x_{122} \\ x_{212} & x_{222} \end{pmatrix}$$

EXAMPLE 3.5.– For a fifth-order tensor $\mathcal{X} \in \mathbb{K}^{I_5}$, we can define matrix slices by fixing three indices, or tensor slices by fixing one or two indices, such as:

$$\mathbf{X}_{i_1,\bullet,i_3,\bullet,i_5} \in \mathbb{K}^{I_2 \times I_4} \;;\; \mathcal{X}_{i_1,\bullet,\bullet,\bullet,\bullet} \in \mathbb{K}^{I_2 \times I_3 \times I_4 \times I_5} \;;\; \mathcal{X}_{\bullet,\bullet,i_3,\bullet,i_5} \in \mathbb{K}^{I_1 \times I_2 \times I_4}.$$

3.4. Mode combination

The mode combination operation plays a very important role in tensor calculus. It can be viewed as a contraction of a tensor of order N to a tensor of order $N_1 < N$. Various contractions are possible, depending on how the modes are combined. For example, suppose that the set $\langle N \rangle = \{1, \ldots, N\}$ is partitioned into N_1 disjoint ordered subsets, with $1 \leq N_1 \leq N - 1$, with each subset $\mathbb{S}_{n_1}$ being composed of $p(n_1)$ modes, with $n_1 \in \langle N_1 \rangle$ and $\sum_{n_1=1}^{N_1} p(n_1) = N$.

The mode combination associated with $\mathbb{S}_{n_1}$ has dimension $J_{n_1} = \prod_{n \in \mathbb{S}_{n_1}} I_n$. Thus, we can rewrite the tensor $\mathcal{X} \in \mathbb{K}^{I_1 \times \cdots \times I_N}$ of order N, defined in [3.2], as the tensor $\mathcal{Y} \in \mathbb{K}^{J_1 \times \cdots \times J_{N_1}}$ of order N_1 such that:

$$\mathcal{Y} = \sum_{j_1=1}^{J_1} \cdots \sum_{j_{N_1}=1}^{J_{N_1}} y_{j_1,\cdots,j_{N_1}} \overset{N_1}{\underset{n_1=1}{\circ}} \mathbf{e}_{j_{n_1}}^{(J_{n_1})} \quad \text{with} \quad \mathbf{e}_{j_{n_1}}^{(J_{n_1})} = \underset{n \in \mathbb{S}_{n_1}}{\otimes} \mathbf{e}_{i_n}^{(I_n)}. \quad [3.13]$$

A mode combination is also called a reshaping operation. It transforms a tensor of order N into a reduced order tensor containing the same elements. Two particular types of mode combination corresponding to the operations of matricization and vectorization will be presented in sections 3.7 and 3.9, respectively.

EXAMPLE 3.6.– Case of a fourth-order tensor $\mathcal{X} \in \mathbb{K}^{I \times J \times K \times L}$.

The following expressions represent combinations of four, three and two modes, corresponding to a vectorization, a matricization and a contraction to a third-order tensor, respectively, starting from the tensor $\mathcal{X}$:

$$\mathbf{y} \triangleq \mathbf{x}_{IJKL} = \sum_{m=1}^{M} y_m \, \mathbf{e}_m^{(M)}$$

$$\mathbf{Y} \triangleq \mathbf{X}_{I \times JKL} = \sum_{i=1}^{I} \sum_{p=1}^{P} y_{i,p} \, \mathbf{e}_i^{(I)} \circ \mathbf{e}_p^{(P)}$$

$$\mathcal{Y} \triangleq \mathcal{X}_{I \times J \times KL} = \sum_{i=1}^{I} \sum_{j=1}^{J} \sum_{q=1}^{Q} y_{i,j,q} \, \mathbf{e}_i^{(I)} \circ \mathbf{e}_j^{(J)} \circ \mathbf{e}_q^{(Q)},$$

$$M = IJKL \, , \ P = JKL \, , \ Q = KL. \quad\quad\quad [3.14]$$

These three mode combinations give three tensors of reduced order, namely a vector $\mathbf{y} = \mathbf{x}_{IJKL} \in \mathbb{K}^M$, a matrix $\mathbf{Y} = \mathbf{X}_{I \times JKL} \in \mathbb{K}^{I \times P}$ and a third-order tensor $\mathcal{Y} = \mathcal{X}_{I \times J \times KL} \in \mathbb{K}^{I \times J \times Q}$, such that, for $i \in \langle I \rangle, j \in \langle J \rangle, k \in \langle K \rangle, l \in \langle L \rangle$, we have:

$$y_m = x_{i,j,k,l} \ \text{with} \ m = l + (k-1)L + (j-1)KL + (i-1)JKL \in \langle M \rangle$$

$$y_{i,p} = x_{i,j,k,l} \ \text{with} \ p = l + (k-1)L + (j-1)KL \in \langle P \rangle$$

$$y_{i,j,q} = x_{i,j,k,l} \ \text{with} \ q = l + (k-1)L \in \langle Q \rangle,$$

where the dimensions M, P and Q are defined by [3.14]. In the following, we will also use the notation:

$$m \triangleq \overline{ijkl} \, , \ p \triangleq \overline{jkl} \, , \ q \triangleq \overline{kl}, \tag{3.15}$$

and in general:

$$\overline{i_1 i_2 \cdots i_P} \triangleq i_P + \sum_{p=1}^{P-1} (i_p - 1) \prod_{k=p+1}^{P} I_k. \tag{3.16}$$

Recall that, by convention, the order of the dimensions in a product $\prod_{p=1}^{P} I_p \triangleq I_1 \cdots I_P$ associated with the index combination $\overline{i_1 i_2 \cdots i_P}$ follows the order of variation of the indices $\mathbf{i}_P = (i_1, \cdots, i_P)$, with i_1 varying more slowly than i_2, which in turn varies more slowly than i_3, etc.

Thus, for $\mathcal{X} \in \mathbb{K}^{\mathcal{I}_P}$, lexicographical vectorization gives the vector $\mathbf{y} \triangleq \mathbf{x}_{I_1 \cdots I_P}$, with element $x_{\mathbf{i}_P}$ at the position $m = \overline{i_1 i_2 \cdots i_P}$ of $\mathbf{y}$, i.e. $y_m = x_{i_1, \cdots, i_P} = x_{\mathbf{i}_P}$.

For example, for $\mathbb{K}^{IJK}$, the index i varies more slowly than j, which in turn varies more slowly than k. The elements of the vector $\mathbf{y} = \mathbf{x}_{IJK}$ are therefore arranged in such a way that $x_{i,j,k}$ is located at the position $m = k + (j-1)K + (i-1)JK$ in $\mathbf{y}$, so $y_m = x_{i,j,k}$. For $I = J = K = 2$, we have: $\mathbf{y} = \mathbf{x}_{IJK} = [x_{111} \ x_{112} \ x_{121} \ x_{122} \ x_{211} \ x_{212} \ x_{221} \ x_{222}]^T$.

REMARK 3.7.– If we choose to vary i_1 (the quickest) and i_P (the slowest), the position of $y_m = x_{\mathbf{i}_P}$ in $\mathbf{y} = \mathbf{x}_{I_1 \cdots I_P}$ becomes:

$$m = i_1 + \sum_{p=2}^{P} (i_p - 1) \prod_{k=1}^{p-1} I_k \triangleq \overline{i_P \cdots i_2 i_1}. \tag{3.17}$$

3.5. Partitioned tensors or block tensors

In the same way that a vector $\mathbf{x} \in \mathbb{K}^I$ and a matrix $\mathbf{X} \in \mathbb{K}^{I \times J}$ can be partitioned into M sub-vectors $\mathbf{x}_m \in \mathbb{K}^{I_m}$ and MN sub-matrices $\mathbf{X}_{m,n} \in \mathbb{K}^{I_m \times J_n}$, respectively, with $m \in \langle M \rangle$, $n \in \langle N \rangle$, $\sum_{m=1}^{M} I_m = I$ and $\sum_{n=1}^{N} J_n = J$, a tensor $\mathcal{X} \in \mathbb{K}^{I_1 \times \cdots \times I_N}$ of order $N > 2$ can be partitioned into $\prod_{n=1}^{N} M_n$ sub-tensors $\mathcal{X}_{m_1, \cdots, m_N} \in \mathbb{K}^{J_{m_1}^{(1)} \times \cdots \times J_{m_N}^{(N)}}$ of order N, with $m_n \in \langle M_n \rangle$ and $\sum_{m_n=1}^{M_n} J_{m_n}^{(n)} = I_n$, for $n \in \langle N \rangle$.

This tensor partitioning is obtained by partitioning each dimension I_n into M_n subsets of dimensions $J_{m_n}^{(n)}$ so that each subset is associated with indices

$j_{m_n}^{(n)} \in \langle J_{m_n}^{(n)} \rangle$, and $\sum_{m_n=1}^{M_n} J_{m_n}^{(n)} = I_n$. Thus, the element $(\mathscr{X}_{m_1,\cdots,m_N})_{j_{m_1}^{(1)},\cdots,j_{m_N}^{(N)}}$ of the sub-tensor $\mathscr{X}_{m_1,\cdots,m_N}$ corresponds to the element $x_{i_1,\cdots,i_N}$ of the tensor $\mathscr{X}$ such that:

$$i_n = \sum_{k=1}^{m_n-1} J_k^{(n)} + j_{m_n}^{(n)} \; , \; \forall n \in \langle N \rangle.$$

REMARK 3.8.– In the matrix case, the element $(\mathbf{X}_{m,n})_{i_m,j_n}$ of the sub-matrix $\mathbf{X}_{m,n} \in \mathbb{K}^{I_m \times J_n}$ corresponds to the element $x_{i,j}$ of $\mathbf{X} \in \mathbb{K}^{I \times J}$ such that:

$$(\mathbf{X}_{m,n})_{i_m,j_n} = x_{i,j} \; \text{with} \; i = \sum_{k=1}^{m-1} I_k + i_m \; , \; j = \sum_{k=1}^{n-1} J_k + j_n.$$

The matrix and tensor slices of a tensor introduced in section 3.3.2 are special partitionings. Thus, a third-order tensor $\mathscr{X} \in \mathbb{K}^{I \times J \times K}$ can be partitioned into K (matrix) sub-tensors of size $I \times J$, denoted $\mathbf{X}_{..k}$ with $k \in \langle K \rangle$, corresponding to the frontal slices of $\mathscr{X}$. Similarly, a fourth-order tensor $\mathscr{X} \in \mathbb{K}^{I_1 \times I_2 \times I_3 \times I_4}$ can be partitioned into $I_1 I_2$ (matrix) sub-tensors of size $I_3 \times I_4$, denoted $\mathbf{X}_{i_1,i_2,..}$ with $i_1 \in \langle I_1 \rangle$ and $i_2 \in \langle I_2 \rangle$.

In general, a tensor $\mathscr{X}$ of order N partitioned into sub-tensors of order N will be called a block tensor, with each block having a size smaller than or equal to the size of $\mathscr{X}$. Tensor blocks generalize matrix blocks to orders greater than two. Like for matrix blocks, some operations such as transposition, addition and vectorization can be performed block-wise. See section 3.10.2 for block transposition.

EXAMPLE 3.9.– A cubic tensor $\mathscr{X} \in \mathbb{K}^{I_1 \times I_2 \times I_3}$, with $I_1 = I_2 = I_3 = 6$, contains 216 elements. It can, for example, be partitioned into 27 cubic sub-tensors of size $2 \times 2 \times 2$, by choosing $M_1 = M_2 = M_3 = 3$ and $J_{m_1}^{(1)} = J_{m_2}^{(2)} = J_{m_3}^{(3)} = 2$, for $m_n \in \langle 3 \rangle$ and $n \in \langle 3 \rangle$. The tensor $\mathscr{X}$ is then partitioned into sub-tensors $\mathscr{X}_{m_1,m_2,m_3} \in \mathbb{K}^{2 \times 2 \times 2}$, each containing eight elements. We can therefore view the tensor $\mathscr{X}$ with this partitioning as a cubic block tensor of size $3 \times 3 \times 3$, where each block is itself a cubic tensor of size $2 \times 2 \times 2$.

For this partitioning, the element $x_{2,3,4}$ of the tensor $\mathscr{X}$ is located in the sub-tensor $\mathscr{X}_{1,2,2}$, at the position $(2,1,2)$, namely the element $(\mathscr{X}_{1,2,2})_{2,1,2}$. Similarly, $x_{4,5,6}$ corresponds to the element $(\mathscr{X}_{2,3,3})_{2,1,2}$. Note that the sub-tensor $\mathscr{X}_{2,3,3}$ is formed by the set of elements x_{i_1,i_2,i_3} such that $3 \leq i_1 \leq 4$ and $5 \leq i_2, i_3 \leq 6$.

EXAMPLE 3.10.– A tensor $\mathscr{X} \in \mathbb{K}^{9 \times 6 \times 4}$ can be partitioned into 12 sub-tensors, with $M_1 = 3$, $M_2 = M_3 = 2$, $J_1^{(1)} = J_2^{(1)} = 4, J_3^{(1)} = 1, J_1^{(2)} = J_2^{(2)} = 3, J_1^{(3)} = J_2^{(3)} = 2$. The element $x_{7,2,3}$ is located at the position $(3,2,1)$ of the sub-tensor

$\mathcal{X}_{2,1,2}$, i.e. $x_{7,2,3} = (\mathcal{X}_{2,1,2})_{3,2,1}$, where the sub-tensor $\mathcal{X}_{2,1,2}$ consists of the set of elements x_{i_1,i_2,i_3}, such that $i_1 \in \{J_1^{(1)} + 1, \cdots, J_1^{(1)} + J_2^{(1)}\}$, $i_2 \in \{1, \cdots, J_1^{(2)}\}$ and $i_3 \in \{J_1^{(3)} + 1, \cdots, J_1^{(3)} + J_2^{(3)}\}$, i.e. $\{x_{i_1,i_2,i_3}; 5 \le i_1 \le 8, 1 \le i_2 \le 3,$ and $3 \le i_3 \le 4\}$.

Block tensors can also be obtained by concatenating column block or row block tensors. Thus, if we consider the tensors $\mathcal{A} \in \mathbb{K}^{\underline{L}_P \times \underline{J}_N}$, $\mathcal{B} \in \mathbb{K}^{\underline{L}_P \times \underline{K}_N}$, $\mathcal{C} \in \mathbb{K}^{\underline{M}_P \times \underline{J}_N}$, we can define column block and row block tensors as follows:

$$\mathcal{X} = \left[\begin{array}{ccc} \mathcal{A} & \vdots & \mathcal{B} \end{array} \right] \in \mathbb{K}^{\underline{L}_P \times \underline{L}_N} \text{ with } L_n = J_n + K_n, \, n \in \langle N \rangle \quad [3.18]$$

$$x_{\underline{\mathbf{i}}_P, \underline{\mathbf{l}}_N} = \begin{cases} a_{\underline{\mathbf{i}}_P, \underline{\mathbf{j}}_N} & \text{for } l_n = j_n \in \langle J_n \rangle, \, n \in \langle N \rangle \\ b_{\underline{\mathbf{i}}_P, \underline{\mathbf{k}}_N} & \text{for } l_n = J_n + k_n \in \{J_n + 1, \cdots J_n + K_n\}, \, n \in \langle N \rangle \end{cases}$$

and

$$\mathcal{Y} = \left[\begin{array}{c} \mathcal{A} \\ \cdots \\ \mathcal{C} \end{array} \right] \in \mathbb{K}^{\underline{L}_P \times \underline{J}_N} \text{ with } L_p = I_p + M_p, \, p \in \langle P \rangle \quad [3.19]$$

$$y_{\underline{\mathbf{l}}_P, \underline{\mathbf{j}}_N} = \begin{cases} a_{\underline{\mathbf{i}}_P, \underline{\mathbf{j}}_N} & \text{for } l_p = i_p \in \langle I_p \rangle, \, p \in \langle P \rangle \\ b_{\underline{\mathbf{k}}_P, \underline{\mathbf{j}}_N} & \text{for } l_p = I_p + k_p \in \{I_p + 1, \cdots I_p + K_p\}, \, p \in \langle P \rangle \end{cases}$$

3.6. Diagonal tensors

Below, we present several types of diagonal and partially diagonal tensors according to which class they belong to.

3.6.1. *Case of a tensor* $\mathcal{X} \in \mathbb{K}^{[N;I]}$

Recall that a hypercubic tensor $\mathcal{X} \in \mathbb{K}^{[N;I]}$ of order N and dimensions I is diagonal if $x_{i_1,i_2\cdots,i_N} \neq 0$ only holds when $i_1 = i_2 = \cdots = i_N$.

We can also define partially diagonal tensors, i.e. tensors that are diagonal with respect to two or more modes (Rezghi and Elder 2011).

Let $\mathbb{T} = \{i_{n_1}, \cdots, i_{n_T}\}$ be a subset of $\{i_1, \cdots, i_N\}$, with $T \le N$, and $I_{n_1} = \cdots = I_{n_T}$. A tensor $\mathcal{X} \in \mathbb{K}^{I_1 \times \cdots \times I_N}$ is said to be diagonal with respect to the set of modes associated with $\mathbb{T}$ if $x_{i_1,\cdots,i_N} \neq 0$ only holds when $i_{n_1} = i_{n_2} = \cdots = i_{n_T}$.

For example, a hypercubic tensor $\mathcal{X} \in \mathbb{K}^{[N;I]}$ of order N and dimensions I is said to be row-diagonal if it is diagonal with respect to its $N-1$ last modes (Liu *et al.* 2018).

This means that every tensor slice $\mathcal{X}_{i_1\ldots}$ obtained by fixing the first mode ($i_1 \in \langle I \rangle$) is diagonal, i.e. the coefficients $x_{i_1,i_2,\cdots,i_N}$ are only non-zero if $i_2 = \cdots = i_N$.

The tensor $\mathcal{X} \in \mathbb{K}^{I_1 \times \cdots \times I_N}$ is diagonal with respect to two disjoint subsets $\mathbb{T} = \{i_{n_1}, \cdots, i_{n_T}\}$ and $\mathbb{Q} = \{i_{p_1}, \cdots, i_{p_Q}\}$, with $I_{n_1} = \cdots = I_{n_T}$ and $I_{p_1} = \cdots = I_{p_Q}$, if $x_{i_1,\cdots,i_N} \neq 0$ only holds when $i_{n_1} = \cdots = i_{n_T}$ and $i_{p_1} = \cdots = i_{p_Q}$.

EXAMPLE 3.11.– A third-order tensor $\mathcal{X} \in \mathbb{K}^{I \times J \times K}$, with $J = K$, is diagonal with respect to its last two modes if $x_{ijk} = 0, \forall j \neq k$. The horizontal slices $\mathbf{X}_{i..}$, for $i \in \langle I \rangle$, are then diagonal matrices, and the matrix unfolding $\mathbf{X}_{J \times IK} = [\mathbf{X}_{1..} \cdots \mathbf{X}_{I..}]$ is a column block matrix, whose column blocks are diagonal matrices.

Similarly, in the case where $I = J$, the tensor $\mathcal{X} \in \mathbb{K}^{I \times I \times K}$ is said to be diagonal with respect to the first two modes if the frontal slices $\mathbf{X}_{..k}$, for $k \in \langle K \rangle$, are diagonal matrices, i.e. if $x_{ijk} = 0, \forall i \neq j$.

EXAMPLE 3.12.– For a fourth-order tensor $\mathcal{X} \in \mathbb{K}^{I \times J \times K \times L}$, with $I = J$ and $K = L$, we say that it is diagonal with respect to the two index subsets $\{i, j\}$ and $\{k, l\}$ if $x_{ijkl} \neq 0$ only holds when $i = j$ and $k = l$. Note that, in this case, the matrix unfolding $\mathbf{X}_{\{1,3\};\{2,4\}} = \mathbf{X}_{IK \times JL}$ is a diagonal matrix.

3.6.2. *Case of a square tensor*

In the case of a square tensor $\mathcal{X} \in \mathbb{K}^{\underline{I}_N \times \underline{I}_N}$ of order $2N$, we say that it is block diagonal if its elements satisfy (Brazell *et al.* 2013):

$$x_{\underline{\mathbf{i}}_N, \underline{\mathbf{j}}_N} = \begin{cases} \alpha_{\underline{\mathbf{i}}_N, \underline{\mathbf{j}}_N} & \text{if } i_n = j_n, \ \forall n \in \langle N \rangle \\ 0 & \text{otherwise} \end{cases} \qquad [3.20]$$

where the $\alpha_{\underline{\mathbf{i}}_N, \underline{\mathbf{j}}_N}$ are arbitrary scalars. This means that every element of $\mathcal{X}$ is zero except the elements $x_{i_1,\cdots,i_N,i_1,\cdots,i_N}$, with $i_n \in I_n$ for $n \in \langle N \rangle$.

EXAMPLE 3.13.– For a tensor $\mathcal{X} \in \mathbb{K}^{3 \times 3 \times 3 \times 3}$, which is a fourth-order hypercubic tensor, we say that $\mathcal{X}$ is block diagonal if the only non-zero coefficients are as follows:

$$x_{1111}, x_{1212}, x_{1313}, x_{2121}, x_{2222}, x_{2323}, x_{3131}, x_{3232}, x_{3333}. \qquad [3.21]$$

Note that $\mathcal{X}$ is diagonal only if the non-zero coefficients are: x_{1111}, x_{2222}, and x_{3333}. We can therefore conclude that any diagonal tensor is also block diagonal. It is clear that the converse is not true.

We define the identity block tensor of order $2N$, denoted $\mathcal{J}_{2N} \in \mathbb{R}^{\underline{I}_N \times \underline{I}_N}$, as the tensor whose elements satisfy:

$$(\mathcal{J})_{\underline{i}_N, \underline{j}_N} = \prod_{n=1}^{N} \delta_{i_n, j_n} \; , \quad i_n, j_n \in I_n, \tag{3.22}$$

namely:

$$(\mathcal{J})_{\underline{i}_N, \underline{j}_N} = \begin{cases} 1 & \text{if} \quad i_n = j_n \; , \; \forall n \in \langle N \rangle \\ 0 & \text{otherwise} \end{cases} . \tag{3.23}$$

Note that the matrix unfolding $\mathbf{J}_{I_1 \cdots I_N \times J_1 \cdots J_N}$, with $I_n = J_n$ for $n \in \langle N \rangle$, of this identity block tensor of order $2N$ is the identity matrix $\mathbf{I}_{\prod_{n=1}^{N} I_n}$ of order $\prod_{n=1}^{N} I_n$. In other words:

$$\mathbf{J}_{\prod_{n=1}^{N} I_n \times \prod_{n=1}^{N} I_n} = \mathbf{I}_{\prod_{n=1}^{N} I_n}. \tag{3.24}$$

Indeed, the diagonal elements $(\mathcal{J})_{\underline{i}_N, \underline{i}_N}$ of this unfolding are equal to 1, whereas the other elements $(\mathcal{J})_{\underline{i}_N, \underline{j}_N}$, where $i_n \neq j_n$ for at least one index $n \in \langle N \rangle$, are zero.

EXAMPLE 3.14.– When $N = 2$ and $I_n = 3$, for $n \in \langle 2 \rangle$, the unfolding corresponding to the identity block tensor $\mathcal{J}_4$ is given by:

$$(\mathcal{J})_{I_1 I_2 \times J_1 J_2} = \text{diag}(j_{1111}, j_{1212}, j_{1313}, j_{2121}, j_{2222}, j_{2323}, j_{3131}, j_{3232}, j_{3333}) = \mathbf{I}_9,$$

with $j_{i_1, i_2, i_1, i_2} = 1$ for $i_1, i_2 \in \langle 3 \rangle$.

3.6.3. *Case of a rectangular tensor*

Analogously to the case of a rectangular matrix, a rectangular tensor $\mathcal{X} \in \mathbb{K}^{\underline{I}_P \times \underline{J}_N}$ is said to be pseudo-diagonal if (Behera *et al.* 2019):

$$x_{\underline{i}_P, \underline{j}_N} = 0 \; \text{for} \; \overline{i_1 \cdots i_P} \neq \overline{j_1 \cdots j_N}, \tag{3.25}$$

where $\overline{i_1 \cdots i_P}$ and $\overline{j_1 \cdots j_N}$ are defined as in [3.16].

3.7. Matricization

This section presents the operation of matricization of a tensor, which plays a very important role in the processing of data tensors. In Chapter 5, this operation will be detailed for various tensor decompositions.

First, we will define the various matricized forms of a third-order tensor. We will then use the index convention in an original way to prove an expression for these matrix unfoldings in terms of matrix slices, as well as to establish compact and general expressions for the matricization of a tensor of order N and for tensor matricization by index blocks.

3.7.1. *Matricization of a third-order tensor*

Different matricizations can be defined by stacking the matrix slices. Thus, for a third-order tensor $\mathcal{X} \in \mathbb{K}^{I \times J \times K}$, there are six flat unfoldings, denoted $\mathbf{X}_{I \times JK}, \mathbf{X}_{I \times KJ}, \mathbf{X}_{J \times KI}, \mathbf{X}_{J \times IK}, \mathbf{X}_{K \times IJ}, \mathbf{X}_{K \times JI}$, and six tall unfoldings $\mathbf{X}_{JK \times I}, \mathbf{X}_{KJ \times I}, \mathbf{X}_{KI \times J}, \mathbf{X}_{IK \times J}, \mathbf{X}_{IJ \times K}, \mathbf{X}_{JI \times K}$, which are transposes of the flat unfoldings.

For example, by stacking the mode-1, mode-2 and mode-3 matrix slices defined in [3.7]–[3.9] as column blocks, we obtain the flat unfoldings:

$$\mathbf{X}_{I \times KJ} = [\mathbf{X}_{..1} \cdots \mathbf{X}_{..K}] \in \mathbb{K}^{I \times KJ} \tag{3.26}$$

$$\mathbf{X}_{J \times IK} = [\mathbf{X}_{1..} \cdots \mathbf{X}_{I..}] \in \mathbb{K}^{J \times IK} \tag{3.27}$$

$$\mathbf{X}_{K \times JI} = [\mathbf{X}_{.1.} \cdots \mathbf{X}_{.J.}] \in \mathbb{K}^{K \times JI} \tag{3.28}$$

or alternatively:

$$\mathbf{X}_{I \times JK} = \left[\mathbf{X}_{.1.}^T \cdots \mathbf{X}_{.J.}^T \right] \in \mathbb{K}^{I \times JK} \tag{3.29}$$

$$\mathbf{X}_{J \times KI} = \left[\mathbf{X}_{..1}^T \cdots \mathbf{X}_{..K}^T \right] \in \mathbb{K}^{J \times KI} \tag{3.30}$$

$$\mathbf{X}_{K \times IJ} = \left[\mathbf{X}_{1..}^T \cdots \mathbf{X}_{I..}^T \right] \in \mathbb{K}^{K \times IJ}. \tag{3.31}$$

This gives us matrices partitioned into column blocks.

REMARK 3.15.– Writing the dimensions as a subscript allows us to distinguish between mode combinations more easily.

Noting that $\mathbf{X}_{..k} = [\mathbf{x}_{.1k} \cdots \mathbf{x}_{.Jk}] \in \mathbb{K}^{I \times J}$, with $\mathbf{x}_{.jk} \in \mathbb{K}^I$, for $j \in \langle J \rangle$ and $k \in \langle K \rangle$, the matrix $\mathbf{X}_{I \times KJ}$ defined in [3.26] can also be partitioned into KJ columns:

$$\mathbf{X}_{I \times KJ} = \left[\begin{array}{ccccccccc} \mathbf{x}_{.11} & \cdots & \mathbf{x}_{.J1} & \mathbf{x}_{.12} & \cdots & \mathbf{x}_{.J2} & \cdots & \mathbf{x}_{.1K} & \cdots & \mathbf{x}_{.JK} \end{array} \right].$$

EXAMPLE 3.16.– For $I = J = K = 2$, we have:

$$\mathbf{X}_{I \times KJ} = \begin{bmatrix} x_{111} & x_{121} & x_{112} & x_{122} \\ x_{211} & x_{221} & x_{212} & x_{222} \end{bmatrix} \in \mathbb{K}^{2 \times 4}$$

$$\mathbf{X}_{I \times JK} = \begin{bmatrix} x_{111} & x_{112} & x_{121} & x_{122} \\ x_{211} & x_{212} & x_{221} & x_{222} \end{bmatrix} \in \mathbb{K}^{2 \times 4}$$

$$\mathbf{X}_{J \times KI} = \begin{bmatrix} x_{111} & x_{211} & x_{112} & x_{212} \\ x_{121} & x_{221} & x_{122} & x_{222} \end{bmatrix} \in \mathbb{K}^{2 \times 4}$$

$$\mathbf{X}_{K \times IJ} = \begin{bmatrix} x_{111} & x_{121} & x_{211} & x_{221} \\ x_{112} & x_{122} & x_{212} & x_{222} \end{bmatrix} \in \mathbb{K}^{2 \times 4}.$$

Similarly, we can define vertical stacks of the matrix slices [3.7]–[3.9] as matrices partitioned into row blocks such that:

$$\mathbf{X}_{JK \times I} = \begin{bmatrix} \mathbf{X}_{.1.} \\ \vdots \\ \mathbf{X}_{.J.} \end{bmatrix} = \mathbf{X}_{I \times JK}^T \qquad [3.32]$$

$$\mathbf{X}_{KI \times J} = \begin{bmatrix} \mathbf{X}_{..1} \\ \vdots \\ \mathbf{X}_{..K} \end{bmatrix} = \mathbf{X}_{J \times KI}^T \qquad [3.33]$$

and

$$\mathbf{X}_{IJ \times K} = \begin{bmatrix} \mathbf{X}_{1..} \\ \vdots \\ \mathbf{X}_{I..} \end{bmatrix} = \mathbf{X}_{K \times IJ}^T, \qquad [3.34]$$

where $\mathbf{X}_{I \times JK}, \mathbf{X}_{J \times KI}$ and $\mathbf{X}_{K \times IJ}$ are defined in [3.29]–[3.31].

This implies, for example:

$$x_{ijk} = [\mathbf{X}_{I \times KJ}]_{i,(k-1)J+j} = [\mathbf{X}_{J \times IK}]_{j,(i-1)K+k} = [\mathbf{X}_{K \times JI}]_{k,(j-1)I+i}.$$

3.7.2. Matrix unfoldings and index convention

We can use the index convention to express matrix unfoldings in terms of the matrix slices [3.7]–[3.9]. Thus, for example, using the expression [3.12] of $\mathbf{X}_{..k}$, we have:

$$\mathbf{X}_{I \times KJ} = x_{ijk} \mathbf{e}_i^{kj} = x_{ijk} (\mathbf{e}_i \otimes \mathbf{e}^k \otimes \mathbf{e}^j) \qquad [3.35]$$

$$= \mathbf{e}^k \otimes (x_{ijk} \mathbf{e}_i^j) = \mathbf{e}^k \otimes \mathbf{X}_{..k} \qquad [3.36]$$

$$= \begin{bmatrix} \mathbf{X}_{..1} & \cdots & \mathbf{X}_{..K} \end{bmatrix} \in \mathbb{K}^{I \times KJ}. \qquad [3.37]$$

Thus, we recover the unfolding [3.26]. Similarly, noting that $\mathrm{vec}(\mathbf{X}_{i..}) = x_{ijk} \mathbf{e}_{kj}$ and $\mathrm{vec}^T(\mathbf{X}_{i..}) = x_{ijk} \mathbf{e}^{kj}$ by [2.105], the unfolding $\mathbf{X}_{I \times KJ}$ can also be written as:

$$\mathbf{X}_{I \times KJ} = \mathbf{e}_i \otimes (x_{ijk} \mathbf{e}^{kj}) = \mathbf{e}_i \otimes \mathrm{vec}^T(\mathbf{X}_{i..}) = \begin{bmatrix} \mathrm{vec}^T(\mathbf{X}_{1..}) \\ \vdots \\ \mathrm{vec}^T(\mathbf{X}_{I..}) \end{bmatrix} \in \mathbb{K}^{I \times KJ}.$$

Likewise, using the equations [3.10]–[3.12], we obtain:

$$\mathbf{X}_{JK \times I} = x_{ijk}\mathbf{e}_{jk}^i = \mathbf{e}_j \otimes (x_{ijk}\mathbf{e}_k^i) = \mathbf{e}_j \otimes \mathbf{X}_{.j.} = \begin{bmatrix} \mathbf{X}_{.1.} \\ \vdots \\ \mathbf{X}_{.J.} \end{bmatrix}$$

$$\mathbf{X}_{KI \times J} = x_{ijk}\mathbf{e}_{ki}^j = \mathbf{e}_k \otimes (x_{ijk}\mathbf{e}_i^j) = \mathbf{e}_k \otimes \mathbf{X}_{..k} = \begin{bmatrix} \mathbf{X}_{..1} \\ \vdots \\ \mathbf{X}_{..K} \end{bmatrix}$$

$$\mathbf{X}_{IJ \times K} = x_{ijk}\mathbf{e}_{ij}^k = \mathbf{e}_i \otimes (x_{ijk}\mathbf{e}_j^k) = \mathbf{e}_i \otimes \mathbf{X}_{i..} = \begin{bmatrix} \mathbf{X}_{1..} \\ \vdots \\ \mathbf{X}_{I..} \end{bmatrix}.$$

Thus, we recover the unfoldings [3.32]–[3.34].

3.7.3. Matricization of a tensor of order N

Let us consider a tensor $\mathcal{X} \in \mathbb{K}^{I_N}$ of order N, together with a partitioning of the set of modes $\{1, \cdots, N\}$ into two disjoint ordered subsets $\mathbb{S}_1$ and $\mathbb{S}_2$, composed of p and $N - p$ modes, respectively, with $p \in \langle N - 1 \rangle$. A general matrix unfolding formula was given by Favier and de Almeida (2014a) as follows:

$$\mathbf{X}_{\mathbb{S}_1;\mathbb{S}_2} = \sum_{i_1=1}^{I_1} \cdots \sum_{i_N=1}^{I_N} x_{i_1,\cdots,i_N} \left(\underset{n \in \mathbb{S}_1}{\otimes} \mathbf{e}_{i_n}^{(I_n)} \right) \left(\underset{n \in \mathbb{S}_2}{\otimes} \mathbf{e}_{i_n}^{(I_n)} \right)^T \in \mathbb{K}^{J_1 \times J_2},$$

[3.38]

where $J_{n_1} = \prod_{n \subset \mathbb{S}_{n_1}} I_n$, for $n_1 = 1$ and 2. We say that $\mathbf{X}_{\mathbb{S}_1;\mathbb{S}_2}$ is a matrix unfolding of $\mathcal{X}$ along the modes of $\mathbb{S}_1$ for the rows and along the modes of $\mathbb{S}_2$ for the columns, with $\mathbb{S}_1 \cap \mathbb{S}_2 = \emptyset$ and $\mathbb{S}_1 \cup \mathbb{S}_2 = \langle N \rangle$.

Using the index convention, the unfolding $\mathbf{X}_{\mathbb{S}_1;\mathbb{S}_2}$ defined in [3.38] can be written concisely as:

$$\mathbf{X}_{\mathbb{S}_1;\mathbb{S}_2} = x_{i_1,\cdots,i_N} \mathbf{e}_{\mathbb{I}_1}^{\mathbb{I}_2} \in \mathbb{K}^{J_1 \times J_2},$$

[3.39]

where $\mathbb{I}_1$ and $\mathbb{I}_2$ represent the index combinations associated with the mode subsets $\mathbb{S}_1$ and $\mathbb{S}_2$, respectively.

From this unfolded form $\mathbf{X}_{\mathbb{S}_1;\mathbb{S}_2}$, we can deduce the following expression for the element $x_{\mathbf{i}_N}$:

$$x_{\mathbf{i}_N} = x_{i_1,\cdots,i_N} = \left(\bigotimes_{n\in\mathbb{S}_1} \mathbf{e}_{i_n}^{(I_n)} \right)^T \mathbf{X}_{\mathbb{S}_1;\mathbb{S}_2} \left(\bigotimes_{n\in\mathbb{S}_2} \mathbf{e}_{i_n}^{(I_n)} \right)$$

$$= \mathbf{e}^{\mathbb{I}_1} \mathbf{X}_{\mathbb{S}_1;\mathbb{S}_2} \mathbf{e}_{\mathbb{I}_2}. \qquad [3.40]$$

In the case of a tensor $\mathcal{X} \in \mathbb{K}^{\underline{I}_N}$ of order N, we define two particular matrix unfoldings: the flat mode-n unfolding, denoted $\mathbf{X}_n$, and the unfolding $\mathbf{X}_{\langle n \rangle}$ corresponding to $\mathbb{S}_1 = \{1,\cdots,n\}$, $\mathbb{S}_2 = \{n+1,\cdots,N\}$, such that:

$$\mathbf{X}_n = \mathbf{X}_{I_n \times I_{n+1}\cdots I_N I_1 \cdots I_{n-1}} \qquad [3.41]$$

$$\mathbf{X}_{\langle n \rangle} = \mathbf{X}_{\{1,\cdots,n\};\{n+1,\cdots,N\}} = \mathbf{X}_{I_1\cdots I_n \times I_{n+1}\cdots I_N}. \qquad [3.42]$$

REMARK 3.17.– The element $x_{\mathbf{i}_N} \triangleq x_{i_1,\cdots,i_N}$ is placed at position (i_n, j) in $\mathbf{X}_n$, and at (m, l) in $\mathbf{X}_{\langle n \rangle}$, with:

$$j = i_{n-1} + (i_{n-2} - 1)I_{n-1} + \cdots + (i_1 - 1) \prod_{k=2}^{n-1} I_k + (i_N - 1) \prod_{k=1}^{n-1} I_k + \cdots$$

$$+ (i_{n+1} - 1) \prod_{k=1,k\neq n}^{N} I_k$$

$$m = i_n + (i_{n-1} - 1)I_n + \cdots + (i_1 - 1) \prod_{k=2}^{n} I_k = \overline{i_1 i_2 \cdots i_n}$$

$$l = i_N + (i_{N-1} - 1)I_N + \cdots + (i_{n+1} - 1) \prod_{k=n+2}^{N} I_k = \overline{i_{n+1} i_{n+2} \cdots i_N}.$$

REMARK 3.18.– Some authors define the flat mode-n unfolding as having size $I_n \times I_1 \cdots I_{n-1} I_{n+1} \cdots I_N$, which is equivalent to combining the modes $1,\cdots,$ $n-1, n+1, \cdots, N$ without reversing the order.

REMARK 3.19.– In the case of a tensor $\mathcal{X} \in \mathbb{K}^{I\times J\times K}$, the unfoldings [3.29]–[3.31] correspond to:

$$\mathbf{X}_{I\times JK} = \mathbf{X}_1 \ , \ \ \mathbf{X}_{J\times KI} = \mathbf{X}_2 \ , \ \ \mathbf{X}_{K\times IJ} = \mathbf{X}_3.$$

For a fourth-order tensor $\mathcal{X} \in \mathbb{K}^{I\times J\times K\times L}$, the matrix $\mathbf{X}_{IJ\times KL}$ can be written as a column block matrix with KL columns of size IJ. Indeed, using the index convention, the unfolding $\mathbf{X}_{IJ\times KL}$ can be written as:

$$\mathbf{X}_{IJ\times KL} = x_{i,j,k,l}\, \mathbf{e}_{ij}^{kl} = \mathbf{e}^{kl} \otimes x_{i,j,k,l}\, \mathbf{e}_{ij} = \mathbf{e}^{kl} \otimes \mathbf{y}_{.kl}$$

where $\mathbf{y}_{.kl} \in \mathbb{K}^{IJ}, k \in \langle K \rangle, l \in \langle L \rangle$ is the mode-1 fiber of the third-order tensor $\mathcal{Y} \in \mathbb{K}^{M \times K \times L}$, with $M = IJ$, obtained by combining the first two modes of $\mathcal{X}$, satisfying $y_{m,k,l} = x_{i,j,k,l}$, with $m = j + (i-1)J$. Thus, we obtain:

$$\mathbf{X}_{IJ \times KL} = \begin{bmatrix} \mathbf{y}_{.11} & \cdots & \mathbf{y}_{.1L} & \mathbf{y}_{.21} & \cdots & \mathbf{y}_{.2L} & \cdots & \mathbf{y}_{.K1} & \cdots & \mathbf{y}_{.KL} \end{bmatrix}$$
$$\in \mathbb{K}^{IJ \times KL},$$

with $\mathbf{y}_{.kl} = [x_{1,1,k,l} \cdots x_{1,J,k,l} \cdots x_{I,1,k,l} \cdots x_{I,J,k,l}]^T$, which gives:

$$\mathbf{X}_{IJ \times KL} = \begin{bmatrix} x_{1111} & \cdots & x_{111L} & \cdots & x_{11K1} & \cdots & x_{11KL} \\ \vdots & \cdots & \vdots & \cdots & \vdots & \cdots & \vdots \\ x_{1J11} & \cdots & x_{1J1L} & \cdots & x_{1JK1} & \cdots & x_{1JKL} \\ \vdots & \cdots & \vdots & \cdots & \vdots & \cdots & \vdots \\ x_{I111} & \cdots & x_{I11L} & \cdots & x_{I1K1} & \cdots & x_{I1KL} \\ \vdots & \cdots & \vdots & \cdots & \vdots & \cdots & \vdots \\ x_{IJ11} & \cdots & x_{IJ1L} & \cdots & x_{IJK1} & \cdots & x_{IJKL} \end{bmatrix}.$$

Similarly, by writing:

$$\mathbf{X}_{IJ \times KL} = \mathbf{e}_{ij} \otimes x_{i,j,k,l} \mathbf{e}^{kl} = \mathbf{e}_{ij} \otimes \mathbf{v}_{ij.}^T.$$

the unfolding $\mathbf{X}_{IJ \times KL}$ can be decomposed into IJ row blocks $\mathbf{v}_{ij.}^T \in \mathbb{K}^{1 \times KL}$, for $i \in \langle I \rangle$ and $j \in \langle J \rangle$, where $\mathbf{v}_{ij.}$ is the mode-3 fiber of the tensor $\mathcal{V} \in \mathbb{K}^{I \times J \times N}$, with $N = KL$, obtained by combining the last two modes of $\mathcal{X}$, satisfying:

$$\mathbf{v}_{ij.}^T = \begin{bmatrix} x_{i,j,1,1} & \cdots & x_{i,j,1,L} & \cdots x_{i,j,K,1} & \cdots & x_{i,j,K,L} \end{bmatrix}.$$

EXAMPLE 3.20.– For $I = J = K = L = 2$, we have:

$$\mathbf{X}_{IJ \times KL} = [\mathbf{y}_{.11} \, \mathbf{y}_{.12} \, \mathbf{y}_{.21} \, \mathbf{y}_{.22}] \quad \text{with} \quad \mathbf{y}_{.kl} = \begin{bmatrix} x_{11kl} \\ x_{12kl} \\ x_{21kl} \\ x_{22kl} \end{bmatrix},$$

which gives: $\mathbf{X}_{IJ \times KL} = \begin{bmatrix} x_{1111} & x_{1112} & x_{1121} & x_{1122} \\ x_{1211} & x_{1212} & x_{1221} & x_{1222} \\ x_{2111} & x_{2112} & x_{2121} & x_{2122} \\ x_{2211} & x_{2212} & x_{2221} & x_{2222} \end{bmatrix} \in \mathbb{K}^{4 \times 4}.$

Similarly, we have:

$$\mathbf{X}_{IJ \times KL} = \begin{bmatrix} \mathbf{v}_{11.}^T \\ \mathbf{v}_{12.}^T \\ \mathbf{v}_{21.}^T \\ \mathbf{v}_{22.}^T \end{bmatrix} \quad \text{with} \quad \mathbf{v}_{ij.}^T = \begin{bmatrix} x_{i,j,1,1} & x_{i,j,1,2} & x_{i,j,2,1} & x_{i,j,2,2} \end{bmatrix}.$$

Writing $\mathbf{Y} = \mathbf{X}_{IJ \times KL} \in \mathbb{K}^{M \times N}$, with $M = IJ$ and $N = KL$, the element $x_{i,j,k,l}$ of the tensor $\mathcal{X}$ is located at the position (m, n) of $\mathbf{Y}$, such that:

$$y_{m,n} = x_{i,j,k,l} \text{ with } m = j + (i-1)J \text{ and } n = l + (k-1)L.$$

Table 3.5 summarizes the main notation used.

Notation	Definitions
$\mathcal{X} \in \mathbb{K}^{\underline{I}_N}$	Tensor of order N, size $I_1 \times \cdots \times I_N$
$x_{\underline{i}_N} \triangleq x_{i_1,\cdots,i_N} = (\mathcal{X})_{i_1,\cdots,i_N}$	Element $(i_1, \cdots, i_N)$ of $\mathcal{X}$
$\mathcal{X}_{i_1,\cdots,i_{n-1},\bullet,i_{n+1},\cdots,i_N}$	Mode-n fiber of size I_n
$\mathcal{X}_{\cdots i_n \cdots}$	Mode-n tensor slice, of size $I_{n+1} \times \cdots \times I_N \times I_1 \times \cdots \times I_{n-1}$.
$\mathbf{X}_n \triangleq \mathbf{X}_{I_n \times I_{n+1} \cdots I_N I_1 \cdots I_{n-1}}$	Mode-n flat matrix unfolding of size $I_n \times I_{n+1} \cdots I_N I_1 \cdots I_{n-1}$
$\mathbf{X}_{\mathbb{S}_1;\mathbb{S}_2}$	Matrix unfolding with respect to the mode combinations $\mathbb{S}_1$ and $\mathbb{S}_2$
$\mathbf{X}_{\langle n \rangle} \triangleq \mathbf{X}_{\{1,\cdots,n\};\{n+1,\cdots,N\}}$	Unfolding with respect to the mode combinations $\{1, \cdots, n\}$ and $\{n+1, \cdots, N\}$ for the rows and columns of size $I_1 \cdots I_n \times I_{n+1} \cdots I_N$

Table 3.5. *Notation for tensors*

3.7.4. *Tensor matricization by index blocks*

In section 3.13.5, we will use an important special case of matrix unfolding of a tensor $\mathcal{X} \in \mathbb{K}^{\underline{I}_P \times \underline{J}_N}$ with the Einstein product, namely the unfolding $\mathbf{X}_{I \times J} \triangleq \mathbf{X}_{I_1 \cdots I_P \times J_1 \cdots J_N}$, with $I = I_1 \cdots I_P$ and $J = J_1 \cdots J_N$, corresponding to a combination of the P first and N last modes of $\mathcal{X}$ for the rows and columns of $\mathbf{X}_{I \times J}$, respectively.

Using the definition [3.16], the element $x_{\underline{i}_P, \underline{j}_N}$ is located in $\mathbf{X}_{I \times J}$ at position $(\overline{i_1 \cdots i_P}, \overline{j_1 \cdots j_N})$, with:

$$\overline{j_1 \cdots j_N} \triangleq j_N + \sum_{n=1}^{N-1} (j_n - 1) \prod_{k=n+1}^{N} J_k. \tag{3.43}$$

From the expression [2.68] of a matrix using the index convention, the matrix unfolding $\mathbf{X}_{I \times J}$ can be written as:

$$\mathbf{X}_{I \times J} = x_{\underline{i}_P, \underline{j}_N} \, \mathbf{e}_{\overline{i_1 \cdots i_P}}^{\overline{j_1 \cdots j_N}}, \tag{3.44}$$

with:

$$\overline{\mathbf{e}_{i_1\cdots i_P}^{j_1\cdots j_N}} = \overset{P}{\underset{p=1}{\otimes}} \mathbf{e}_{i_p}^{(I_p)} \Big(\overset{N}{\underset{n=1}{\otimes}} \mathbf{e}_{j_n}^{(J_n)} \Big)^T.$$ [3.45]

REMARK 3.21.– In the special case of a hypercubic tensor $\mathcal{X} \in \mathbb{K}^{I_P \times I_P}$ with $I_p = I$ for $p \in \langle P \rangle$, the unfolding $\mathbf{X} \triangleq \mathbf{X}_{I_1\cdots I_P \times I_1\cdots I_P} \in \mathbb{K}^{I^P \times I^P}$ satisfies:

$$\big(\mathbf{X}\big)_{\overline{i_1\cdots i_P},\overline{j_1\cdots j_P}} = x_{\underline{i}_P,\underline{j}_P},$$ [3.46]

where $i_p, j_p \in \langle I \rangle$, $\overline{i_1 \cdots i_P}$, and $\overline{j_1 \cdots j_P}$ are defined similarly as:

$$\overline{j_1 \cdots j_P} \triangleq j_P + (j_{P-1} - 1)I + \cdots + (j_2 - 1)I^{P-2} + (j_1 - 1)I^{P-1}.$$ [3.47]

Table 3.6 summarizes the matricizations for three particular classes of tensors, whose index sets are divided into either two blocks of the same or different dimensions or into index pairs, i.e. the sets $\mathbb{K}^{I_P \times I_P}$, $\mathbb{K}^{I_P \times J_N}$ and $\mathbb{K}^{I_1 \times J_1 \times \cdots \times I_P \times J_P}$, respectively.

Space of $\mathcal{X}$	Unfolding $\mathbf{X}_{I \times I}/\mathbf{X}_{I \times J}$	Space of $\mathbf{X}$	Dimensions I ; J
$\mathbb{K}^{I_P \times I_P}$	$(\mathbf{X}_{I \times I})_{\underline{i}_P,\underline{j}_P} = x_{\underline{i}_P,\underline{j}_P}$	$\mathbb{K}^{I \times I}$	$I = \prod_{p=1}^{P} I_p$
$\mathbb{K}^{I_P \times J_N}$	$(\mathbf{X}_{I \times J})_{\underline{i}_P,\underline{j}_N} = x_{\underline{i}_P,\underline{j}_N}$	$\mathbb{K}^{I \times J}$	$I = \prod_{p=1}^{P} I_p$; $J = \prod_{n=1}^{N} J_n$
$\mathbb{K}^{I_1 \times J_1 \times \cdots \times I_P \times J_P}$	$(\mathbf{X}_{I \times J})_{\underline{i}_P,\underline{j}_N} = x_{i_1,j_1,\cdots,i_P,j_P}$	$\mathbb{K}^{I \times J}$	$I = \prod_{p=1}^{P} I_p$; $J = \prod_{n=1}^{N} J_n$

Table 3.6. *Matricization by index blocks for different sets of tensors*

3.8. Subspaces associated with a tensor and multilinear rank

For a tensor $\mathcal{X} \in \mathbb{K}^{I \times J \times K}$, the column vectors of the unfoldings $\mathbf{X}_{I \times JK}$, $\mathbf{X}_{J \times KI}$ and $\mathbf{X}_{K \times IJ}$ are called the mode-1, mode-2 and mode-3 vectors of $\mathcal{X}$, respectively. They span three linear spaces whose dimensions are called the mode-1, mode-2 and mode-3 ranks, respectively, denoted $R_1 = \mathrm{r}(\mathbf{X}_{I \times JK}) = \mathrm{r}_1(\mathcal{X})$, $R_2 = \mathrm{r}(\mathbf{X}_{J \times KI}) = \mathrm{r}_2(\mathcal{X})$ and $R_3 = \mathrm{r}(\mathbf{X}_{K \times IJ}) = \mathrm{r}_3(\mathcal{X})$. The triplet (R_1, R_2, R_3) is called the trilinear rank of $\mathcal{X}$. It satisfies $R_1 \leq I, R_2 \leq J$ and $R_3 \leq K$.

In the case of a tensor $\mathcal{X} \in \mathbb{K}^{I_N}$, the column vectors of the matricized form $\mathbf{X}_n$ are the mode-n vectors, and $R_n = r(\mathbf{X}_n) = \mathrm{r}_n(\mathcal{X}) \leq I_n$ is the mode-n rank of $\mathcal{X}$. The N-tuple $(R_1, ..., R_N)$ is called the multilinear (or N-linear) rank of $\mathcal{X}$.

The multilinear rank generalizes the row and column ranks of a matrix to the case of tensors of order greater than two. Note that, for a matrix, the row and column ranks are both equal to the rank of the matrix. However, this does not necessarily hold for $N > 2$, and the modal ranks $R_n, n \in \langle N \rangle$ may differ, unless $\mathcal{X}$ is a symmetric tensor.

3.9. Vectorization

In this section, we present a general formula for the vectorization of a tensor of arbitrary order using the index convention. This formula extends the vectorization results presented in Chapter 2 for matrices, to the case of tensors of order greater than two. The inverse operation of vectorization, denoted *unvec*, transforms a vectorized form into the original tensor.

3.9.1. *Vectorization of a tensor of order N*

By [3.2], using the index convention, a tensor $\mathcal{X} = [x_{i_1,\cdots,i_N}]$ of order N can be written as follows with respect to the canonical basis:

$$\mathcal{X} = x_{i_1,\cdots,i_N} \overset{N}{\underset{n=1}{\circ}} \mathbf{e}_{i_n}^{(I_n)}. \tag{3.48}$$

There are $N!$ vectorizations in the form of a column vector, each associated with a different mode combination. Thus, for lexicographical order, the vectorization is given as:

$$\mathbf{x}_{I_1 I_2 \cdots I_N} = \sum_{i_1=1}^{I_1} \cdots \sum_{i_N=1}^{I_N} x_{i_1,\cdots,i_N} (\mathbf{e}_{i_1}^{(I_1)} \otimes \mathbf{e}_{i_2}^{(I_2)} \otimes \cdots \otimes \mathbf{e}_{i_N}^{(I_N)}) \in \mathbb{K}^{I_1 I_2 \cdots I_N},$$

or alternatively, using the index convention:

$$\mathbf{x}_{I_1 \cdots I_N} = x_{i_1,\cdots,i_N} \overset{N}{\underset{n=1}{\otimes}} \mathbf{e}_{i_n}^{(I_n)} = x_{i_1,\cdots,i_N} \mathbf{e}_{i_1 \cdots i_N}^{(I_1 \cdots I_N)}. \tag{3.49}$$

By comparing [3.48] with [3.49], we conclude that the vectorization of a tensor of order N can be translated into $(N-1)$ Kronecker products of the N basis vectors.

In general, the vectorization of an outer product of N vectors is obtained by replacing the outer products with the Kronecker products of the vectors.

If we consider the set $\$ = \{n_1, \ldots, n_N\}$ associated with a permutation of the modes $\{1, \cdots, N\}$, the vectorization of $\mathcal{X}$ according to this new ordering of the modes is:

$$\mathbf{x}_{I_{n_1} I_{n_2} \cdots I_{n_N}} = x_{i_1,\cdots,i_N} \mathbf{e}_{i_{n_1} i_{n_2} \cdots i_{n_N}}^{(I_{n_1} I_{n_2} \cdots I_{n_N})} \in \mathbb{K}^{I_{n_1} I_{n_2} \cdots I_{n_N}}. \tag{3.50}$$

The transformation from the lexicographical vectorization to the vectorization associated with the set $\$$ is given by the following equation, corresponding to a permutation of the rows of $\mathbf{x}_{I_1 I_2 \cdots I_N}$:

$$\mathbf{x}_{I_{n_1} I_{n_2} \cdots I_{n_N}} = (\mathbf{e}_{i_{n_1}}^{i_1} \otimes \mathbf{e}_{i_{n_2}}^{i_2} \otimes \cdots \otimes \mathbf{e}_{i_{n_N}}^{i_N}) \mathbf{x}_{I_1 I_2 \cdots I_N}. \tag{3.51}$$

Note that, since $(i_{n_1}, \cdots, i_{n_N})$ is a permutation of $(i_1, \cdots, i_N)$, the formula [3.51] does indeed involve a repetition of each index, which implies that the index convention is being used. The 1 of each column n of the permutation matrix is obtained by computing $\mathbf{e}_{i_{n_1}}^{i_1} \otimes \mathbf{e}_{i_{n_2}}^{i_2} \otimes \cdots \otimes \mathbf{e}_{i_{n_N}}^{i_N}$ for the N-tuple of indices $(i_1, \cdots, i_N)$ of the element $x_{i_1, \cdots, i_N}$, located on the nth row of $\mathbf{x}_{I_1 I_2 \cdots I_N}$. The next section gives an example for a third-order tensor.

3.9.2. *Vectorization of a third-order tensor*

Given three vectors $\mathbf{u} \in \mathbb{K}^I, \mathbf{v} \in \mathbb{K}^J, \mathbf{w} \in \mathbb{K}^K$, their outer product defines a rank-one, third-order tensor such that:

$$\mathcal{X} = \mathbf{u} \circ \mathbf{v} \circ \mathbf{w} \in \mathbb{K}^{I \times J \times K} \Leftrightarrow x_{ijk} = u_i v_j w_k. \qquad [3.52]$$

The lexicographical vectorization is given by:

$$\mathbf{x}_{IJK} = u_i v_j w_k \mathbf{e}_{ijk} \in \mathbb{K}^{IJK}. \qquad [3.53]$$

EXAMPLE 3.22.– For $I = J = K = 2$, we have:

$$\mathbf{x}_{JIK} = u_i v_j w_k \mathbf{e}_{jik} = (\mathbf{e}_{ji}^{ij} \otimes \mathbf{e}_k^k) \mathbf{x}_{IJK} = (\mathbf{e}_{ji}^{ij} \otimes \mathbf{I}_K) \mathbf{x}_{IJK}, \qquad [3.54]$$

with the permutation matrix $\mathbf{e}_{ji}^{ij}$, for $I = J = 2$, defined in [2.167], which gives:

$$\mathbf{e}_{ji}^{ij} \otimes \mathbf{I}_2 = \begin{bmatrix} 1 & 0 & 0 & 0 \\ 0 & 0 & 1 & 0 \\ 0 & 1 & 0 & 0 \\ 0 & 0 & 0 & 1 \end{bmatrix} \otimes \begin{bmatrix} 1 & 0 \\ 0 & 1 \end{bmatrix} = \begin{bmatrix} \mathbf{I}_2 & \mathbf{0} & \mathbf{0} & \mathbf{0} \\ \mathbf{0} & \mathbf{0} & \mathbf{I}_2 & \mathbf{0} \\ \mathbf{0} & \mathbf{I}_2 & \mathbf{0} & \mathbf{0} \\ \mathbf{0} & \mathbf{0} & \mathbf{0} & \mathbf{I}_2 \end{bmatrix},$$

where $\mathbf{0} = \mathbf{0}_{2 \times 2}$ is the zero square matrix of order two. The vectorization $\mathbf{x}_{JIK}$ can therefore also be deduced from $\mathbf{x}_{IJK}$ using the following equation:

$$\mathbf{x}_{JIK} = \begin{bmatrix} u_1 v_1 w_1 \\ u_1 v_1 w_2 \\ u_2 v_1 w_1 \\ u_2 v_1 w_2 \\ u_1 v_2 w_1 \\ u_1 v_2 w_2 \\ u_2 v_2 w_1 \\ u_2 v_2 w_2 \end{bmatrix} = \begin{bmatrix} 1 & 0 & 0 & 0 & 0 & 0 & 0 & 0 \\ 0 & 1 & 0 & 0 & 0 & 0 & 0 & 0 \\ 0 & 0 & 0 & 0 & 1 & 0 & 0 & 0 \\ 0 & 0 & 0 & 0 & 0 & 1 & 0 & 0 \\ 0 & 0 & 1 & 0 & 0 & 0 & 0 & 0 \\ 0 & 0 & 0 & 1 & 0 & 0 & 0 & 0 \\ 0 & 0 & 0 & 0 & 0 & 0 & 1 & 0 \\ 0 & 0 & 0 & 0 & 0 & 0 & 0 & 1 \end{bmatrix} \begin{bmatrix} u_1 v_1 w_1 \\ u_1 v_1 w_2 \\ u_1 v_2 w_1 \\ u_1 v_2 w_2 \\ u_2 v_1 w_1 \\ u_2 v_1 w_2 \\ u_2 v_2 w_1 \\ u_2 v_2 w_2 \end{bmatrix}.$$

It can be checked that the 1 of the fourth column, associated with the element $u_1 v_2 w_2$ of the vector $\mathbf{x}_{IJK}$, and hence with the triplet $(i, j, k) = (1, 2, 2)$, is obtained by computing:

$$\mathbf{e}_{21}^{12} \otimes \mathbf{e}_2^2 = \begin{bmatrix} 0 \\ 0 \\ 1 \\ 0 \end{bmatrix} \begin{bmatrix} 0 & 1 & 0 & 0 \end{bmatrix} \otimes \begin{bmatrix} 0 & 0 \\ 0 & 1 \end{bmatrix},$$

which gives:

$$
\mathbf{e}_{21}^{12} \otimes \mathbf{e}_2^2 =
\begin{bmatrix}
0 & 0 & 0 & 0 & 0 & 0 & 0 & 0 \\
0 & 0 & 0 & 0 & 0 & 0 & 0 & 0 \\
0 & 0 & 0 & 0 & 0 & 0 & 0 & 0 \\
0 & 0 & 0 & 0 & 0 & 0 & 0 & 0 \\
0 & 0 & 0 & 0 & 0 & 0 & 0 & 0 \\
0 & 0 & 0 & 1 & 0 & 0 & 0 & 0 \\
0 & 0 & 0 & 0 & 0 & 0 & 0 & 0 \\
0 & 0 & 0 & 0 & 0 & 0 & 0 & 0
\end{bmatrix}.
$$

EXAMPLE 3.23.– To illustrate the vectorization of a third-order tensor, consider the tensor $\mathcal{X} \in \mathbb{K}^{2 \times 2 \times 2}$, corresponding to $I = J = K = 2$. We have:

$$
\mathbf{x}_{IJK} = \begin{bmatrix} x_{111} & x_{112} & x_{121} & x_{122} & x_{211} & x_{212} & x_{221} & x_{222} \end{bmatrix}^T
$$

$$
\mathbf{x}_{IKJ} = \begin{bmatrix} x_{111} & x_{121} & x_{112} & x_{122} & x_{211} & x_{221} & x_{212} & x_{222} \end{bmatrix}^T
$$

$$
\mathbf{x}_{KIJ} = \begin{bmatrix} x_{111} & x_{121} & x_{211} & x_{221} & x_{112} & x_{122} & x_{212} & x_{222} \end{bmatrix}^T
$$

$$
\mathbf{x}_{JKI} = \begin{bmatrix} x_{111} & x_{211} & x_{112} & x_{212} & x_{121} & x_{221} & x_{122} & x_{222} \end{bmatrix}^T.
$$

3.10. Transposition

3.10.1. *Definition of a transpose tensor*

The transposition of a tensor $\mathcal{X} \in \mathbb{K}^{I_N}$ of order N amounts to reordering the indices $(i_1, \cdots, i_N)$, i.e. performing a permutation π of the modes $(1, \cdots, N)$.

The transpose tensors of $\mathcal{X}$ can therefore be written as:

$$
y_{i_{j_1}, \cdots, i_{j_N}} = x_{i_1, \cdots, i_N} \text{ with } j_n = \pi(n), n \in \langle N \rangle, \tag{3.55}
$$

where the set $(j_1, \cdots, j_N)$ is obtained by applying the permutation $\pi \in S_N$ to the set $\langle N \rangle$, where S_N is the symmetric permutation group of order N. The transpose tensor $\mathcal{Y}$, denoted $\mathcal{X}^{T,\pi}$, therefore has the entries $y_{i_{\pi(1)}, \cdots, i_{\pi(N)}}$ and the size $I_{\pi(1)} \times \cdots \times I_{\pi(N)}$.

Using the notations from Tables 3.1 and 3.2, equation [3.55] can also be written as:

$$
y_{\mathbf{i}_{\pi(N)}} = x_{\mathbf{i}_N}, \tag{3.56}
$$

with $y_{\mathbf{i}_{\pi(N)}} = y_{i_{\pi(1)}, \cdots, i_{\pi(N)}}$ and $x_{\mathbf{i}_N} = x_{i_1, \cdots, i_N}$, and the size of the tensor $\mathcal{Y}$ is $\underline{I}_{\pi(N)} \triangleq I_{\pi(1)} \times \cdots \times I_{\pi(N)}$.

For a tensor of order N, there are $N! - 1$ transpose tensors. Thus, for $N = 3$, there are five transpose tensors of $\mathcal{X} \in \mathbb{K}^{I \times J \times K}$, of sizes $I \times K \times J, J \times I \times K, J \times K \times I, K \times I \times J$, and $K \times J \times I$.

We can distinguish between total transposition and partial transposition. A total transposition corresponds to a permutation π, such that every index i_n is no longer in its original position after applying the permutation, i.e. $\pi(n) \neq n, \forall n \in \langle N \rangle$.

Thus, in the case of a third-order tensor, there are two total transpositions, corresponding to the transpose tensors of size $J \times K \times I$ and $K \times I \times J$. The three other transpose tensors, of size $I \times K \times J, K \times J \times I$, and $J \times I \times K$, correspond to partial transpositions associated with permutations of only two modes.

3.10.2. *Properties of transpose tensors*

The transposition operation satisfies the following properties (Brazell *et al.* 2013; Pan 2014):

– Composition of two transpositions associated with the mode permutations π_1 and π_2 is such that:

$$(\mathcal{X}^{T, \pi_1})^{T, \pi_2} = \mathcal{X}^{T, \pi_2 \pi_1}, \tag{3.57}$$

where $\pi_2 \pi_1 = \pi_2 \circ \pi_1$ represents the composition of the two permutations. In particular, denoting by π^{-1} the inverse of the permutation π, we have $(\mathcal{X}^{T, \pi})^{T, \pi^{-1}} = \mathcal{X}$.

– Given two tensors $\mathcal{A}$ and $\mathcal{B}$ of the same size, we have:

$$\langle \mathcal{A}^{T, \pi}, \mathcal{B}^{T, \pi} \rangle = \langle \mathcal{A}, \mathcal{B} \rangle, \tag{3.58}$$

in other words, the inner product of two tensors of the same size is preserved if the two tensors are transposed using the same permutation of modes (see section 3.16 for the definition of the inner product).

– For a tensor $\mathcal{A} \in \mathbb{K}^{I_N}$ of order N, a sequence of N transpositions associated with a cyclic permutation $(\pi_1, \pi_2, \cdots, \pi_{N-1}, \pi_N)$ of the indices $(i_1, \cdots, i_N)$, of length N, leaves the tensor unchanged. Indeed, since the composition of the permutations $\pi = \pi_N \circ \pi_{N-1} \circ \cdots \circ \pi_2 \circ \pi_1$ is the identity mapping, i.e. the mapping that transforms the N-tuple $(i_1, \cdots, i_N)$ into itself, we have:

$$\left(\cdots ((\mathcal{A}^{T, \pi_1})^{T, \pi_2}) \cdots \right)^{T, \pi_N} = \mathcal{A}. \tag{3.59}$$

EXAMPLE 3.24.– For $N = 3$, with $\mathcal{A} \in \mathbb{K}^{I \times J \times K}$, there are two cyclic permutations of length three, which satisfy:

$$\pi_1 : (i, j, k) \to (j, k, i)\,;\ \pi_2 : (j, k, i) \to (k, i, j)\,;\ \pi_3 : (k, i, j) \to (i, j, k)$$

$$\pi_1 : (i, k, j) \to (k, j, i)\,;\ \pi_2 : (k, j, i) \to (j, i, k)\,;\ \pi_3 : (j, i, k) \to (i, k, j)$$

and we therefore have: $\left(\left(\mathcal{A}^{T, \pi_1} \right)^{T, \pi_2} \right)^{T, \pi_3} = \mathcal{A}^{T, \pi_3 \pi_2 \pi_1} = \mathcal{A}.$

– In the matrix case, the properties [3.58] and [3.59] become:

$$\langle \mathbf{A}^T, \mathbf{B}^T \rangle = \langle \mathbf{A}, \mathbf{B} \rangle \ , \ (\mathbf{A}^T)^T = \mathbf{A}. \tag{3.60}$$

– The transpose of a tensor $\mathcal{X}$ of order N can be obtained from its vectorization using the following formula:

$$\mathrm{vec}(\mathcal{X}^{T, \pi}) = \mathbf{e}^{i_1 \cdots i_N}_{i_{\pi(1)} \cdots i_{\pi(N)}} \, \mathrm{vec}(\mathcal{X}), \tag{3.61}$$

where $\mathbf{e}^{i_1 \cdots i_N}_{i_{\pi(1)} \cdots i_{\pi(N)}}$ is the permutation matrix associated with the permutation π of the set $\langle N \rangle$ that transforms the indices $(i_1, \cdots, i_N)$ of the tensor $\mathcal{X}$ into the indices $(i_{\pi(1)}, \cdots, i_{\pi(N)})$ of the transpose tensor $\mathcal{X}^{T, \pi}$. See section 2.7.4 for this type of permutation.

EXAMPLE 3.25.– Consider the tensor $\mathcal{X} \in \mathbb{K}^{I \times J \times K \times L}$ and the permutation π that transforms the quadruple $\{i, k, j, l\}$ into the quadruple $\{i, j, k, l\}$. The formula [3.61] that allows us to compute the transpose tensor $\mathcal{Y} = \mathcal{X}^{T, \pi}$ such that $y_{ikjl} = x_{ijkl}$ is then given by:

$$\mathrm{vec}(\mathcal{X}^{T, \pi}) = (\mathbf{I}_I \otimes \mathbf{e}^{jk}_{kj} \otimes \mathbf{I}_L) \mathrm{vec}(\mathcal{X}),$$

where the permutation matrix $\mathbf{e}^{jk}_{kj}$ is defined as in [2.167].

– In the case of a tensor $\mathcal{X} \in \mathbb{R}^{\underline{I}_P \times \underline{J}_N}$ of order $P+N$, we can define two particular types of partial block-wise transposition/transconjugation with respect to the P first and the N last modes, depending on whether or not the indices associated with each block are permuted using some permutations π and σ, respectively. The corresponding transpose tensors, denoted $\mathcal{X}^T$ and $\mathcal{X}^{T, \pi, \sigma}$, are given in Table 3.7, with the following notation for the index sets after permutation, and the associated dimensions:

$$\underline{\mathbf{i}}_{\pi(P)} = \{i_{\pi(1)}, \cdots, i_{\pi(P)}\} \ , \ \underline{\mathbf{j}}_{\sigma(N)} = \{j_{\sigma(1)}, \cdots, j_{\sigma(N)}\} \tag{3.62}$$

$$\underline{\mathbf{I}}_{\pi(P)} = \{I_{\pi(1)}, \cdots, I_{\pi(P)}\} \ , \ \underline{\mathbf{J}}_{\sigma(N)} = \{J_{\sigma(1)}, \cdots, J_{\sigma(N)}\} \tag{3.63}$$

$$\underline{I}_{\pi(P)} = I_{\pi(1)} \times \cdots \times I_{\pi(P)} \ , \ \underline{J}_{\sigma(N)} = J_{\sigma(1)} \times \cdots \times J_{\sigma(P)}. \tag{3.64}$$

The block-wise transposition of even-order tensors characterized by two mode sets of same dimension $\underline{I}_P$ and belonging to the space $\mathbb{K}^{\underline{I}_P \times \underline{I}_P}$ is also defined. This tensor space will be considered in section 3.13.5, in connection with the Einstein product.

Properties	Notation	Conditions	Dimensions
	$\mathcal{X} \in \mathbb{K}^{\underline{I}_P \times \underline{J}_N}$		$\underline{I}_P \times \underline{J}_N$
Block transpose	$\mathcal{Y} = \mathcal{X}^T$	$y_{\underline{j}_N,\underline{i}_P} = x_{\underline{i}_P,\underline{j}_N}$	$\underline{J}_N \times \underline{I}_P$
Block transconjugate	$\mathcal{Y} = \mathcal{X}^H$	$y_{\underline{j}_N,\underline{i}_P} = x^*_{\underline{i}_P,\underline{j}_N}$	$\underline{J}_N \times \underline{I}_P$
Block transpose with permutations	$\mathcal{Y} = \mathcal{X}^{T,\pi,\sigma}$	$y_{\underline{j}_{\sigma(N)},\underline{i}_{\pi(P)}} = x_{\underline{i}_P,\underline{j}_N}$	$\underline{J}_{\sigma(N)} \times \underline{I}_{\pi(P)}$
	$\mathcal{X} \in \mathbb{K}^{\underline{I}_P \times \underline{I}_P}$		$\underline{I}_P \times \underline{I}_P$
Block transpose	$\mathcal{Y} = \mathcal{X}^T$	$y_{\underline{i}_P,\underline{i}_P} = x_{\underline{i}_P,\underline{i}_P}$	$\underline{I}_P \times \underline{I}_P$
Block transconjugate	$\mathcal{Y} = \mathcal{X}^H$	$y_{\underline{i}_P,\underline{i}_P} = x^*_{\underline{i}_P,\underline{i}_P}$	$\underline{I}_P \times \underline{I}_P$
Block transpose with permutations	$\mathcal{Y} = \mathcal{X}^{T,\pi}$	$y_{\underline{i}_{\pi(P)},\underline{i}_{\pi(P)}} = x_{\underline{i}_P,\underline{j}_P}$	$\underline{I}_{\pi(P)} \times \underline{I}_{\pi(P)}$

Table 3.7. *Block-wise transposition of $\mathcal{X} \in \mathbb{K}^{\underline{I}_P \times \underline{J}_N}$ and $\mathcal{X} \in \mathbb{K}^{\underline{I}_P \times \underline{I}_P}$*

EXAMPLE 3.26.– Let $\mathcal{X} \in \mathbb{C}^{I_1 \times I_2 \times J_1 \times J_2}$, with $I_1 = 3$ and $I_2 = J_1 = J_2 = 2$. The block-wise transconjugate tensor $\mathcal{Y} = \mathcal{X}^H \in \mathbb{C}^{J_1 \times J_2 \times I_1 \times I_2}$ in the unfolded form $\mathbf{Y}_{J_1 J_2 \times I_1 I_2}$ is given by:

$$\mathbf{Y}_{J_1 J_2 \times I_1 I_2} = \begin{bmatrix} x^*_{1111} & x^*_{1211} & x^*_{2111} & x^*_{2211} & x^*_{3111} & x^*_{3211} \\ x^*_{1112} & x^*_{1212} & x^*_{2112} & x^*_{2212} & x^*_{3112} & x^*_{3212} \\ x^*_{1121} & x^*_{1221} & x^*_{2121} & x^*_{2221} & x^*_{3121} & x^*_{3221} \\ x^*_{1122} & x^*_{1222} & x^*_{2122} & x^*_{2222} & x^*_{3122} & x^*_{3222} \end{bmatrix}.$$

If we choose the permutations π and σ to be identical and such that $(1,2) \to (2,1)$, the block-wise transpose tensor with permutations $\mathcal{Y} = \mathcal{X}^{T,\pi,\sigma}$ is given by:

$$\mathbf{Y}_{J_2 J_1 \times I_2 I_1} = \begin{bmatrix} x_{1111} & x_{2111} & x_{3111} & x_{1211} & x_{2211} & x_{3211} \\ x_{1121} & x_{2121} & x_{3121} & x_{1221} & x_{2221} & x_{3221} \\ x_{1112} & x_{2112} & x_{3112} & x_{1212} & x_{2212} & x_{3212} \\ x_{1122} & x_{2122} & x_{3122} & x_{1222} & x_{2222} & x_{3222} \end{bmatrix}. \qquad [3.65]$$

PROPOSITION 3.27.– The block-wise transpose tensor with permutations $\mathcal{Y} = \mathcal{X}^{T,\pi,\sigma}$ satisfies the following relation with the tensor $\mathcal{X}$ in terms of matrix unfoldings:

$$\mathbf{Y}_{J_{\sigma(1)} \cdots J_{\sigma(N)} \times I_{\pi(1)} \cdots I_{\pi(P)}} = \mathbf{P}_\sigma (\mathbf{X}_{I \times J})^T \mathbf{P}_\pi, \qquad [3.66]$$

where $I = \prod_{p=1}^P I_p, J = \prod_{n=1}^N J_n$, and $\mathbf{P}_\sigma$ and $\mathbf{P}_\pi$ are permutation matrices defined as follows (see section 2.7 for commutation matrices):

$$\mathbf{P}_\sigma = \mathbf{e}^{j_1 \cdots j_N}_{j_{\sigma(1)} \cdots j_{\sigma(N)}} \qquad [3.67]$$

$$\mathbf{P}_\pi = \mathbf{e}^{i_{\pi(1)} \cdots i_{\pi(P)}}_{i_1 \cdots i_P} \qquad [3.68]$$

EXAMPLE 3.28.– For the tensor $\mathcal{X} \in \mathbb{C}^{I_1 \times I_2 \times J_1 \times J_2}$, with $I_1 = 3$ and $I_2 = J_1 = J_2 = 2$ from the previous example, the unfolded form [3.65] can be recovered using the formulae [3.66]–[3.68], with:

$$\mathbf{X}_{I \times J} = \mathbf{X}_{I_1 I_2 \times J_1 J_2} = \begin{bmatrix} x_{1111} & x_{1112} & x_{1121} & x_{1122} \\ x_{1211} & x_{1212} & x_{1221} & x_{1222} \\ x_{2111} & x_{2112} & x_{2121} & x_{2122} \\ x_{2211} & x_{2212} & x_{2221} & x_{2222} \\ x_{3111} & x_{3112} & x_{3121} & x_{3122} \\ x_{3211} & x_{3212} & x_{3221} & x_{3222} \end{bmatrix},$$

$$\mathbf{P}_\sigma = \mathbf{e}_{j_2 j_1}^{j_1 j_2} = \begin{bmatrix} 1 & 0 & 0 & 0 \\ 0 & 0 & 1 & 0 \\ 0 & 1 & 0 & 0 \\ 0 & 0 & 0 & 1 \end{bmatrix}, \quad \mathbf{P}_\pi = \mathbf{e}_{i_1 i_2}^{i_2 i_1} = \begin{bmatrix} 1 & 0 & 0 & 0 & 0 & 0 \\ 0 & 0 & 0 & 1 & 0 & 0 \\ 0 & 1 & 0 & 0 & 0 & 0 \\ 0 & 0 & 0 & 0 & 1 & 0 \\ 0 & 0 & 1 & 0 & 0 & 0 \\ 0 & 0 & 0 & 0 & 0 & 1 \end{bmatrix}.$$

3.10.3. Transposition and tensor contraction

The transposition operation can be used to perform a tensor contraction. For example, using index notation, the contraction defined by:

$$y_{i,j,k} = a_{i,l,j,m,n} b_{m,k,l,n}, \tag{3.69}$$

corresponding to triple summation over the indices (l, m, n), can be obtained by multiplying two matrix unfoldings obtained by combining the modes of the transpose tensors of $\mathcal{A} \in \mathbb{K}^{I \times L \times J \times M \times N}$ and $\mathcal{B} \in \mathbb{K}^{M \times K \times L \times N}$, as summarized below:

1) transposition of the tensor $\mathcal{A}$: $a_{i,l,j,m,n} \mapsto x_{i,j,l,m,n}$;
2) matrix unfolding of the transpose tensor $\mathcal{X}$ of $\mathcal{A}$: $\mathbf{X}_{IJ \times LMN}$;
3) transposition of the tensor $\mathcal{B}$: $b_{m,k,l,n} \mapsto z_{l,m,n,k}$;
4) matrix unfolding of the transpose tensor $\mathcal{Z}$ of $\mathcal{B}$: $\mathbf{Z}_{LMN \times K}$;
5) matrix multiplication: $\mathbf{Y}_{IJ \times K} = \mathbf{X}_{IJ \times LMN} \mathbf{Z}_{LMN \times K}$;
6) reconstruction of the tensor $\mathcal{Y}$: $\mathbf{Y}_{IJ \times K} \mapsto y_{i,j,k}$.

For tensors with very large dimensions, efficient algorithms can be used to determine the transpose tensors, in particular with the goal of minimizing the computational complexity of tensor contractions (see, for example, Lyakh (2015)).

3.11. Symmetric/partially symmetric tensors

3.11.1. *Symmetric tensors*

A hypercubic tensor $\mathcal{A} \in \mathbb{K}^{[N;I]}$ of order N and dimensions I is said to be symmetric if it is invariant under any permutation π of its modes, i.e.:

$$a_{\pi(i_1,i_2,\cdots,i_N)} \triangleq a_{i_{\pi(1)},i_{\pi(2)},\cdots,i_{\pi(N)}} = a_{i_1,i_2,\cdots,i_N} \qquad [3.70]$$

or equivalently:

$$\mathcal{A} = \mathcal{A}^{T,\pi} , \ \forall \pi \in \mathcal{S}_N. \qquad [3.71]$$

A tensor is therefore symmetric if it is equal to all of its transpose tensors. If so, the symmetry is said to be total or complete. Note that in the case of a complex tensor ($\mathbb{K} = \mathbb{C}$), Hermitian symmetry will be considered in section 3.11.2.

The set of symmetric real or complex tensors of order N and dimensions I will be denoted $\mathbb{K}_S^{[N;I]}$. Another notation used in literature is $\$^N(\mathbb{K}^I)$. Some authors call these tensors super-symmetric tensors (Kofidis and Regalia 2002; Qi 2005).

EXAMPLE 3.29.– High-order moments and cumulants of random variables form symmetric tensors. See the property [A1.31] in the Appendix.

In the case of a third-order tensor $\mathcal{A} \in \mathbb{K}_S^{I \times J \times K}$, with $I = J = K$, (total) symmetry implies the following equalities for every $i, j, k \in \langle I \rangle$:

$$a_{ijk} = a_{ikj} = a_{jik} = a_{jki} = a_{kij} = a_{kji}. \qquad [3.72]$$

Like for a symmetric matrix, a symmetric tensor $\mathcal{A} \in \mathbb{K}_S^{[N;I]}$ is uniquely determined by the set of entries located in its upper triangular part, i.e. $\{a_{i_1,\cdots,i_N}; 1 \le i_1 \le \cdots \le i_N \le I\}$.

A symmetric rank-one tensor of order N and dimensions I can be written as $(N-1)$ outer products of some vector $\mathbf{x} \in \mathbb{K}^I$ with itself (see section 3.13.1 for the definition of the outer product):

$$\mathcal{A} = \underbrace{\mathbf{x} \circ \cdots \circ \mathbf{x}}_{N \text{ terms}} = \mathbf{x}^{\circ N} \in \mathbb{K}_S^{[N;I]}. \qquad [3.73]$$

3.11.2. *Partially symmetric/Hermitian tensors*

In the same way that we defined partially diagonal tensors, we can also define partially symmetric tensors, i.e. tensors that are symmetric with respect to subsets of modes. There are different ways to define partial symmetry according to the partitioning being considered for the index set. For example, for a cubic tensor $\mathcal{A} \in \mathbb{K}^{I \times I \times I}$, there are three partial symmetries with respect to two modes (i, j), (i, k), or (j, k), which correspond to the following three conditions, respectively: (1) $x_{ijk} = x_{jik}$, (2) $x_{ijk} = x_{kji}$ and (3) $x_{ijk} = x_{ikj}$, satisfied for every $i, j, k \in \langle I \rangle$. Note that if two of these partial symmetries are satisfied, then this implies the third is also satisfied.

In the following, we will consider partial block-wise symmetries for three particular sets of even-order tensors: $\mathbb{K}^{\underline{I}_P \times \underline{I}_P}$, $\mathbb{K}^{[2P;I]}$ and $\mathbb{R}_p^{[2N;\underline{I}_n \times \underline{J}_n]} \triangleq \mathbb{R}^{I_1 \times J_1 \times \cdots \times I_N \times J_N}$.

A complex square tensor $\mathcal{A} = \left[a_{i_1, \cdots, i_P, j_1, \cdots, j_P} \right] = \left[a_{\mathbf{i}_P, \mathbf{j}_P} \right] \in \mathbb{C}^{\underline{I}_P \times \underline{I}_P}$ of order $2P$ is said to be Hermitian (by blocks of order P), written as $\mathcal{A}^H = \mathcal{A}$, if it is invariant under transconjugation by blocks of order P (Panigrahy and Mishra 2018; Ni 2019):

$$a_{\mathbf{j}_P, \mathbf{i}_P} = a_{\mathbf{i}_P, \mathbf{j}_P}^* , \quad \forall i_p \text{ and } j_p \in \langle I_p \rangle, \ p \in \langle P \rangle. \tag{3.74}$$

If so, the diagonal coefficients $a_{\mathbf{i}_P, \mathbf{i}_P}$ are real numbers.

PROPOSITION 3.30.– Consider the tensor $\mathcal{A} \in \mathbb{C}^{\underline{I}_P \times \underline{I}_P}$ and the matrix unfolding $\mathbf{A} \in \mathbb{C}^{\Pi I_P \times \Pi I_P}$, with $\Pi I_P \triangleq \prod_{p=1}^{P} I_p$, defined as:

$$(\mathbf{A})_{\overline{i_1 \cdots i_P}, \overline{j_1 \cdots j_P}} = a_{\mathbf{i}_P, \mathbf{j}_P}, \tag{3.75}$$

where $\overline{i_1 \cdots i_P}$ is defined as in [3.16], with $i_p, j_p \in \langle I_p \rangle$, $\forall p \in \langle P \rangle$. If $\mathcal{A}$ is Hermitian by blocks of order P, then the matrix unfolding $\mathbf{A}$ defined above is a Hermitian matrix, and we therefore have the following equivalence:

$$\mathcal{A}^H = \mathcal{A} \iff \mathbf{A}^H = \mathbf{A}. \tag{3.76}$$

EXAMPLE 3.31.– Consider the tensor $\mathcal{A} \in \mathbb{K}^{2 \times 2 \times 2 \times 2}$, with $I_1 = I_2 = 2$, assumed to be Hermitian by blocks of order two. Its unfolding $\mathbf{A}_{I_1 I_2 \times I_1 I_2}$ can then be written as:

$$\mathbf{A}_{I_1 I_2 \times I_1 I_2} = \begin{bmatrix} a_{1111} & a_{1211}^* & a_{2111}^* & a_{2211}^* \\ a_{1211} & a_{1212} & a_{2112}^* & a_{2212}^* \\ a_{2111} & a_{2112} & a_{2121} & a_{2221}^* \\ a_{2211} & a_{2212} & a_{2221} & a_{2222} \end{bmatrix}, \tag{3.77}$$

where the elements $a_{1111}, a_{1212}, a_{2121}$ and a_{2222} are real numbers.

Similarly, using a transposition/transconjugation by blocks of order P, we can define a skew-Hermitian, symmetric and skew-symmetric tensor, as outlined in Table 3.8.

Properties	Notation	Conditions
$\mathcal{A} \in \mathbb{K}^{\underline{I}_P \times \underline{I}_P}$		
Hermitian	$\mathcal{A}^H = \mathcal{A}$	$a_{\underline{j}_P, \underline{i}_P} = a^*_{\underline{i}_P, \underline{j}_P}$
Skew-Hermitian	$\mathcal{A}^H = -\mathcal{A}$	$a_{\underline{j}_P, \underline{i}_P} = -a^*_{\underline{i}_P, \underline{j}_P}$
Symmetric	$\mathcal{A}^T = \mathcal{A}$	$a_{\underline{j}_P, \underline{i}_P} = a_{\underline{i}_P, \underline{j}_P}$
Skew-symmetric	$\mathcal{A}^T = -\mathcal{A}$	$a_{\underline{j}_P, \underline{i}_P} = -a_{\underline{i}_P, \underline{j}_P}$

Table 3.8. *Hermitian/symmetric tensors by blocks of order P*

PROPOSITION 3.32.– Like for matrices, any real tensor $\mathcal{A} \in \mathbb{R}^{\underline{I}_P \times \underline{I}_P}$ has a unique decomposition as the sum of a symmetric tensor and a skew-symmetric tensor by blocks of order P:

$$\mathcal{A} = \frac{1}{2}(\mathcal{A} + \mathcal{A}^T) + \frac{1}{2}(\mathcal{A} - \mathcal{A}^T), \qquad [3.78]$$

where $\mathcal{A}^T$ is defined as in Table 3.8. In the case of a complex tensor $\mathcal{A} \in \mathbb{C}^{\underline{I}_P \times \underline{I}_P}$, replacing $\mathcal{A}^T$ by $\mathcal{A}^H$ in [3.78] gives a unique decomposition of $\mathcal{A}$ as the sum of a Hermitian tensor and a skew-Hermitian tensor by blocks of order P.

We say that a real (respectively, complex) even-order hypercubic tensor $\mathcal{A} \in \mathbb{R}^{[2P;I]}$ is twice block-wise symmetric (respectively, Hermitian) if it is invariant under any permutation by blocks of order P and under any permutation π of the indices $\{i_1, \cdots, i_P\}$ and $\{j_1, \cdots, j_P\}$ of each block, i.e.:

$$a_{\pi(\underline{j}_P), \pi(\underline{i}_P)} = a_{\underline{i}_P, \underline{j}_P}, \quad \left(\text{resp. } a^*_{\pi(\underline{j}_P), \pi(\underline{i}_P)} = a_{\underline{i}_P, \underline{j}_P} \right) \qquad [3.79]$$

$$i_p, j_p \in \langle I \rangle, \, p \in \langle P \rangle. \qquad [3.80]$$

We can also define a partial symmetry by blocks of order P such that:

$$a_{\pi(\underline{i}_P), \pi(\underline{j}_P)} = a_{\underline{i}_P, \underline{j}_P}, \quad i_p, j_p \in \langle I \rangle, \, p \in \langle P \rangle. \qquad [3.81]$$

This partial symmetry is equivalent to a symmetry with respect to the P first modes combined with a symmetry with respect to the P last modes.

A real even-order paired tensor $\mathcal{A} \in \mathbb{R}^{I_1 \times J_1 \times \cdots \times I_N \times J_N}$, with $I_n = J_n$ for $n \in \langle N \rangle$, is said to be paired symmetric if it is invariant under any permutation of adjacent indices $\{i_n, j_n\}$ for $n \in \langle N \rangle$ (Huang and Qi 2018):

$$a_{i_1 j_1 i_2 j_2 \cdots i_N j_N} = a_{j_1 i_1 i_2 j_2 \cdots i_N j_N} = a_{i_1 j_1 j_2 i_2 \cdots i_N j_N} = \cdots = a_{i_1 j_1 i_2 j_2 \cdots j_N i_N}.$$

$$[3.82]$$

We say that $\mathcal{A} \in \mathbb{R}^{I_1 \times J_1 \times \cdots \times I_N \times J_N}$, with $I_n = I$ and $J_n = J$ for $n \in \langle N \rangle$, is circularly symmetric if it is invariant under paired circular permutation of indices (Huang and Qi 2018):

$$a_{i_1 j_1 i_2 j_2 \cdots i_N j_N} = a_{i_2 j_2 i_3 j_3 \cdots i_1 j_1} = \cdots = a_{i_N j_N i_1 j_1 \cdots i_{N-1} j_{N-1}}. \qquad [3.83]$$

EXAMPLE 3.33.– For a fourth-order tensor $\mathcal{A} = [a_{ijkl}] \in \mathbb{R}^{[4\,;\,I]}$, with $i, j, k, l, \in \langle I \rangle$, paired symmetry means that the following constraints are satisfied[1]:

$$a_{ijkl} = a_{jikl} = a_{ijlk} = a_{jilk} \ \forall i, j, k, l \in \langle I \rangle. \qquad [3.84]$$

One example of this type of tensor is given by the fourth-order elasticity tensor[2] $\mathcal{C} \in \mathbb{R}^{[4;3]}$, which satisfies the following symmetries (Olive and Auffray 2013):

$$c_{ijkl} = c_{jikl} = c_{ijlk} \ \forall i, j, k, l \in \langle 3 \rangle \qquad [3.85]$$

$$c_{ijkl} = c_{klij} \ \forall i, j, k, l \in \langle 3 \rangle. \qquad [3.86]$$

The paired symmetries corresponding to the constraints [3.85] are called minor symmetries, whereas the symmetry by blocks of order two [3.86] is described as major. The symmetry relations [3.85] reduce the number of independent coefficients from $3^4 = 81$ to 36, and the symmetries [3.86] further reduce this number to 21.

3.11.3. *Multilinear forms with Hermitian symmetry and Hermitian tensors*

In the next two propositions, we will prove that, given certain symmetry conditions on the associated tensor, the multilinear form [3.5], for $P = N$ with $I_p = J_p, \forall p \in \langle P \rangle$, is Hermitian, and the multi-Hermitian form deduced from this multilinear form is real-valued.

PROPOSITION 3.34.– Consider the complex multilinear form deduced from [3.5] by choosing $P = N$, written using the index convention:

$$f\left((\mathbf{y}^{(1)})^*, \cdots, (\mathbf{y}^{(P)})^*, \mathbf{x}^{(1)}, \cdots, \mathbf{x}^{(P)}\right) = a_{\mathbf{i}_P \mathbf{j}_P} \prod_{p=1}^{P} (y_{i_p}^{(p)})^* \prod_{p=1}^{P} x_{j_p}^{(p)}, \quad [3.87]$$

---

1 Note that the third equality immediately follows from the first two.

2 In the case of small deformations of an elastic solid under stress, Hooke's experimental law (1676) can be written as follows with the index convention: $t_{ij} = c_{ijkl}\,\epsilon_{kl}$, where $[t_{ij}]$, $[\epsilon_{kl}]$, and $[c_{ijkl}]$ represent the stress, strain and rigidity tensors, respectively. The rigidity tensor is also called the tensor of elastic constants or the elasticity tensor. The symmetry of the stress tensor $(t_{ij} = t_{ji})$ and the strain tensor $(\epsilon_{kl} = \epsilon_{lk})$ implies that the rigidity tensor must also be symmetric.

with $\mathbf{x}^{(p)}, \mathbf{y}^{(p)} \in \mathbb{C}^{I_p}, \forall p \in \langle P \rangle$ and $\mathcal{A} \in \mathbb{C}^{\underline{I}_P \times \underline{I}_P}$. This multilinear form satisfies the property of Hermitian symmetry in the sense of the following equality:

$$f\big((\mathbf{x}^{(1)})^*, \cdots, (\mathbf{x}^{(P)})^*, \mathbf{y}^{(1)}, \cdots, \mathbf{y}^{(P)}\big) = f^*\big((\mathbf{y}^{(1)})^*, \cdots, (\mathbf{y}^{(P)})^*, \mathbf{x}^{(1)}, \cdots, \mathbf{x}^{(P)}\big)$$

[3.88]

if and only if $\mathcal{A}$ is Hermitian, as defined in Table 3.8, i.e. $\mathcal{A}^H = \mathcal{A}$.

PROOF.– Consider the matrix unfolding $\mathbf{A} \in \mathbb{C}^{\Pi I_P \times \Pi I_P}$ of the tensor $\mathcal{A} \in \mathbb{C}^{\underline{I}_P \times \underline{I}_P}$, with $\Pi I_P \triangleq \prod_{p=1}^{P} I_p$, as defined in [3.75], as well as the following vectors:

$$\mathbf{x} = \underbrace{\mathbf{x}^{(1)} \otimes \cdots \otimes \mathbf{x}^{(P)}}_{P \text{ terms}} \in \mathbb{C}^{\Pi I_P} \qquad\qquad [3.89]$$

$$\mathbf{y} = \underbrace{\mathbf{y}^{(1)} \otimes \cdots \otimes \mathbf{y}^{(P)}}_{P \text{ terms}} \in \mathbb{C}^{\Pi I_P}. \qquad\qquad [3.90]$$

The complex multilinear form [3.87] can then be rewritten as:

$$f(\mathbf{y}^*, \mathbf{x}) = \mathbf{y}^H \mathbf{A} \mathbf{x}. \qquad\qquad [3.91]$$

Using the property [3.76], the hypothesis of Hermitian symmetry of the tensor $\mathcal{A}$ implies that the matrix unfolding $\mathbf{A}$ is Hermitian, i.e. $\mathbf{A}^H = \mathbf{A}$. Hence, we can deduce that:

$$f(\mathbf{x}^*, \mathbf{y}) = \mathbf{x}^H \mathbf{A} \mathbf{y} = (\mathbf{x}^H \mathbf{A} \mathbf{y})^T \text{ (by transposition of a scalar quantity)}$$

$$= \mathbf{y}^T \mathbf{A}^T \mathbf{x}^* = (\mathbf{y}^H \mathbf{A}^H \mathbf{x})^* = (\mathbf{y}^H \mathbf{A} \mathbf{x})^* \text{ (by the hypothesis } \mathbf{A}^H = \mathbf{A})$$

$$= f^*(\mathbf{y}^*, \mathbf{x}). \qquad\qquad [3.92]$$

From this equality, we can deduce the Hermitian symmetry property [3.88], and conversely, [3.92] implies $\mathbf{A}^H = \mathbf{A}$. $\qquad\square$

PROPOSITION 3.35.– Consider the complex multilinear form defined in Table 3.4:

$$f\big(\underbrace{\mathbf{x}^*, \cdots, \mathbf{x}^*}_{P \text{ terms}}, \underbrace{\mathbf{x}, \cdots, \mathbf{x}}_{P \text{ terms}}\big) = a_{\underline{i}_P, \underline{j}_P} \prod_{p=1}^{P} x_{i_p}^* \prod_{n=1}^{P} x_{j_n}. \qquad\qquad [3.93]$$

This multilinear form is Hermitian in the sense of the following equality:

$$f^*\big(\mathbf{x}^*, \cdots, \mathbf{x}^*, \mathbf{x}, \cdots, \mathbf{x}\big) = f\big(\mathbf{x}^*, \cdots, \mathbf{x}^*, \mathbf{x}, \cdots, \mathbf{x}\big) \qquad\qquad [3.94]$$

if and only if $\mathcal{A}$ is Hermitian ($\mathcal{A}^H = \mathcal{A}$), in which case it is real-valued.

PROOF.– Note that the multilinear form [3.93] can be deduced from the multilinear form [3.87] by choosing $\mathbf{x}^{(p)} = \mathbf{y}^{(p)} = \mathbf{x} \in \mathbb{C}^I$, for $\forall p \in \langle P \rangle$, and $\mathcal{A} \in \mathbb{C}^{[2P;I]}$. We can therefore deduce from the previous proposition that if $\mathcal{A}$ is Hermitian, then the multilinear form [3.93] is multi-Hermitian in the sense of the equality [3.94], and therefore it is real-valued, since it is equal to its conjugate. □

REMARK 3.36.– Similarly, we can show that the multilinear form [3.93] is skew-Hermitian if $\mathcal{A}$ is skew-Hermitian ($\mathcal{A}^H = -\mathcal{A}$).

3.11.4. Symmetrization of a tensor

Like for matrices, it is possible to symmetrize a tensor, either as a tensor with the same size as the original tensor, or as a block tensor whose blocks are either transposed forms of the original tensor or zero tensors.

In the matrix case, for a square matrix $\mathbf{A} \in \mathbb{R}^{I \times I}$, a symmetrized form is given by:

$$\mathbf{Y} = \frac{1}{2}(\mathbf{A} + \mathbf{A}^T).$$ [3.95]

The next proposition extends the above symmetrization formula to the case of a hypercubic tensor of order N.

PROPOSITION 3.37.– Any real hypercubic tensor $\mathcal{A} \in \mathbb{R}^{[N;I]}$ of order N and dimensions I can be uniquely symmetrized using the following formula:

$$\mathcal{Y} = \frac{1}{N!} \sum_{\pi_n \in S_N} \mathcal{A}^{T,\pi_n},$$ [3.96]

where π_n denotes one of the $N!$ permutations of S_N, the symmetric group defined over the set $\langle N \rangle$ of the N first integers.

EXAMPLE 3.38.– Consider the cubic tensor $\mathcal{A} \in \mathbb{R}^{I \times I \times I}$, with $I = 2$, whose vectorization, according to lexicographical order, is the following vector:

$$\mathbf{a}_{IJK} = [1, 1, 2, 3, 3, 2, 4, 3]^T \in \mathbb{R}^8,$$

which corresponds to the following mode-1 flat matrix unfolding:

$$\mathbf{A}_{I \times JK} = \begin{bmatrix} 1 & 1 & | & 2 & 3 \\ 3 & 2 & | & 4 & 3 \end{bmatrix}.$$

Using the formula [3.96], the symmetrized tensor is given by:

$$\mathbf{y}_{IJK} = \begin{bmatrix} 1 & 2 & 2 & 3 & 2 & 3 & 3 & 3 \end{bmatrix}^T$$

$$\mathbf{Y}_{I \times JK} = \begin{bmatrix} 1 & 2 & | & 2 & 3 \\ 2 & 3 & | & 3 & 3 \end{bmatrix}.$$

Note that this symmetrized form $\mathcal{Y}$ of the tensor $\mathcal{A}$ does indeed satisfy the symmetry conditions [3.72], namely: $y_{112} = y_{121} = y_{211} = 2$, and $y_{122} = y_{212} = y_{221} = 3$.

In the matrix case, a block symmetrized form of a rectangular matrix $\mathbf{A} \in \mathbb{R}^{I_1 \times I_2}$, denoted $\mathbf{X}$, is given by:

$$\mathbf{X} = \begin{bmatrix} \mathbf{0}_{I_1 \times I_1} & \mathbf{A} \\ \mathbf{A}^T & \mathbf{0}_{I_2 \times I_2} \end{bmatrix} \in \mathbb{R}^{(I_1 + I_2) \times (I_1 + I_2)}, \tag{3.97}$$

which implies $\mathbf{X} = \mathbf{X}^T$.

PROPOSITION 3.39.– The block symmetric matrix $\mathbf{X}$ can be used to determine the SVD of $\mathbf{A}$ from the EVD of $\mathbf{X}$.

PROOF.– To prove this result, recall equations [1.42] and [1.43], which show that the right singular vectors $(\mathbf{v}_k)$ and left singular vectors $(\mathbf{u}_k)$ associated with the singular value σ_k are the eigenvectors of the symmetric matrices $\mathbf{A}^T \mathbf{A}$ and $\mathbf{A} \mathbf{A}^T$, respectively, i.e.:

$$\mathbf{A} \mathbf{v}_k = \sigma_k \mathbf{u}_k \iff \mathbf{A}^T \mathbf{A} \mathbf{v}_k = \sigma_k^2 \mathbf{v}_k \tag{3.98}$$

$$\mathbf{A}^T \mathbf{u}_k = \sigma_k \mathbf{v}_k \iff \mathbf{A} \mathbf{A}^T \mathbf{u}_k = \sigma_k^2 \mathbf{u}_k. \tag{3.99}$$

Using the block symmetrized form [3.97] of $\mathbf{A}$, the equations defining the above singular vectors can be rewritten as:

$$\begin{bmatrix} \mathbf{0}_{I_1 \times I_1} & \mathbf{A} \\ \mathbf{A}^T & \mathbf{0}_{I_2 \times I_2} \end{bmatrix} \begin{bmatrix} \mathbf{u}_k \\ \mathbf{v}_k \end{bmatrix} = \sigma_k \begin{bmatrix} \mathbf{u}_k \\ \mathbf{v}_k \end{bmatrix}, \tag{3.100}$$

which corresponds to the equation of the EVD of $\mathbf{X}$ defined in [3.97]. Assuming that $\mathbf{A}$ has rank R, the SVD of $\mathbf{A}$ (written $\mathbf{U}\boldsymbol{\Sigma}\mathbf{V}^T$) can therefore be determined by using the EVD [3.100] of the block symmetrized form [3.97], with:

$$\mathbf{U} = [\mathbf{u}_1, \cdots, \mathbf{u}_R], \ \mathbf{V} = [\mathbf{v}_1, \cdots, \mathbf{v}_R], \ \boldsymbol{\Sigma} = \text{diag}(\sigma_1, \cdots, \sigma_R), \tag{3.101}$$

where the σ_r, with $r \in \langle R \rangle$, are the R non-zero singular values associated with the R non-zero eigenvalues of $\mathbf{X}$, and the corresponding eigenvectors provide the singular vectors $(\mathbf{u}_k, \mathbf{v}_k)$ of $\mathbf{A}$. $\qquad \square$

In the next proposition, we extend the block symmetrization [3.97] to the case of a tensor $\mathcal{A} \in \mathbb{R}^{I_N}$ (Ragnarsson and Van Loan 2013). As we have just seen, in the case of a matrix $\mathbf{A} \in \mathbb{R}^{I_1 \times I_2}$, the block symmetrized matrix $\mathbf{X} \in \mathbb{R}^{(I_1 + I_2) \times (I_1 + I_2)}$ is formed of the blocks $\mathbf{A}$ and $\mathbf{A}^T$, as well as two zero blocks. Analogously to the matrix case, a tensor $\mathcal{A} \in \mathbb{R}^{\underline{I}_N}$ can be symmetrized as a block hypercubic tensor $\mathcal{X} \in \mathbb{R}^{[N,I]}$, with $I = \sum_{n=1}^{N} I_n$, whose blocks are either the $N!$ transpose tensors of $\mathcal{A}$ or zero tensors.

Recall the notation $\mathcal{A}^{T,\pi_p}$, with $p \in \langle N! \rangle$, of the transpose tensor of $\mathcal{A}$ associated with the permutation π_p of the set $\langle N \rangle$ of modes, such that:

$$\mathcal{A} \ni (\mathcal{A})_{\underline{i}_N} \overset{T,\pi_p}{\longmapsto} \left(\mathcal{A}^{T,\pi_p} \right)_{\underline{i}_{\pi_p\langle N \rangle}} \in \mathcal{A}^{T,\pi_p}. \qquad [3.102]$$

The multi-index $\pi_p\langle N \rangle \triangleq \{\pi_p(1), \cdots, \pi_p(N)\}$ is used to define the position of the transpose tensor $\mathcal{A}^{T,\pi_p}$ in the block tensor $\mathcal{X}$, as detailed in the next proposition.

PROPOSITION 3.40.– Any tensor $\mathcal{A} \in \mathbb{R}^{\underline{I}_N}$ can be symmetrized as a hypercubic block tensor $\mathcal{X} \in \mathbb{R}^{[N,I]}$, with $I = \sum_{n=1}^{N} I_n$, defined as:

$$[\mathcal{X}]_{\underline{\mathbf{j}}_N} = \begin{cases} \mathcal{A}^{T,\pi_p} & \text{for } \underline{\mathbf{j}}_N = \pi_p\langle N \rangle \\ \mathcal{O}_{\underline{\mathbf{j}}_N} & \text{for } \underline{\mathbf{j}}_N \neq \pi_p\langle N \rangle \quad \forall p \in \langle N! \rangle \end{cases} \qquad [3.103]$$

where $[\mathcal{X}]_{\pi_p\langle N \rangle} \in \mathbb{R}^{\underline{I}_{\pi_p\langle N \rangle}}$, with $\underline{I}_{\pi_p\langle N \rangle} = I_{\pi_p(1)} \times \cdots \times I_{\pi_p(N)}$, denotes the block of $\mathcal{X}$ located at position $(\pi_p(1), \cdots, \pi_p(N))$, and the zero block $\mathcal{O}_{\underline{\mathbf{j}}_N} \in \mathbb{R}^{I_{j_1} \times \cdots \times I_{j_N}}$ is located at position $(j_1, \cdots, j_N)$ in $\mathcal{X}$. The block $\mathcal{O}_{\underline{\mathbf{j}}_N}$ will also be denoted $\mathcal{O}_{I_{j_1} \times \cdots \times I_{j_N}}$, with $j_n \in \langle N \rangle$, $n \in \langle N \rangle$, in order to specify its dimensions.

REMARK 3.41.– The zero tensor $\mathcal{O}_{\underline{\mathbf{j}}_N}$ is positioned in $\mathcal{X}$ using the multi-index $\underline{\mathbf{j}}_N = \{j_1, \cdots, j_N\}$, such that $\underline{\mathbf{j}}_N \neq \pi_p\langle N \rangle$ for $\forall p \in \langle N! \rangle$, i.e. an N-tuple $\underline{\mathbf{j}}_N$ that cannot be associated with any permutation π_p of the set $\langle N \rangle$. This means that at least two of the N indices j_n are equal.

Before proving that the block tensor $\mathcal{X}$ thus constructed is symmetric, we will illustrate the block symmetrization formula [3.103] for a third-order tensor (Ragnarsson and Van Loan 2013).

For $\mathcal{A} \in \mathbb{R}^{I_1 \times I_2 \times I_3}$, there exist six transpose tensors associated with the six permutations π_p, with $p \in \langle 6 \rangle$, of the set $\langle 3 \rangle$. The blocks $[\mathcal{X}]_{\pi_p\langle 3 \rangle}$ of the tensor $\mathcal{X}$ containing these six transpose tensors are summarized in Table 3.9, together with their dimensions.

π_p	**Permutations** $\langle 3 \rangle \overset{\pi_p}{\longmapsto} \pi_p\langle 3 \rangle$	$[\mathcal{X}]_{\pi_p\langle 3 \rangle}$	**Dimensions of** $\mathcal{A}^{T,\pi_p}$
π_1	$(1,2,3) \to (1,2,3)$	$[\mathcal{X}]_{123} = \mathcal{A}^{T,\pi_1}$	$I_1 \times I_2 \times I_3$
π_2	$(1,2,3) \to (1,3,2)$	$[\mathcal{X}]_{132} = \mathcal{A}^{T,\pi_2}$	$I_1 \times I_3 \times I_2$
π_3	$(1,2,3) \to (2,1,3)$	$[\mathcal{X}]_{213} = \mathcal{A}^{T,\pi_3}$	$I_2 \times I_1 \times I_3$
π_4	$(1,2,3) \to (2,3,1)$	$[\mathcal{X}]_{231} = \mathcal{A}^{T,\pi_4}$	$I_2 \times I_3 \times I_1$
π_5	$(1,2,3) \to (3,1,2)$	$[\mathcal{X}]_{312} = \mathcal{A}^{T,\pi_5}$	$I_3 \times I_1 \times I_2$
π_6	$(1,2,3) \to (3,2,1)$	$[\mathcal{X}]_{321} = \mathcal{A}^{T,\pi_6}$	$I_3 \times I_2 \times I_1$

Table 3.9. *Block symmetrized tensor $\mathcal{X}$ of a third-order tensor $\mathcal{A} \in \mathbb{R}^{I_1 \times I_2 \times I_3}$*

The equations below indicate the positions of the six transpose tensors $\mathcal{A}^{T,\pi_p}$, for $p \in \langle 6 \rangle$, in the tensor slices $(\mathcal{X})_{..k} \in \mathbb{R}^{I \times I \times I_k}$, with $k \in \langle 3 \rangle$ and $I = I_1 + I_2 + I_3$.

$$(\mathcal{X})_{..1} = \begin{bmatrix} \mathcal{O}_{I_1 \times I_1 \times I_1} & \vdots & \mathcal{O}_{I_1 \times I_2 \times I_1} & \vdots & \mathcal{O}_{I_1 \times I_3 \times I_1} \\ \cdots & & \cdots & & \cdots \\ \mathcal{O}_{I_2 \times I_1 \times I_1} & \vdots & \mathcal{O}_{I_2 \times I_2 \times I_1} & \vdots & \mathcal{A}^{T,\pi_4} \\ \cdots & & \cdots & \cdots & \\ \mathcal{O}_{I_3 \times I_1 \times I_1} & \vdots & \mathcal{A}^{T,\pi_6} & \vdots & \mathcal{O}_{I_3 \times I_3 \times I_1} \end{bmatrix} \in \mathbb{R}^{I \times I \times I_1}$$

$$(\mathcal{X})_{..2} = \begin{bmatrix} \mathcal{O}_{I_1 \times I_1 \times I_2} & \vdots & \mathcal{O}_{I_1 \times I_2 \times I_2} & \vdots & \mathcal{A}^{T,\pi_2} \\ \cdots & & \cdots & & \cdots \\ \mathcal{O}_{I_2 \times I_1 \times I_2} & \vdots & \mathcal{O}_{I_2 \times I_2 \times I_2} & \vdots & \mathcal{O}_{I_2 \times I_3 \times I_2} \\ \cdots & & \cdots & & \cdots \\ \mathcal{A}^{T,\pi_5} & \vdots & \mathcal{O}_{I_3 \times I_2 \times I_2} & \vdots & \mathcal{O}_{I_3 \times I_3 \times I_2} \end{bmatrix} \in \mathbb{R}^{I \times I \times I_2}$$

$$(\mathcal{X})_{..3} = \begin{bmatrix} \mathcal{O}_{I_1 \times I_1 \times I_3} & \vdots & \mathcal{A}^{T,\pi_1} & \vdots & \mathcal{O}_{I_1 \times I_3 \times I_3} \\ \cdots & & \cdots & & \cdots \\ \mathcal{A}^{T,\pi_3} & \vdots & \mathcal{O}_{I_2 \times I_2 \times I_3} & \vdots & \mathcal{O}_{I_2 \times I_3 \times I_3} \\ \cdots & & \cdots & & \cdots \\ \mathcal{O}_{I_3 \times I_1 \times I_3} & \vdots & \mathcal{O}_{I_3 \times I_2 \times I_3} & \vdots & \mathcal{O}_{I_3 \times I_3 \times I_3} \end{bmatrix} \in \mathbb{R}^{I \times I \times I_3}$$

REMARK 3.42.– We can make the following remarks:

– there are zero tensors at the positions (j_1, j_2, j_3) of the slices $(\mathcal{X})_{..k}$ whenever the triplet (j_1, j_2, j_3) does not correspond to a permutation of the set $\langle 3 \rangle$, i.e. when two or three of the indices j_1, j_2, and j_3 are equal. For example, in the slice $(\mathcal{X})_{..2}$, the block $[\mathcal{X}]_{122}$ is the zero tensor $\mathcal{O}_{I_1 \times I_2 \times I_2} \in \mathbb{R}^{I_1 \times I_2 \times I_2}$ due to the repetition of the last two indices ($j_2 = j_3 = 2$);

– all blocks in the same slice $(\mathcal{X})_{..k}$ have the same third dimension I_k.

We will now prove that the block tensor $\mathcal{X}$ is symmetric, which means that it is equal to all of its transpose tensors.

Let $\mathcal{Y} = \mathcal{X}^{T, \pi_q}$ be the transpose tensor of $\mathcal{X}$ associated with the permutation π_q of the set $\langle N \rangle$. Since the block tensor $\mathcal{X}$ is hypercubic with dimensions I, its transpose tensors are themselves hypercubic block tensors with dimensions I, i.e. $\mathcal{Y} \in \mathbb{R}^{[N, I]}$, with $I = \sum_{n=1}^{N} I_n$.

The blocks of $\mathcal{Y}$ are composed of the transposes of the blocks $[\mathcal{X}]_{\underline{\mathbf{j}}_{\langle N \rangle}}$, with positions $\pi_p(\underline{\mathbf{j}}_{\langle N \rangle}) = (\pi_p(j_1), \cdots, \pi_p(j_N))$. Hence, by [3.103], we can conclude that the blocks of $\mathcal{Y}$ are equal to either transposed forms of the blocks $\mathcal{A}^{T, \pi_p}$, or to transposed zero tensors. Thus, after transposition, the block $[\mathcal{X}]_{\pi_p \langle N \rangle} = \mathcal{A}^{T, \pi_p}$ of $\mathcal{X}$ is transformed into the block $(\mathcal{A}^{T, \pi_p})^{T, \pi_q} = \mathcal{A}^{T, \pi_q \pi_p}$, at the position $\pi_q \pi_p \langle N \rangle$ in $\mathcal{Y}$:

$$[\mathcal{Y}]_{\pi_q \pi_p \langle N \rangle} = [\mathcal{X}_{\pi_p \langle N \rangle}]^{T, \pi_q} = (\mathcal{A}^{T, \pi_p})^{T, \pi_q} = \mathcal{A}^{T, \pi_q \pi_p} = [\mathcal{X}]_{\pi_q \pi_p \langle N \rangle},$$

[3.104]

which proves that the block $[\mathcal{Y}]_{\pi_q \pi_p \langle N \rangle}$ is equal to $\mathcal{A}^{T, \pi_q \pi_p}$, and hence equal to $[\mathcal{X}]_{\pi_q \pi_p \langle N \rangle}$ by [3.103].

To illustrate this result, consider the symmetrized form defined in [3.103] of the third-order tensor $\mathcal{A}$, with $p = 6, q = 2$. Noting that the composition of the permutations $\pi_2 \pi_6$ transforms the triplet $(1, 2, 3)$ into $(3, 1, 2)$, we have $\pi_2 \pi_6 = \pi_5$, and therefore:

$$(\mathcal{A}^{T, \pi_6})^{T, \pi_2} = \mathcal{A}^{T, \pi_2 \pi_6} = \mathcal{A}^{T, \pi_5} = [\mathcal{X}]_{\pi_5 \langle 3 \rangle} = [\mathcal{X}]_{312},$$

as can be checked in the tensor slice $(\mathcal{X})_{..2}$ detailed above.

The transpose tensor $\mathcal{X}^{T, \pi_2}$ is therefore such that the block $[\mathcal{X}]_{321} = \mathcal{A}^{T, \pi_6}$ of $\mathcal{X}$ is transformed into $[\mathcal{Y}]_{312} = \mathcal{A}^{T, \pi_2 \pi_6} = \mathcal{A}^{T, \pi_5} = [\mathcal{X}]_{312}$ in $\mathcal{Y} = \mathcal{X}^{T, \pi_2}$.

Now consider a zero block $\mathcal{O}_{\underline{\mathbf{j}}_N} \in \mathbb{R}^{I_{j_1} \times \cdots \times I_{j_N}}$ in $\mathcal{X}$, for $\underline{\mathbf{j}}_N = \{j_1, \cdots, j_N\}$, $j_n \in \langle N \rangle$, and $n \in \langle N \rangle$, with at least two identical indices j_i and j_k, i.e. $\underline{\mathbf{j}}_N \neq \pi_p \langle N \rangle$ for $\forall p \in \langle N! \rangle$. The transpose of the block $\mathcal{O}_{\underline{\mathbf{j}}_N}$ in $\mathcal{X}^{T, \pi_q}$ is given by:

$$[\mathcal{Y}]_{\pi_q(\underline{\mathbf{j}}_N)} = (\mathcal{O}_{\underline{\mathbf{j}}_N})^{T, \pi_q} = \mathcal{O}_{\pi_q(\underline{\mathbf{j}}_N)} = [\mathcal{X}]_{\pi_q(\underline{\mathbf{j}}_N)} \in \mathbb{R}^{I_{\pi_q(j_1)} \times \cdots \times I_{\pi_q(j_N)}}.$$

For example, for the tensor $\mathcal{A} \in \mathbb{R}^{I_1 \times I_2 \times I_3}$, after the transposition associated with the permutation π_2, the zero block $[\mathcal{X}]_{121} = \mathcal{O}_{I_1 \times I_2 \times I_1}$ of $\mathcal{X}$ is transformed into:

$$[\mathcal{Y}]_{\pi_2(1,2,1)} = [\mathcal{Y}]_{112} = [\mathcal{X}]_{121}^{T,\pi_2} = (\mathcal{O})_{121}^{T,\pi_2} = (\mathcal{O})_{\pi_2(1,2,1)} = \mathcal{O}_{I_1 \times I_1 \times I_2}.$$

From the above results, we can write the transpose tensor $\mathcal{Y} = \mathcal{X}^{T,\pi_q}$ as:

$$[\mathcal{Y}]_{\underline{\mathbf{j}}_N} = [\mathcal{X}^{T,\pi_q}]_{\underline{\mathbf{j}}_N} = \begin{cases} \mathcal{A}^{T,\pi_q \pi_p} & \text{for } \underline{\mathbf{j}}_N = \pi_q \pi_p \langle N \rangle \\ (\mathcal{O}_{\underline{\mathbf{i}}_N})^{T,\pi_q} & \text{for } \underline{\mathbf{j}}_N = \pi_q(\underline{\mathbf{i}}_N) \end{cases}$$

where $\underline{\mathbf{i}}_N \neq \pi_p \langle N \rangle$ for $\forall p \in \langle N! \rangle$.

In summary, we conclude that $\mathcal{X}^{T,\pi_q} = \mathcal{X}$ for every $q \in \langle N! \rangle$, i.e. regardless of the transposition of $\mathcal{X}$, which implies that $\mathcal{X}$ is symmetric.

3.12. Triangular tensors

Let $\mathcal{A} \in \mathbb{K}^{\underline{I}_N \times \underline{I}_N}$. This tensor is said to be triangular if its matrix unfolding $\mathbf{A}_{I \times I}$, with $I = \prod_{n=1}^{N} I_n$, is itself triangular. More precisely, noting that:

$$(\mathbf{A}_{I \times I})_{\overline{i_1 \cdots i_N}, \overline{j_1 \cdots j_N}} = a_{\underline{\mathbf{i}}_N \underline{\mathbf{j}}_N}, \qquad [3.105]$$

where $\overline{i_1 \cdots i_N}$ is defined as in [3.16], with $i_n, j_n \in \langle I_n \rangle$, $\forall n \in \langle N \rangle$, the tensor $\mathcal{A}$ is upper (respectively, lower) triangular if $\mathbf{A}_{I \times I}$ is upper (respectively, lower) triangular, which is equivalent to $a_{\underline{\mathbf{i}}_N \underline{\mathbf{j}}_N} = 0$ for $\overline{i_1 \cdots i_N} > \overline{j_1 \cdots j_N}$ (respectively, $\overline{j_1 \cdots j_N} > \overline{i_1 \cdots i_N}$).

Similarly, we say that $\mathcal{A}$ is unit upper (lower) triangular if $\mathbf{A}_{I \times I}$ is unit upper (lower) triangular, i.e. with diagonal terms equal to 1.

EXAMPLE 3.43.– Let $\mathcal{A} \in \mathbb{K}^{2 \times 2 \times 2 \times 2}$, with $I_1 = I_2 = 2$. Its matrix unfolding $\mathbf{A}_{I_1 I_2 \times I_1 I_2}$ can be written as:

$$\mathbf{A}_{I_1 I_2 \times I_1 I_2} = \begin{bmatrix} a_{1111} & a_{1112} & a_{1121} & a_{1122} \\ a_{1211} & a_{1212} & a_{1221} & a_{1222} \\ a_{2111} & a_{2112} & a_{2121} & a_{2122} \\ a_{2211} & a_{2212} & a_{2221} & a_{2222} \end{bmatrix}. \qquad [3.106]$$

The tensor $\mathcal{A}$ is unit upper triangular if: $a_{1111} = a_{1212} = a_{2121} = a_{2222} = 1$ and $a_{1211} = a_{2111} = a_{2112} = a_{2211} = a_{2212} = a_{2221} = 0$, and the other terms can take any values.

3.13. Multiplication operations

There are several types of multiplication with tensors. These operations can be classified as follows:

– products whose result is a tensor of the same order as the tensors being multiplied, like the Hadamard product (section 3.18);

– products that give a tensor of order higher than the tensors being multiplied, like the outer product (section 3.13.1) and the Kronecker product (section 3.18);

– contracted products that correspond to a contraction of tensors being multiplied over one or several modes, like the mode-p and mode-(p,n) products, denoted $\times_p$ and $\times_p^n$, respectively (sections 3.13.2 and 3.13.4), or the Einstein product, denoted $\star_N$ (section 3.13.5). These products involve summation over one or several indices corresponding to the modes shared by both tensors being multiplied.

Note that mode-p multiplication, also known as the Tucker product, is used for tensor decompositions into a linear combination of rank-one tensors, like the PARAFAC and Tucker decompositions that will be presented in Chapter 5, whereas the Einstein product is used for developing tensor factorizations, like the SVD and full-rank decomposition, which will be described in section 3.15. The Einstein product is also used to define certain tensor systems (section 3.17), and to develop methods for solving them based on the notions of inverse and pseudo-inverse tensors (section 3.14).

These various multiplication operations are presented below. First, we will describe three fundamental operations, namely the outer product of tensors, tensor-matrix multiplication and tensor-vector multiplication. We will then present the more general operation of tensor contraction, i.e. tensor–tensor multiplication based on the mode-(p,n) and Einstein products. In section 3.16, the inner product of two tensors will be defined using the Einstein product. The notions of the Frobenius norm and trace of a tensor will also be defined. In section 3.17, these multiplication operations will be used to establish a connection between tensors and tensor systems. Finally, the Hadamard and Kronecker products of tensors will be introduced in section 3.18.

REMARK 3.44.– It should be noted that other types of tensor multiplication exist, such as the t-product of third-order tensors $\mathcal{A}, \mathcal{B}$, denoted $\mathcal{A} \star \mathcal{B}$ (Kilmer and Martin 2011). This type of tensor product is based on the matrix multiplication of a block circulant matrix constructed from frontal slices of $\mathcal{A}$, with a block vectorized form of $\mathcal{B}$ also constructed using frontal slices. Unlike the mode-p and Einstein products, the t-product preserves the order of the tensors being multiplied. Furthermore, the t-product can be computed using the fast Fourier transform (FFT). The t-product operation was used to develop the t-SVD, t-polar, t-LU, t-QR and t-Schur decompositions of third-order tensors (Kilmer and Martin 2011; Hao et al. 2013; Miao et al. 2020).

3.13.1. *Outer product of tensors*

The outer product, also called the tensor product, of two non-zero vectors $\mathbf{u} \in \mathbb{K}^I$ and $\mathbf{v} \in \mathbb{K}^J$, denoted $\mathbf{u} \circ \mathbf{v}$, defines a rank-one matrix of size $I \times J$ such that:

$$(\mathbf{u} \circ \mathbf{v})_{ij} = u_i v_j \quad \Rightarrow \quad \mathbf{u} \circ \mathbf{v} \in \mathbb{K}^{I \times J}. \tag{3.107}$$

The outer product of P non-zero vectors $\mathbf{u}^{(p)} \in \mathbb{K}^{I_p}$, $p \in \langle P \rangle$, gives a rank-one tensor of order P and size $\underline{I}_P$ such that:

$$\left(\underset{p=1}{\overset{P}{\circ}} \mathbf{u}^{(p)} \right)_{\underline{i}_P} = \prod_{p=1}^{P} u_{i_p}^{(p)} \quad \Rightarrow \quad \underset{p=1}{\overset{P}{\circ}} \mathbf{u}^{(p)} \in \mathbb{K}^{\underline{I}_P}. \tag{3.108}$$

In particular, the outer product of three non-zero vectors $\mathbf{u} \in \mathbb{K}^I, \mathbf{v} \in \mathbb{K}^J$, and $\mathbf{w} \in \mathbb{K}^K$ gives a rank-one, third-order tensor $\mathbf{u} \circ \mathbf{v} \circ \mathbf{w}$ of size $I \times J \times K$ such that:

$$(\mathbf{u} \circ \mathbf{v} \circ \mathbf{w})_{ijk} = u_i v_j w_k \ , \quad i \in \langle I \rangle, j \in \langle J \rangle, k \in \langle K \rangle. \tag{3.109}$$

In the case where the vectors $\mathbf{u}^{(p)}$ are all identical and equal to $\mathbf{u} \in \mathbb{K}^I$, the definition [3.108] gives a symmetric hypercubic tensor of order P and dimensions I, as defined in [3.70]:

$$\underbrace{\mathbf{u} \circ \cdots \circ \mathbf{u}}_{P \text{ terms}} \triangleq \mathbf{u}^{\circ P} = \Big[\prod_{p=1}^{P} u_{i_p} \Big] \in \mathbb{K}_S^{[P;I]}. \tag{3.110}$$

Table 3.10 presents a few examples of outer products of matrices and tensors, indicating the space which the tensors resulting from these products belong to, as well as their order.

Matrices/tensors being multiplied	Outer products	Spaces	Orders
$\mathbf{A}^{(p)} \in \mathbb{K}^{I_p \times J_p}$	$\underset{p=1}{\overset{P}{\circ}} \mathbf{A}^{(p)}$	$\mathbb{K}^{[2P;I_p \times J_p]}$	$2P$
$\mathcal{A} \in \mathbb{K}^{\underline{I}_P}, \mathcal{B} \in \mathbb{K}^{\underline{I}_P}$	$\mathcal{A} \circ \mathcal{B}$	$\mathbb{K}^{\underline{I}_P \times \underline{I}_P}$	$2P$
$\mathcal{A} \in \mathbb{K}^{\underline{I}_P}, \mathcal{B} \in \mathbb{K}^{\underline{J}_N}$	$\mathcal{A} \circ \mathcal{B}$	$\mathbb{K}^{\underline{I}_P \times \underline{J}_N}$	$P + N$
$\mathcal{A}^{(p)} \in \mathbb{K}^{\underline{J}_{N_p}}$	$\underset{p=1}{\overset{P}{\circ}} \mathcal{A}^{(p)}$	$\mathbb{K}^{\underline{J}_{N_1} \times \cdots \times \underline{J}_{N_P}}$	$\sum_{p=1}^{P} N_p$

Table 3.10. *Outer products of matrices and tensors*

REMARK 3.45.– Note that the outer product of P matrices $\mathbf{A}^{(p)} \in \mathbb{K}^{I_p \times J_p}$ gives an even-order paired tensor belonging to the space $\mathbb{K}^{I_1 \times J_1 \cdots \times I_P \times J_P}$, also denoted $\mathbb{K}^{[2P;I_p \times J_p]}$. Moreover, the outer product of two tensors $\mathcal{A}$ and $\mathcal{B} \in \mathbb{K}^{\underline{I}_P}$ gives a square tensor of the space $\mathbb{K}^{\underline{I}_P \times \underline{I}_P}$, whereas the outer product of $\mathcal{A} \in \mathbb{K}^{\underline{I}_P}$ with $\mathcal{B} \in \mathbb{K}^{\underline{J}_N}$ gives a rectangular tensor of order $P + N$, belonging to the space $\mathbb{K}^{\underline{I}_P \times \underline{J}_N}$.

Matrices/tensors	Outer products	Elements of $\mathcal{X}$	Indices
$\mathbf{A}^{(p)} \in \mathbb{K}^{I_p \times J_p}$	$\mathcal{X} = \overset{P}{\underset{p=1}{\circ}} \mathbf{A}^{(p)}$	$x_{i_1 j_1 \cdots i_P j_P} = \prod_{p=1}^{P} a_{i_p j_p}^{(p)}$	$i_p \in \langle I_p \rangle \,,\ p \in \langle P \rangle$
			$j_p \in \langle J_p \rangle \,,\ p \in \langle P \rangle$
$\mathcal{A} \in \mathbb{K}^{\underline{I}_P}, \mathcal{B} \in \mathbb{K}^{\underline{I}_P}$	$\mathcal{X} = \mathcal{A} \circ \mathcal{B}$	$x_{\underline{\mathbf{i}}_P, \underline{\mathbf{j}}_P} = a_{\underline{\mathbf{i}}_P} b_{\underline{\mathbf{j}}_P}$	$i_p, j_p \in \langle I_p \rangle \,,\ p \in \langle P \rangle$
$\mathcal{A} \in \mathbb{K}^{\underline{I}_P}, \mathcal{B} \in \mathbb{K}^{\underline{J}_N}$	$\mathcal{X} = \mathcal{A} \circ \mathcal{B}$	$x_{\underline{\mathbf{i}}_P, \underline{\mathbf{j}}_N} = a_{\underline{\mathbf{i}}_P} b_{\underline{\mathbf{j}}_N}$	$i_p \in \langle I_p \rangle \,,\ p \in \langle P \rangle$
			$j_n \in \langle J_n \rangle \,,\ n \in \langle N \rangle$
$\mathcal{A}^{(p)} \in \mathbb{K}^{\underline{J}_{N_p}}$	$\mathcal{X} = \overset{P}{\underset{p=1}{\circ}} \mathcal{A}^{(p)}$	$x_{\underline{\mathbf{j}}_{N_1}, \cdots, \underline{\mathbf{j}}_{N_P}} = \prod_{p=1}^{P} a_{\underline{\mathbf{j}}_{N_p}}^{(p)}$	$\underline{\mathbf{j}}_{N_p} = (j_1, \cdots, j_{N_p})$
			$p \in \langle P \rangle$

Table 3.11. *Scalar elements of outer products*

Table 3.11 gives expressions for the scalar elements of each tensor resulting from the outer products in Table 3.10.

The next proposition presents the properties of the outer product.

PROPOSITION 3.46.– The outer product is associative and distributive with respect to tensor addition but not commutative. For every $\mathcal{A}, \mathcal{A}_1, \mathcal{A}_2 \in \mathbb{K}^{\underline{I}_P}$, $\mathcal{B}, \mathcal{B}_1, \mathcal{B}_2 \in \mathbb{K}^{\underline{J}_N}$, $\mathcal{C} \in \mathbb{K}^{\underline{K}_L}$ and $\lambda \in \mathbb{K}$, we have:

$$\mathcal{A} \circ (\mathcal{B} \circ \mathcal{C}) = (\mathcal{A} \circ \mathcal{B}) \circ \mathcal{C} = \mathcal{A} \circ \mathcal{B} \circ \mathcal{C} \in \mathbb{K}^{\underline{I}_P \times \underline{J}_N \times \underline{K}_L} \qquad [3.111]$$

$$(\mathcal{A}_1 + \mathcal{A}_2) \circ \mathcal{B} = \mathcal{A}_1 \circ \mathcal{B} + \mathcal{A}_2 \circ \mathcal{B} \in \mathbb{K}^{\underline{I}_P \times \underline{J}_N} \qquad [3.112]$$

$$\mathcal{A} \circ (\mathcal{B}_1 + \mathcal{B}_2) = \mathcal{A} \circ \mathcal{B}_1 + \mathcal{A} \circ \mathcal{B}_2 \in \mathbb{K}^{\underline{I}_P \times \underline{J}_N} \qquad [3.113]$$

$$(\lambda \mathcal{A}) \circ \mathcal{B} = \mathcal{A} \circ (\lambda \mathcal{B}) = \lambda (\mathcal{A} \circ \mathcal{B}) \in \mathbb{K}^{\underline{I}_P \times \underline{J}_N}. \qquad [3.114]$$

From the properties [3.112]–[3.114], we can conclude that the outer product is a bilinear mapping from $\mathbb{R}^{\underline{I}_P} \times \mathbb{R}^{\underline{J}_N}$ to $\mathbb{R}^{\underline{I}_P \times \underline{J}_N}$:

$$\mathbb{R}^{\underline{I}_P} \times \mathbb{R}^{\underline{J}_N} \ni (\mathcal{A}, \mathcal{B}) \longmapsto \mathcal{A} \circ \mathcal{B} \in \mathbb{R}^{\underline{I}_P \times \underline{J}_N}.$$

The next proposition gives general matricized and vectorized forms for a rank-one tensor of order P.

PROPOSITION 3.47.– Let $\mathcal{X} = \overset{P}{\underset{p=1}{\circ}} \mathbf{u}^{(p)} \in \mathbb{K}^{I_P}$ be the rank-one tensor of order P defined in [3.108]. The general matricized form [3.39] of $\mathcal{X}$ is given by:

$$\mathbf{X}_{\mathbb{S}_1;\mathbb{S}_2} = x_{i_1,\cdots,i_P} \mathbf{e}_{\mathbb{I}_1}^{\mathbb{I}_2} = \left(\underset{p \in \mathbb{S}_1}{\otimes} u_{i_p}^{(p)} \mathbf{e}_{i_p}^{(I_p)} \right) \left(\underset{p \in \mathbb{S}_2}{\otimes} u_{i_p}^{(p)} \mathbf{e}_{i_p}^{(I_p)} \right)^T$$

$$= \left(\underset{p \in \mathbb{S}_1}{\otimes} \mathbf{u}^{(p)} \right) \left(\underset{p \in \mathbb{S}_2}{\otimes} \mathbf{u}^{(p)} \right)^T \in \mathbb{K}^{J_1 \times J_2} \qquad [3.115]$$

where $J_{p_1} = \prod_{p \in \mathbb{S}_{p_1}} I_p$, for $p_1 = 1$ and 2.

Applying the relation [2.118] to [3.115], with $\mathbf{A} = \underset{p \in \mathbb{S}_1}{\otimes} \mathbf{u}^{(p)}, \mathbf{B} = \left(\underset{p \in \mathbb{S}_2}{\otimes} \mathbf{u}^{(p)} \right)^T$ and $J = 1$, gives the following general vectorized form:

$$\text{vec}(\mathbf{X}_{\mathbb{S}_1;\mathbb{S}_2}) = \left(\underset{p \in \mathbb{S}_2}{\otimes} \mathbf{u}^{(p)} \right) \otimes \left(\underset{p \in \mathbb{S}_1}{\otimes} \mathbf{u}^{(p)} \right). \qquad [3.116]$$

We can therefore conclude that the vectorization of a rank-one tensor transforms the outer product of the vectors $\mathbf{u}^{(p)}$ into a Kronecker product of these same vectors. In particular, the vectorization according to lexicographical order is given by:

$$\mathbf{x}_{I_1 \cdots I_P} = \text{vec}\left(\overset{P}{\underset{p=1}{\circ}} \mathbf{u}^{(p)} \right) = \overset{P}{\underset{p=1}{\otimes}} \mathbf{u}^{(p)}, \qquad [3.117]$$

where the element $x_{i_1,\cdots,i_P} = \prod_{p=1}^{P} u_{i_p}^{(p)}$ of $\mathcal{X}$ is located at the position i in the vectorized form $\mathbf{x}_{I_1 \cdots I_P}$ given by (see [3.43]):

$$i \triangleq \overline{i_1 \cdots i_P} \triangleq i_P + \sum_{p=1}^{P-1} (i_p - 1) \prod_{k=p+1}^{P} I_k. \qquad [3.118]$$

Table 3.12 illustrates the matricization and vectorization formulae for a rank-one, third-order tensor. The Frobenius norm is also given. The proof of this formula is given in section 3.16.2.

3.13.2. *Tensor-matrix multiplication*

3.13.2.1. *Definition of the mode-p product*

The mode-p product of a tensor $\mathcal{X} \in \mathbb{K}^{I_P}$ of order P with a matrix $\mathbf{A} \in \mathbb{K}^{J_p \times I_p}$, denoted $\mathcal{X} \times_p \mathbf{A}$, gives the tensor $\mathcal{Y}$ of order P and size $I_1 \times \cdots \times I_{p-1} \times J_p \times I_{p+1} \times \cdots \times I_P$ such that (Carroll *et al.* 1980):

$$y_{i_1,\dots,i_{p-1},j_p,i_{p+1},\dots,i_P} = \sum_{i_p=1}^{I_p} a_{j_p,i_p} x_{i_1,\dots,i_{p-1},i_p,i_{p+1},\dots,i_P}. \qquad [3.119]$$

$\mathcal{X} = \mathbf{u} \circ \mathbf{v} \circ \mathbf{w} \in \mathbb{K}^{I \times J \times K}$

Matricized forms

$$\mathbf{X}_{I \times JK} = \mathbf{u}(\mathbf{v} \otimes \mathbf{w})^T \; , \; \mathbf{X}_{J \times KI} = \mathbf{v}(\mathbf{w} \otimes \mathbf{u})^T \; , \; \mathbf{X}_{K \times IJ} = \mathbf{w}(\mathbf{u} \otimes \mathbf{v})^T$$

Vectorized forms

$$\mathbf{x}_{IJK} = \mathbf{u} \otimes \mathbf{v} \otimes \mathbf{w} \; , \; \mathbf{x}_{JKI} = \mathbf{v} \otimes \mathbf{w} \otimes \mathbf{u} \; , \; \mathbf{x}_{KIJ} = \mathbf{w} \otimes \mathbf{u} \otimes \mathbf{v}$$

Frobenius norm

$$\| \mathbf{u} \circ \mathbf{v} \circ \mathbf{w} \|_F = \| \mathbf{u} \|_2 \| \mathbf{v} \|_2 \| \mathbf{w} \|_2$$

Table 3.12. *Matricized and vectorized forms of a rank-one third-order tensor*

The operation $\times_p$ therefore corresponds to summation over the index i_p associated with the mode p of the tensor $\mathcal{X}$, which is equivalent to performing a contraction over the mode p of $\mathcal{X}$ and the mode 2 of $\mathbf{A}$. This operation is also called the Tucker product, since it is used to write the Tucker decomposition of a tensor (see section 5.2.1.2). It can be expressed using the mode-p unfoldings of the tensors $\mathcal{X}$ and $\mathcal{Y}$ as follows:

$$\mathbf{Y}_p = \mathbf{A}\mathbf{X}_p. \qquad [3.120]$$

This operation can be interpreted in terms of a linear mapping from the mode-p space of $\mathcal{X}$ to the mode-p space of $\mathcal{Y}$, associated with the matrix $\mathbf{A}$. Note that [3.120] implies that the mode shared by $\mathcal{X}$ and $\mathbf{A}$ corresponds to the second index of $\mathbf{A}$.

Table 3.13 gives three examples of the mode-p product of a third-order tensor $\mathcal{X} \in \mathbb{K}^{I \times J \times K}$ with a matrix $\mathbf{A} \in \mathbb{K}^{P \times I}, \mathbf{B} \in \mathbb{K}^{Q \times J}$ and $\mathbf{C} \in \mathbb{K}^{R \times K}$ for $p \in \langle 3 \rangle$.

Tensor $\mathcal{Y}$	Entries of $\mathcal{Y}$	Dimensions of $\mathcal{Y}$	Unfoldings of $\mathcal{Y}$
$\mathcal{X} \times_1 \mathbf{A}$	$y_{pjk} = \sum_{i=1}^I a_{pi} x_{ijk}$	$P \times J \times K$	$\mathbf{Y}_{P \times JK} = \mathbf{A}\mathbf{X}_{I \times JK}$
$\mathcal{X} \times_2 \mathbf{B}$	$y_{iqk} = \sum_{j=1}^J b_{qj} x_{ijk}$	$I \times Q \times K$	$\mathbf{Y}_{Q \times KI} = \mathbf{B}\mathbf{X}_{J \times KI}$
$\mathcal{X} \times_3 \mathbf{C}$	$y_{ijr} = \sum_{k=1}^K c_{rk} x_{ijk}$	$I \times J \times R$	$\mathbf{Y}_{R \times IJ} = \mathbf{C}\mathbf{X}_{K \times IJ}$

Table 3.13. *Mode-p products of a third-order tensor with a matrix*

If we consider an ordered set $\$ = \{n_1, \ldots, n_P\}$ obtained by permuting the elements of the set $\langle P \rangle = \{1, \ldots, P\}$, a series of mode-$n_p$ products of $\mathcal{X} \in \mathbb{K}^{I_P}$ with $\mathbf{A}^{(n_p)} \in \mathbb{K}^{J_{n_p} \times I_{n_p}}, p \in \langle P \rangle$, will be written concisely as follows:

$$\mathcal{X} \times_{n_1} \mathbf{A}^{(n_1)} \cdots \times_{n_P} \mathbf{A}^{(n_P)} \triangleq \mathcal{X} \overset{n_P}{\underset{n=n_1}{\times}} \mathbf{A}^{(n)}. \qquad [3.121]$$

In the case where the tensor-matrix product is performed for every mode $\{1, \ldots, P\}$, de Silva and Lim (2008) proposed another way to concisely express [3.121]:

$$\mathcal{X} \times_{n_1} \mathbf{A}^{(n_1)} \cdots \times_{n_P} \mathbf{A}^{(n_P)} \triangleq \left(\mathbf{A}^{(n_1)}, \cdots, \mathbf{A}^{(n_P)} \right) . \mathcal{X}, \qquad [3.122]$$

where the order of the matrices $\mathbf{A}^{(n_1)}, \cdots, \mathbf{A}^{(n_P)}$ is identical to the order of the mode-n_p products, i.e. $\times_{n_1}, \cdots, \times_{n_P}$.

REMARK 3.48.– The mode-p product of a tensor can be viewed as a generalization of left (mode-1) and right (mode-2) multiplication of a matrix by another matrix, as illustrated in Table 3.14.

Matrices	Matrix products
$\mathbf{A} \in \mathbb{K}^{K \times J}, \mathbf{B} \in \mathbb{K}^{I \times K}$	$\mathbf{BA} = \mathbf{A} \times_1 \mathbf{B} = \mathbf{B} \times_2 \mathbf{A}^T$
$\mathbf{A} \in \mathbb{K}^{K \times J}, \mathbf{B} \in \mathbb{K}^{I \times K}, \mathbf{C} \in \mathbb{K}^{J \times L}$	$\mathbf{BAC} = \mathbf{A} \times_1 \mathbf{B} \times_2 \mathbf{C}^T$
$\mathbf{U} \in \mathbb{K}^{I \times R}, \mathbf{V} \in \mathbb{K}^{J \times R}, \mathbf{\Sigma} \in \mathbb{K}^{R \times R}$	$\mathbf{U \Sigma V}^T = \mathbf{\Sigma} \times_1 \mathbf{U} \times_2 \mathbf{V}$

Table 3.14. *Matrix products and mode-p product*

3.13.2.2. *Properties*

PROPOSITION 3.49.– The mode-p product satisfies the following properties:

– For $\mathcal{X} \in \mathbb{K}^{I_P}, \alpha_m \in \mathbb{K}$ and $\mathbf{A}^{(m)} \in \mathbb{K}^{J_p \times I_p}$, for $m \in \langle M \rangle$, we have:

$$\mathcal{X} \times_p \left(\alpha_1 \mathbf{A}^{(1)} + \cdots + \alpha_M \mathbf{A}^{(M)} \right) = \mathcal{X} \times_p \left(\sum_{m=1}^{M} \alpha_m \mathbf{A}^{(m)} \right)$$

$$= \sum_{m=1}^{M} \alpha_m \left(\mathcal{X} \times_p \mathbf{A}^{(m)} \right). \qquad [3.123]$$

– Let $\mathcal{X} \in \mathbb{K}^{I_P}$ and $\mathbf{A}^{(p)} \in \mathbb{K}^{J_p \times I_p}$, $p \in \langle P \rangle$. For any permutation π of the elements of the set $\langle P \rangle$, such that $n_p = \pi(p)$, we have:

$$\mathcal{X} \overset{n_P}{\underset{n=n_1}{\times}} \mathbf{A}^{(n)} = \mathcal{X} \overset{P}{\underset{p=1}{\times}} \mathbf{A}^{(p)}.$$

This means that the order of the mode-p products does not matter when the modes are all distinct. Using the index convention, we have:

$$\left(\mathcal{X} \overset{P}{\underset{p=1}{\times}} \mathbf{A}^{(p)} \right)_{\underline{\mathbf{j}}_P} = \sum_{i_1=1}^{I_1} \cdots \sum_{i_P=1}^{I_P} x_{i_1, \cdots, i_P} \prod_{p=1}^{P} a_{j_p, i_p}^{(p)} = x_{\underline{\mathbf{i}}_P} \prod_{p=1}^{P} a_{j_p, i_p}^{(p)}. \qquad [3.124]$$

– For two products of $\mathcal{X} \in \mathbb{K}^{\underline{I}_P}$ along the mode p, with $\mathbf{A} \in \mathbb{K}^{J_p \times I_p}$ and $\mathbf{B} \in \mathbb{K}^{K_p \times J_p}$, we have (de Lathauwer 1997):

$$\mathcal{Y} = \mathcal{X} \times_p \mathbf{A} \times_p \mathbf{B} = \mathcal{X} \times_p (\mathbf{BA}) \in \mathbb{K}^{I_1 \times \cdots \times I_{p-1} \times K_p \times I_{p+1} \times \cdots \times I_P}. \qquad [3.125]$$

PROOF.– Defining $\mathcal{Z} = \mathcal{X} \times_p \mathbf{A}$, we have $\mathcal{Y} = \mathcal{Z} \times_p \mathbf{B}$, and using the definition [3.119] of the mode-p product, we have:

$$z_{i_1,\ldots,i_{p-1},j_p,i_{p+1},\ldots,i_P} = \sum_{i_p=1}^{I_p} a_{j_p,i_p} x_{i_1,\ldots,i_{p-1},i_p,i_{p+1},\ldots,i_P} \qquad [3.126]$$

$$y_{i_1,\ldots,i_{p-1},k_p,i_{p+1},\ldots,i_P} = \sum_{j_p=1}^{J_p} b_{k_p,j_p} z_{i_1,\ldots,i_{p-1},j_p,i_{p+1},\ldots,i_P}. \qquad [3.127]$$

By substituting [3.126] into [3.127] and reversing the order of summation, we obtain:

$$y_{i_1,\ldots,i_{p-1},k_p,i_{p+1},\ldots,i_P} = \sum_{j_p=1}^{J_p} b_{k_p,j_p} \sum_{i_p=1}^{I_p} a_{j_p,i_p} x_{i_1,\ldots,i_{p-1},i_p,i_{p+1},\ldots,i_P}.$$

$$= \sum_{i_p=1}^{I_p} \Big[\sum_{j_p=1}^{J_p} b_{k_p,j_p} a_{j_p,i_p} \Big] x_{i_1,\ldots,i_{p-1},i_p,i_{p+1},\ldots,i_P}.$$

The sum in square brackets gives the current element c_{k_p,i_p} of the matrix $\mathbf{C} = \mathbf{BA}$, which proves the property [3.125]. $\qquad \square$

From this property, we can conclude that the double mode-p product is commutative only if the matrices $\mathbf{A}$ and $\mathbf{B}$ commute ($\mathbf{AB} = \mathbf{BA}$). If so, we have:

$$\mathcal{Y} = \mathcal{X} \times_p \mathbf{A} \times_p \mathbf{B} = \mathcal{X} \times_p (\mathbf{BA}) = \mathcal{X} \times_p (\mathbf{AB}) = \mathcal{X} \times_p \mathbf{B} \times_p \mathbf{A}. \qquad [3.128]$$

– For $\mathcal{X} \in \mathbb{K}^{\underline{I}_P}$, $\mathbf{A}^{(p)} \in \mathbb{K}^{J_p \times I_p}$, and $\mathbf{B}^{(p)} \in \mathbb{K}^{K_p \times J_p}$, with $p \in \langle P \rangle$, we have:

$$\mathcal{Y} = \mathcal{X} \mathop{\times}_{p=1}^{P} \mathbf{A}^{(p)} \mathop{\times}_{p=1}^{P} \mathbf{B}^{(p)} = \mathcal{X} \mathop{\times}_{p=1}^{P} (\mathbf{B}^{(p)} \mathbf{A}^{(p)}) \in \mathbb{K}^{\underline{K}_P}. \qquad [3.129]$$

– For $\mathcal{X} \in \mathbb{K}^{\underline{I}_P}$ and factors $\mathbf{A}^{(p)} \in \mathbb{K}^{J_p \times I_p}$, $p \in \langle P \rangle$, of full column rank, we have:

$$\mathcal{Y} = \mathcal{X} \mathop{\times}_{p=1}^{P} \mathbf{A}^{(p)} \Leftrightarrow \mathcal{X} = \mathcal{Y} \mathop{\times}_{p=1}^{P} \mathbf{A}^{(p)\dagger}, \qquad [3.130]$$

where $\mathbf{A}^{(p)\dagger}$ denotes the Moore–Penrose pseudo-inverse of $\mathbf{A}^{(p)}$.

PROOF.– Using the property [3.129], we have:

$$\mathcal{Y} \underset{p=1}{\overset{P}{\times}} \mathbf{A}^{(p)\dagger} = \left(\mathcal{X} \underset{p=1}{\overset{P}{\times}} \mathbf{A}^{(p)} \right) \underset{p=1}{\overset{P}{\times}} \mathbf{A}^{(p)\dagger} = \mathcal{X} \underset{p=1}{\overset{P}{\times}} (\mathbf{A}^{(p)\dagger} \mathbf{A}^{(p)}) = \mathcal{X}$$

because $\mathbf{A}^{(p)\dagger} \mathbf{A}^{(p)} = \left[\mathbf{A}^{(p)T} \mathbf{A}^{(p)} \right]^{-1} \mathbf{A}^{(p)T} \mathbf{A}^{(p)} = \mathbf{I}_{I_p}$. $\qquad\square$

PROPOSITION 3.50.– The mode-p product of $\mathcal{X} \in \mathbb{K}^{I_P}$ with a matrix $\mathbf{A}^{(p)} \in \mathbb{K}^{J_p \times I_p}$ of full column rank does not change the mode-p rank of $\mathcal{X}$. Consequently, the multiple mode-p product with matrices $\mathbf{A}^{(p)}$ of full column rank, for $p \in \langle P \rangle$, does not change the multilinear rank $(R_1, \cdots, R_P)$ of $\mathcal{X}$.

PROOF.– Let $\mathcal{Y} = \mathcal{X} \times_p \mathbf{A}^{(p)}$. Using the expression [3.120] of the mode-p unfolding of $\mathcal{Y}$, we have:

$$\mathbf{Y}_p = \mathbf{A}^{(p)} \mathbf{X}_p. \tag{3.131}$$

Since the rank of a matrix is preserved when it is pre-multiplied by a matrix of full column rank, we deduce that $r(\mathbf{Y}_p) = r(\mathbf{X}_p) = R_p$, and hence the mode-$p$ rank of $\mathcal{Y}$ is identical to that of $\mathcal{X}$. It is now easy to deduce that the multilinear rank of $\mathcal{Y} = \mathcal{X} \underset{p=1}{\overset{P}{\times}} \mathbf{A}^{(p)}$ is identical to that of $\mathcal{X}$ if the matrices $\mathbf{A}^{(p)}$ all have full column rank. $\qquad\square$

3.13.3. *Tensor–vector multiplication*

3.13.3.1. *Definition*

The mode-p product of $\mathcal{X} \in \mathbb{K}^{I_P}$ with the row vector $\mathbf{u}^T \in \mathbb{K}^{1 \times I_p}$, denoted by $\mathcal{X} \times_p \mathbf{u}^T$, gives a tensor $\mathcal{Y}$ of order $P - 1$ and size $I_1 \times \cdots \times I_{p-1} \times I_{p+1} \times \cdots \times I_P$ such that:

$$y_{i_1, \cdots, i_{p-1}, i_{p+1}, \cdots, i_P} = \sum_{i_p=1}^{I_p} u_{i_p} x_{i_1, \cdots, i_{p-1}, i_p, i_{p+1}, \cdots, i_P},$$

which can be written in vectorized form as $\mathrm{vec}^T(\mathcal{Y}) = \mathbf{u}^T \mathbf{X}_p \in \mathbb{K}^{1 \times I_{p+1} \cdots I_P I_1 \cdots I_{p-1}}$, with $\mathbf{X}_p$ defined as in [3.41].

From this relation, we can deduce that the i_pth mode-p tensor slice of $\mathcal{X}$, defined in Table 3.5, is given by $\mathcal{X}_{\cdots i_p \cdots} = \mathcal{X} \times_p \mathbf{e}_{i_p}^{(I_p)T}$. Thus, by multiplying the tensor $\mathcal{X}$ with the P basis vectors $\mathbf{e}_{i_p}^{(I_p)}$ along the P modes, we obtain the entry $x_{i_1, \cdots, i_P} = \mathcal{X} \underset{p=1}{\overset{P}{\times}} \left[\mathbf{e}_{i_p}^{(I_p)} \right]^T$.

3.13.3.2. *Case of a third-order tensor*

For $\mathcal{X} \in \mathbb{K}^{I \times J \times K}, \mathbf{u} \in \mathbb{K}^I, \mathbf{v} \in \mathbb{K}^J, \mathbf{w} \in \mathbb{K}^K$, with the index convention, we have:

$$\mathcal{X} \times_1 \mathbf{u}^T = \sum_{i=1}^{I} u_i \mathbf{X}_{i..} \in \mathbb{K}^{J \times K} \quad , \quad (\mathcal{X} \times_1 \mathbf{u}^T)_{jk} = u_i x_{ijk}$$

$$\mathcal{X} \times_2 \mathbf{v}^T = \sum_{j=1}^{J} v_j \mathbf{X}_{.j.} \in \mathbb{K}^{K \times I} \quad , \quad (\mathcal{X} \times_2 \mathbf{v}^T)_{ki} = v_j x_{ijk}$$

$$\mathcal{X} \times_3 \mathbf{w}^T = \sum_{k=1}^{K} w_k \mathbf{X}_{..k} \in \mathbb{K}^{I \times J} \quad , \quad (\mathcal{X} \times_3 \mathbf{w}^T)_{ij} = w_k x_{ijk}.$$

Similarly, we have:

$$\mathbf{y} = \mathcal{X} \times_1 \mathbf{u}^T \times_2 \mathbf{v}^T = \sum_{i=1}^{I} \sum_{j=1}^{J} u_i v_j \mathbf{X}_{ij.} \in \mathbb{K}^K \quad , \quad y_k = u_i v_j x_{ijk} \qquad [3.132]$$

$$\mathbf{y} = \mathcal{X} \times_1 \mathbf{u}^T \times_3 \mathbf{w}^T = \sum_{i=1}^{I} \sum_{k=1}^{K} u_i w_k \mathbf{X}_{i.k} \in \mathbb{K}^J \quad , \quad y_j = u_i w_k x_{ijk} \qquad [3.133]$$

$$\mathbf{y} = \mathcal{X} \times_2 \mathbf{v}^T \times_3 \mathbf{w}^T = \sum_{j=1}^{J} \sum_{k=1}^{K} v_j w_k \mathbf{X}_{.jk} \in \mathbb{K}^I \quad , \quad y_i = v_j w_k x_{ijk} \qquad [3.134]$$

and:

$$\mathcal{X} \times_1 \mathbf{u}^T \times_2 \mathbf{v}^T \times_3 \mathbf{w}^T = \sum_{i=1}^{I} \sum_{j=1}^{J} \sum_{k=1}^{K} u_i v_j w_k x_{ijk} = u_i v_j w_k x_{ijk} \in \mathbb{K}. \quad [3.135]$$

REMARK 3.51.– The product [3.132] of $\mathcal{X}$ with $\mathbf{u}^T$ and $\mathbf{v}^T$ can be performed in three different ways, as illustrated by the following three expressions:

$$\mathcal{X} \times_1 \mathbf{u}^T \times_2 \mathbf{v}^T = (\mathcal{X} \times_2 \mathbf{v}^T) \times_1 \mathbf{u}^T = (\mathcal{X} \times_1 \mathbf{u}^T) \times_1 \mathbf{v}^T. \qquad [3.136]$$

The first equality means that the mode-1 and mode-2 products can be performed consecutively, regardless of their order, whereas, for the third expression, the product $\mathcal{X} \times_1 \mathbf{u}^T \in \mathbb{K}^{J \times K}$ is performed first, which eliminates the first mode of $\mathcal{X}$ associated with the index i to replace it with the mode associated with the index j. This explains the mode-1 product of $(\mathcal{X} \times_1 \mathbf{u}^T)$ with $\mathbf{v}^T$.

For $\mathcal{X} \in \mathbb{K}^{I \times J \times K}$, the mode-$p$ products with canonical basis vectors give:

$$\mathbf{X}_{i..} = \mathcal{X} \times_1 \mathbf{e}_i^{(I)^T}, \quad \mathbf{X}_{.j.} = \mathcal{X} \times_2 \mathbf{e}_j^{(J)^T}, \quad \mathbf{X}_{..k} = \mathcal{X} \times_3 \mathbf{e}_k^{(K)^T}$$

$$\mathbf{X}_{ij.} = \mathcal{X} \times_1 \mathbf{e}_i^{(I)^T} \times_2 \mathbf{e}_j^{(J)^T} \in \mathbb{K}^K$$

$$\mathbf{X}_{.jk} = \mathcal{X} \times_2 \mathbf{e}_j^{(J)^T} \times_3 \mathbf{e}_k^{(K)^T} \in \mathbb{K}^I$$

$$\mathbf{X}_{i.k} = \mathcal{X} \times_1 \mathbf{e}_i^{(I)^T} \times_3 \mathbf{e}_k^{(K)^T} \in \mathbb{K}^J$$

$$x_{ijk} = \mathcal{X} \times_1 \mathbf{e}_i^{(I)^T} \times_2 \mathbf{e}_j^{(J)^T} \times_3 \mathbf{e}_k^{(K)^T}.$$

3.13.4. *Mode-(p, n) product*

The mode-p product of a tensor with a matrix or a vector can be extended to a product with tensors of arbitrary orders (Gelfand *et al.* 1992; Lee and Cichocki 2014). If we consider the tensors $\mathcal{A} \in \mathbb{K}^{\underline{I}_P}$ and $\mathcal{B} \in \mathbb{K}^{\underline{J}_N}$, with $I_P = J_1$, the mode-$(P, 1)$ product of $\mathcal{A}$ with $\mathcal{B}$, denoted $\mathcal{A} \times_P^1 \mathcal{B}$, gives a tensor $\mathcal{C}$ of order $P + N - 2$ and size $I_1 \times \cdots \times I_{P-1} \times J_2 \times \cdots \times J_N$, with entries:

$$c_{i_1,\cdots,i_{P-1},j_2,\cdots,j_N} = \sum_{k=1}^{I_P} a_{i_1,\cdots,i_{P-1},k}\, b_{k,j_2,\cdots,j_N}. \qquad [3.137]$$

This contraction operation can be performed using a matrix multiplication involving tall mode-P and flat mode-1 matrix unfoldings of $\mathcal{A}$ and $\mathcal{B}$, respectively:

$$\mathbf{C}_{I_1 \cdots I_{P-1} \times J_2 \cdots J_N} = \mathbf{A}_{I_1 \cdots I_{P-1} \times I_P} \mathbf{B}_{I_P \times J_2 \cdots J_N}. \qquad [3.138]$$

REMARK 3.52.– In the matrix case with $\mathbf{A} \in \mathbb{K}^{I \times K}$ and $\mathbf{B} \in \mathbb{K}^{K \times J}$, where the number of columns of $\mathbf{A}$ is equal to the number of rows of $\mathbf{B}$, we have:

$$\mathbf{C} = \mathbf{A} \times_2^1 \mathbf{B} = \mathbf{AB} \quad \Leftrightarrow \quad c_{ij} = \sum_{k=1}^{K} a_{ik}b_{kj}, \qquad [3.139]$$

i.e. the standard matrix product.

This contraction operation can also be performed for two arbitrary modes (p, n) of $\mathcal{A} \in \mathbb{K}^{\underline{I}_P}$ and $\mathcal{B} \in \mathbb{K}^{\underline{J}_N}$, with $I_p = J_n = K$. In this case, it is denoted $\mathcal{A} \times_p^n \mathcal{B}$ and gives a tensor $\mathcal{C}$ of order $P + N - 2$ and size $I_1 \times \cdots \times I_{p-1} \times I_{p+1} \times \cdots \times I_P \times J_1 \times \cdots \times J_{n-1} \times J_{n+1} \times \cdots \times J_N$ such that:

$$c_{i_1, \cdots, i_{p-1}, i_{p+1}, \cdots, i_P, j_1, \cdots, j_{n-1}, j_{n+1}, \cdots, j_N} = \sum_{k=1}^{K} a_{i_1, \cdots, i_{p-1}, k, i_{p+1}, \cdots, i_P}$$

[3.140]

$$\times b_{j_1, \cdots, j_{n-1}, k, j_{n+1}, \cdots, j_N}.$$

This type of contraction, of which [3.137] is a special case, was introduced by Gelfand *et al.* (1992) under the name of convolution (or product), denoted $\mathcal{A} \star_{p,n} \mathcal{B}$.

PROPOSITION 3.53.– The contracted product $\times_p^n$ is associative; in other words, for any tensors $\mathcal{A} \in \mathbb{K}^{\underline{I}_P}$, $\mathcal{B} \in \mathbb{K}^{\underline{J}_N}$ and $\mathcal{C} \in \mathbb{K}^{\underline{K}_Q}$ such that $I_p = J_n$ and $J_m = K_q$, with $m \neq n$, we have:

$$(\mathcal{A} \times_p^n \mathcal{B}) \times_m^q \mathcal{C} = \mathcal{A} \times_p^n (\mathcal{B} \times_m^q \mathcal{C}) = \mathcal{A} \times_p^n \mathcal{B} \times_m^q \mathcal{C}.$$

[3.141]

This double contracted product gives a tensor of order $P + N + Q - 4$.

This property is easy to prove using the definition of the product $\times_p^n$.

REMARK 3.54.– In the matrix case, with $\mathbf{A} \in \mathbb{K}^{I \times J}$, $\mathbf{B} \in \mathbb{K}^{J \times K}$ and $\mathbf{C} \in \mathbb{K}^{K \times L}$, the property [3.141] becomes:

$$(\mathbf{A} \times_2^1 \mathbf{B}) \times_2^1 \mathbf{C} = \mathbf{A} \times_2^1 (\mathbf{B} \times_2^1 \mathbf{C}) = \mathbf{A} \times_2^1 \mathbf{B} \times_2^1 \mathbf{C},$$

[3.142]

or equivalently:

$$(\mathbf{AB})\mathbf{C} = \mathbf{A}(\mathbf{BC}) = \mathbf{ABC}.$$

[3.143]

Thus, we recover the associativity property of the standard matrix product.

EXAMPLE 3.55.– The mode-(p, n) product can be used to define the TT decomposition, introduced in Table I.3, that will be studied in more detail in the next volume. This decomposition represents an Nth-order tensor $\mathcal{X} \in \mathbb{K}^{\underline{I}_P}$ as a train of two matrices $\mathbf{G}^{(1)}$ and $\mathbf{G}^{(P)}$, and $(P - 2)$ third-order tensors $\mathcal{G}^{(p)}, p \in \{2, \cdots, P-1\}$, with the following sizes:

$$\mathbf{G}^{(1)} \in \mathbb{K}^{I_1 \times R_1} \ ; \ \mathbf{G}^{(P)} \in \mathbb{K}^{R_{P-1} \times I_P} \ ; \ \mathcal{G}^{(p)} \in \mathbb{K}^{R_{P-1} \times I_P \times R_P}.$$

The factors $\mathcal{G}^{(p)}, p \in \langle P \rangle$, are called the TT-cores, and the integers $R_p, p \in \langle P - 1 \rangle$, are called the TT-ranks. Note that, for $p = 1$ and $p = P$, the TT-cores are two matrices.

The TT decomposition can be written as:

$$\mathcal{X} = \mathbf{G}^{(1)} \times_2^1 \mathcal{G}^{(2)} \times_3^1 \mathcal{G}^{(3)} \times_4^1 \cdots \times_{P-1}^1 \mathcal{G}^{(P-1)} \times_P^1 \mathbf{G}^{(P)}, \qquad [3.144]$$

or in scalar form, with the index convention:

$$x_{\mathbf{i}_P} = \sum_{r_1=1}^{R_1} \cdots \sum_{r_{P-1}=1}^{R_{P-1}} g_{i_1,r_1}^{(1)} g_{r_1,i_2,r_2}^{(2)} g_{r_2,i_3,r_3}^{(3)} \cdots g_{r_{P-2},i_{P-1},r_{P-1}}^{(P-1)} g_{r_{P-1},i_P}^{(P)}$$

$$= \prod_{p=1}^{P} g_{r_{p-1},i_p,r_p}^{(p)} \qquad [3.145]$$

with $g_{r_0,i_1,r_1}^{(1)} = g_{i_1,r_1}^{(1)}$ and $g_{r_{P-1},i_P,r_P}^{(P)} = g_{r_{P-1},i_P}^{(P)}$. This entry-wise representation can also be written using matrix–vector and matrix–matrix products as:

$$x_{\mathbf{i}_P} = \mathbf{g}_{i_1,\bullet}^{(1)} \mathbf{G}_{\bullet,i_2,\bullet}^{(2)} \mathbf{G}_{\bullet,i_3,\bullet}^{(3)} \cdots \mathbf{G}_{\bullet,i_{P-1},\bullet}^{(P-1)} \mathbf{g}_{\bullet,i_P}^{(P)} \qquad [3.146]$$

where $\mathbf{g}_{i_1,\bullet}^{(1)} \in \mathbb{K}^{1 \times R_1}$ and $\mathbf{g}_{\bullet,i_P}^{(P)} \in \mathbb{K}^{R_{P-1}}$ are the i_1th row vector of $\mathbf{G}^{(1)}$, and the i_Pth column vector of $\mathbf{G}^{(P)}$, respectively, whereas $\mathbf{G}_{\bullet,i_p,\bullet}^{(p)} \in \mathbb{K}^{R_{p-1} \times R_p}$ is the i_pth lateral slice of $\mathcal{G}^{(p)}$, for $p \in \{2, \cdots, P-1\}$.

The contraction operation can also be applied to several arbitrary pairs of indices corresponding to modes with the same dimension. Thus, for example, for $\mathcal{A} \in \mathbb{K}^{\underline{I}_P}$ and $\mathcal{B} \in \mathbb{K}^{\underline{J}_N}$, with $I_p = J_n$, $I_s = J_r$, and $p < s$, $n < r$, it is possible to sum over the index i_p of $\mathcal{A}$ and the index j_n of $\mathcal{B}$ on one hand, and over the index i_s of $\mathcal{A}$ and the index j_r of $\mathcal{B}$ on the other hand. This contraction is defined as follows:

$$\langle \mathcal{A}, \mathcal{B} \rangle_{p:n;s:r} = \sum_{k=1}^{I_p} \sum_{l=1}^{I_s} a_{i_1,\cdots,k,\cdots,l,\cdots i_P} b_{j_1,\cdots,j_{n-1},k,j_{n+1}\cdots,j_{r-1},l,j_{r+1},\cdots j_N}.$$

Like in the example [3.138], this contraction can be performed by multiplying the matrices $\mathbf{A}_{I_1 \cdots I_{p-1} I_{p+1} \cdots I_{s-1} I_{s+1} \cdots I_P \times I_p I_s}$ and $\mathbf{B}_{I_p I_s \times J_1 \cdots J_{n-1} J_{n+1} \cdots J_{r-1} J_{r+1} \cdots J_N}$ to obtain the unfolding $\mathbf{C}_{I_1 \cdots I_{p-1} I_{p+1} \cdots I_{s-1} I_{s+1} \cdots I_P \times J_1 \cdots J_{n-1} J_{n+1} \cdots J_{r-1} J_{r+1} \cdots J_N}$ of the tensor $\mathcal{C}$ resulting from a double contraction.

3.13.5. *Einstein product*

We will now define the Einstein product, denoted $\star_N$, which generalizes the product [3.137]. We will then prove that the space $\mathbb{K}^{\underline{I}_N \times \underline{I}_N}$ equipped with the product $\star_N$ has a group structure, and will show that there is an isomorphism of groups between this space $\mathbb{K}^{\underline{I}_N \times \underline{I}_N}$ and the general linear group GL_I, where $I = \prod_{n=1}^{N} I_n$, i.e. the group of invertible square matrices of order I under the operation of standard matrix multiplication. This isomorphism will allow us to define the inverse of a tensor $\mathcal{X} \in \mathbb{K}^{\underline{I}_N \times \underline{I}_N}$ via the inverse of its matrix unfolding $\mathbf{X}_{I \times I} \in \mathbb{K}^{I \times I}$ (Brazell *et al.* 2013). Note that the Einstein product will be used to define the properties of orthogonality and idempotence for a tensor, as well as certain tensor factorizations (see sections 3.13.5.2 and 3.15).

3.13.5.1. *Definition and properties*

The product [3.137] can be generalized by defining the Einstein product of the tensor $\mathcal{A} \in \mathbb{K}^{\underline{I}_P \times \underline{J}_N}$ of order $P + N$ with the tensor $\mathcal{B} \in \mathbb{K}^{\underline{J}_N \times \underline{K}_Q}$ of order $N + Q$ along the N shared indices $\mathbf{j}_N = \{j_1, \cdots, j_N\}$, corresponding to the N last modes of $\mathcal{A}$ and the N first modes of $\mathcal{B}$. This product gives the tensor $\mathcal{C} = \mathcal{A} \star_N \mathcal{B} \in \mathbb{K}^{\underline{I}_P \times \underline{K}_Q}$ of order $P + Q$ such that:

$$c_{\underline{\mathbf{i}}_P, \underline{\mathbf{k}}_Q} = \sum_{\underline{\mathbf{j}}_N = \underline{1}}^{\underline{\mathbf{J}}_N} a_{\underline{\mathbf{i}}_P, \underline{\mathbf{j}}_N} b_{\underline{\mathbf{j}}_N, \underline{\mathbf{k}}_Q} = a_{\underline{\mathbf{i}}_P, \underline{\mathbf{j}}_N} b_{\underline{\mathbf{j}}_N, \underline{\mathbf{k}}_Q}, \qquad [3.147]$$

where the last equality follows by using the index convention, which implies summation over the repeated indices $\mathbf{j}_N$, explaining the name of Einstein product. Recall the notation defined earlier in Table 3.1:

$$\underline{\mathbf{i}}_P = \{i_1, \cdots, i_P\} \;, \; \underline{\mathbf{k}}_Q = \{k_1, \cdots, k_Q\} \;, \; \underline{\mathbf{J}}_N = \{J_1, \cdots, J_N\}.$$

For the Einstein product [3.147], the modes associated with the P first indices and the N last indices of $a_{\underline{\mathbf{i}}_P, \underline{\mathbf{j}}_N}$ are called the row modes and column modes of $\mathcal{A}$, respectively.

EXAMPLE 3.56.– For $N = 1$, with $\mathcal{A} \in \mathbb{K}^{\underline{I}_P \times J}$ and $\mathcal{B} \in \mathbb{K}^{J \times \underline{K}_Q}$, we have:

$$(\mathcal{A} \star_1 \mathcal{B})_{\underline{\mathbf{i}}_P, \underline{\mathbf{k}}_Q} = a_{\underline{\mathbf{i}}_P, j} b_{j, \underline{\mathbf{k}}_Q} \qquad [3.148]$$

and hence:

$$\mathcal{A} \star_1 \mathcal{B} = \mathcal{A} \times_{P+1}^1 \mathcal{B} \in \mathbb{K}^{\underline{I}_P \times \underline{K}_Q}. \qquad [3.149]$$

EXAMPLE 3.57.– For $\mathcal{A} \in \mathbb{K}^{I \times J \times K}$, $\mathbf{X} \in \mathbb{K}^{J \times K}$ and $\mathbf{x} \in \mathbb{K}^K$, the $\star_2$ product of $\mathcal{A}$ with $\mathbf{X}$ and the $\star_1$ product of $\mathcal{A}$ with $\mathbf{x}$ give a vector $\mathbf{y} \in \mathbb{K}^I$ and a matrix $\mathbf{Y} \in \mathbb{K}^{I \times J}$, respectively, such that:

$$y_i = [\mathcal{A} \star_2 \mathbf{X}]_i = \sum_{j,k=1}^{J,K} a_{i,j,k} x_{j,k} = a_{i,j,k} x_{j,k} \qquad [3.150]$$

$$y_{i,j} = [\mathcal{A} \star_1 \mathbf{x}]_{i,j} = \sum_{k=1}^{K} a_{i,j,k} x_k = a_{i,j,k} x_k. \qquad [3.151]$$

EXAMPLE 3.58.– For $\mathcal{A} \in \mathbb{K}^{I \times J \times K \times L}$ and $\mathcal{B} \in \mathbb{K}^{K \times L \times M \times N}$, the $\star_2$ product of $\mathcal{A}$ with $\mathcal{B}$ gives the fourth-order tensor $\mathcal{C} \in \mathbb{K}^{I \times J \times M \times N}$ such that:

$$c_{i,j,m,n} = [\mathcal{A} \star_2 \mathcal{B}]_{i,j,m,n} = \sum_{k,l=1}^{K,L} a_{i,j,k,l} b_{k,l,m,n} = a_{i,j,k,l} b_{k,l,m,n}. \qquad [3.152]$$

PROPOSITION 3.59.– The Einstein product is associative and distributive with respect to addition, i.e. for $\mathcal{A} \in \mathbb{K}^{\underline{I}_P \times \underline{J}_N}$, $\mathcal{B}, \mathcal{D} \in \mathbb{K}^{\underline{J}_N \times \underline{K}_M}$ and $\mathcal{C} \in \mathbb{K}^{\underline{K}_M \times \underline{L}_Q}$, we have:

$$(\mathcal{A} \star_N \mathcal{B}) \star_M \mathcal{C} = \mathcal{A} \star_N (\mathcal{B} \star_M \mathcal{C}) = \mathcal{A} \star_N \mathcal{B} \star_M \mathcal{C}. \qquad [3.153]$$

This double contracted product gives a tensor $\mathcal{P} \in \mathbb{K}^{\underline{I}_P \times \underline{L}_Q}$ of order $P + Q$.

Moreover, we have:

$$\mathcal{A} \star_N (\mathcal{B} + \mathcal{D}) = (\mathcal{A} \star_N \mathcal{B}) + (\mathcal{A} \star_N \mathcal{D}) \in \mathbb{K}^{\underline{I}_P \times \underline{K}_M} \qquad [3.154]$$

$$(\mathcal{B} + \mathcal{D}) \star_M \mathcal{C} = (\mathcal{B} \star_M \mathcal{C}) + (\mathcal{D} \star_M \mathcal{C}) \in \mathbb{K}^{\underline{J}_N \times \underline{L}_Q}. \qquad [3.155]$$

REMARK 3.60.–

– The Einstein product is not commutative.

– For $\mathbf{A} \in \mathbb{K}^{I \times K}$ and $\mathbf{B} \in \mathbb{K}^{K \times J}$, the $\star_1$ product of $\mathbf{A}$ with $\mathbf{B}$ gives the matrix $\mathbf{C} \in \mathbb{K}^{I \times J}$, such that:

$$c_{i,j} = [\mathbf{A} \star_1 \mathbf{B}]_{i,j} = \sum_{k=1}^{K} a_{i,k} b_{k,j} = a_{i,k} b_{k,j}. \qquad [3.156]$$

The $\star_1$ product of two matrices is therefore equivalent to the standard matrix product, and we have:

$$\mathbf{A} \star_1 \mathbf{B} = \mathbf{A} \times_2^1 \mathbf{B} = \mathbf{B} \times_1 \mathbf{A} = \mathbf{A}\mathbf{B}. \qquad [3.157]$$

– The Einstein product is also defined for even-order paired tensors $\mathcal{A} \in \mathbb{K}^{I_1 \times K_1 \cdots I_P \times K_P}$ and $\mathcal{B} \in \mathbb{K}^{K_1 \times J_1 \cdots K_P \times J_P}$, giving a tensor $\mathcal{C} \in \mathbb{K}^{I_1 \times J_1 \cdots I_P \times J_P}$ that is also an even-order paired tensor of order $2P$ (Chen *et al.* 2019):

$$c_{i_1 j_1 \cdots i_P j_P} = \sum_{\underline{\mathbf{k}}_N = \underline{\mathbf{1}}}^{\underline{\mathbf{K}}_N} a_{i_1 k_1 \cdots i_P k_P} b_{k_1 j_1 \cdots k_P j_P}. \qquad [3.158]$$

In this case, the summations apply to alternating indices, i.e. every other index.

– The Einstein product can also be used with block tensors, such as those defined in section 3.5, as illustrated using the following examples.

For $\mathcal{A} \in \mathbb{K}^{\underline{I}_P \times \underline{J}_N}, \mathcal{B} \in \mathbb{K}^{\underline{I}_P \times \underline{K}_N}, \mathcal{C} \in \mathbb{K}^{\underline{M}_P \times \underline{J}_N}, \mathcal{D} \in \mathbb{K}^{\underline{J}_N \times \underline{K}_Q}, \mathcal{F} \in \mathbb{K}^{\underline{J}_N \times \underline{R}_Q}$, and $\mathcal{X} \in \mathbb{K}^{\underline{H}_Q \times \underline{I}_P}, \mathcal{Y} \in \mathbb{K}^{\underline{J}_N \times \underline{G}_Q}$, we have:

$$\mathcal{X} \star_P \begin{bmatrix} \mathcal{A} & \vdots & \mathcal{B} \end{bmatrix} = \begin{bmatrix} \mathcal{X} \star_P \mathcal{A} & \vdots & \mathcal{X} \star_P \mathcal{B} \end{bmatrix} \in \mathbb{K}^{\underline{H}_Q \times \underline{L}_N}$$

[3.159]

with $\underline{L}_N = \underline{J}_N \times \underline{K}_N$

$$\begin{bmatrix} \mathcal{A} \\ \cdots \\ \mathcal{C} \end{bmatrix} \star_N \mathcal{Y} = \begin{bmatrix} \mathcal{A} \star_N \mathcal{Y} \\ \cdots \\ \mathcal{C} \star_N \mathcal{Y} \end{bmatrix} \in \mathbb{K}^{\underline{L}_P \times \underline{G}_Q}$$

[3.160]

with $\underline{L}_P = \underline{I}_P \times \underline{M}_P$

$$\begin{bmatrix} \mathcal{A} \\ \cdots \\ \mathcal{C} \end{bmatrix} \star_N \begin{bmatrix} \mathcal{D} & \vdots & \mathcal{F} \end{bmatrix} = \begin{bmatrix} \mathcal{A} \star_N \mathcal{D} & \vdots & \mathcal{A} \star_N \mathcal{F} \\ \mathcal{C} \star_N \mathcal{D} & \vdots & \mathcal{C} \star_N \mathcal{F} \end{bmatrix} \in \mathbb{K}^{\underline{L}_P \times \underline{S}_Q}$$

[3.161]

with $\underline{L}_P = \underline{I}_P \times \underline{M}_P$, $\underline{S}_Q = \underline{K}_Q \times \underline{R}_Q$.

Table 3.15 summarizes the key results about multiplication with tensors, using the notations from Table 3.1 and the index convention.

Tensors	Operations	Definitions
$\mathcal{X} \in \mathbb{K}^{\underline{I}_P}, \mathbf{A} \in \mathbb{K}^{J \times I_P}$	$\mathcal{X} \times_p \mathbf{A}$	$\sum_{i_p} a_{j,i_p} x_{\underline{i}_P} = a_{j,i_p} x_{\underline{i}_P}$
$\mathcal{X} \in \mathbb{K}^{\underline{I}_P}, \mathbf{u} \in \mathbb{K}^{I_P}$	$\mathcal{X} \times_p \mathbf{u}^T$	$\sum_{i_p} u_{i_p} x_{\underline{i}_P} = u_{i_p} x_{\underline{i}_P}$
$\mathcal{X} \in \mathbb{K}^{\underline{I}_P}, \mathcal{Y} \in \mathbb{K}^{\underline{J}_N}$ with $I_P = J_1$	$\mathcal{X} \times_P^1 \mathcal{Y}$	$\sum_{k=1}^{I_P} x_{i_1,\cdots,i_{P-1},k}\, y_{k,j_2,\cdots,j_N}$ $= x_{i_1,\cdots,i_{P-1},k}\, y_{k,j_2,\cdots,j_N}$
$\mathcal{X} \in \mathbb{K}^{\underline{I}_P \times \underline{J}_N}, \mathcal{Y} \in \mathbb{K}^{\underline{J}_N \times \underline{K}_Q}$	$\mathcal{X} \star_N \mathcal{Y}$	$\sum_{\underline{j}_N=\underline{1}}^{\underline{J}_N} x_{\underline{i}_P,\underline{j}_N}\, y_{\underline{j}_N,\underline{k}_Q} = x_{\underline{i}_P,\underline{j}_N}\, y_{\underline{j}_N,\underline{k}_Q}$

Table 3.15. *Different types of multiplication with tensors*

3.13.5.2. *Orthogonality and idempotence properties*

Using the Einstein product, we can define the properties of orthogonality[3] and idempotence for an even-order tensor $\mathcal{X} \in \mathbb{K}^{\underline{I}_N \times \underline{I}_N}$, as outlined in Table 3.16, where $\mathcal{J}_{2N}$ is the block identity tensor defined in [3.22], and the tensors $\mathcal{X}^T$ and $\mathcal{X}^H$ are

---

3 The orthogonality property based on the Einstein product was introduced by Brazell *et al.* (2013) for a fourth-order tensor.

defined as in Table 3.7, replacing the transposition operation with that of transconjugation for $\mathcal{X}^H$. The properties of orthogonality and idempotence for a square matrix $\mathbf{X} \in \mathbb{K}^{I \times I}$ are also recalled.

Properties	Conditions for a matrix	Conditions for a tensor
	$\mathbf{X} \in \mathbb{K}^{I \times I}$	$\mathcal{X} \in \mathbb{K}^{\underline{I}_N \times \underline{I}_N}$
$\mathbf{X}, \mathcal{X}$ orthogonal	$\mathbf{X}^T \mathbf{X} = \mathbf{X}\mathbf{X}^T = \mathbf{I}_I$	$\mathcal{X}^T \star_N \mathcal{X} = \mathcal{X} \star_N \mathcal{X}^T = \mathcal{J}_{2N}$
$\mathbf{X}, \mathcal{X}$ unitary	$\mathbf{X}^H \mathbf{X} = \mathbf{X}\mathbf{X}^H = \mathbf{I}_I$	$\mathcal{X}^H \star_N \mathcal{X} = \mathcal{X} \star_N \mathcal{X}^H = \mathcal{J}_{2N}$
$\mathbf{X}, \mathcal{X}$ idempotent	$\mathbf{X}^2 = \mathbf{X}$	$\mathcal{X} \star_N \mathcal{X} = \mathcal{X}$

Table 3.16. *Orthogonality and idempotence properties*

3.13.5.3. *Isomorphism of tensor and matrix groups*

Below, we consider the space $(\mathbb{K}^{\underline{I}_N \times \underline{I}_N}, \star_N)$, i.e. the space of square tensors of size $\underline{I}_N \times \underline{I}_N$, equipped with the $\star_N$ product, and the matrix unfolding $\mathbf{X}_{I \times I}$, with $I = \prod_{n=1}^{N} I_n$, of $\mathcal{X} \in \mathbb{K}^{\underline{I}_N \times \underline{I}_N}$, as defined in [3.105]. In the next proposition, we begin by showing two preliminary results concerning the closure property of the space $(\mathbb{K}^{\underline{I}_N \times \underline{I}_N}, \star_N)$ under the operation $\star_N$ and the existence of a neutral (also called identity) element for this operation.

PROPOSITION 3.61.– The $\star_N$ product satisfies the following properties:

– For any tensors $\mathcal{X}, \mathcal{Y} \in \mathbb{K}^{\underline{I}_N \times \underline{I}_N}$, the product $\mathcal{Z} = \mathcal{X} \star_N \mathcal{Y}$ gives an element of $\mathbb{K}^{\underline{I}_N \times \underline{I}_N}$. This means that the space $(\mathbb{K}^{\underline{I}_N \times \underline{I}_N}, \star_N)$ is closed under the operation $\star_N$. The unfolding of the product $\mathcal{Z}$ can be computed using the product of the unfoldings of $\mathcal{X}$ and $\mathcal{Y}$, i.e.:

$$\mathbf{Z}_{I \times I} = (\mathcal{X} \star_N \mathcal{Y})_{I \times I} = \mathbf{X}_{I \times I} \mathbf{Y}_{I \times I}. \qquad [3.162]$$

From this property and the definition of an idempotent tensor given in Table 3.16, we can conclude that the tensor $\mathcal{X} \in \mathbb{K}^{\underline{I}_N \times \underline{I}_N}$ is idempotent whenever its unfolding $\mathbf{X}_{I \times I}$ is idempotent itself, i.e. $\mathbf{X}_{I \times I}^2 = \mathbf{X}_{I \times I}$.

PROOF.– The closure property follows from the definition [3.147] of the product $\star_N$:

$$\mathcal{Z} = \mathcal{X} \star_N \mathcal{Y} \ \Rightarrow \ z_{\underline{\mathbf{i}}_N, \underline{\mathbf{k}}_N} = x_{\underline{\mathbf{i}}_N, \underline{\mathbf{j}}_N} y_{\underline{\mathbf{j}}_N, \underline{\mathbf{k}}_N} \ \Rightarrow \ \mathcal{Z} \in \mathbb{K}^{\underline{I}_N \times \underline{I}_N}, \qquad [3.163]$$

with $i_n, j_n, k_n \in \langle I_n \rangle$, $\forall n \in \langle N \rangle$. Using the expression [2.68] of a matrix with the index convention, the matrix unfolding $\mathbf{Z}_{I \times I}$ can be written as:

$$\mathbf{Z}_{I \times I} = z_{\underline{\mathbf{i}}_N, \underline{\mathbf{k}}_N} \mathbf{e}_{i_1 \cdots i_N}^{\overline{k_1 \cdots k_N}}. \qquad [3.164]$$

Replacing $z_{\underline{i}_N,\underline{k}_N}$ with its expression [3.163], using the expression [2.81] of the matrix product and the index convention, we obtain:

$$\mathbf{Z}_{I \times I} = x_{\underline{i}_N \underline{j}_N} y_{\underline{j}_N,\underline{k}_N} \, e_{\overline{i_1 \cdots i_N}}^{\overline{k_1 \cdots k_N}} \qquad [3.165]$$

$$= \mathbf{X}_{I \times I} \mathbf{Y}_{I \times I}, \qquad [3.166]$$

with $I = \prod_{n=1}^{N} I_n$. This proves the relation [3.162]. $\qquad \square$

– The identity tensor $\mathcal{J}_{2N}$ defined in [3.22] is the neutral element for the $\star_N$ product, i.e. for any $\mathcal{X} \in \mathbb{K}^{\underline{I}_N \times \underline{I}_N}$, we have:

$$\mathcal{X} \star_N \mathcal{J}_{2N} = \mathcal{J}_{2N} \star_N \mathcal{X} = \mathcal{X}. \qquad [3.167]$$

PROOF.– This result follows from the definition of the $\star_N$ product:

$$(\mathcal{X} \star_N \mathcal{J}_{2N})_{\underline{i}_N \underline{j}_N} = \sum_{\underline{k}_N = \underline{1}}^{\underline{I}_N} x_{\underline{i}_N,\underline{k}_N} \prod_{n=1}^{N} \delta_{k_n,j_n} = x_{\underline{i}_N \underline{j}_N}, \qquad [3.168]$$

with $i_n, j_n \in \langle I_n \rangle, \forall n \in \langle N \rangle$, which proves that $\mathcal{X} \star_N \mathcal{J}_{2N} = \mathcal{X}$. The equality $\mathcal{J}_{2N} \star_N \mathcal{X} = \mathcal{X}$ can be proven in the same way. We conclude that the space $(\mathbb{K}^{\underline{I}_N \times \underline{I}_N}, \star_N)$ admits $\mathcal{J}_{2N}$ as a neutral element. This result can also be recovered using the properties [3.162] and [3.24]:

$$(\mathcal{X} \star_N \mathcal{J}_{2N})_{I \times I} = \mathbf{X}_{I \times I} \mathbf{I}_I = \mathbf{X}_{I \times I}, \qquad [3.169]$$

which completes the proof of [3.167]. $\qquad \square$

EXAMPLE 3.62.– For $\mathcal{X}, \mathcal{Y} \in \mathbb{K}^{2 \times 1 \times 2 \times 1}$, i.e. for $I_1 = 2, I_2 = 1$, the product $\mathcal{Z} = \mathcal{X} \star_2 \mathcal{Y}$ satisfies:

$$z_{ijmn} = \sum_{k=1}^{2} \sum_{l=1}^{1} x_{ijkl} \, y_{klmn}, \quad i, m \in \{1, 2\}, \; j, n = 1, \qquad [3.170]$$

which gives:

$$z_{1111} = x_{1111} y_{1111} + x_{1121} y_{2111}, \quad z_{1121} = x_{1111} y_{1121} + x_{1121} y_{2121}$$

$$z_{2111} = x_{2111} y_{1111} + x_{2121} y_{2111}, \quad z_{2121} = x_{2111} y_{1121} + x_{2121} y_{2121}.$$

This result can be found using equation [3.162]:

$$(\mathcal{Z})_{I \times I} = \begin{bmatrix} z_{1111} & z_{1121} \\ z_{2111} & z_{2121} \end{bmatrix} = \mathbf{X}_{I \times I} \mathbf{Y}_{I \times I}$$

$$= \begin{bmatrix} x_{1111} & x_{1121} \\ x_{2111} & x_{2121} \end{bmatrix} \begin{bmatrix} y_{1111} & y_{1121} \\ y_{2111} & y_{2121} \end{bmatrix}.$$

To show the group isomorphism, we define the set $(\mathbb{T}_N, \star_N)$, equipped with the Einstein product as a binary operation, composed of the square tensors $\mathcal{X} \in \mathbb{K}^{\underline{I}_N \times \underline{I}_N}$, whose unfolding $\mathbf{X}_{I \times I}$, with $I = \prod_{n=1}^{N} I_n$, is a non-singular matrix, i.e.:

$$\mathbb{T}_N = \{\mathcal{X} \in \mathbb{K}^{\underline{I}_N \times \underline{I}_N} \ / \ \det(\mathbf{X}_{I \times I}) \neq 0\}. \tag{3.171}$$

Since the matrix $\mathbf{X}_{I \times I}$ is invertible, it is an element of the general linear group GL_I, whose elements are the invertible square matrices of order I. Note that the invertibility of $\mathbf{X}_{I \times I}$ implies that this matrix has full rank: $r(\mathbf{X}_{I \times I}) = I$.

Now, let us define the mapping f that transforms a tensor $\mathcal{X} \in \mathbb{T}_N$ into its matrix unfolding $\mathbf{X}_{I \times I}$:

$$f : \mathbb{T}_N \ni \mathcal{X} \longmapsto f(\mathcal{X}) = \mathbf{X}_{I \times I} \in \mathbb{K}^{I \times I} \text{ with } I = \prod_{n=1}^{N} I_n. \tag{3.172}$$

This mapping is bijective. Indeed, bijectivity follows from the relation [3.105], which associates each element $x_{\mathbf{i}_N \mathbf{j}_N}$ of $\mathcal{X}$ with the element $(\mathbf{X}_{I \times I})_{\overline{i_1 \cdots i_N}, \overline{j_1 \cdots j_N}}$ of the unfolding $\mathbf{X}_{I \times I}$, via the bijective mapping g defined as:

$$\underline{I}_N \times \underline{I}_N \ni (\mathbf{i}_N, \mathbf{j}_N) \longmapsto g(\mathbf{i}_N, \mathbf{j}_N) = (\overline{i_1 \cdots i_N}, \overline{j_1 \cdots j_N}) \in \prod_{n=1}^{N} I_n \times \prod_{n=1}^{N} I_n. \tag{3.173}$$

The property [3.162] can be rewritten using the mapping f as follows:

$$f(\mathcal{X} \star_N \mathcal{Y}) = f(\mathcal{X})f(\mathcal{Y}) = \mathbf{X}_{I \times I}\mathbf{Y}_{I \times I}. \tag{3.174}$$

This formula allows us to determine the tensor $\mathcal{Z} = \mathcal{X} \star_N \mathcal{Y}$ by computing its matrix unfolding $\mathbf{Z}_{I \times I} = \mathbf{X}_{I \times I}\mathbf{Y}_{I \times I}$.

Since the mapping f is bijective, it admits an inverse f^{-1} such that:

$$\mathcal{X} \star_N \mathcal{Y} = f^{-1}[f(\mathcal{X})f(\mathcal{Y})]. \tag{3.175}$$

The relation [3.175] can also be expressed using matrix unfoldings as follows:

$$f^{-1}(\mathbf{X}_{I \times I}) \star_N f^{-1}(\mathbf{Y}_{I \times I}) = f^{-1}(\mathbf{X}_{I \times I}\mathbf{Y}_{I \times I}). \tag{3.176}$$

Furthermore, the property [3.24] can be written as:

$$f(\mathcal{J}_{2N}) = \mathbf{I}_I \text{ with } I = \prod_{n=1}^{N} I_n, \tag{3.177}$$

which means that the image of the neutral element of $\mathbb{T}_N$ under f is the neutral element of the group GL_I.

Using the properties [3.174], [3.175] and [3.177], we can deduce the following proposition, which shows that the inverse of a square tensor $\mathcal{X} \in \mathbb{K}^{\underline{L}_N \times \underline{L}_N}$ belonging to the set $\mathbb{T}_N$, defined in [3.171], can be determined from the inverse of its matrix unfolding $f(\mathcal{X}) = \mathbf{X}_{I \times I}$.

PROPOSITION 3.63.– Any tensor $\mathcal{X} \in \mathbb{T}_N$ admits a unique inverse, denoted $\mathcal{X}^{-1} \in \mathbb{T}_N$, with respect to the Einstein product, satisfying (Brazell *et al.* 2013):

$$\mathcal{X} \star_N \mathcal{X}^{-1} = \mathcal{X}^{-1} \star_N \mathcal{X} = \mathcal{J}_{2N}. \qquad [3.178]$$

Its matrix unfolding is $f(\mathcal{X}^{-1}) = \mathbf{X}_{I \times I}^{-1}$, i.e.

$$\mathcal{X}^{-1} = f^{-1}(\mathbf{X}_{I \times I}^{-1}). \qquad [3.179]$$

PROOF.– Using the properties [3.174], [3.175] and [3.177], the equality $\mathcal{X} \star_N \mathcal{X}^{-1} = \mathcal{J}_{2N}$ gives us:

$$f(\mathcal{X} \star_N \mathcal{X}^{-1}) = f(\mathcal{X})f(\mathcal{X}^{-1}) = \mathbf{X}_{I \times I}f(\mathcal{X}^{-1}), = f(\mathcal{J}_{2N}) = \mathbf{I}_I \qquad [3.180]$$

from which we deduce:

$$f(\mathcal{X}^{-1}) = \mathbf{X}_{I \times I}^{-1} \Leftrightarrow \mathcal{X}^{-1} = f^{-1}(\mathbf{X}_{I \times I}^{-1}). \qquad [3.181]$$

The same result can be obtained from the equality $\mathcal{X}^{-1} \star_N \mathcal{X} = \mathcal{J}_{2N}$. □

In summary, from the properties stated in propositions 3.59, 3.61 and 3.63, we conclude that:

– the set $(\mathbb{T}_N, \star_N)$ is closed under the operation $\star_N$;

– the operation $\star_N$ is associative;

– there exists a multiplicative neutral element, namely the identity tensor $\mathcal{J}_{2N}$;

– every element $\mathcal{X} \in (\mathbb{T}_N, \star_N)$ admits an inverse.

These properties define a group structure on the set $(\mathbb{T}_N, \star_N)$. The property [3.174] then defines a morphism of groups. Furthermore, since f is bijective, it is an isomorphism from the tensor group $(\mathbb{T}_N, \star_N)$ to the general linear group GL_I.

REMARK 3.64.– It is easy to check that the mapping f maps a symmetric matrix $\mathbf{X}_{I \times I}$ to a tensor $\mathcal{X} \in \mathbb{R}^{\underline{L}_N \times \underline{L}_N}$ that is symmetric by blocks of order N (see Table 3.8).

Moreover, if $\mathcal{X} \in \mathbb{T}_N$ is orthogonal (respectively, unitary) according to the definition given in Table 3.16, its matrix unfolding $\mathbf{X}_{I \times I}$ is orthogonal itself (respectively, unitary).

3.14. Inverse and pseudo-inverse tensors

The concept of the inverse tensor, introduced by Brazell *et al.* (2013) for square tensors, and therefore the invertibility conditions of a tensor depend on several factors simultaneously: the structure of the tensor, defined in terms of order and dimensions; the definition of a multiplication operation ($\bullet$) between tensors; and an identity tensor $\mathcal{J}$ for this operation, satisfying $\mathcal{X} \bullet \mathcal{X}^{-1} = \mathcal{X}^{-1} \bullet \mathcal{X} = \mathcal{J}$, where $\mathcal{X}^{-1}$ is the inverse of $\mathcal{X}$.

In general, inversion is based on exploiting an isomorphism f between the tensor space $(\mathbb{T}, \bullet)$ containing $\mathcal{X}$, equipped with the product $\bullet$ and a space $(\mathbb{M}, \cdot)$ of invertible matrices that contains a particular matrix unfolding of the tensor $\mathcal{X}$, equipped with the standard matrix product, whose multiplication symbol $(\cdot)$ is typically omitted from the notation:

$$f : \mathbb{T} \ni \mathcal{X} \longrightarrow \mathbf{X} = f(\mathcal{X}) \in \mathbb{M}. \tag{3.182}$$

Since the mapping f is an isomorphism, the inverse mapping f^{-1} allows us to determine the inverse of $\mathcal{X}$ from that of its unfolding $\mathbf{X}$, i.e.:

$$\mathcal{X}^{-1} = f^{-1}\big[[f(\mathcal{X})]^{-1}\big] = f^{-1}[\mathbf{X}^{-1}]. \tag{3.183}$$

This result is illustrated by the formula [3.179], which gives the inverse of a tensor $\mathcal{X} \in \mathbb{T}_N$.

Table 3.17 gives inverses for orthogonal and unitary tensors.

Properties of $\mathcal{X} \in \mathbb{K}^{I_N \times I_N}$	Definitions	Inverses
Orthogonal	$\mathcal{X}^T \star_N \mathcal{X} = \mathcal{X} \star_N \mathcal{X}^T = \mathcal{J}_{2N}$	$\mathcal{X}^{-1} = \mathcal{X}^T = f^{-1}[\mathbf{X}_{I \times I}^T]$
Unitary	$\mathcal{X}^H \star_N \mathcal{X} = \mathcal{X} \star_N \mathcal{X}^H = \mathcal{J}_{2N}$	$\mathcal{X}^{-1} = \mathcal{X}^H = f^{-1}[\mathbf{X}_{I \times I}^H]$

Table 3.17. *Inverses of orthogonal/unitary tensors*

In the next proposition, we show how the inverse of the tensor $\mathcal{Z} = \mathcal{X} \star_N \mathcal{Y}$ can be determined via its matrix unfolding, i.e. $\mathcal{Z}^{-1} = f^{-1}\big[\mathbf{Z}_{I \times I}^{-1}\big]$.

PROPOSITION 3.65.– Consider the tensors $\mathcal{X}, \mathcal{Y} \in \mathbb{T}_N$. The inverse of their product $\mathcal{Z} = \mathcal{X} \star_N \mathcal{Y}$ can be obtained from its matrix unfolding $\mathbf{Z}_{I \times I}^{-1} = \mathbf{Y}_{I \times I}^{-1} \mathbf{X}_{I \times I}^{-1}$, where $\mathbf{X}_{I \times I}$ and $\mathbf{Y}_{I \times I}$ are the unfoldings of $\mathcal{X}$ and $\mathcal{Y}$, i.e.:

$$\mathcal{Z}^{-1} = f^{-1}(\mathbf{Z}_{I \times I}^{-1}) = f^{-1}(\mathbf{Y}_{I \times I}^{-1} \mathbf{X}_{I \times I}^{-1}) \tag{3.184}$$

$$= f^{-1}\big[f(\mathcal{Y}^{-1})f(\mathcal{X}^{-1})\big] \tag{3.185}$$

$$= \mathcal{Y}^{-1} \star_N \mathcal{X}^{-1} \quad \text{by [3.175].} \tag{3.186}$$

REMARK 3.66.– This result requires $\mathbf{X}_{I\times I}$ and $\mathbf{Y}_{I\times I}$ to be invertible, which is the case by the hypothesis $\mathcal{X}, \mathcal{Y} \in \mathbb{T}_N$.

EXAMPLE 3.67.– Consider two tensors $\mathcal{X}, \mathcal{Y} \in \mathbb{K}^{2\times2\times2\times2}$, whose unfoldings $\mathbf{X}_{I\times I}$ and $\mathbf{Y}_{I\times I}$, with $I = I_1 I_2$ and $I_1 = I_2 = 2$, are given by:

$$\mathbf{X}_{I\times I} = \begin{bmatrix} 1 & 0 & 0 & 0 \\ 1 & 1 & 0 & 0 \\ 0 & 1 & 1 & 0 \\ 0 & 0 & 1 & 1 \end{bmatrix}, \quad \mathbf{Y}_{I\times I} = \begin{bmatrix} 1 & 1 & 0 & 0 \\ 0 & 1 & 1 & 0 \\ 0 & 0 & 1 & 1 \\ 0 & 0 & 0 & 1 \end{bmatrix},$$

which corresponds to $x_{ijkl} = 0$ for all (i,j,k,l) such that $l + 2(k-1) > j + 2(i-1)$ and $y_{ijkl} = 0$ for all (i,j,k,l) such that $j + 2(i-1) > l + 2(k-1)$, respectively.

It is easy to check that:

$$\mathbf{X}_{I\times I}^{-1} = \begin{bmatrix} 1 & 0 & 0 & 0 \\ -1 & 1 & 0 & 0 \\ 1 & -1 & 1 & 0 \\ -1 & 1 & -1 & 1 \end{bmatrix}, \quad \mathbf{Y}_{I\times I}^{-1} = \begin{bmatrix} 1 & -1 & 1 & -1 \\ 0 & 1 & -1 & 1 \\ 0 & 0 & 1 & -1 \\ 0 & 0 & 0 & 1 \end{bmatrix},$$

and hence the unfoldings $\mathbf{Z}_{I\times I}$ and $\mathbf{Z}_{I\times I}^{-1}$ are given by:

$$\mathbf{Z}_{I\times I} = \mathbf{X}_{I\times I}\mathbf{Y}_{I\times I} = \begin{bmatrix} 1 & 1 & 0 & 0 \\ 1 & 2 & 1 & 0 \\ 0 & 1 & 2 & 1 \\ 0 & 0 & 1 & 2 \end{bmatrix}$$

$$\mathbf{Z}_{I\times I}^{-1} = \mathbf{Y}_{I\times I}^{-1}\mathbf{X}_{I\times I}^{-1} = \begin{bmatrix} 4 & -3 & 2 & -1 \\ -3 & 3 & -2 & 1 \\ 2 & -2 & 2 & -1 \\ -1 & 1 & -1 & 1 \end{bmatrix}.$$

The latter matrix does indeed correspond to the unfolding of the inverse of the tensor $\mathcal{Z}$, as can be checked by computing the product $\mathbf{Z}_{I\times I}\mathbf{Z}_{I\times I}^{-1}$, which must give the unfolding of the identity tensor $\mathcal{J}_4$ defined in [3.22], namely the identity matrix $\mathbf{I}_4$, by the relation [3.24].

Indeed, we have:

$$\mathbf{Z}_{I\times I}\mathbf{Z}_{I\times I}^{-1} = \begin{bmatrix} 1 & 1 & 0 & 0 \\ 1 & 2 & 1 & 0 \\ 0 & 1 & 2 & 1 \\ 0 & 0 & 1 & 2 \end{bmatrix} \begin{bmatrix} 4 & -3 & 2 & -1 \\ -3 & 3 & -2 & 1 \\ 2 & -2 & 2 & -1 \\ -1 & 1 & -1 & 1 \end{bmatrix} = \mathbf{I}_4.$$

In section 3.15, we will show how the group isomorphism introduced above can be exploited to develop tensor decompositions in a factorized form, such as the SVD and full-rank factorization.

The next proposition presents a generalization of the inversion formula [3.184] for $\mathcal{Z} = \mathcal{X}^T \star_P \mathcal{X}$, with $\mathcal{X} \in \mathbb{K}^{\underline{I}_P \times \underline{J}_N}$.

PROPOSITION 3.68.– Let $\mathcal{X} \in \mathbb{K}^{\underline{I}_P \times \underline{J}_N}$ and $\mathcal{X}^T \in \mathbb{K}^{\underline{J}_N \times \underline{I}_P}$. The inverse of the Einstein product $\mathcal{Z} = \mathcal{X}^T \star_P \mathcal{X} \in \mathbb{K}^{\underline{J}_N \times \underline{J}_N}$ admits the matrix unfolding $\left(\mathbf{X}_{I \times J}^T \mathbf{X}_{I \times J}\right)^{-1}$, where $\mathbf{X}_{I \times J}$ is the matrix unfolding of the tensor $\mathcal{X}$, with $I = \prod_{p=1}^{P} I_p$ and $J = \prod_{n=1}^{N} J_n$. Hence, the necessary condition for $\mathcal{Z}$ to be invertible is that $\mathbf{X}_{I \times J}$ must have full column rank, i.e. $r(\mathbf{X}_{I \times J}) = J = \prod_{n=1}^{N} J_n \leq I = \prod_{p=1}^{P} I_p$.

PROOF.– By defining the mapping f that transforms a tensor $\mathcal{X} \in \mathbb{K}^{\underline{I}_P \times \underline{J}_N}$ into its matrix unfolding $f(\mathcal{X}) = \mathbf{X}_{I \times J}$ with $I = \prod_{p=1}^{P} I_p$ and $J = \prod_{n=1}^{N} J_n$, as defined in [3.44]–[3.45], we have:

$$\mathcal{Z} = \mathcal{X}^T \star_P \mathcal{X} \in \mathbb{K}^{\underline{J}_N \times \underline{J}_N} \tag{3.187}$$

and:

$$f(\mathcal{Z}) = \mathbf{X}_{I \times J}^T \mathbf{X}_{I \times J} = f(\mathcal{X}^T) f(\mathcal{X}) = \mathbf{Z}_{J \times J} \tag{3.188}$$

$$\mathcal{Z}^{-1} = f^{-1}(\mathbf{Z}_{J \times J}^{-1}) = f^{-1}\left[(\mathbf{X}_{I \times J}^T \mathbf{X}_{I \times J})^{-1}\right]. \tag{3.189}$$

We can therefore conclude that $\mathcal{Z}$ is invertible if $\mathbf{X}_{I \times J}^T \mathbf{X}_{I \times J}$ is invertible, which implies the necessary condition that $\mathbf{X}_{I \times J}$ must have full column rank, i.e. $r(\mathbf{X}_{I \times J}) = J$.   □

EXAMPLE 3.69.– Consider the fourth-order rectangular tensor $\mathcal{X} \in R^{3 \times 2 \times 2 \times 2}$ corresponding to $(I_1, I_2, J_1, J_2) = (3, 2, 2, 2)$ and hence $(I, J) = (6, 4)$, with the following matrix unfolding:

$$\mathbf{X}_{I \times J} = \begin{bmatrix} 1 & 0 & 0 & 0 \\ 1 & 1 & 0 & 0 \\ 0 & 1 & 1 & 0 \\ 0 & 0 & 1 & 1 \\ 0 & 0 & 0 & 1 \\ 0 & 0 & 0 & 0 \end{bmatrix}. \tag{3.190}$$

From this matrix unfolding, we can deduce the elements of the tensor $\mathcal{X}$, such as: $(\mathbf{X}_{I \times J})_{i_2+(i_1-1)I_2, j_2+(j_1-1)J_2} = x_{i_1,i_2,j_1,j_2}$. For example, $x_{1221} = (\mathbf{X}_{I \times J})_{23} = 0$, and $x_{3122} = (\mathbf{X}_{I \times J})_{54} = 1$. Using the formulae [3.188] and [3.189], we obtain:

$$\mathbf{Z}_{J \times J} = \begin{bmatrix} 2 & 1 & 0 & 0 \\ 1 & 2 & 1 & 0 \\ 0 & 1 & 2 & 1 \\ 0 & 0 & 1 & 2 \end{bmatrix}, \quad \mathbf{Z}_{J \times J}^{-1} = \begin{bmatrix} 0.8 & -0.6 & 0.4 & -0.2 \\ -0.6 & 1.2 & -0.8 & 0.4 \\ 0.4 & -0.8 & 1.2 & -0.6 \\ -0.2 & 0.4 & -0.6 & 0.8 \end{bmatrix},$$

with, for example:

$$z_{1211} = (\mathcal{Z})_{1211} = \sum_{i_1=1}^{3} \sum_{i_2=1}^{2} = x_{i_1,i_2,1,2}\, x_{i_1,i_2,1,1} = (\mathbf{Z}_{J \times J})_{21} = 1$$

$$(\mathcal{Z}^{-1})_{2212} = (\mathbf{Z}_{J \times J}^{-1})_{42} = 0.4.$$

It is easy to check that $\mathbf{Z}_{J \times J}\mathbf{Z}_{J \times J}^{-1} = \mathbf{Z}_{J \times J}^{-1}\mathbf{Z}_{J \times J} = \mathbf{I}_4$.

In the following, we will introduce the notions of the generalized inverse and Moore–Penrose pseudo-inverse of a tensor $\mathcal{A} \in \mathbb{K}^{\underline{I}_P \times \underline{J}_N}$ of order $P + N$, using the Einstein product (Sun $et\ al.$ 2016; Behera and Mishra 2017; Liang and Zheng 2018). Generalized inverses and the Moore–Penrose pseudo-inverse of tensors are in particular used to solve certain tensor systems (Brazell $et\ al.$ 2013; Bu $et\ al.$ 2014; Sun $et\ al.$ 2016; Behera and Mishra 2017; Jin $et\ al.$ 2018; Liang and Zheng 2018b), as will be illustrated in section 3.17.

Table 3.18 states the four conditions[4] that define different types of generalized inverses of a matrix $\mathbf{A} \in \mathbb{K}^{I \times J}$ and a tensor $\mathcal{A} \in \mathbb{K}^{\underline{I}_P \times \underline{J}_N}$ of order $P + N$, denoted $\mathbf{A}^\sharp$ and $\mathcal{A}^\sharp$, respectively. When these four conditions are satisfied[5], we have the Moore–Penrose pseudo-inverse, denoted $\mathbf{A}^\dagger$ in the matrix case and $\mathcal{A}^\dagger$ in the tensor case.

Conditions	$\mathbf{A} \in \mathbb{K}^{I \times J}$, $\mathbf{A}^\sharp \in \mathbb{K}^{J \times I}$	$\mathcal{A} \in \mathbb{K}^{\underline{I}_P \times \underline{J}_N}$, $\mathcal{A}^\sharp \in \mathbb{K}^{\underline{J}_N \times \underline{I}_P}$
(1)	$\mathbf{A}\mathbf{A}^\sharp\mathbf{A} = \mathbf{A}$	$\mathcal{A} \star_N \mathcal{A}^\sharp \star_P \mathcal{A} = \mathcal{A}$
(2)	$\mathbf{A}^\sharp\mathbf{A}\mathbf{A}^\sharp = \mathbf{A}^\sharp$	$\mathcal{A}^\sharp \star_P \mathcal{A} \star_N \mathcal{A}^\sharp = \mathcal{A}^\sharp$
(3)	$(\mathbf{A}\mathbf{A}^\sharp)^H = \mathbf{A}\mathbf{A}^\sharp$	$(\mathcal{A} \star_N \mathcal{A}^\sharp)^H = \mathcal{A} \star_N \mathcal{A}^\sharp$
(4)	$(\mathbf{A}^\sharp\mathbf{A})^H = \mathbf{A}^\sharp\mathbf{A}$	$(\mathcal{A}^\sharp \star_P \mathcal{A})^H = \mathcal{A}^\sharp \star_P \mathcal{A}$

Table 3.18. *Generalized inverses*

There are several types of generalized inverse for a matrix $\mathbf{A}$ or a tensor $\mathcal{A}$, depending on which conditions are satisfied in Table 3.18. For example, Burns $et\ al.$ (1974) and Sun $et\ al.$ (2016) define the following generalized inverses for matrices and tensors, respectively:

– for (1): inner inverse, denoted $\mathcal{A}^{\{1\}}$;

– for (2): outer inverse, denoted $\mathcal{A}^{\{2\}}$;

---

4 The conditions (1)–(4) are called the general, reflexivity, normalization and inverse normalization conditions, respectively.

5 In the real case ($\mathbb{K} = \mathbb{R}$), replace the operation of tranconjugation with that of transposition in the conditions (3) and (4).

– for (1) and (2): reflexive generalized inverse, denoted $\mathcal{A}^{\{1,2\}}$;

– for (1)–(3): weak (or normalized) generalized inverse, denoted $\mathcal{A}^{\{1,2,3\}}$;

– for (1)–(4): Moore–Penrose pseudo-inverse, denoted $\mathcal{A}^{\dagger}$.

Based on the isomorphism f defined in [3.172], Brazell *et al.* (2013) presented the SVD of a real fourth-order square tensor. A generalization of the SVD to the case of a rectangular tensor $\mathcal{X} \in \mathbb{C}^{\underline{I}_P \times \underline{J}_N}$, via the Einstein product, was established by Liang and Zheng (2018), using the mapping f defined as:

$$f : \mathbb{C}^{\underline{I}_P \times \underline{J}_N} \ni \mathcal{X} \longmapsto f(\mathcal{X}) = \mathbf{X}_{I \times J} \in \mathbb{C}^{I \times J} \text{ with } I = \prod_{p=1}^{P} I_p, \ J = \prod_{n=1}^{N} J_n.$$

[3.191]

Like the mapping f defined in [3.172], the mapping defined by the above equation is bijective.

Before establishing the Moore–Penrose pseudo-inverse and the SVD of a rectangular tensor, let us present a preliminary result in the following lemma (Stanimirovic *et al.* 2020):

LEMMA 3.70.– Consider the rectangular tensors $\mathcal{X} \in \mathbb{C}^{\underline{I}_P \times \underline{J}_N}$ and $\mathcal{Y} \in \mathbb{C}^{\underline{J}_N \times \underline{K}_L}$, and their respective matrix unfoldings $\mathbf{X}_{I \times J} \in \mathbb{C}^{I \times J}$ and $\mathbf{Y}_{J \times K} \in \mathbb{C}^{J \times K}$, with $I = \prod_{p=1}^{P} I_p$, $J = \prod_{n=1}^{N} J_n$, and $K = \prod_{l=1}^{L} K_l$. The mapping f defined in [3.191] satisfies the following properties:

$$f(\mathcal{X} \star_N \mathcal{Y}) = \mathbf{X}_{I \times J} \mathbf{Y}_{J \times K} = f(\mathcal{X}) f(\mathcal{Y}) \in \mathbb{C}^{I \times K}$$

[3.192]

$$\mathcal{X} \star_N \mathcal{Y} = f^{-1}(\mathbf{X}_{I \times J} \mathbf{Y}_{J \times K}) = f^{-1}(\mathbf{X}_{I \times J}) \star_N f^{-1}(\mathbf{Y}_{J \times K})$$

[3.193]

and:

$$f^{-1}(\mathbf{X}_{I \times J}^T) = \mathcal{X}^T \ , \ f^{-1}(\mathbf{X}_{I \times J}^H) = \mathcal{X}^H.$$

[3.194]

REMARK 3.71.– From the above results, we can deduce the expression of the transconjugate of an Einstein product of tensors. Let $\mathcal{X} \in \mathbb{C}^{\underline{I}_P \times \underline{J}_N}$, $\mathcal{Y} \in \mathbb{C}^{\underline{J}_N \times \underline{K}_M}$, and $\mathcal{Z} \in \mathbb{C}^{\underline{K}_M \times \underline{L}_Q}$. We have:

$$(\mathcal{X} \star_N \mathcal{Y})^H = \mathcal{Y}^H \star_N \mathcal{X}^H \in \mathbb{C}^{\underline{K}_M \times \underline{I}_P}$$

[3.195]

$$(\mathcal{X} \star_N \mathcal{Y} \star_M \mathcal{Z})^H = \mathcal{Z}^H \star_M \mathcal{Y}^H \star_N \mathcal{X}^H \in \mathbb{C}^{\underline{L}_Q \times \underline{I}_P}.$$

[3.196]

Using the isomorphism defined in [3.191], we can deduce the following proposition.

PROPOSITION 3.72.– Any rectangular tensor $\mathcal{X} \in \mathbb{C}^{\underline{I}_P \times \underline{J}_N}$ whose matrix unfolding $\mathbf{X}_{I \times J}$ has full column rank admits a Moore–Penrose pseudo-inverse, defined as:

$$\mathcal{X}^{\dagger} = f^{-1}[\mathbf{X}_{I \times J}^{\dagger}] = f^{-1}\big[(\mathbf{X}_{I \times J}^{T}\mathbf{X}_{I \times J})^{-1}\mathbf{X}_{I \times J}^{T}\big]. \qquad [3.197]$$

Table 3.19 summarizes the key results established for the Einstein product over the previous sections.

Mappings/Properties	Classes of tensors and operations with the Einstein product	Eq.
	Square tensors: $\mathbb{T}_N = \{\mathcal{X} \in \mathbb{K}^{\underline{I}_N \times \underline{I}_N} \;/\; \det(\mathbf{X}_{I \times I}) \neq 0\}$ $\mathcal{X}, \mathcal{Y} \in \mathbb{K}^{\underline{I}_N \times \underline{I}_N}$, $I = \prod_{n=1}^{N} I_n$	(3.171)
Isomorphism: f	$\mathbb{T}_N \ni \mathcal{X} \longmapsto f(\mathcal{X}) = \mathbf{X}_{I \times I} \in \mathbb{K}^{I \times I}$	(3.172)
Product	$f(\mathcal{X} \star_N \mathcal{Y}) = \mathbf{X}_{I \times I}\mathbf{Y}_{I \times I} = f(\mathcal{X})f(\mathcal{Y})$	(3.174)
Inverse of $\mathcal{X}$	$\mathcal{X}^{-1} = f^{-1}(\mathbf{X}_{I \times I}^{-1})$	(3.179)
Inverse of $\mathcal{X} \star_N \mathcal{Y}$	$(\mathcal{X} \star_N \mathcal{Y})^{-1} = f^{-1}[\mathbf{Y}_{I \times I}^{-1}\mathbf{X}_{I \times I}^{-1}] = \mathcal{Y}^{-1} \star_N \mathcal{X}^{-1}$	(3.186)
	Rectangular tensors: $\mathcal{X} \in \mathbb{C}^{\underline{I}_P \times \underline{J}_N}, \mathcal{Y} \in \mathbb{C}^{\underline{J}_N \times \underline{K}_L}$ $I = \prod_{p=1}^{P} I_p$, $J = \prod_{n=1}^{N} J_n$, $K = \prod_{l=1}^{L} K_l$	
Mapping: f	$\mathbb{C}^{\underline{I}_P \times \underline{J}_N} \ni \mathcal{X} \longmapsto f(\mathcal{X}) = \mathbf{X}_{I \times J} \in \mathbb{C}^{I \times J}$	(3.191)
Product	$f(\mathcal{X} \star_N \mathcal{Y}) = \mathbf{X}_{I \times J}\mathbf{Y}_{J \times K} = f(\mathcal{X})f(\mathcal{Y})$	(3.192)
	$\mathcal{X} \star_N \mathcal{Y} = f^{-1}(\mathbf{X}_{I \times J}) \star_N f^{-1}(\mathbf{Y}_{J \times K})$	(3.193)
Inverse of $\mathcal{X}^T \star_P \mathcal{X}$	$(\mathcal{X}^T \star_P \mathcal{X})^{-1} = f^{-1}\big[(\mathbf{X}_{I \times J}^{T}\mathbf{X}_{I \times J})^{-1}\big]$	(3.189)
Moore-Penrose pseudo-inverse	$\mathcal{X}^{\dagger} = f^{-1}[\mathbf{X}_{I \times J}^{\dagger}] = f^{-1}\big[(\mathbf{X}_{I \times J}^{T}\mathbf{X}_{I \times J})^{-1}\mathbf{X}_{I \times J}^{T}\big]$	(3.197)

Table 3.19. *Operations with the Einstein product*

Table 3.20 presents various properties of the Moore–Penrose pseudo-inverse of a tensor $\mathcal{X} \in \mathbb{C}^{\underline{I}_P \times \underline{J}_N}$, in parallel to those of the Moore–Penrose pseudo-inverse of a matrix (Favier, 2019), with the following hypotheses for specific properties:

– Property (xv): $\mathbf{A} \in \mathbb{K}^{I \times J}$ of full column rank, in the matrix case; $\mathcal{A} \in \mathbb{K}^{\underline{I}_P \times \underline{J}_N}$, with the unfolding $\mathbf{A}_{I \times J}$ of full column rank.

– Property (xvi): $\mathbf{A} \in \mathbb{K}^{I \times J}$ of full row rank, in the matrix case; $\mathcal{A} \in \mathbb{K}^{\underline{I}_P \times \underline{J}_N}$, with the unfolding $\mathbf{A}_{I \times J}$ of full row rank.

– Property (xvii): $\mathbf{A} \in \mathbb{K}^{I \times K}, \mathbf{B} \in \mathbb{K}^{K \times J}$ and $\mathbf{C} \in \mathbb{K}^{K \times K}$ of full rank, in the matrix case; $\mathcal{A} \in \mathbb{K}^{\underline{I}_P \times \underline{J}_N}, \mathcal{B} \in \mathbb{K}^{\underline{J}_N \times \underline{K}_Q}$ and $\mathcal{C} \in \mathbb{K}^{\underline{J}_N \times \underline{J}_N}$, in the tensor case, with $\mathcal{C}$ assumed to be invertible in the sense of the definition [3.179].

– Property (xviii): $\mathbf{A} \in \mathbb{C}^{I \times I}$ and $\mathbf{B} \in \mathbb{C}^{J \times J}$ in the matrix case, and $\mathcal{A} \in \mathbb{C}^{\underline{I}_P \times \underline{I}_P}$ and $\mathcal{B} \in \mathbb{C}^{\underline{J}_N \times \underline{J}_N}$ in the tensor case, are assumed to be unitary, with $\mathbf{C} \in \mathbb{K}^{I \times J}$ and $\mathcal{C} \in \mathbb{K}^{\underline{I}_P \times \underline{J}_N}$.

	Matrix case	Tensor case
	$\mathbf{A} \in \mathbb{C}^{I \times J}$	$\mathcal{A} \in \mathbb{C}^{\underline{I}_P \times \underline{J}_N}$
SVD	$\mathbf{A} = \mathbf{U}\mathbf{D}\mathbf{V}^H$ $\mathbf{U} \in \mathbb{C}^{I \times I}, \mathbf{V} \in \mathbb{C}^{J \times J}$ unitary	$\mathcal{A} = \mathcal{U} \star_P \mathcal{D} \star_N \mathcal{V}^H$ $\mathcal{U} \in \mathbb{C}^{\underline{I}_P \times \underline{I}_P}, \mathcal{V} \in \mathbb{C}^{\underline{J}_N \times \underline{J}_N}$ unitary
Pseudo- **inverse**	$\mathbf{A}^\dagger = \mathbf{V}\mathbf{D}^\dagger \mathbf{U}^H \in \mathbb{C}^{J \times I}$	$\mathcal{A}^\dagger = \mathcal{V} \star_N \mathcal{D}^\dagger \star_P \mathcal{U}^H \in \mathbb{C}^{\underline{J}_N \times \underline{I}_P}$
(i)	$(\mathbf{A}^\dagger)^\dagger = \mathbf{A}$	$(\mathcal{A}^\dagger)^\dagger = \mathcal{A}$
(ii)	$(\mathbf{A}^H)^\dagger = (\mathbf{A}^\dagger)^H$	$(\mathcal{A}^H)^\dagger = (\mathcal{A}^\dagger)^H$
(iii)	$(\mathbf{A}\mathbf{A}^\dagger)^2 = \mathbf{A}\mathbf{A}^\dagger$	$(\mathcal{A} \star_N \mathcal{A}^\dagger) \star_P (\mathcal{A} \star_N \mathcal{A}^\dagger) = \mathcal{A} \star_N \mathcal{A}^\dagger$
(iv)	$(\mathbf{A}^\dagger\mathbf{A})^2 = \mathbf{A}^\dagger\mathbf{A}$	$(\mathcal{A}^\dagger \star_P \mathcal{A}) \star_N (\mathcal{A}^\dagger \star_P \mathcal{A}) = \mathcal{A}^\dagger \star_P \mathcal{A}$
(v)	$(\mathbf{A}\mathbf{A}^\dagger)^\dagger = \mathbf{A}\mathbf{A}^\dagger$	$(\mathcal{A} \star_N \mathcal{A}^\dagger)^\dagger = \mathcal{A} \star_N \mathcal{A}^\dagger$
(vi)	$(\mathbf{A}^\dagger\mathbf{A})^\dagger = \mathbf{A}^\dagger\mathbf{A}$	$(\mathcal{A}^\dagger \star_P \mathcal{A})^\dagger = \mathcal{A}^\dagger \star_P \mathcal{A}$
(vii)	$(\mathbf{A}\mathbf{A}^H)^\dagger = (\mathbf{A}^\dagger)^H\mathbf{A}^\dagger$	$(\mathcal{A} \star_N \mathcal{A}^H)^\dagger = (\mathcal{A}^\dagger)^H \star_N \mathcal{A}^\dagger$
(viii)	$(\mathbf{A}^H\mathbf{A})^\dagger = \mathbf{A}^\dagger(\mathbf{A}^\dagger)^H$	$(\mathcal{A}^H \star_P \mathcal{A})^\dagger = \mathcal{A}^\dagger \star_P (\mathcal{A}^\dagger)^H$
(ix)	$\mathbf{A}^H = \mathbf{A}^H\mathbf{A}\mathbf{A}^\dagger = \mathbf{A}^\dagger\mathbf{A}\mathbf{A}^H$	$\mathcal{A}^H = \mathcal{A}^H \star_P \mathcal{A} \star_N \mathcal{A}^\dagger = \mathcal{A}^\dagger \star_P \mathcal{A} \star_N \mathcal{A}^H$
(x)	$\mathbf{A}^\dagger = \mathbf{A}^H(\mathbf{A}^\dagger)^H\mathbf{A}^\dagger$ $= \mathbf{A}^\dagger(\mathbf{A}^\dagger)^H\mathbf{A}^H$	$\mathcal{A}^\dagger = \mathcal{A}^H \star_P (\mathcal{A}^\dagger)^H \star_N \mathcal{A}^\dagger$ $= \mathcal{A}^\dagger \star_P (\mathcal{A}^\dagger)^H \star_N \mathcal{A}^H$
(xi)	$\mathbf{A} = \mathbf{A}\mathbf{A}^H(\mathbf{A}^\dagger)^H$ $= (\mathbf{A}^\dagger)^H\mathbf{A}^H\mathbf{A}$	$\mathcal{A} = \mathcal{A} \star_N \mathcal{A}^H \star_P (\mathcal{A}^\dagger)^H$ $= (\mathcal{A}^\dagger)^H \star_N \mathcal{A}^H \star_P \mathcal{A}$
(xii)	$\mathbf{A}^\dagger = (\mathbf{A}^H\mathbf{A})^\dagger\mathbf{A}^H$ $= \mathbf{A}^H(\mathbf{A}\mathbf{A}^H)^\dagger$	$\mathcal{A}^\dagger = (\mathcal{A}^H \star_P \mathcal{A})^\dagger \star_N \mathcal{A}^H$ $= \mathcal{A}^H \star_P (\mathcal{A} \star_N \mathcal{A}^H)^\dagger$
(xiii)	$\mathbf{A}\mathbf{A}^\dagger = (\mathbf{A}^H)^\dagger\mathbf{A}^H$	$\mathcal{A} \star_N \mathcal{A}^\dagger = (\mathcal{A}^H)^\dagger \star_N \mathcal{A}^H$
(xiv)	$\mathbf{A}^\dagger\mathbf{A} = \mathbf{A}^H(\mathbf{A}^H)^\dagger$	$\mathcal{A}^\dagger \star_P \mathcal{A} = \mathcal{A}^H \star_P (\mathcal{A}^H)^\dagger$
(xv)	$\mathbf{A}^\dagger = (\mathbf{A}^H\mathbf{A})^{-1}\mathbf{A}^H$	$\mathcal{A}^\dagger = (\mathcal{A}^H \star_P \mathcal{A})^{-1} \star_N \mathcal{A}^H$
(xvi)	$\mathbf{A}^\dagger = \mathbf{A}^H(\mathbf{A}\mathbf{A}^H)^{-1}$	$\mathcal{A}^\dagger = \mathcal{A}^H \star_P (\mathcal{A} \star_N \mathcal{A}^H)^{-1}$
(xvii)	$(\mathbf{A}\mathbf{C}\mathbf{B})^\dagger = \mathbf{B}^\dagger\mathbf{C}^{-1}\mathbf{A}^\dagger$	$(\mathcal{A} \star_N \mathcal{C} \star_N \mathcal{B})^\dagger = \mathcal{B}^\dagger \star_N \mathcal{C}^{-1} \star_N \mathcal{A}^\dagger$
(xviii)	$(\mathbf{A}\mathbf{C}\mathbf{B})^\dagger = \mathbf{B}^H\mathbf{C}^\dagger\mathbf{A}^H$	$(\mathcal{A} \star_P \mathcal{C} \star_N \mathcal{B})^\dagger = \mathcal{B}^H \star_N \mathcal{C}^\dagger \star_P \mathcal{A}^H$
(xix)	$(\mathbf{0}_{I \times J})^\dagger = \mathbf{0}_{J \times I}$	$(\mathbf{0}_{\underline{I}_P \times \underline{J}_N})^\dagger = \mathbf{0}_{\underline{J}_N \times \underline{I}_P}$

Table 3.20. *Properties of the Moore-Penrose pseudo-inverse*

REMARK 3.73.– We can make the following remarks:

– In the case of an invertible square matrix $\mathbf{A} \in \mathbb{K}^{I \times I}$ and an invertible square tensor $\mathcal{A} \in \mathbb{K}^{\underline{I}_P \times \underline{I}_P}$, we have $\mathbf{A}^\dagger = \mathbf{A}^{-1}$ and $\mathcal{A}^\dagger = \mathcal{A}^{-1}$, as defined in [3.179].

– The relations (iii) and (iv) mean that $\mathbf{A}\mathbf{A}^\dagger$ and $\mathbf{A}^\dagger\mathbf{A}$, in the matrix case, and $\mathcal{A} \star_N \mathcal{A}^\dagger$ and $\mathcal{A}^\dagger \star_P \mathcal{A}$, in the tensor case, are idempotent. To prove these properties, it suffices to replace the pseudo-inverses $\mathbf{A}^\dagger$ and $\mathcal{A}^\dagger$ by the expressions provided by the definition equations (1) and (2) in Table 3.18.

– The expressions (xv) and (xvi) follow directly from the relations (xii), because the pseudo-inverses on both right-hand sides can be replaced with inverses, due to the hypotheses of full column rank and full row rank, respectively.

Some of the results presented in Table 3.20 are proven in Panigrahy *et al.* (2017) for a tensor $\mathcal{A} \in \mathbb{C}^{\underline{I}_P \times \underline{J}_P}$, i.e., for $N = P$. In general, the proof of these results can be established using the equations defining the Moore–Penrose pseudo-inverse given in Table 3.18, together with the properties [3.192]–[3.193] and [3.197].

3.15. Tensor decompositions in the form of factorizations

In the next three sections, we present the eigendecomposition of a symmetric square tensor, the SVD of a rectangular tensor and the full-rank decomposition of a tensor of arbitrary order.

3.15.1. *Eigendecomposition of a symmetric square tensor*

PROPOSITION 3.74.– Let $\mathcal{A} \in \mathbb{R}^{\underline{I}_P \times \underline{I}_P}$ be a block symmetric square tensor in the sense of the definition in Table 3.8. Then there exists an orthogonal tensor $\mathcal{U} \in \mathbb{R}^{\underline{I}_P \times \underline{I}_P}$ and a diagonal tensor $\mathcal{D} \in \mathbb{R}^{\underline{I}_P \times \underline{I}_P}$, such that:

$$\mathcal{A} = \mathcal{U} \star_P \mathcal{D} \star_P \mathcal{U}^T, \qquad [3.198]$$

where the diagonal tensor $\mathcal{D}$ is defined as in [3.20]:

$$d_{\underline{i}_P, \underline{j}_P} = \begin{cases} \sigma_{\overline{i_1 \cdots i_P}} & \text{if } i_p = j_p, \ \forall p \in \langle P \rangle \\ 0 & \text{otherwise} \end{cases} \qquad [3.199]$$

This factorization is called the eigendecomposition of $\mathcal{A}$. It is expressed in terms of the tensors $(\mathcal{U}, \mathcal{D})$ associated with the matrices $(\mathbf{U}, \mathbf{D})$ of the eigendecomposition of the unfolding $\mathbf{A}_{I \times I}$. It constitutes an extension of the eigendecomposition of a fourth-order tensor introduced by Brazell *et al.* (2013).

PROOF.– Since the tensor $\mathcal{A}$ is symmetric, its unfolding $\mathbf{A}_{I \times I}$ is a symmetric matrix. This matrix therefore admits an eigendecomposition of the form:

$$\mathbf{A}_{I \times I} = \mathbf{U} \mathbf{D} \mathbf{U}^T, \qquad [3.200]$$

where $\mathbf{U} \in \mathbb{R}^{I \times I}$ is orthogonal and $\mathbf{D} \in \mathbb{R}^{I \times I} = \text{diag}(\sigma_1 \cdots \sigma_I)$ is diagonal, with $I = \prod_{p=1}^{P} I_p$. Using the inverse isomorphism f^{-1} and based on the property [3.193], it is easy to deduce [3.198] from [3.200], with $\mathcal{U} = f^{-1}(\mathbf{U})$ and $\mathcal{D} = f^{-1}(\mathbf{D})$.

Moreover, using the property [3.193] and the relation [3.177] once again, we conclude that the tensor $\mathcal{U}$ is itself orthogonal. Indeed, from $\mathbf{U} \mathbf{U}^T = \mathbf{I}_I$, we deduce:

$$f^{-1}(\mathbf{U} \mathbf{U}^T) = f^{-1}(\mathbf{U}) \star_P f^{-1}(\mathbf{U}^T) = \mathcal{U} \star_P \mathcal{U}^T = f^{-1}(\mathbf{I}_I) = \mathcal{I}_{2P}. \quad [3.201]$$

Similarly, we can show that $\mathcal{U}^T \star_P \mathcal{U} = \mathcal{I}_{2P}$, which implies that the tensor $\mathcal{U}$ is orthogonal. $\qquad \square$

3.15.2. *SVD decomposition of a rectangular tensor*

PROPOSITION 3.75.– Any rectangular tensor $\mathcal{X} \in \mathbb{C}^{\underline{I}_P \times \underline{J}_N}$ whose matrix unfolding $\mathbf{X}_{I \times J}$ is of rank R admits an SVD of the following form (Liang and Zheng 2018; Behera *et al.* 2019):

$$\mathcal{X} = \mathcal{U} \star_P \mathcal{D} \star_N \mathcal{V}^H, \tag{3.202}$$

where $\mathcal{U} \in \mathbb{C}^{\underline{I}_P \times \underline{I}_P}$ and $\mathcal{V} \in \mathbb{C}^{\underline{J}_N \times \underline{J}_N}$ are unitary tensors as defined in Table 3.16, and $\mathcal{D} \in \mathbb{C}^{\underline{I}_P \times \underline{J}_N}$ is a diagonal tensor as defined in [3.20]:

$$(\mathcal{D})_{\underline{i}_P, \underline{j}_N} = \begin{cases} \sigma_r > 0 & \text{if} \quad \overline{i_1 \cdots i_P} = \overline{j_1 \cdots j_N} = r \in \langle R \rangle \\ 0 & \text{otherwise} \end{cases}. \tag{3.203}$$

PROOF.– If we assume that the matrix unfolding $\mathbf{X}_{I \times J}$ is of rank R, with $I = \prod_{p=1}^{P} I_p$ and $J = \prod_{n=1}^{N} J_n$, its SVD can be written as follows:

$$\mathbf{X}_{I \times J} = \mathbf{U}\mathbf{D}\mathbf{V}^H, \tag{3.204}$$

where $\mathbf{U} \in \mathbb{C}^{I \times I}$ and $\mathbf{V} \in \mathbb{C}^{J \times J}$ are unitary matrices, and $\mathbf{D} \in \mathbb{C}^{I \times J}$ is a pseudo-diagonal matrix such that:

$$(\mathbf{D})_{\overline{i_1, \cdots, i_P}, \overline{j_1, \cdots, j_N}} = \begin{cases} \sigma_r > 0 & \text{if} \quad \overline{i_1 \cdots i_P} = \overline{j_1 \cdots j_N} = r \in \langle R \rangle \\ 0 & \text{otherwise} \end{cases}.$$

$$\tag{3.205}$$

Using the property [3.193], it is easy to deduce the expression [3.202] of the SVD of the tensor $\mathcal{X}$ from the SVD [3.204] of $\mathbf{X}_{I \times J}$. Furthermore, proceeding in the same way as in the previous section, but replacing the operation of transposition with that of transconjugation, we can show that the tensors $\mathcal{U}$ and $\mathcal{V}$ are unitary, i.e.:

$$\mathcal{U} \star_P \mathcal{U}^H = \mathcal{U}^H \star_P \mathcal{U} = \mathcal{I}_{2P} \tag{3.206}$$

$$\mathcal{V} \star_N \mathcal{V}^H = \mathcal{V}^H \star_N \mathcal{V} = \mathcal{I}_{2N}. \tag{3.207}$$

This completes the proof of the proposition. $\qquad\qquad\square$

REMARK 3.76.– The SVD decomposition presented above is a generalization of the SVD proposed by Sun *et al.* (2016) for tensors $\mathcal{X} \in \mathbb{C}^{\underline{I}_P \times \underline{J}_P}$, i.e. corresponding to $N = P$.

3.15.3. *Connection between SVD and HOSVD*

In this section, we will establish the link between the SVD [3.202] of the tensor $\mathcal{X} \in \mathbb{C}^{\underline{I}_P \times \underline{J}_N}$ and its HOSVD decomposition, deduced from [5.15]:

$$\mathcal{X} = \mathcal{G} \overset{P}{\underset{p=1}{\times}} \mathbf{U}^{(p)} \overset{N}{\underset{n=1}{\times}} \mathbf{U}^{(p+n)}, \tag{3.208}$$

where $\mathbf{U}^{(p)} \in \mathbb{C}^{I_p \times I_p}$, for $p \in \langle P \rangle$ and $\mathbf{U}^{(p+n)} \in \mathbb{C}^{J_n \times J_n}$, for $n \in \langle N \rangle$, are unitary matrices, and $\mathcal{G} \in \mathbb{C}^{\underline{I}_P \times \underline{J}_N}$ is the core tensor. This connection is the object of the next proposition, which generalizes the relationship established by Cui *et al.* (2015) for a square tensor to the case of a rectangular tensor.

PROPOSITION 3.77.– Consider the tensor $\mathcal{X} \in \mathbb{C}^{\underline{I}_P \times \underline{J}_N}$. Its SVD [3.202] can be rewritten using the matrix unfolding defined in [3.191], with $I = \prod_{p=1}^{P} I_p$ and $J = \prod_{n=1}^{N} J_n$:

$$\mathbf{X} \triangleq \mathbf{X}_{I \times J} = f(\mathcal{X}) = \mathbf{U} \mathbf{\Sigma} \mathbf{V}^H, \qquad [3.209]$$

where the unitary matrices $\mathbf{U} = f(\mathcal{U})$ and $\mathbf{V} = f(\mathcal{V})$ and the pseudo-diagonal matrix $\mathbf{\Sigma} = f(\mathcal{D})$ of singular values are directly related to the matrices of the HOSVD [3.208] of $\mathcal{X}$, as well as to the matrices of the EVDs of $\mathbf{G}\mathbf{G}^H$ and $\mathbf{G}^H\mathbf{G}$, where $\mathbf{G} \triangleq \mathbf{G}_{I \times J}$ is the matrix unfolding of the core tensor $\mathcal{G}$, by the following relations:

$$\mathbf{U} = \left(\overset{P}{\underset{p=1}{\otimes}} \mathbf{U}^{(p)} \right) \mathbf{Q} \qquad [3.210]$$

$$\mathbf{V} = \left(\overset{N}{\underset{n=1}{\otimes}} (\mathbf{U}^{(p+n)})^* \right) \mathbf{P} \qquad [3.211]$$

$$\mathbf{G}\mathbf{G}^H = \mathbf{Q}\mathbf{D}_I\mathbf{Q}^H \; ; \; \mathbf{G}^H\mathbf{G} = \mathbf{P}\mathbf{D}_J\mathbf{P}^H \qquad [3.212]$$

$$(\mathbf{\Sigma})_{\overline{i_1, \cdots, i_P}, \overline{j_1, \cdots, j_N}} = \begin{cases} \sqrt{\lambda_r} & \text{if } \overline{i_1 \cdots i_P} = \overline{j_1 \cdots j_N} = r \in \langle R \rangle \\ 0 & \text{otherwise} \end{cases}$$
$$[3.213]$$

where the λ_j are the non-zero eigenvalues of $\mathbf{G}\mathbf{G}^H$ and $\mathbf{G}^H\mathbf{G}$, and R is the rank of $\mathbf{G}$.

PROOF.– Let $\mathbf{X} \triangleq \mathbf{X}_{I \times J}$ be the unfolding of $\mathcal{X}$, as defined in [3.191]. Using the general matricization formula [5.5] for the Tucker model, this matrix unfolding is given by:

$$\mathbf{X} = \left(\overset{P}{\underset{p=1}{\otimes}} \mathbf{U}^{(p)} \right) \mathbf{G} \left(\overset{N}{\underset{n=1}{\otimes}} (\mathbf{U}^{(p+n)})^T \right), \qquad [3.214]$$

where $\mathbf{G} \triangleq \mathbf{G}_{I \times J}$. Taking into account the fact that matrices $\mathbf{U}^{(p+n)}$ are unitary for $n \in \langle N \rangle$, i.e. $(\mathbf{U}^{(p+n)})^T (\mathbf{U}^{(p+n)})^* = \mathbf{I}_{J_n}$, we deduce that:

$$\mathbf{X}\mathbf{X}^H = \left(\overset{P}{\underset{p=1}{\otimes}} \mathbf{U}^{(p)} \right) \mathbf{G}\mathbf{G}^H \left(\overset{P}{\underset{p=1}{\otimes}} \mathbf{U}^{(p)} \right)^H \in \mathbb{C}^{I \times I}. \qquad [3.215]$$

Since the matrix $\mathbf{G}\mathbf{G}^H$ is Hermitian positive semi-definite, it admits the following EVD:

$$\mathbf{G}\mathbf{G}^H = \mathbf{Q}\mathbf{D}_I\mathbf{Q}^H, \qquad [3.216]$$

where $\mathbf{Q} \in \mathbb{C}^{I \times I}$ is unitary and $\mathbf{D}_I \in \mathbb{C}^{I \times I}$ is diagonal positive semi-definite.

After replacing $\mathbf{GG}^H$ with its EVD, equation [3.215] becomes:

$$\mathbf{XX}^H = \left(\overset{P}{\underset{p=1}{\otimes}} \mathbf{U}^{(p)} \right) \mathbf{QD}_I \mathbf{Q}^H \left(\overset{P}{\underset{p=1}{\otimes}} \mathbf{U}^{(p)} \right)^H = \mathbf{UD}_I \mathbf{U}^H \qquad [3.217]$$

$$\mathbf{U} \triangleq \left(\overset{P}{\underset{p=1}{\otimes}} \mathbf{U}^{(p)} \right) \mathbf{Q}. \qquad [3.218]$$

Note that $\mathbf{U}$ is a unitary matrix because the matrices $\mathbf{U}^{(p)}$ and $\mathbf{Q}$ are themselves unitary, which implies:

$$\mathbf{UU}^H = \left(\overset{P}{\underset{p=1}{\otimes}} \mathbf{U}^{(p)} \right) \mathbf{QQ}^H \left(\overset{P}{\underset{p=1}{\otimes}} \mathbf{U}^{(p)} \right)^H$$

$$= \overset{P}{\underset{p=1}{\otimes}} \mathbf{U}^{(p)} (\mathbf{U}^{(p)})^H = \mathbf{I}_I.$$

Similarly, we have $\mathbf{U}^H \mathbf{U} = \mathbf{I}_I$. Proceeding in the same way, we have:

$$\mathbf{X}^H \mathbf{X} = \left(\overset{N}{\underset{n=1}{\otimes}} (\mathbf{U}^{(p+n)})^* \right) \mathbf{G}^H \mathbf{G} \left(\overset{N}{\underset{n=1}{\otimes}} (\mathbf{U}^{(p+n)})^T \right) \in \mathbb{C}^{J \times J}. \qquad [3.219]$$

Defining the EVD of $\mathbf{G}^H \mathbf{G}$ as: $\mathbf{G}^H \mathbf{G} = \mathbf{PD}_J \mathbf{P}^H$, with $\mathbf{P} \in \mathbb{C}^{J \times J}$ unitary and $\mathbf{D}_J \in \mathbb{C}^{J \times J}$ positive semi-definite and diagonal, equation [3.219] becomes:

$$\mathbf{X}^H \mathbf{X} = \mathbf{VD}_J \mathbf{V}^H \qquad [3.220]$$

$$\mathbf{V} \triangleq \left(\overset{N}{\underset{n=1}{\otimes}} (\mathbf{U}^{(p+n)})^* \right) \mathbf{P}. \qquad [3.221]$$

Like $\mathbf{U}$, the matrix $\mathbf{V}$ is unitary. Furthermore, as we saw in Volume 1 (Favier 2019), the matrices $\mathbf{GG}^H$ and $\mathbf{G}^H \mathbf{G}$ have the same (positive) non-zero eigenvalues. Suppose that $I \geq J$, and R is the rank of $\mathbf{X}$, so that $\text{r}(\mathbf{X}) = R \leq \min(I, J) = J$, and write $\mathbf{D} \in \mathbb{C}^{I \times J}$ for the diagonal matrix (if $I = J$ or pseudo-diagonal if $I > J$), whose diagonal (or pseudo-diagonal) terms satisfy:

$$d_{jj} = \begin{cases} \lambda_j > 0 & \text{for} \quad j \in \langle R \rangle \\ 0 & \text{otherwise} \end{cases} \qquad [3.222]$$

where the λ_j are the non-zero eigenvalues of $\mathbf{GG}^H$ and $\mathbf{G}^H \mathbf{G}$, i.e. the squares of the singular values (σ_j) of $\mathbf{G}$.

The SVD of the matrix unfolding $\mathbf{X}$ is then given by:

$$\mathbf{X} = \mathbf{U\Sigma V}^H, \qquad [3.223]$$

where the unitary matrices $\mathbf{U}$ and $\mathbf{V}$ are defined in [3.218] and [3.221], respectively, and $\mathbf{\Sigma}$ is defined in [3.213]. This completes the proof of the proposition. $\qquad \square$

3.15.4. *Full-rank decomposition*

Before presenting the full-rank decomposition for tensors, we will recall an elementary result from matrix algebra, concerning the factorization of any matrix as the product of a matrix of full column rank with a matrix of full row rank. This is called the full-rank factorization.

PROPOSITION 3.78.– Any matrix $\mathbf{A} \in \mathbb{K}^{I \times J}$ of rank R can be expressed as the product of a matrix $\mathbf{F} \in \mathbb{K}^{I \times R}$ of full column rank with a matrix $\mathbf{G} \in \mathbb{K}^{R \times J}$ of full row rank:

$$\mathbf{A} = \mathbf{FG}. \tag{3.224}$$

PROOF.– Let $\mathbf{F} \in \mathbb{K}^{I \times R}$ be a matrix whose columns form a basis of the column space $\mathcal{C}(\mathbf{A})$ of $\mathbf{A}$. Each column $j \in \langle J \rangle$ of $\mathbf{A}$ can then be uniquely written as a linear combination of the R columns of $\mathbf{F}$. Hence, by denoting the matrix by $\mathbf{G} \in \mathbb{K}^{R \times J}$, such that each column j contains the R coefficients of the linear combination for the jth column of $\mathbf{A}$, we obtain the factorization [3.224]. Furthermore, since the matrix $\mathbf{F}$ has full column rank, we have:

$$r(\mathbf{A}) = r(\mathbf{FG}) = r(\mathbf{G}) = R, \tag{3.225}$$

which implies that $\mathbf{G}$ has full row rank.   □

Full-rank matrix decomposition can be extended to tensors $\mathcal{A} \in \mathbb{K}^{\underline{I}_P \times \underline{J}_N}$ using a new definition of the rank of the tensor $\mathcal{A}$ as the rank of its matrix unfolding $\mathbf{A}_{I \times J}$, i.e. $r(\mathcal{A}) = r[f(\mathcal{A})] = r(\mathbf{A}_{I \times J})$. The tensor $\mathcal{A}$ is then said to have full column (respectively, row) rank if its unfolding $\mathbf{A}_{I \times J}$ has full column (respectively, row) rank itself (Liang and Zheng 2018). This decomposition is the object of the next proposition, which uses the Einstein product.

PROPOSITION 3.79.– Consider a tensor $\mathcal{A} \in \mathbb{K}^{\underline{I}_P \times \underline{J}_N}$ and its matrix unfolding $\mathbf{A}_{I \times J}$, as defined in [3.191]. If we assume that $r(\mathbf{A}_{I \times J}) = K = \prod_{r=1}^{R} K_r$, and consider the full-rank decomposition of $\mathbf{A}_{I \times J}$ as in [3.224], with $I = \prod_{p=1}^{P} I_p$ and $J = \prod_{n=1}^{N} J_n$:

$$\mathbf{A}_{I \times J} = \mathbf{FG}, \tag{3.226}$$

with $\mathbf{F} \in \mathbb{K}^{I \times K}$ and $\mathbf{G} \in \mathbb{K}^{K \times J}$, then the full-rank decomposition of the tensor $\mathcal{A}$ is given by Liang and Zheng (2018) and Behera *et al.* (2020):

$$\mathcal{A} = \mathcal{F} \star_R \mathcal{G}, \tag{3.227}$$

where $\mathcal{F} = f^{-1}(\mathbf{F}) \in \mathbb{K}^{\underline{I}_P \times \underline{K}_R}$ is a tensor of full column rank and $\mathcal{G} = f^{-1}(\mathbf{G}) \in \mathbb{K}^{\underline{K}_R \times \underline{J}_N}$ is a tensor of full row rank.

PROOF.– This result can be shown using equation [3.193]:

$$f^{-1}(\mathbf{FG}) = f^{-1}(\mathbf{F}) \star_R f^{-1}(\mathbf{G}) = \mathcal{F} \star_R \mathcal{G}.$$

Furthermore, since the matrices $\mathbf{F}$ and $\mathbf{G}$ are the factors of the full-rank decomposition [3.226] of $\mathbf{A}_{I \times J}$, they have full column rank and full row rank, respectively. Hence, given that these matrices are the matrix unfoldings of the tensors $\mathcal{F}$ and $\mathcal{G}$, we may conclude that these tensors also have full column rank and full row rank, respectively. □

In the next proposition, we derive a formula for computing the Moore–Penrose pseudo-inverse of the tensor $\mathcal{A} \in \mathbb{K}^{\underline{I}_P \times \underline{J}_N}$, based on the full-rank decomposition [3.226] of $\mathbf{A}_{I \times J}$. Our proof of this formula is different to the one proposed by Liang and Zheng (2018).

PROPOSITION 3.80.– Consider a tensor $\mathcal{A} \in \mathbb{K}^{\underline{I}_P \times \underline{J}_N}$ whose unfolding $\mathbf{A}_{I \times J}$ is assumed to have full column rank. Its Moore–Penrose pseudo-inverse, defined in [3.197], can be computed using the factors $\mathcal{F} \in \mathbb{K}^{\underline{I}_P \times \underline{K}_R}$ and $\mathcal{G} \in \mathbb{K}^{\underline{K}_R \times \underline{J}_N}$ in its full-rank factorization [3.227], thanks to the following formula:

$$\mathcal{A}^{\dagger} = \mathcal{G}^H \star_R (\mathcal{F}^H \star_P \mathcal{A} \star_N \mathcal{G}^H)^{-1} \star_R \mathcal{F}^H. \tag{3.228}$$

PROOF.– From [3.197] and [3.226], we have:

$$\mathcal{A}^{\dagger} = f^{-1}(\mathbf{A}_{I \times J}^{\dagger}) = f^{-1}[(\mathbf{FG})^{\dagger}]. \tag{3.229}$$

Since the matrices $\mathbf{F}$ and $\mathbf{G}$ have full column rank and full row rank, respectively, the properties (xv) and (xvi) in Table 3.20 give us:

$$(\mathbf{FG})^{\dagger} = \mathbf{G}^{\dagger}\mathbf{F}^{\dagger} = \mathbf{G}^H(\mathbf{GG}^H)^{-1}(\mathbf{F}^H\mathbf{F})^{-1}\mathbf{F}^H, \tag{3.230}$$

where $\mathbf{F}^H\mathbf{F}$ and $\mathbf{GG}^H$ have full rank, and are therefore invertible. From [3.226], we deduce:

$$\mathbf{F}^H \mathbf{A}_{I \times J} \mathbf{G}^H = (\mathbf{F}^H\mathbf{F})(\mathbf{GG}^H) \Rightarrow (\mathbf{GG}^H)^{-1}(\mathbf{F}^H\mathbf{F})^{-1} = (\mathbf{F}^H \mathbf{A}_{I \times J} \mathbf{G}^H)^{-1}.$$

Substituting this last expression into [3.230], then into [3.229], we obtain:

$$\mathcal{A}^{\dagger} = f^{-1}[\mathbf{G}^H(\mathbf{F}^H \mathbf{A}_{I \times J} \mathbf{G}^H)^{-1}\mathbf{F}^H]. \tag{3.231}$$

Since the matrix $\mathbf{F}^H \mathbf{A}_{I \times J} \mathbf{G}^H$ is invertible and of size $R \times R$, using the properties [3.183], [3.193] and [3.194] leads to:

$$f^{-1}[(\mathbf{F}^H \mathbf{A}_{I \times J} \mathbf{G}^H)^{-1}] = (\mathcal{F}^H \star_P \mathcal{A} \star_N \mathcal{G}^H)^{-1}. \tag{3.232}$$

Hence, from [3.231] and [3.232], it is easy to deduce the formula [3.228]. □

3.16. Inner product, Frobenius norm and trace of a tensor

3.16.1. *Inner product of two tensors*

In the case of two tensors $\mathcal{A}, \mathcal{B} \in \mathbb{R}^{I_N}$ of order N and same size, a contraction on every pair of indices is equivalent to the $\star_N$ product of the two tensors, which corresponds to their inner product:

$$\langle \mathcal{A}, \mathcal{B} \rangle = \mathcal{A} \star_N \mathcal{B} = \sum_{i_1=1}^{I_1} \cdots \sum_{i_N=1}^{I_N} a_{i_1, \cdots, i_N} b_{i_1, \cdots, i_N} = \sum_{\mathbf{i}_N=\mathbf{1}}^{\mathbf{I}_N} a_{\mathbf{i}_N} b_{\mathbf{i}_N}. \qquad [3.233]$$

Using the index convention, the inner product can be concisely written as:

$$\langle \mathcal{A}, \mathcal{B} \rangle = a_{\mathbf{i}_N} b_{\mathbf{i}_N}. \qquad [3.234]$$

We can also write it using the Euclidean inner product of vectorized forms of $\mathcal{A}$ and $\mathcal{B}$:

$$\langle \mathcal{A}, \mathcal{B} \rangle = \text{vec}^T(\mathcal{B}) \, \text{vec}(\mathcal{A}) = \text{vec}^T(\mathcal{A}) \, \text{vec}(\mathcal{B}), \qquad [3.235]$$

where $\text{vec}(\mathcal{A})$ and $\text{vec}(\mathcal{B})$ are vectorizations of $\mathcal{A}$ and $\mathcal{B}$ associated with the same mode combination.

In the case of complex-valued tensors, the Hermitian inner product is obtained by conjugating $b_{i_1, \cdots, i_N}$ in the definition [3.233], i.e.:

$$\langle \mathcal{A}, \mathcal{B} \rangle = a_{\mathbf{i}_N} b_{\mathbf{i}_N}^* = \text{vec}^H(\mathcal{B}) \, \text{vec}(\mathcal{A}). \qquad [3.236]$$

PROPOSITION 3.81.– For $\mathcal{A} \in \mathbb{R}^{I_P}$ and $\mathcal{B} = \overset{P}{\underset{p=1}{\circ}} \mathbf{x}^{(p)} \in \mathbb{R}^{I_P}$, a rank-one tensor of order P, their inner product is given by:

$$\langle \mathcal{A}, \overset{P}{\underset{p=1}{\circ}} \mathbf{x}^{(p)} \rangle = a_{\mathbf{i}_P} \prod_{p=1}^{P} x_{i_p}^{(p)} = \mathcal{A} \overset{P}{\underset{p=1}{\times}} \mathbf{x}^{(P)T}. \qquad [3.237]$$

The inner product of the tensor $\mathcal{A}$ with a rank-one tensor is therefore equal to the multiple mode-p product of $\mathcal{A}$, with P vectors that define the rank-one tensor.

PROPOSITION 3.82.– Let $\mathcal{A} \in \mathbb{R}^{I_P}$ and $\mathbf{x}^{(p)} \in \mathbb{R}^{I_p}$ for $p \in \langle P \rangle$. The inner product of the vector $\mathbf{y} = \mathcal{A} \overset{P}{\underset{\substack{q=1 \\ q \neq p}}{\times}} \mathbf{x}^{(q)T} \in \mathbb{R}^{I_p}$ with $\mathbf{x}^{(p)}$ is equivalent to the multiple mode-p product of $\mathcal{A}$ with the P vectors $\mathbf{x}^{(p)}$, and therefore to the inner product of $\mathcal{A}$ with the rank-one tensor $\overset{P}{\underset{p=1}{\circ}} \mathbf{x}^{(P)}$, by [3.237]:

$$\langle \mathcal{A} \overset{P}{\underset{\substack{q=1 \\ q \neq p}}{\times}} \mathbf{x}^{(q)T}, \mathbf{x}^{(p)} \rangle = \mathcal{A} \overset{P}{\underset{p=1}{\times}} \mathbf{x}^{(P)T} = \langle \mathcal{A}, \overset{P}{\underset{p=1}{\circ}} \mathbf{x}^{(p)} \rangle. \qquad [3.238]$$

PROOF.– Since the i_pth component of the vector $\mathbf{y} = \mathcal{A} \underset{\substack{q=1 \\ q \neq p}}{\overset{P}{\times}} \mathbf{x}^{(q)}{}^{T}$ is:

$$y_{i_p} = \sum_{i_1=1}^{I_1} \cdots \sum_{i_{p-1}=1}^{I_{p-1}} \sum_{i_{p+1}=1}^{I_{p+1}} \cdots \sum_{i_P=1}^{I_P} a_{i_1,\cdots,i_p,\cdots,i_P} \prod_{\substack{q=1 \\ q \neq p}}^{P} x_{i_q}^{(q)}, \qquad [3.239]$$

we have:

$$\langle \mathbf{y}, \mathbf{x}^{(p)} \rangle = \sum_{i_p=1}^{I_p} y_{i_p} x_{i_p}^{(p)} = a_{\underline{\mathbf{i}}_p} \prod_{p=1}^{P} x_{i_p}^{(p)} \qquad [3.240]$$

$$= \mathcal{A} \underset{p=1}{\overset{P}{\times}} \mathbf{x}^{(p)}{}^{T} = \langle \mathcal{A}, \underset{p=1}{\overset{P}{\circ}} \mathbf{x}^{(p)} \rangle, \qquad [3.241]$$

which proves [3.238] $\square$

3.16.2. *Frobenius norm of a tensor*

The Frobenius norm of the tensor $\mathcal{A} \in \mathbb{K}^{I_N}$ is the square root of the inner product of the tensor with itself, i.e.:

$$\|\mathcal{A}\|_F = \langle \mathcal{A}, \mathcal{A} \rangle^{1/2} = \sqrt{\sum_{i_1=1}^{I_1} \cdots \sum_{i_N=1}^{I_N} |a_{i_1,\cdots,i_N}|^2} = \sqrt{\sum_{\underline{\mathbf{i}}_N=\underline{\mathbf{1}}}^{\underline{\mathbf{I}}_N} |a_{\underline{\mathbf{i}}_N}|^2}, \quad [3.242]$$

where $|.|$ represents the absolute value or the modulus, depending on whether $\mathcal{A}$ is real or complex.

Since the Frobenius norm is equal to the square root of the sum of the squares of the absolute value or modulus of all elements of the tensor, it is also equal to the following expressions:

$$\|\mathcal{A}\|_F = \|\text{vec}(\mathcal{A})\|_2 = \|\mathbf{A}_{\mathbb{S}_1;\mathbb{S}_2}\|_F = \|\mathbf{A}_n\|_F \ , \ \forall n \in \langle N \rangle. \qquad [3.243]$$

The Frobenius norm of a tensor can therefore be defined as the Euclidean norm of one of its vectorized forms, or as the Frobenius norm of one of its matrix unfolding $\mathbf{A}_{\mathbb{S}_1;\mathbb{S}_2}$ or mode-n unfolding $\mathbf{A}_n$, defined in [3.39] and [3.41], respectively.

PROPOSITION 3.83.– For a rank-one tensor $\mathcal{A} = \overset{P}{\underset{p=1}{\circ}} \mathbf{u}^{(p)} \in \mathbb{K}^{\underline{I}_P}$ of order P, applying the result [3.243] with the vectorized form [3.117] leads to:

$$\| \overset{P}{\underset{p=1}{\circ}} \mathbf{u}^{(p)} \|_F^2 = \| \mathrm{vec}(\overset{P}{\underset{p=1}{\circ}} \mathbf{u}^{(p)}) \|_2^2 = \| \overset{P}{\underset{p=1}{\otimes}} \mathbf{u}^{(p)} \|_2^2 \qquad [3.244]$$

$$= (\overset{P}{\underset{p=1}{\otimes}} \mathbf{u}^{(p)})^H (\overset{P}{\underset{p=1}{\otimes}} \mathbf{u}^{(p)}) = \overset{P}{\underset{p=1}{\otimes}} (\mathbf{u}^{(p)})^H \mathbf{u}^{(p)} \qquad [3.245]$$

$$= \prod_{p=1}^{P} \| \mathbf{u}^{(p)} \|_2^2. \qquad [3.246]$$

From this relation, we deduce that the Frobenius norm of an outer product of P vectors is equal to the product of the Hermitian (respectively, Euclidean) norms of these vectors in the complex (respectively, real) case.

Like for vectors and matrices, we can define other norms for tensors of order greater than two, such as the Höder or l_p norm:

$$\| \mathcal{A} \|_p = \Big(\sum_{\underline{i}_N=\underline{1}}^{\underline{I}_N} |a_{\underline{i}_N}|^p \Big)^{1/p} , \ 1 \leq p \leq \infty, \qquad [3.247]$$

which has the norms l_1 and l_∞ as special cases:

$$\| \mathcal{A} \|_1 = \sum_{\underline{i}_N=\underline{1}}^{\underline{I}_N} |a_{\underline{i}_N}| \qquad [3.248]$$

$$\| \mathcal{A} \|_\infty = \max_{i_1,\cdots,i_N} |a_{i_1,\cdots,i_N}|. \qquad [3.249]$$

The Frobenius norm of a complex (respectively, real) tensor is preserved when the tensor is multiplied mode-p with a unitary (respectively, orthogonal) matrix, as shown in the next proposition.

PROPOSITION 3.84.– Let $\mathcal{X} \in \mathbb{C}^{\underline{I}_P}$, and let $\mathbf{U}^{(p)} \in \mathbb{C}^{I_p \times I_p}$ be a unitary matrix, i.e. $\mathbf{U}^{(p)}(\mathbf{U}^{(p)})^H = (\mathbf{U}^{(p)})^H \mathbf{U}^{(p)} = \mathbf{I}_{I_p}$. The mode-$p$ product of $\mathcal{X}$ with $\mathbf{U}^{(p)}$ does not change the Frobenius norm of $\mathcal{X}$.

PROOF.– Let $\mathcal{Y} = \mathcal{X} \times_p \mathbf{U}^{(p)}$. Using the results [3.243] and [3.120], as well as the expression of the Frobenius norm of a matrix in terms of the trace ($\| \mathbf{A} \|_F = \mathrm{tr}(\mathbf{A}^H \mathbf{A})$), we have:

$$\| \mathcal{Y} \|_F = \| \mathbf{Y}_p \|_F = \mathrm{tr}(\mathbf{Y}_p^H \mathbf{Y}_p) = \mathrm{tr}(\mathbf{X}_p^H (\mathbf{U}^{(p)})^H \mathbf{U}^{(p)} \mathbf{X}_p \|_F, \qquad [3.250]$$

and taking into account the hypothesis that $\mathbf{U}^{(p)}$ is unitary, we deduce that:

$$\|\mathcal{Y}\|_F = \operatorname{tr}(\mathbf{X}_p^H \mathbf{X}_p) = \|\mathbf{X}_p\|_F = \|\mathcal{X}\|_F, \qquad [3.251]$$

which proves the stated property.   □

Like in the matrix case, this property means that any unitary transformation applied to the mode-p subspace of a tensor does not change the norm of the tensor. It is easy to prove that this result may be generalized to the case of a multiple mode-p product, i.e. for $\mathcal{Y} = \mathcal{X} \overset{P}{\underset{p=1}{\times}} \mathbf{U}^{(p)}$, where the matrices $\mathbf{U}^{(p)}$ are unitary. We then have:

$$\|\mathcal{X} \overset{P}{\underset{p=1}{\times}} \mathbf{U}^{(p)}\|_F = \|\mathcal{X}\|_F. \qquad [3.252]$$

As we will see in section 5.2.1, the equation $\mathcal{Y} = \mathcal{X} \overset{P}{\underset{p=1}{\times}} \mathbf{U}^{(p)}$ corresponds to a Tucker model of the tensor $\mathcal{Y}$, with unitary factors $\mathbf{U}^{(p)}$, where the tensor $\mathcal{X}$ is called the core tensor. Equation [3.252] therefore implies that the Frobenius norm of a tensor represented using a Tucker model with unitary factors is equal to the Frobenius norm of the core tensor.

REMARK 3.85.– In the case of a real tensor $\mathcal{X} \in \mathbb{R}^{\underline{I}_P}$, proposition 3.84 holds with an orthogonal matrix $\mathbf{U}^{(p)} \in \mathbb{R}^{I_p \times I_p}$.

Below, we give a formula for computing the Frobenius norm of an Einstein product of two tensors.

PROPOSITION 3.86.– Consider the Einstein product $\mathcal{C} = \mathcal{A} \star_N \mathcal{B} \in \mathbb{R}^{\underline{I}_P \times \underline{K}_Q}$ of the tensors $\mathcal{A} \in \mathbb{R}^{\underline{I}_P \times \underline{J}_N}$ and $\mathcal{B} \in \mathbb{R}^{\underline{J}_N \times \underline{K}_Q}$. By [3.243], the Frobenius norm of $\mathcal{C}$ is equal to the Frobenius norm of its matrix unfolding $\mathbf{C}_{I \times K}$, where $I \triangleq \prod_{p=1}^{P} I_p$, and $K \triangleq \prod_{q=1}^{Q} K_q$. It can therefore be computed using the Frobenius norm of the product of the matrix unfoldings of $\mathcal{A}$ and $\mathcal{B}$ as follows:

$$\|\mathcal{C}\|_F = \|\mathcal{A} \star_N \mathcal{B}\|_F = \|\mathbf{A}_{I \times J} \mathbf{B}_{J \times K}\|_F, \qquad [3.253]$$

with $J \triangleq \prod_{n=1}^{N} J_n$.

3.16.3. *Trace of a tensor*

The next two propositions show that the traces of a hypercubic tensor of order N and a square tensor of order $2N$ can be obtained as the inner products of these tensors with identity tensors.

PROPOSITION 3.87.– Let $\mathcal{A} \in \mathbb{K}^{[N;I]}$ be a hypercubic tensor of order N and dimensions I. Its inner product with the identity tensor of order N gives the trace of $\mathcal{A}$, denoted $\mathrm{tr}(\mathcal{A})$:

$$\langle \mathcal{A}, \mathcal{I}_{N,I} \rangle = \mathcal{A} \star_N \mathcal{I}_{N,I} = \sum_{i_1,\cdots,i_N=1}^{I} a_{i_1,\cdots,i_N} \delta_{i_1,\cdots,i_N} \qquad [3.254]$$

$$= a_{\underline{i}_N} \delta_{\underline{i}_N} = \sum_{i=1}^{I} a_{i,\cdots,i} = \mathrm{tr}(\mathcal{A}). \qquad [3.255]$$

REMARK 3.88.– In the matrix case, for $\mathbf{A} \in \mathbb{K}^{I \times I}$, we have:

$$\langle \mathbf{A}, \mathbf{I} \rangle = \sum_{i=1}^{I} a_{i,i} = \mathrm{tr}(\mathbf{A}). \qquad [3.256]$$

PROPOSITION 3.89.– Let $\mathcal{A} \in \mathbb{K}^{\underline{I}_N \times \underline{I}_N}$ be a square tensor of order $2N$ and size $\underline{I}_N \times \underline{I}_N$. Its trace is given by the inner product of $\mathcal{A}$ with the identity block tensor of order $2N$, defined in [3.22]–[3.23]:

$$\mathrm{tr}(\mathcal{A}) = \sum_{\underline{i}_N=\underline{1}}^{\mathbf{I}_N} a_{\underline{i}_N,\underline{i}_N} = \langle \mathcal{A}, \mathcal{J}_{2N} \rangle. \qquad [3.257]$$

Indeed, we have:

$$\langle \mathcal{A}, \mathcal{J}_{2N} \rangle = \mathcal{A} \star_{2N} \mathcal{J}_{2N} = \sum_{\underline{i}_N=\underline{1}}^{\mathbf{I}_N} \sum_{\underline{j}_N=\underline{1}}^{\mathbf{I}_N} a_{\underline{i}_N,\underline{j}_N} \prod_{n=1}^{N} \delta_{i_n,j_n} \qquad [3.258]$$

$$= a_{\underline{i}_N,\underline{i}_N} = \sum_{i_1=1}^{I_1} \cdots \sum_{i_N=1}^{I_N} a_{i_1,\cdots,i_N,i_1,\cdots,i_N} = \mathrm{tr}(\mathcal{A}). \qquad [3.259]$$

3.17. Tensor systems and homogeneous polynomials

The tensor products $\times_n$ and $\star_N$ can be used to define (linear and multilinear) tensor systems, as well as homogeneous polynomials.

3.17.1. *Multilinear systems based on the mode-n product*

The bilinear form ϕ, with the associated matrix $\mathbf{A} \in \mathbb{R}^{I \times J}$, such that:

$$\phi : \mathbb{R}^I \times \mathbb{R}^J \ni (\mathbf{x}, \mathbf{y}) \mapsto \phi(\mathbf{x}, \mathbf{y}) = \mathbf{x}^T \mathbf{A} \mathbf{y} = b \in \mathbb{R}, \qquad [3.260]$$

can be written as follows using the mode-n product:

$$\mathbf{A} \times_1 \mathbf{x}^T \times_2 \mathbf{y}^T = \sum_{i=1}^{I} \sum_{j=1}^{J} a_{ij} x_i y_j = a_{ij} x_i y_j = b, \qquad [3.261]$$

i.e., as a homogeneous polynomial of degree 2 in the components x_i and y_j, with $i \in \langle I \rangle$ and $j \in \langle J \rangle$, of the vectors $\mathbf{x}$ and $\mathbf{y}$. Since the quantity b is scalar, it is equal to its transpose, and using the properties [2.149] and [2.28] gives us:

$$b = \text{tr}(\mathbf{x}^T \mathbf{A} \mathbf{y}) = \text{tr}(\mathbf{A} \mathbf{y} \mathbf{x}^T)$$

$$= \text{vec}^T(\mathbf{A}^T) \text{vec}(\mathbf{y} \mathbf{x}^T) = \text{vec}^T(\mathbf{A}^T)(\mathbf{x} \otimes \mathbf{y}). \qquad [3.262]$$

This expression clearly shows the bilinear character of the system with respect to the vectors $\mathbf{x}$ and $\mathbf{y}$. Similarly, from equation [3.135], we have:

$$\mathcal{G} \times_1 \mathbf{x}^T \times_2 \mathbf{y}^T \times_3 \mathbf{z}^T = \sum_{i=1}^{I} \sum_{j=1}^{J} \sum_{k=1}^{K} g_{ijk} x_i y_j z_k = g_{ijk} x_i y_j z_k = b. \qquad [3.263]$$

This tensor equation, which is trilinear with respect to the vectors $(\mathbf{x}, \mathbf{y}, \mathbf{z})$, corresponds to a third-order Tucker model, denoted $[\![\mathcal{G}; \mathbf{x}, \mathbf{y}, \mathbf{z}]\!]$, with $\mathcal{G} \in \mathbb{R}^{I \times J \times K}$, $\mathbf{x} \in \mathbb{R}^I, \mathbf{y} \in \mathbb{R}^J, \mathbf{z} \in \mathbb{R}^K$ (see section 5.2.1.6). Using the vectorization formula given in Table 5.1, and replacing $(\mathbf{A}, \mathbf{B}, \mathbf{C})$ with $(\mathbf{x}^T, \mathbf{y}^T, \mathbf{z}^T)$, we can rewrite this trilinear system as:

$$b = \text{vec}^T(\mathcal{G})(\mathbf{x} \otimes \mathbf{y} \otimes \mathbf{z}), \qquad [3.264]$$

with $\text{vec}(\mathcal{G}) \in \mathbb{R}^{IJK}$. This equation generalizes the bilinear system [3.262] to the trilinear case. It should be noted that equation [3.263] is associated with the following trilinear form:

$$\phi : \mathbb{R}^I \times \mathbb{R}^J \times \mathbb{R}^I \ni (\mathbf{x}, \mathbf{y}, \mathbf{z}) \mapsto \phi(\mathbf{x}, \mathbf{y}, \mathbf{z}) = \mathcal{G} \times_1 \mathbf{x}^T \times_2 \mathbf{y}^T \times_3 \mathbf{z}^T \in \mathbb{R}.$$

$$[3.265]$$

The generalization to the case of a P-linear form, with $P \geq 2$, is given by:

$$\phi : \mathop{\times}_{p=1}^{P} \mathbb{R}^{I_p} \ni (\mathbf{x}^{(1)}, \cdots, \mathbf{x}^{(P)}) \mapsto \phi(\mathbf{x}^{(1)}, \cdots, \mathbf{x}^{(P)}) = \mathcal{G} \mathop{\times}_{p=1}^{P} \mathbf{x}^{(p)T}$$

$$= \sum_{i_1=1}^{I_1} \cdots \sum_{i_P=1}^{I_P} g_{i_1, \cdots, i_P} x_{i_1}^{(1)} \cdots x_{i_P}^{(P)} \in \mathbb{R}, \qquad [3.266]$$

where the tensor $\mathcal{G} \in \mathbb{R}^{I_P}$ of order P is called the core tensor in the Tucker model.

The bilinear form [3.261] and the trilinear form [3.263] can be obtained as special cases of the P-linear form [3.266] for $P = 2$ and $P = 3$, respectively.

In the case of a hypercubic tensor $\mathcal{A} \in \mathbb{R}^{[P;I]}$ of order P and dimensions I, the mode-p multiproduct of $\mathcal{A}$ with the vector $\mathbf{x} \in \mathbb{R}^I$ gives a P-linear form that can be written as:

$$\phi : \mathbb{R}^I \ni \mathbf{x} \mapsto \phi(\mathbf{x}) = \mathcal{A} \overset{P}{\underset{p=1}{\times}} \mathbf{x}^T \qquad [3.267]$$

$$= \sum_{i_1, \cdots, i_P = 1}^{I} a_{i_1, \cdots, i_P} x_{i_1} \cdots x_{i_P} = a_{\underline{i}_P} \prod_{p=1}^{P} x_{i_p} \in \mathbb{R}. \qquad [3.268]$$

This P-linear form is often written using the notation $\mathcal{A}\mathbf{x}^P$, which was introduced by Qi (2005). It corresponds to a homogeneous polynomial of degree P in the variables x_i, with $i \in \langle I \rangle$, whose coefficients are the entries $a_{i_1, \cdots, i_P}$ of the tensor $\mathcal{A}$. It can also be written as the inner product of the tensor $\mathcal{A}$ with the rank-one tensor $\mathbf{x}^{\circ P} \triangleq \overset{P}{\underset{p=1}{\circ}} \mathbf{x} = [x_{i_1} \cdots x_{i_P}]$, or as the Euclidean inner product of the vectors $\mathbf{x}$ and $\mathcal{A}\mathbf{x}^{\circ(P-1)}$, i.e.:

$$\mathcal{A}\mathbf{x}^P \triangleq \mathcal{A} \overset{P}{\underset{p=1}{\times}} \mathbf{x}^T = \langle \mathcal{A}, \mathbf{x}^{\circ P} \rangle = \langle \mathcal{A}\mathbf{x}^{\circ(P-1)}, \mathbf{x} \rangle = \mathbf{x}^T(\mathcal{A}\mathbf{x}^{\circ(P-1)}). \qquad [3.269]$$

REMARK 3.90.– In the matrix case ($P = 2$), with $\mathbf{A} \in \mathbb{R}^{I \times I}$, we obtain a homogeneous polynomial of degree 2 in the variables x_i, with $i \in \langle I \rangle$:

$$\phi : \mathbb{R}^I \ni \mathbf{x} \mapsto \phi(\mathbf{x}) = \mathbf{A}\mathbf{x}^2 \triangleq \mathbf{A} \times_1 \mathbf{x}^T \times_2 \mathbf{x}^T = \mathbf{x}^T \mathbf{A}\mathbf{x} = \sum_{i,j=1}^{I} a_{i,j} x_i x_j \in \mathbb{R},$$

i.e., a quadratic form.

REMARK 3.91.– If $\mathcal{A}$ is symmetric and satisfies a PARAFAC decomposition of rank R, i.e. $\mathcal{A} = \sum_{r=1}^{R} \mathbf{u}_r^{\circ P}$, with $\mathbf{u}_r \in \mathbb{K}^I$ (see equation [5.31]), equation [3.269] becomes:

$$\mathcal{A}\mathbf{x}^P = \left(\sum_{r=1}^{R} \mathbf{u}_r^{\circ P} \right) \overset{P}{\underset{p=1}{\times}} \mathbf{x}^T = \sum_{r=1}^{R} (\mathbf{x}^T \mathbf{u}_r)^P.$$

We then obtain a homogeneous polynomial of degree P in I variables x_i, expressed as a sum of powers of linear forms, which is directly connected to the Waring problem.

EXAMPLE 3.92.– An example of this type of nonlinear system is given by a Volterra series, whose homogeneous term of order P can be written as[6]:

$$y_k = \sum_{i_1,\cdots,i_P=1}^{M_P} h_{i_1,\cdots,i_P} u_{k-i_1} \cdots u_{k-i_P}, \qquad [3.270]$$

where u_k and y_k represent the input and output of the system at the sampling instant kT, where T is the sampling period. The tensor $h_{i_1,\cdots,i_P}$ is the Volterra kernel of order P, and M_P is the memory of this kernel.

Writing the kernel of order P as the Pth-order tensor $\mathcal{H}_P \in \mathbb{R}^{M_P \times \cdots \times M_P}$, and defining the input vector $\mathbf{u}_k^T = [u_{k-1}, \cdots, u_{k-M_P}]$, equation [3.270] of the system output at time kT can also be written as:

$$y_k = \mathcal{H}_P \underset{p=1}{\overset{P}{\times}} \mathbf{u}_k^T = \langle \mathcal{H}_P, \mathbf{u}_k^{\circ P} \rangle, \qquad [3.271]$$

which is a homogeneous polynomial of degree P in the components of the input vector $\mathbf{u}_k$. In order to reduce the complexity of this type of model, a PARAFAC decomposition of the symmetrized kernel was exploited by Favier *et al.* (2012a).

More generally, for a hypercubic tensor $\mathcal{A} \in \mathbb{K}^{[P;I]}$ of order P and dimensions I, we define the multiple mode-$(Q+1, \cdots, P)$ product, with $0 \leq Q < P$, of $\mathcal{A}$ with $\mathbf{x}$, as follows:

$$\phi : \mathbb{K}^I \ni \mathbf{x} \mapsto \phi(\mathbf{x}) = \mathcal{A} \underset{p=Q+1}{\overset{P}{\times}} \mathbf{x}^T = \mathcal{A}\mathbf{x}^{P-Q}. \qquad [3.272]$$

We then obtain a tensor $\mathcal{B}$ of order Q and dimensions I such that:

$$b_{i_1,\cdots,i_Q} = \sum_{i_{Q+1},\cdots,i_P=1}^{I} a_{i_1,\cdots,i_Q,i_{Q+1},\cdots,i_P} x_{i_{Q+1}} \cdots x_{i_P}. \qquad [3.273]$$

In particular, for $Q = 1$ and $Q = 2$, we obtain:

– a vector $\mathbf{b} = \mathcal{A} \underset{p=2}{\overset{P}{\times}} \mathbf{x}^T = \mathcal{A}\mathbf{x}^{P-1} \in \mathbb{K}^I$, for which each component is a homogeneous polynomial of degree $P - 1$ in the variables x_i, with $i \in \langle I \rangle$, that can be deduced from [3.268] by fixing the index $i_1 = i$ of the mode-1 of $\mathcal{A}$:

$$\left(\mathcal{A}\mathbf{x}^{P-1}\right)_i = \sum_{i_2,i_3,\cdots,i_P=1}^{I} a_{i,i_2,i_3,\cdots,i_P} x_{i_2} x_{i_3} \cdots x_{i_P} \in \mathbb{K} \text{ , with } i \in \langle I \rangle;$$

$$[3.274]$$

---

6 Here, we assume that the system has a pure delay of one sampling period. If this is not the case, the lower bound on the sum over the indices must be considered 0.

– a matrix $\mathbf{B} = \mathcal{A} \overset{P}{\underset{p=3}{\times}} \mathbf{x}^T = \mathcal{A}\mathbf{x}^{P-2} \in \mathbb{K}^{I \times I}$ for which each element is given by:

$$\left(\mathcal{A}\mathbf{x}^{P-2}\right)_{ij} = \sum_{i_3, i_4, \cdots, i_P = 1}^{I} a_{i,j,i_3,\cdots,i_P} x_{i_3} \cdots x_{i_P} \in \mathbb{K} \,, \text{ with } i, j \in \langle I \rangle.$$

[3.275]

EXAMPLE 3.93.– For a third-order tensor $\mathcal{A} \in \mathbb{C}^{I \times I \times I}$ and $\mathbf{x} \in \mathbb{C}^I$ with $I = 2$, the system of equations ([3.274] becomes $\mathcal{A}\mathbf{x}^2 = b$, or in expanded form:

$$a_{111}x_1^2 + (a_{112} + a_{121})x_1 x_2 + a_{122}x_2^2 = b_1$$

$$a_{211}x_1^2 + (a_{212} + a_{221})x_1 x_2 + a_{222}x_2^2 = b_2.$$

Thus, we obtain a system of two polynomial equations consisting of two homogeneous polynomials of degree two in the variables (x_1, x_2), whereas equation [3.275] can be written as: $\left(\mathcal{A}\mathbf{x}\right)_{ij} = \sum_{k=1}^{2} a_{ijk} x_k$, which corresponds to four linear equations in the variables (x_1, x_2).

In the case of a complex hypercubic tensor $\mathcal{A} \in \mathbb{C}^{[2P;I]}$, which is a square tensor of order $2P$, we define the following complex multilinear form (Fu *et al.* 2018):

$$\phi : \mathbb{C}^I \ni \mathbf{x} \mapsto \phi(\mathbf{x}^*, \mathbf{x}) = \mathcal{A} \overset{P}{\underset{p=1}{\times}} \mathbf{x}^H \overset{2P}{\underset{p=P+1}{\times}} \mathbf{x}^T = a_{\mathbf{i}_{2P}} \prod_{p=1}^{P} x_{i_p}^* \prod_{n=P+1}^{2P} x_{i_n}.$$

[3.276]

This multilinear form can also be written as the Hermitian inner product of the tensor $\mathcal{A}$ with the rank-one tensor $\underbrace{\mathbf{x} \circ \cdots \circ \mathbf{x}}_{P \text{ terms}} \circ \underbrace{\mathbf{x}^* \circ \cdots \circ \mathbf{x}^*}_{P \text{ terms}} = \mathbf{x}^{\circ P} \circ (\mathbf{x}^*)^{\circ P}$:

$$\phi(\mathbf{x}^*, \mathbf{x}) = \langle \mathcal{A}, \mathbf{x}^{\circ P} \circ (\mathbf{x}^*)^{\circ P} \rangle.$$

[3.277]

The multi-products presented above will be used in Chapter 4 to define the notions of eigenvalue and eigenvector of a tensor.

3.17.2. *Tensor systems based on the Einstein product*

We can also use the Einstein product to define multilinear and linear tensor systems. Thus, a bilinear tensor system associated with a bilinear mapping, characterized by the tensor $\mathcal{A} \in \mathbb{R}^{I \times J \times K \times L}$ is defined in Brazell $et\ al.$ (2013) as follows:

$$\phi : \mathbb{R}^{K \times L} \times \mathbb{R}^{I \times J} \ni (\mathbf{X}, \mathbf{Y}) \mapsto \phi(\mathbf{X}, \mathbf{Y}) = \mathcal{A} \star_2 \mathbf{X} \star_2 \mathbf{Y} = b \in \mathbb{R}. \quad [3.278]$$

Similarly, for a bilinear tensor system whose variables are the tensors $(\mathcal{X}, \mathcal{Y})$, and which is characterized by the tensor $\mathcal{A} \in \mathbb{R}^{I \times J \times K \times L \times M \times N}$, we can write:

$$\phi : \mathbb{R}^{M \times N \times P} \times \mathbb{R}^{K \times L \times P} \ni (\mathcal{X}, \mathcal{Y}) \mapsto \phi(\mathcal{X}, \mathcal{Y}) = \mathcal{A} \star_2 \mathcal{X} \star_3 \mathcal{Y} = \mathbf{B} \in \mathbb{R}^{I \times J}.$$

Another example of a tensor system using the Einstein product is given by Sylvester's tensor equation, which can be stated as follows (Behera and Mishra 2017):

$$\mathcal{T} \star_P \mathcal{X} + \mathcal{X} \star_N \mathcal{S} = \mathcal{C}, \quad [3.279]$$

with $\mathcal{T} \in \mathbb{R}^{\underline{I}_P \times \underline{I}_P}$, $\mathcal{S} \in \mathbb{R}^{\underline{J}_N \times \underline{J}_N}$ and $\mathcal{C}, \mathcal{X} \in \mathbb{R}^{\underline{I}_P \times \underline{J}_N}$. This equation generalizes Sylvester's matrix equation [2.285] to the tensor case. It can be rewritten using the Einstein product of block tensors as follows:

$$\begin{bmatrix} \mathcal{T} & \mathcal{I}_{2P} \end{bmatrix} \star_P \begin{bmatrix} \mathcal{X} & \mathcal{O}_{P \times N} \\ \mathcal{O}_{P \times N} & \mathcal{X} \end{bmatrix} \star_N \begin{bmatrix} \mathcal{I}_{2N} \\ \mathcal{S} \end{bmatrix} = \mathcal{C}, \quad [3.280]$$

where $\mathcal{O}_{P \times N}$ is the zero tensor of size $\underline{I}_P \times \underline{J}_N$, and $\mathcal{I}_{2P}$ and $\mathcal{I}_{2N}$ are the identity tensors of size $\underline{I}_P \times \underline{I}_P$ and $\underline{J}_N \times \underline{J}_N$, respectively, as defined in [3.22].

Two other examples of basic tensor equations are as follows:

$$\mathcal{A} \star_N \mathcal{X} = \mathcal{C} \ ; \ \mathcal{X} \star_N \mathcal{B} = \mathcal{D}, \quad [3.281]$$

with $\mathcal{A} \in \mathbb{R}^{\underline{I}_P \times \underline{J}_N}$, $\mathcal{B} \in \mathbb{R}^{\underline{J}_N \times \underline{I}_P}$, $\mathcal{C}, \mathcal{D} \in \mathbb{R}^{\underline{I}_P}$ and $\mathcal{X} \in \mathbb{R}^{\underline{J}_N}$.

This type of system was called multilinear by Brazell $et\ al.$ (2013), but in our view it seems more appropriate to speak of a system of linear tensor equations (or a tensor system) with respect to the unknown tensor $\mathcal{X}$.

REMARK 3.94.– Using the definition [3.147] of the Einstein product and the results [3.266] and [3.273], we can prove the following proposition, which establishes a link between tensor systems based on the mode-p and Einstein products.

PROPOSITION 3.95.– Let $\mathcal{A} \in \mathbb{K}^{\underline{I}_P \times \underline{J}_N}$ and $\mathbf{x}^{(n)} \in \mathbb{K}^{J_n}$, $n \in \langle N \rangle$. We have:

$$\mathcal{B} = \mathcal{A} \times_{p+1} \mathbf{x}^{(1)} \times_{p+2} \cdots \times_{p+N} \mathbf{x}^{(N)} = \mathcal{A} \star_N \mathcal{C} \quad [3.282]$$

$$\mathcal{C} = \overset{N}{\underset{n=1}{\circ}} \mathbf{x}^{(n)} = [c_{\underline{\mathbf{j}}_N}] = \left[\prod_{n=1}^{N} x_{j_n}^{(n)} \right] \in \mathbb{K}^{\underline{J}_N}. \quad [3.283]$$

PROOF.– By the definition of the multiple mode-p product, we have:

$$\mathcal{B} = \mathcal{A} \times_{p+1} \mathbf{x}^{(1)} \times_{p+2} \cdots \times_{p+N} \mathbf{x}^{(N)} \;\Rightarrow\; b_{\underline{\mathbf{i}}_P} = \sum_{\underline{\mathbf{j}}_N=\underline{\mathbf{1}}}^{\underline{\mathbf{J}}_N} a_{\underline{\mathbf{i}}_P,\underline{\mathbf{j}}_N} x_{j_1}^{(1)} \cdots x_{j_N}^{(N)}.$$

Taking into account the definition [3.283] of the tensor $\mathcal{C}$, we deduce that:

$$b_{\underline{\mathbf{i}}_P} = \sum_{\underline{\mathbf{j}}_N=\underline{\mathbf{1}}}^{\underline{\mathbf{J}}_N} a_{\underline{\mathbf{i}}_P,\underline{\mathbf{j}}_N} c_{\underline{\mathbf{j}}_N} \;\Rightarrow\; \mathcal{B} = \mathcal{A} \star_N \mathcal{C},$$

which proves the relation [3.282]. $\qquad\square$

3.17.3. Solving tensor systems using LS

Using the results from sections 3.13.5 and 3.14, it is possible to use the LS method to solve systems of tensor equations expressed in terms of the Einstein product.

EXAMPLE 3.96.– Consider the following tensor system:

$$\mathcal{A} \star_2 \mathbf{X} = \mathbf{B}, \qquad\qquad [3.284]$$

with $\mathcal{A} \in \mathbb{R}^{I \times J \times K \times L}, \mathbf{X} \in \mathbb{R}^{K \times L}$ and $\mathbf{B} \in \mathbb{R}^{I \times J}$. The solution of this system in the sense of minimizing the LS criterion $\min_{\mathbf{X}} \|\mathcal{A} \star_2 \mathbf{X} - \mathbf{B}\|_F^2$ can be obtained by rewriting equation [3.284] using the unfolding $\mathbf{A}_{IJ \times KL}$ of the tensor $\mathcal{A}$ and the vectorized forms $\mathbf{x}_{KL}$ and $\mathbf{b}_{IJ}$ of the matrices $\mathbf{X}$ and $\mathbf{B}$. The LS criterion then becomes:

$$\min_{\mathbf{x}_{KL}} \|\mathbf{A}_{IJ \times KL} \mathbf{x}_{KL} - \mathbf{b}_{IJ}\|_2^2. \qquad\qquad [3.285]$$

Minimizing this criterion leads us to the normal equations:

$$(\mathbf{A}_{IJ \times KL}^T \mathbf{A}_{IJ \times KL})\hat{\mathbf{x}}_{KL} = \mathbf{A}_{IJ \times KL}^T \mathbf{b}_{IJ} \qquad\qquad [3.286]$$

$$\Downarrow \qquad\qquad [3.287]$$

$$\hat{\mathbf{x}}_{KL} = (\mathbf{A}_{IJ \times KL}^T \mathbf{A}_{IJ \times KL})^{-1} \mathbf{A}_{IJ \times KL}^T \mathbf{b}_{IJ} \qquad [3.288]$$

if the matrix $\mathbf{A}_{IJ \times KL}^T \mathbf{A}_{IJ \times KL}$ is invertible, i.e. if $\mathbf{A}_{IJ \times KL}$ has full column rank, which implies the necessary but not sufficient condition that $IJ \geq KL$.

The normal equations and LS solution can be rewritten using the Einstein product as follows:

$$\mathcal{A}^T \star_2 \mathcal{A} \star_2 \hat{\mathbf{X}} = \mathcal{A}^T \star_2 \mathbf{B} \;\Rightarrow\; \hat{\mathbf{X}} = (\mathcal{A}^T \star_2 \mathcal{A})^{-1} \star_2 \mathcal{A}^T \star_2 \mathbf{B}, \qquad [3.289]$$

with $\mathcal{A}^T \in \mathbb{R}^{K \times L \times I \times J}$.

EXAMPLE 3.97.– The following tensor system was considered by Brazell *et al.* (2013):

$$\mathcal{A} \star_2 \mathcal{X} = \mathcal{B}, \tag{3.290}$$

with $\mathcal{A} \in \mathbb{R}^{I \times J \times K \times L}$, $\mathcal{X} \in \mathbb{R}^{K \times L \times M \times N}$ and $\mathcal{B} \in \mathbb{R}^{I \times J \times M \times N}$. The LS solution of this system of equations is obtained by minimizing the criterion:

$$\min_{\mathcal{X}} \| \mathcal{A} \star_2 \mathcal{X} - \mathcal{B} \|_F^2. \tag{3.291}$$

Using proposition 3.86, this criterion can also be written as:

$$\min_{\mathbf{X}_{KL \times MN}} \| \mathbf{A}_{IJ \times KL} \mathbf{X}_{KL \times MN} - \mathbf{B}_{IJ \times MN} \|_F^2. \tag{3.292}$$

Its minimization then leads us to the following normal equations and solution:

$$\mathbf{A}_{IJ \times KL}^T \mathbf{A}_{IJ \times KL} \hat{\mathbf{X}}_{KL \times MN} = \mathbf{A}_{IJ \times KL}^T \mathbf{B}_{IJ \times MN} \tag{3.293}$$

$$\Downarrow$$

$$\hat{\mathbf{X}}_{KL \times MN} = (\mathbf{A}_{IJ \times KL}^T \mathbf{A}_{IJ \times KL})^{-1} \mathbf{A}_{IJ \times KL}^T \mathbf{B}_{IJ \times MN} \tag{3.294}$$

if $\mathbf{A}_{IJ \times KL}^T \mathbf{A}_{IJ \times KL}$ is invertible, which implies the necessary but not sufficient condition that $IJ \geq KL$. After defining $\mathcal{A}^T \in \mathbb{R}^{K \times L \times I \times J}$, the normal equations and the LS solution can be rewritten in terms of the Einstein product as follows:

$$\mathcal{A}^T \star_2 \mathcal{A} \star_2 \hat{\mathcal{X}} = \mathcal{A}^T \star_2 \mathcal{B} \quad \Rightarrow \quad \hat{\mathcal{X}} = (\mathcal{A}^T \star_2 \mathcal{A})^{-1} \star_2 \mathcal{A}^T \star_2 \mathcal{B}. \tag{3.295}$$

Thus, we recover the result of Brazell *et al.* (2013) much more straightforwardly because of our use of the property [3.253] and by rewriting the LS criterion in the form [3.292]. Minimizing this criterion then leads us to the matrix normal equations [3.293], from which it is easy to deduce the tensor normal equations.

Note that the solution [3.295] is of the same form as [3.289], with tensors $\mathcal{X}$ and $\mathcal{B}$ instead of matrices $\mathbf{X}$ and $\mathbf{B}$.

The solutions [3.289] and [3.295] are expressed using the Moore–Penrose pseudo-inverse of the tensor $\mathcal{A}$, which is given by $\mathcal{A}^\dagger = (\mathcal{A}^T \star_2 \mathcal{A})^{-1} \star_2 \mathcal{A}^T$, and deduced from the matrix pseudo-inverse $\mathbf{A}_{IJ \times KL}^\dagger = (\mathbf{A}_{IJ \times KL}^T \mathbf{A}_{IJ \times KL})^{-1} \mathbf{A}_{IJ \times KL}^T$ (see proposition 3.72).

Proceeding in the same way, we can solve the tensor system $\mathcal{A} \star_2 \mathbf{X} = \mathbf{b}$, with $\mathcal{A} \in \mathbb{R}^{I \times K \times L}$, $\mathbf{X} \in \mathbb{R}^{K \times L}$ and $\mathbf{b} \in \mathbb{R}^I$. The LS criterion is then given by:

$$\min_{\mathbf{X}} \| \mathcal{A} \star_2 \mathbf{X} - \mathbf{b} \|_2^2 \quad \Leftrightarrow \quad \min_{\mathbf{x}_{KL}} \| \mathbf{A}_{I \times KL} \mathbf{x}_{KL} - \mathbf{b} \|_2^2. \tag{3.296}$$

Minimizing these criteria gives us the following solutions:

$$\hat{\mathbf{x}}_{KL} = (\mathbf{A}_{I \times KL}^T \mathbf{A}_{I \times KL})^{-1} \mathbf{A}_{I \times KL}^T \mathbf{b} \quad \Leftrightarrow \quad \hat{\mathbf{X}} = (\mathcal{A}^T \star_1 \mathcal{A})^{-1} \star_2 \mathcal{A}^T \times_3 \mathbf{b}$$

[3.297]

with $\mathcal{A}^T \in \mathbb{R}^{K \times L \times I}$. This LS solution exists if the matrix $\mathbf{A}_{I \times KL}$ has full column rank, which implies the necessary but not sufficient condition that $I \geq KL$.

The LS solution of the three tensor systems considered here are summarized in Table 3.21.

Tensor systems	LS solutions	Necessary conditions
$\mathcal{A} \star_2 \mathcal{X} = \mathcal{B}$ $\mathcal{A} \in \mathbb{R}^{I \times J \times K \times L}, \mathcal{X} \in \mathbb{R}^{K \times L \times M \times N}$ $\mathcal{B} \in \mathbb{R}^{I \times J \times M \times N}$	$\hat{\mathcal{X}} = (\mathcal{A}^T \star_2 \mathcal{A})^{-1} \star_2 \mathcal{A}^T \star_2 \mathcal{B}$	$IJ \geq KL$
$\mathcal{A} \star_2 \mathbf{X} = \mathbf{B}$ $\mathcal{A} \in \mathbb{R}^{I \times J \times K \times L}, \mathbf{X} \in \mathbb{R}^{K \times L}$ $\mathbf{B} \in \mathbb{R}^{I \times J}$	$\hat{\mathbf{X}} = (\mathcal{A}^T \star_2 \mathcal{A})^{-1} \star_2 \mathcal{A}^T \star_2 \mathbf{B}$	$IJ \geq KL$
$\mathcal{A} \star_2 \mathbf{X} = \mathbf{b}$ $\mathcal{A} \in \mathbb{R}^{I \times K \times L}, \mathbf{X} \in \mathbb{R}^{K \times L}$ $\mathbf{b} \in \mathbb{R}^I$	$\hat{\mathbf{X}} = (\mathcal{A}^T \star_1 \mathcal{A})^{-1} \star_2 \mathcal{A}^T \times_3 \mathbf{b}$	$I \geq KL$

Table 3.21. *LS solutions of tensor linear systems*

3.18. Hadamard and Kronecker products of tensors

The Hadamard and Kronecker products of matrices can be extended to tensors (Lee and Cichocki 2017), as summarized in Tables 3.22 and 3.23, where we have used the notation [3.16] to denote the combination of indices associated with the Kronecker product:

$$\overline{i_1 i_2 \cdots i_I} \triangleq i_I + (i_{I-1} - 1)I_I + \cdots + (i_1 - 1)I_2 \cdots I_I,$$

[3.298]

with the convention that the first index i_1 varies more slowly than i_2, which itself varies more slowly than i_3, and so on. In particular, $\overline{ki} \triangleq i+(k-1)I$, $\overline{lj} \triangleq j+(l-1)J$.

EXAMPLE 3.98.– The Hadamard product of two tensors appears, in particular, when solving the problem of estimating a tensor model for a data tensor characterized by missing data. Let $\mathcal{X} \in \mathbb{K}^{I_P}$ be a tensor modeled by means of a PARAFAC model $[\![\mathbf{A}^{(1)}, \cdots, \mathbf{A}^{(P)}; R]\!]$ of order P and rank R, as defined in section 5.2.5, and let $\mathcal{W} \in \mathbb{K}^{I_P}$ be the binary tensor such that:

$$w_{i_1, \cdots, i_P} = w_{\mathbf{i}_P} = \begin{cases} 0 & \text{if } x_{\mathbf{i}_P} \text{missing} \\ 1 & \text{if } x_{\mathbf{i}_P} \text{measured} \end{cases}$$

for $\forall i_p \in \langle I_p \rangle$ and $\forall p \in \langle P \rangle$. The quadratic criterion in terms of the difference between the data tensor $\mathcal{X}$ and the model output, taking into account the missing data, can be written as:

$$f(\mathbf{A}^{(1)}, \cdots, \mathbf{A}^{(P)}) = \| \mathcal{W} \odot \left(\mathcal{X} - [\![\mathbf{A}^{(1)}, \cdots, \mathbf{A}^{(P)}; R]\!] \right) \|_F^2. \qquad [3.299]$$

Operations	Symbols and entries	Dimensions
	$\mathbf{A}, \mathbf{B} \in \mathbb{K}^{I \times J}, \mathbf{C} \in \mathbb{K}^{K \times L}$	
Outer product	$(\mathbf{A} \circ \mathbf{C})_{i,j,k,l} = a_{ij} c_{kl}$	$I \times J \times K \times L$
Hadamard product	$(\mathbf{A} \odot \mathbf{B})_{ij} = a_{ij} b_{ij}$	$I \times J$
Kronecker product	$(\mathbf{C} \otimes \mathbf{A})_{\overline{ki}, \overline{lj}} = a_{ij} c_{kl}$	$KI \times LJ$

Table 3.22. *Matrix operations*

Operations	Symbols and entries	Dimensions
	$\mathbf{v} \in \mathbb{K}^J, \mathcal{A}, \mathcal{B} \in \mathbb{K}^{\underline{I}_P}, \mathcal{C} \in \mathbb{K}^{\underline{J}_P}, \mathcal{D} \in \mathbb{K}^{\underline{K}_N}$	
Outer product	$(\mathcal{A} \circ \mathbf{v})_{\underline{i}_P, j} = a_{\underline{i}_P} v_j$	$\underline{I}_P \times J = \overset{P}{\underset{p=1}{\times}} I_p \times J$
Outer product	$(\mathcal{A} \circ \mathcal{D})_{\underline{i}_P, \underline{k}_N} = a_{\underline{i}_P} d_{\underline{k}_N}$	$\underline{I}_P \times \underline{K}_N$
Hadamard product	$(\mathcal{A} \odot \mathcal{B})_{\underline{i}_P} = a_{\underline{i}_P} b_{\underline{i}_P}$	$\underline{I}_P$
Kronecker product	$(\mathcal{C} \otimes \mathcal{A})_{\overline{j_1 i_1}, \cdots, \overline{j_P i_P}} = a_{\underline{i}_P} c_{\underline{j}_P}$	$J_1 I_1 \times \cdots \times J_P I_P$

Table 3.23. *Tensor operations*

Since the rank R is fixed, the criterion needs to be minimized with respect to the factor matrices $(\mathbf{A}^{(1)}, \cdots, \mathbf{A}^{(P)})$ (see Table I.4).

EXAMPLE 3.99.– Let $\mathcal{A}$, $\mathcal{B}$ and $\mathcal{C}$ be three rank-one tensors defined as $\mathcal{A} = \overset{P}{\underset{p=1}{\circ}} \mathbf{a}^{(p)}$, $\mathcal{B} = \overset{P}{\underset{p=1}{\circ}} \mathbf{b}^{(p)}$ and $\mathcal{C} = \overset{P}{\underset{p=1}{\circ}} \mathbf{c}^{(p)}$, with $\mathbf{a}^{(p)}, \mathbf{b}^{(p)} \in \mathbb{K}^{I_p}$ and $\mathbf{c}^{(p)} \in \mathbb{K}^{J_p}, p \in \langle P \rangle$. Their Hadamard and Kronecker products are given by:

$$\mathcal{A} \odot \mathcal{B} = \overset{P}{\underset{p=1}{\circ}} (\mathbf{a}^{(p)} \odot \mathbf{b}^{(p)}) \in \mathbb{K}^{\underline{I}_P} \; ; \; \mathcal{C} \otimes \mathcal{A} = \overset{P}{\underset{p=1}{\circ}} (\mathbf{c}^{(p)} \otimes \mathbf{a}^{(p)}) \in \mathbb{K}^{J_1 I_1 \times \cdots \times J_P I_P}.$$

We can also define partial Hadamard and Kronecker products (Favier and de Almeida 2014a; Lee and Cichocki 2017). The partial Hadamard product was used to model a wireless communication system, represented using a generalized PARATUCK model, whose core tensor is obtained as the partial Hadamard product of the coding tensor with the resource allocation tensor (Favier and de Almeida 2014b).

The partial Hadamard product, denoted $\mathcal{X} \underset{\underline{\mathbf{i}}_P}{\odot} \mathcal{Y}$, along a set $\underline{\mathbf{i}}_P = (i_1, \cdots, i_P)$ of indices shared by the tensors $\mathcal{X} \in \mathbb{K}^{\underline{R}_L \times \underline{I}_P}$ and $\mathcal{Y} \in \mathbb{K}^{\underline{S}_M \times \underline{I}_P}$ is defined as the tensor $\mathcal{D} \in \mathbb{K}^{\underline{R}_L \times \underline{S}_M \times \underline{I}_P}$ whose entries are given by (without using the index convention):

$$d_{\underline{\mathbf{r}}_L, \underline{\mathbf{s}}_M, \underline{\mathbf{i}}_P} = x_{\underline{\mathbf{r}}_L, \underline{\mathbf{i}}_P}\, y_{\underline{\mathbf{s}}_M, \underline{\mathbf{i}}_P}, \qquad [3.300]$$

with $\underline{\mathbf{r}}_L = \{r_1, \cdots, r_L\}$ and $\underline{\mathbf{s}}_M = \{s_1, \cdots, s_M\}$.

EXAMPLE 3.100.– Consider two third-order tensors $\mathcal{X} \in \mathbb{K}^{R \times I_1 \times I_2}$ and $\mathcal{Y} \in \mathbb{K}^{S \times I_1 \times I_2}$. The partial Hadamard product along the indices corresponding to the shared modes 2 and 3, denoted $\mathcal{X} \underset{\{i_1, i_2\}}{\odot} \mathcal{Y}$, gives a fourth-order tensor $\mathcal{D} \in \mathbb{K}^{R \times S \times I_1 \times I_2}$ such that $d_{r,s,i_1,i_2} = x_{r,i_1,i_2} y_{s,i_1,i_2}$.

In the case of two tensors $\mathcal{X} \in \mathbb{K}^{\underline{R}_L \times \underline{I}_P \times \underline{J}_M}$ and $\mathcal{Y} \in \mathbb{K}^{\underline{S}_L \times \underline{I}_P \times \underline{K}_M}$ that share the indices $(i_1, \cdots, i_P)$, we define the partial Kronecker product as the tensor $\mathcal{D}$ such that:

$$\mathcal{D}_{\cdots, i_1, \cdots, i_P, \cdots} = \mathcal{X}_{\cdots, i_1, \cdots, i_P, \cdots} \otimes \mathcal{Y}_{\cdots, i_1, \cdots, i_P, \cdots}$$
$$\in \mathbb{K}^{R_1 S_1 \times \cdots \times R_L S_L \times I_1 \times \cdots \times I_P \times J_1 K_1 \times \cdots \times J_M K_M},$$

where the Kronecker product applies to each pair of indices (r_l, s_l), for $l \in \langle L \rangle$, and (j_m, k_m), for $m \in \langle M \rangle$, not shared by the tensors $\mathcal{X}$ and $\mathcal{Y}$.

3.19. Tensor extension

Tensor extension of a matrix refers to the construction of a tensor of order greater than two from this matrix. For example, given a matrix $\mathbf{B} \in \mathbb{K}^{I \times J}$, we define the tensor $\mathcal{A} \in \mathbb{K}^{I \times J \times K}$ such that $a_{i,j,k} = b_{i,j}$ for $k = 1, \cdots, K$. This amounts to defining $\mathcal{A}$ in such a way that its K frontal slices $\mathbf{A}_{..k}$, $k \in \langle K \rangle$, are all equal to $\mathbf{B}$. Using index notation, the mode-1 flat unfolding $\mathbf{A}_{I \times KJ}$ of $\mathcal{A}$ is given by:

$$\mathbf{A}_{I \times KJ} = a_{i,j,k}\mathbf{e}_i^{kj} = \sum_{k=1}^{K} \mathbf{e}^k \otimes b_{i,j}\mathbf{e}_i^j$$
$$= \mathbf{1}_K^T \otimes \mathbf{B} = \mathbf{B}(\mathbf{1}_K^T \otimes \mathbf{I}_J) = [\underbrace{\mathbf{B} \cdots \mathbf{B}}_{K \text{ terms}}]. \qquad [3.301]$$

Similarly, we can extend the matrix $\mathbf{B}$ to a third-order tensor $\mathcal{A} \in \mathbb{K}^{M \times I \times J}$ such that $a_{m,i,j} = b_{i,j}$ for $m = 1, \cdots, M$, i.e. a tensor whose horizontal slices $\mathbf{A}_{m..}$, for $m \in \langle M \rangle$, are all equal to $\mathbf{B}$. We then have:

$$\mathbf{A}_{MI \times J} = a_{m,i,j}\mathbf{e}_{mi}^j = \sum_{m=1}^{M} \mathbf{e}_m \otimes b_{i,j}\mathbf{e}_i^j$$

$$= \mathbf{1}_M \otimes \mathbf{B} = (\mathbf{1}_M \otimes \mathbf{I}_I)\mathbf{B} = \left.\begin{bmatrix} \mathbf{B} \\ \vdots \\ \mathbf{B} \end{bmatrix}\right\} M \text{ terms.} \tag{3.302}$$

EXAMPLE 3.101.– For the extension of $\mathbf{B} \in \mathbb{K}^{I \times J}$ to $\mathcal{A} \in \mathbb{K}^{M \times I \times J \times K}$ such that:

$a_{m,i,j,k} = b_{i,j}$, $\forall m \in \langle M \rangle, \forall k \in \langle K \rangle$,

combining the formulae [3.301] and [3.302] gives:

$$\mathbf{A}_{MI \times KJ} = (\mathbf{1}_M \otimes \mathbf{I}_I)\mathbf{B}(\mathbf{1}_K^T \otimes \mathbf{I}_J). \tag{3.303}$$

We therefore obtain a matrix partitioned into (M, K) blocks, whose blocks of size $I \times J$ are all equal to the matrix $\mathbf{B}$.

We can also extend a tensor by repeating some of its dimensions. For example, the third-order tensor $\mathcal{A} \in \mathbb{K}^{I_1 \times I_2 \times I_3}$ can be extended to the tensor $\mathcal{B} \in \mathbb{K}^{R_1 I_1 \times R_2 I_2 \times I_3}$ by repeating its first two dimensions R_1 and R_2 times, respectively. The tensor $\mathcal{B}$ can then be written as the following Tucker decomposition:

$$\mathcal{B} = \mathcal{A} \times_1 (\mathbf{1}_{R_1} \otimes \mathbf{I}_{I_1}) \times_2 (\mathbf{1}_{R_2} \otimes \mathbf{I}_{I_2}) \times_3 \mathbf{I}_{I_3} \in \mathbb{K}^{J_1 \times J_2 \times I_3}, \tag{3.304}$$

with $J_1 = R_1 I_1$ and $J_2 = R_2 I_2$, and $\mathcal{A}$ as the core tensor (see section 5.2.1 for the definition of a Tucker model). The tensor $\mathcal{B}$ admits the following mode-1 flat and mode-2 tall unfoldings (see equation [5.5]):

$$\mathbf{B}_{J_1 \times J_2 I_3} = (\mathbf{1}_{R_1} \otimes \mathbf{I}_{I_1})\mathbf{A}_{I_1 \times I_2 I_3}\left((\mathbf{1}_{R_2} \otimes \mathbf{I}_{I_2}) \otimes I_{I_3}\right)^T$$

$$= (\mathbf{1}_{R_1} \otimes \mathbf{I}_{I_1})\mathbf{A}_{I_1 \times I_2 I_3}(\mathbf{1}_{R_2} \otimes \mathbf{I}_{I_2 I_3})^T$$

$$\mathbf{B}_{J_1 J_2 \times I_3} = (\mathbf{1}_{R_1} \otimes \mathbf{I}_{I_1}) \otimes (\mathbf{1}_{R_2} \otimes \mathbf{I}_{I_2})\mathbf{A}_{I_1 I_2 \times I_3}$$

$$= (\mathbf{1}_{R_1 R_2} \otimes \mathbf{I}_{I_1 I_2})\mathbf{A}_{I_1 I_2 \times I_3}.$$

From these equations, we conclude that the unfolding $\mathbf{B}_{J_1 \times J_2 I_3}$ corresponds to a partitioned matrix, with R_1 row blocks and R_2 column blocks, all equal to $\mathbf{A}_{I_1 \times I_2 I_3}$, whereas $\mathbf{B}_{J_1 J_2 \times I_3}$ is a matrix partitioned into $R_1 R_2$ row blocks, all equal to $\mathbf{A}_{I_1 I_2 \times I_3}$, as illustrated by Figure 3.3.

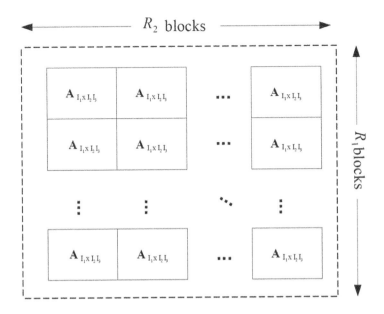

Figure 3.3. *Unfolding* $\mathbf{B}_{J_1 \times J_2 I_3}$.

3.20. Tensorization

Tensorization is the process of constructing data tensors. We distinguish three main types of tensorization approach based on (i) segmentation of data vectors or matrices, (ii) repetition of experiments in different configurations and (iii) system design. The first approach can be viewed as the inverse process to vectorization and matricization. It involves constructing a high-order tensor from a given vector or matrix. For instance, a vector of size $I_1 I_2 \cdots I_P$ partitioned into P sub-vectors of size I_p, $p \in \langle P \rangle$, can be reformatted into a tensor of order P and size $I_1 \times I_2 \times \cdots \times I_P$. Similarly, a partitioned matrix of size $\prod_{p=1}^{P_1} I_p \times \prod_{p=1}^{P_2} J_p$ can be tensorized into a tensor of order $P_1 + P_2$ and size $I_1 \times \cdots \times I_{P_1} \times J_1 \times \cdots \times J_{P_2}$.

Thus, a matrix $\mathbf{A} \in \mathbb{C}^{IK \times JL}$ partitioned into (I, J) blocks of same size $K \times L$ can be tensorized into a fourth-order tensor $\mathcal{B} \in \mathbb{K}^{I \times J \times K \times L}$, such that:

$$b_{ijkl} = a_{k+(i-1)K,\, l+(j-1)L} = a_{\overline{ik},\overline{jl}}$$

using the notation defined in [3.298]. Note that $\mathbf{A}$ can be seen as the matrix unfolding $\mathbf{B}_{IK \times JL}$ of the tensor $\mathcal{B}$ obtained by combining the modes 1 and 3 along the rows and the modes 2 and 4 along the columns.

EXAMPLE 3.102.– For $I = J = 3, K = 2, L = 1$, the matrix $\mathbf{A} \in \mathbb{C}^{6 \times 3}$ is partitioned into (3,3)-blocks of size 2×1, i.e. vectors of size 2:

$$
\mathbf{A} = \begin{bmatrix}
a_{11} & | & a_{12} & | & a_{13} \\
a_{21} & | & a_{22} & | & a_{23} \\
- & & - & & - \\
a_{31} & | & a_{32} & | & a_{33} \\
a_{41} & | & a_{42} & | & a_{43} \\
- & & - & & - \\
a_{51} & | & a_{52} & | & a_{53} \\
a_{61} & | & a_{62} & | & a_{63}
\end{bmatrix} .
$$

The element a_{52} of the matrix $\mathbf{A}$ corresponding to $(i = 3, j = 2, k = 1, l = 1)$ is associated with the element $b_{3,2,1,1}$ of the tensor $\mathcal{B}$. Similarly, a_{43}, which corresponds to $(i = 2, j = 3, k = 2, l = 1)$, is associated with $b_{2,3,2,1}$.

We can also increase the order of a tensor, i.e. transform a tensor $\mathcal{X} \in \mathbb{K}^{I_P}$ of order P into a tensor $\mathcal{Y} \in \mathbb{K}^{J_Q}$ of order Q, with $Q > P$ and $\prod_{q=1}^{Q} J_q = \prod_{p=1}^{P} I_p$. This type of transformation is used in image processing, in particular for the compression and reconstruction of images (Latorre 2005; Bengua *et al.* 2017).

Below, we summarize the idea introduced by Latorre (2005) to represent a color image, viewed as a third-order tensor $\mathcal{X} \in \mathbb{K}^{I_1 \times I_2 \times I_3}$, where i_1 and i_2 are the spatial modes associated with the rows and columns of an image, and i_3 is the color mode (red, green, blue), with $I_3 = 3$.

An image of size $I_1 \times I_2 = 2^N \times 2^N$ can be divided into elementary square blocks of 2×2 pixels, such that each block is associated with a matrix:

$$
\mathbf{Y} = \sum_{j_1=1}^{4} \sum_{i_3=1}^{3} x_{j_1,i_3} \mathbf{e}_{j_1}^{(4)} \circ \mathbf{e}_{i_3}^{(3)},
$$

where x_{j_1,i_3} represents the value of the pixel for the color $i_3 \in \{1, 2, 3\}$, with $j_1 \in \{1, 2, 3, 4\}$ according to the position of the pixel within the block: top left $(j_1 = 1)$, top right $(j_1 = 2)$, bottom left $(j_1 = 3)$ and bottom right $(j_1 = 4)$, respectively. For an image containing $2^N \times 2^N$ pixels, we define a tensor $\mathcal{Y} \in \mathbb{K}^{4 \times \cdots \times 4 \times 3}$ of order $N + 1$ containing all pixels, of the form:

$$
\mathcal{Y} = \sum_{j_1=1}^{4} \cdots \sum_{j_N=1}^{4} \sum_{i_3=1}^{3} y_{j_1,\cdots,j_N,i_3} \mathbf{e}_{j_N}^{(4)} \circ \cdots \circ \mathbf{e}_{j_1}^{(4)} \circ \mathbf{e}_{i_3}^{(3)}.
$$

Each mode $j_n, n \in \langle N \rangle$, is associated with some division of the image into N overlapping elementary square 2×2 blocks, and the set of modes $(j_1, \cdots, j_N)$ defines the position of the pixel of value $y_{j_1,\cdots,j_N,i_3}$ for the color i_3. Thus, for example, in the case $N = 2$, for $j_1 = 3, j_2 = 2$, the position within the image is given by:

$$\mathbf{e}_{j_2}^{(4)} \circ \mathbf{e}_{j_1}^{(4)} = \begin{bmatrix} 0 \\ 1 \\ 0 \\ 0 \end{bmatrix} \begin{bmatrix} 0 & 0 & 1 & 0 \end{bmatrix} = \begin{bmatrix} 0 & 0 & 0 & 0 \\ 0 & 0 & 1 & 0 \\ 0 & 0 & 0 & 0 \\ 0 & 0 & 0 & 0 \end{bmatrix}.$$

The value $j_2 = 2$ is associated with the 2×2 block in the top right, whereas $j_1 = 3$ corresponds to a position of the bottom left pixel of the top right block.

An example of the second tensorization approach that is often used in applications is to stack a set of data matrices of the same size along a third mode to build a third-order data tensor. This type of tensorization can result, for example, from the repetition of some experiments at different time instances (time diversity), the use of multiple sensors as in array processing (space diversity), or the exploitation of various conditions like illumination, views, and expressions in a face recognition system (see Table I.1).

The third tensorization approach is illustrated by wireless communication systems whose design can simultaneously incorporate several diversities, like space-time-frequency-code diversities (Favier and de Almeida 2014). Tensor-based approaches to designing such communication systems lead to tensors of received signals at the receiver that satisfy tensor models with different structures. The multilinear structure of the received signals is then exploited to develop semi-blind receivers to jointly estimate the channels and information symbols. This application will be studied in detail in the next two volumes.

Another tensorization approach that leads to tensor decompositions originated from the use of high-order cumulants to solve BSS and blind system identification problems (see the Appendix). For instance, in the context of wireless communication systems, fourth-order cumulants of the signals, measured at the output of a finite impulse response (FIR) communication channel, satisfy a PARAFAC model that can be exploited to estimate the channel. See section 5.3.3 for a detailed presentation of this application. There are particular tensorization techniques that construct structured tensors, like Cauchy, Toeplitz or Hankel tensors, from a given vector of data. The corresponding Hankelization is briefly described in the next section. For an overview of tensorization techniques, see Debals and de Lathauwer (2017).

3.21. Hankelization

Hankelization is the process of constructing Hankel matrices and Hankel tensors from a given data vector. In the matrix case, the Hankelization maps a vector $\mathbf{u} \in \mathbb{K}^N$ to a Hankel matrix $\mathbf{A} \in \mathbb{K}^{I \times J}$, defined as follows:

$$\mathbb{K}^N \ni \mathbf{u} \mapsto \mathbf{A} \in \mathbb{K}^{I \times J} \ , \quad N = I + J - 1$$

$$a_{ij} = u_{i+j-1} \ , \quad i \in \langle I \rangle \ , \quad j \in \langle J \rangle. \tag{3.305}$$

It is important to note that the Hankelization process introduces redundancy to the information contained in the vector $\mathbf{u}$, due to the repetition of its components in the Hankel matrix. This redundancy allows the performance of estimating the parameters contained in $\mathbf{u}$ to be improved by Hankelization.

EXAMPLE 3.103.– For $N = 5, I = J = 3$, we obtain: $\mathbf{A} = \begin{bmatrix} u_1 & u_2 & u_3 \\ u_2 & u_3 & u_4 \\ u_3 & u_4 & u_5 \end{bmatrix}$.

For certain vectors $\mathbf{u}$, Hankelization provides low rank Hankel matrices or low multilinear rank Hankel tensors. For instance, when the components of $\mathbf{u}$ are exponentials such that $u_n = z^{n-1}, n \in \langle N \rangle$, the Hankel matrix $\mathbf{A}$ is a rank-one matrix that can be factored as:

$$\mathbf{A} = \begin{bmatrix} 1 & z & \cdots & z^{J-1} \\ z & z^2 & \cdots & z^J \\ \vdots & \vdots & & \vdots \\ z^{I-1} & z^I & \cdots & z^{I+J-2} \end{bmatrix} = \begin{bmatrix} 1 \\ z \\ \vdots \\ z^{I-1} \end{bmatrix} \begin{bmatrix} 1 & z & \cdots & z^{J-1} \end{bmatrix}.$$

In the case of a sum of R exponentials $u_n = \sum_{r=1}^{R} z_r^{n-1}, n \in \langle N \rangle$, the Hankelization process gives a Hankel matrix that can be factored as the product of two Vandermonde matrices $\mathbf{A} = \mathbf{U}\mathbf{V}^T$, with:

$$\mathbf{U} = \begin{bmatrix} 1 & 1 & \cdots & 1 \\ z_1 & z_2 & \cdots & z_R \\ \vdots & \vdots & & \vdots \\ z_1^{I-1} & z_2^{I-1} & \cdots & z_R^{I-1} \end{bmatrix}, \quad \mathbf{V} = \begin{bmatrix} 1 & 1 & \cdots & 1 \\ z_1 & z_2 & \cdots & z_R \\ \vdots & \vdots & & \vdots \\ z_1^{J-1} & z_2^{J-1} & \cdots & z_R^{J-1} \end{bmatrix}.$$

If the generators (z_r) are all distinct and the number R of exponentials is less than or equal to $\min\{I, J\}$, the Vandermonde matrices have full column rank, which implies that the Hankel matrix admits a full-rank decomposition. Such a Hankelization is used to solve the BSS problem. In this context, de Lathauwer (2011) and Debals $et\ al.$ (2017) present different examples of matrix Hankelization from data vectors obtained by evaluating various signals corresponding to sums and/or products of exponentials, sinusoids and/or polynomials. The rank of the resulting Hankel matrix is given for each class of signals.

Similarly, the Hankelization of a vector $\mathbf{u} \in \mathbb{K}^N$ to a Hankel tensor $\mathcal{A} \in \mathbb{K}^{I_1 \times \cdots \times I_P}$ of order P is defined as the following transformation (Papy $et\ al.$ 2005):

$$\mathbb{K}^N \ni \mathbf{u} \mapsto \mathcal{A} \in \mathbb{K}^{I_1 \times \cdots \times I_P} \quad , \quad N = \sum_{p=1}^{P} I_p - P + 1$$

$$a_{i_1,\cdots,i_P} = u_{i_1 + \cdots + i_P - P + 1} \quad , \quad i_p \in \langle I_p \rangle \ , \ p \in \langle P \rangle. \tag{3.306}$$

The tensor thus formed is composed of $P - 1$ tensor slices, each with size $I_1 \times \cdots \times I_{P-1}$.

EXAMPLE 3.104.– For $N = 5, P = 3, I_1 = 3, I_2 = I_3 = 2$, the third-order Hankel tensor $\mathcal{A} \in \mathbb{K}^{3 \times 2 \times 2}$ contains two frontal slices with Hankel structure and size $I_1 \times I_2 = 3 \times 2$, defined as follows:

$$\mathbf{A}_{..1} = \begin{bmatrix} u_1 & u_2 \\ u_2 & u_3 \\ u_3 & u_4 \end{bmatrix} \ , \ \mathbf{A}_{..2} = \begin{bmatrix} u_2 & u_3 \\ u_3 & u_4 \\ u_4 & u_5 \end{bmatrix}.$$

For this example, data Hankelization is obtained by stacking two Hankel matrices along the third mode.

Tensor Hankelization was used by de Lathauwer (2011) and Phan $et\ al.$ (2017) to solve the BSS problem. In section 5.3.2, we will briefly present this application for an instantaneous source mixture.

4

Eigenvalues and Singular Values of a Tensor

4.1. Introduction

The notion of determinant plays an important role in matrix computation, and more generally in linear algebra, for example for determining the rank, inverse and eigenvalues of a matrix, as well as for solving systems of linear equations using Cramer formulae. It is a well-known fact that the system of equations $\mathbf{Ax} = \mathbf{b}$, with $\mathbf{A} \in \mathbb{C}^{N \times N}$ and $\mathbf{x}, \mathbf{b} \in \mathbb{C}^N$, is invertible, in the sense that there exists a unique solution, if and only if the matrix $\mathbf{A}$ is invertible, and therefore if and only if the determinant of this matrix is non-zero. The notions of eigenvalue and eigenvector of $\mathbf{A} \in \mathbb{C}^{N \times N}$ are also encountered when solving the system of equations $\mathbf{Ax} = \lambda \mathbf{x}$, where the eigenvalue–eigenvector pair $(\lambda, \mathbf{x}) \in \mathbb{C} \times \{\mathbb{C}^N \backslash \{\mathbf{0}\}\}$ is called an eigenpair of $\mathbf{A}$. The existence of a solution $\mathbf{x} \neq \mathbf{0}$ for the system $(\mathbf{A} - \lambda \mathbf{I})\mathbf{x} = \mathbf{0}$ means that the kernel of $\mathbf{A} - \lambda \mathbf{I}$ must contain non-zero vectors, which implies that the matrix $\mathbf{A} - \lambda \mathbf{I}$ is singular, and therefore $\det(\mathbf{A} - \lambda \mathbf{I}) = 0$, known as the characteristic equation. The polynomial $p_{\mathbf{A}}(\lambda) = \mathbf{A} - \lambda \mathbf{I}$ is called the characteristic polynomial of $\mathbf{A}$, and its roots are the eigenvalues of $\mathbf{A}$. The set V_λ of all eigenvectors associated with the eigenvalue λ is called the eigenspace associated with λ.

The problem of computing eigenvalues is closely related to that of solving ordinary and partial differential equations. Thus, in the case of a system of first-order linear differential equations with constant coefficients such that $\frac{d\mathbf{u}(t)}{dt} = \mathbf{Au}(t)$, with the initial condition $\mathbf{u}(0) = \mathbf{u}_0$, an exponential solution $\mathbf{u}(t) = e^{\lambda t}\mathbf{x}$ leads to the equation $\mathbf{Ax} = \lambda \mathbf{x}$, whose unknowns λ and $\mathbf{x}$ correspond to an eigenpair of $\mathbf{A}$. Given N particular solutions $\mathbf{u}_n(t) = e^{\lambda_n t}\mathbf{x}_n$, $n \in \langle N \rangle$, the superposition principle for linear systems implies that every linear combination $\sum_{n=1}^{N} c_n \mathbf{u}_n(t)$ is also a solution.

Once the N eigenvectors $\mathbf{x}_n$ have been computed, the coefficients c_n of this linear combination can be determined using the initial condition, namely:

$$\sum_{n=1}^{N} c_n \mathbf{u}_n(0) = \sum_{n=1}^{N} c_n \mathbf{x}_n = [\mathbf{x}_1 \cdots \mathbf{x}_N] \begin{bmatrix} c_1 \\ \vdots \\ c_N \end{bmatrix} = \mathbf{u}_0.$$

In the same way that an exponential solution of a first-order differential equation leads to an eigenvalue problem, searching for an exponential solution $\mathbf{u}(t) = e^{j\omega t}\mathbf{x}$ to a system of second-order differential equations $\mathbf{B}\frac{d^2\mathbf{u}(t)}{dt^2} = \mathbf{A}\mathbf{u}(t)$, with $\mathbf{A}, \mathbf{B} \in \mathbb{C}^{N \times N}$, leads to a generalized eigenvalue problem of the form $\mathbf{A}\mathbf{x} = \lambda\mathbf{B}\mathbf{x}$, where $\lambda = (j\omega)^2$. This problem admits a solution $\mathbf{x} \neq \mathbf{0}$ only if the matrix $\mathbf{A} - \lambda\mathbf{B}$ is singular and therefore if $\det(\mathbf{A} - \lambda\mathbf{B}) = 0$. The roots λ of this polynomial are then called the generalized eigenvalues of $\mathbf{A}$ and $\mathbf{B}$, and the problem is called the generalized form of the classical eigenvalue problem. This type of generalized eigenvalue problem is often encountered in solid mechanics. In the case where $\mathbf{B}$ is symmetric positive definite, the above polynomial can be rewritten as $\det(\mathbf{B}^{-1}\mathbf{A} - \lambda\mathbf{I}) = 0$, which amounts to solving a classical eigenvalue problem for $\mathbf{B}^{-1}\mathbf{A}$ corresponding to the equation $\mathbf{B}^{-1}\mathbf{A}\mathbf{x} = \lambda\mathbf{x}$.

Like in the case of a system of first-order matrix differential equations, searching for an exponential solution $\mathbf{u}(t) = e^{\lambda t}\mathbf{x}$ to the system of first-order tensor differential equations defined in terms of the tensors $\mathcal{A}$ and $\mathcal{B}$ as:

$$\frac{d\mathcal{B}(\mathbf{u}(t))^{P-1}}{dt} = \mathcal{A}(\mathbf{u}(t))^{P-1} \tag{4.1}$$

leads to the following generalized eigenvalue problem (Ding and Wei 2015):

$$\mathcal{A}\mathbf{x}^{P-1} = (P-1)\lambda\mathcal{B}\mathbf{x}^{P-1}. \tag{4.2}$$

Another application of the spectral properties of a matrix, i.e. the properties concerning its eigenvectors and eigenvalues, is encountered when studying continuous-time and discrete-time linear dynamic systems represented using state-space models described by the following equations:

$$\begin{cases} \dot{\mathbf{x}}(t) = \mathbf{A}\mathbf{x}(t) + \mathbf{B}\mathbf{u}(t) \\ \mathbf{y}(t) = \mathbf{C}\mathbf{x}(t) + \mathbf{D}\mathbf{u}(t) \end{cases} \text{ and } \begin{cases} \mathbf{x}(k+1) = \mathbf{A}\mathbf{x}(k) + \mathbf{B}\mathbf{u}(k) \\ \mathbf{y}(k) = \mathbf{C}\mathbf{x}(k) + \mathbf{D}\mathbf{u}(k) \end{cases} \tag{4.3}$$

where $\mathbf{x} \in \mathbb{R}^P$, $\mathbf{u} \in \mathbb{R}^M$ and $\mathbf{y} \in \mathbb{R}^Q$ represent the state, the control and the output of the system, with $\mathbf{A} \in \mathbb{R}^{P \times P}$, $\mathbf{B} \in \mathbb{R}^{P \times M}$, $\mathbf{C} \in \mathbb{R}^{Q \times P}$ and $\mathbf{D} \in \mathbb{R}^{Q \times M}$. When the matrices $(\mathbf{A}, \mathbf{B}, \mathbf{C}, \mathbf{D})$ are independent of time t and k, the system is said to be time-invariant or stationary.

The spectral properties of the state-transition matrix $\mathbf{A}$ allow us to study the stability of the system and determine the state–space representation associated with the modal form of $\mathbf{A}$, i.e. its diagonal or Jordan canonical form. It is a well-known fact that the bounded-input bounded-output (BIBO) stability of systems represented using state-space models [4.3] is guaranteed if and only if all eigenvalues of $\mathbf{A}$ have negative real parts in the continuous-time case or modulus less than 1 in the discrete-time case.

The state-space models recalled above can be generalized to the case of tensor systems defined in terms of the Einstein product as follows:

$$\begin{cases} \dot{\mathcal{X}}(t) = \mathcal{A} \star_P \mathcal{X}(t) + \mathcal{B} \star_R \mathcal{U}(t) \\ \mathcal{Y}(t) = \mathcal{C} \star_P \mathcal{X}(t) + \mathcal{D} \star_R \mathcal{U}(t) \end{cases} , \quad \begin{cases} \mathcal{X}(k+1) = \mathcal{A} \star_P \mathcal{X}(k) + \mathcal{B} \star_R \mathcal{U}(k) \\ \mathcal{Y}(k) = \mathcal{C} \star_P \mathcal{X}(k) + \mathcal{D} \star_R \mathcal{U}(k) \end{cases}$$

$$[4.4]$$

where $\mathcal{X} \in \mathbb{R}^{\underline{I}_P}$, $\mathcal{U} \in \mathbb{R}^{\underline{M}_R}$ and $\mathcal{Y} \in \mathbb{R}^{\underline{Q}_s}$ are called the state, input and output tensors, respectively, and $\mathcal{A} \in \mathbb{R}^{\underline{I}_P \times \underline{I}_P}$, $\mathcal{B} \in \mathbb{R}^{\underline{I}_P \times \underline{M}_R}$, $\mathcal{C} \in \mathbb{R}^{\underline{Q}_s \times \underline{I}_P}$ and $\mathcal{D} \in \mathbb{R}^{\underline{Q}_s \times \underline{M}_R}$ represent the dynamics, control and output (or observation) tensors of the tensor state-space model.

Analogously to standard state-space models [4.3], we can study the stability, controllability and observability properties of dynamic systems represented using tensor state-space models [4.4].

Thus, in the case of a discrete-time tensor system where $\mathcal{A} \in \mathbb{R}^{I_1 \times I_1 \times \cdots \times I_P \times I_P}$ is an even-order paired tensor, and with the Einstein product defined as in [3.158], asymptotic stability in the sense of $\|\mathcal{X}(k)\|_F \to 0$ as $k \to \infty$ is guaranteed if and only if every eigenvalue of the dynamics tensor $\mathcal{A}$ defined by the equation $\mathcal{A} \star_P \mathcal{X} = \lambda \mathcal{X}$ has modulus less than 1 (Chen et $al.$ 2019). See equation [4.22] in definition 4.4.

Note that computing matrix eigenvalues and singular values is the key to determining the eigenvalue decomposition (EVD) and the HOSVD for high-order tensors.

Spectral theory has been developed in more detail for specific classes of structured tensors, such as symmetric tensors, non-negative tensors, Toeplitz tensors, Hankel tensors and Cauchy tensors. Two important applications of the spectral theory of non-negative tensors concern:

– hypergraphs, to study the spectral properties of the connectivity hypermatrix (adjacency hypermatrix, Laplacian hypermatrix, etc.) of a hypergraph (Li et $al.$ 2013; Hu and Qi 2014; Pearson and Zhang 2014; Xie and Qi 2016; Banerjee et $al.$ 2017);

– high-order Markov chains, to study the properties of the transition probability tensors (Chang and Zhang 2013b; Li and Ng 2014; Culp et $al.$ 2017).

See the article by Chang *et al.* (2013) for a presentation of different applications of the spectral theory of non-negative tensors.

Unlike the matrix case, there are several ways to define the eigenvalues of a tensor. Solving a polynomial optimization problem plays a key role in each of them. The objective of this chapter is to present the various definitions. The notions of eigenvalue and singular value of a tensor will be defined in sections 4.2 and 4.5, respectively. Positive/negative definite tensors and orthogonally/unitarily similar tensors will also be introduced. The link between eigenvalues and the best rank-one approximation of a tensor will be established in section 4.3, and the notion of orthogonal decomposition of a tensor will be introduced in section 4.4.

4.2. Eigenvalues of a tensor of order greater than two

Following the publication of the articles by Qi (2005) and Lim (2005) on the eigenvalues and singular values of a tensor, there was extensive research concerning the spectral properties of tensors. There are various notions and problems underlying the concept of eigenvalue of a tensor, such as the notions of characteristic polynomial and hyperdeterminant, the Perron–Frobenius theorem for non-negative tensors, the computation of eigenvalues and therefore solving polynomial equations, among many other results that generalize known results from the matrix case. See for example the surveys on the spectral theory of tensors by Qi (2012) and Chang *et al.* (2013).

The notion of eigenvalue for a tensor can be defined using either the mode-p product (Qi 2005) or the Einstein product (Cui *et al.* 2016; Liang *et al.* 2018). In the latter case, we speak of an eigentensor rather than an eigenvector. Five different definitions are introduced in the next section.

4.2.1. *Different definitions of the eigenvalues of a tensor*

The work by Qi (2005) was originally motivated by research into positive definite homogeneous polynomials such as [3.269]. These polynomials play an important role in automatic control when studying the stability of nonlinear autonomous systems of the form $\dot{x}(t) = f(x(t))$ using Lyapunov's direct method[1].

---

1 Consider the continuous-time autonomous nonlinear system $\dot{x}(t) = f(x(t))$, with $f(0) = 0$. Lyapunov's direct method aims to construct a positive definite polynomial function (called a Lyapunov function) $V(x) > 0$ in the neighborhood of the origin, i.e. on a ball B_r of radius r around the origin, with $V(0) = 0$, such that:

$$\dot{V}(x) = (\frac{\partial V}{\partial x})^T \dot{x}(t) = (\frac{\partial V}{\partial x})^T f(x) < 0 \text{ for } \forall x \in \mathbb{R}^n, x \neq 0,$$

They also allow us to determine whether a tensor is positive definite (see the definition in Table 4.2).

The notion of eigenvalue of a real symmetric[2] hypercubic tensor $\mathcal{A} \in \mathbb{R}_S^{[P;I]}$ was defined by Qi (2005) in terms of the multiple tensor–vector product [3.274].

DEFINITION 4.1.– *A pair $(\lambda, \mathbf{x})$ is said to be an eigenpair of $\mathcal{A} \in \mathbb{R}_S^{[P;I]}$ if it satisfies the following system of homogeneous polynomial equations:*

$$\mathcal{A}\mathbf{x}^{P-1} = \lambda\mathbf{x}^{[P-1]}, \tag{4.5}$$

where $\mathbf{x}^{[P-1]} = [x_1^{P-1}, \cdots, x_I^{P-1}]^T$ is the vector whose components are equal to the components of $\mathbf{x}$ raised to the power $P-1$, i.e. $(\mathbf{x}^{[P-1]})_i = x_i^{P-1}$ for $i \in \langle I \rangle$. The system [4.5] is equivalent to the I following polynomial equations:

$$\sum_{i_2=1}^{I} \cdots \sum_{i_P=1}^{I} a_{i,i_2,\cdots,i_P} x_{i_2} \cdots x_{i_P} = \lambda x_i^{P-1} \;,\; i \in \langle I \rangle. \tag{4.6}$$

In the case where the pair $(\lambda, \mathbf{x}) \in \mathbb{R} \times \{\mathbb{R}^I \backslash \{\mathbf{0}\}\}$ is real, it is called an H-eigenpair.

Equation [4.5] implies that $\lambda \in \mathbb{C}$ is an eigenvalue of $\mathcal{A}$ if and only if it is a non-zero root of the characteristic polynomial $\phi(\lambda) = \det(\mathcal{A} - \lambda \mathcal{I}_{P,I})$, where $\mathcal{I}_{P,I} = [\delta_{i_1,\cdots,i_P}]$, with $i_p \in \langle I \rangle$ for $p \in \langle P \rangle$, is the identity tensor of order P.

The hyperdeterminant of the tensor $\mathcal{A}$, denoted $\det(\mathcal{A})$, is defined as the resultant of the I homogeneous polynomials $(\mathcal{A}\mathbf{x}^{P-1})_i$, with $i \in \langle I \rangle$, in the I components x_i of $\mathbf{x}$:

$$\det(\mathcal{A}) = \text{Res}\big((\mathcal{A}\mathbf{x}^{P-1})_1, \cdots, (\mathcal{A}\mathbf{x}^{P-1})_I\big), \tag{4.7}$$

where $(\mathcal{A}\mathbf{x}^{P-1})_i$ is the ith component of the vector $\mathcal{A}\mathbf{x}^{P-1}, i \in \langle I \rangle$.

---

where $\frac{\partial V}{\partial \mathbf{x}} \triangleq [\frac{\partial V}{\partial x_1}, \cdots, \frac{\partial V}{\partial x_n}]^T$ is the gradient of $V(\mathbf{x})$. The origin is then an asymptotically stable equilibrium point with domain of attraction B_r, i.e. $\lim_{t \to +\infty} \mathbf{x}(t) = \mathbf{0}, \forall \mathbf{x}(0) \in B_r$.

In the case of an autonomous linear system $\dot{\mathbf{x}}(t) = \mathbf{A}\mathbf{x}(t)$, a Lyapunov function is given by the quadratic form $V(\mathbf{x}) = \mathbf{x}^T \mathbf{P}\mathbf{x}$, where $\mathbf{P} \in \mathbb{R}^{n \times n}$ is a positive definite symmetric matrix. The Lyapunov condition is then written as:

$$\dot{V}(\mathbf{x}) = \dot{\mathbf{x}}^T \mathbf{P}\mathbf{x} + \mathbf{x}^T \mathbf{P}\dot{\mathbf{x}} = \mathbf{x}^T (\mathbf{A}^T \mathbf{P} + \mathbf{P}\mathbf{A})\mathbf{x} < 0,$$

which leads to the Lyapunov equation $\mathbf{A}^T \mathbf{P} + \mathbf{P}\mathbf{A} = -\mathbf{Q}$, with $\mathbf{Q} > 0$. The autonomous linear system $\dot{\mathbf{x}}(t) = \mathbf{A}\mathbf{x}(t)$ is asymptotically stable at the origin if and only if, for a given positive definite matrix $\mathbf{Q}$, the Lyapunov equation admits precisely one solution $\mathbf{P} > 0$.

2 Recall that $\mathbb{R}_S^{[P;I]}$ denotes the set of real symmetric tensors of order P and dimensions I.

REMARK 4.1.–

– Resultants play an important role in the study of polynomial equations. The notion of resultant, which generalizes that of determinant, is used to establish whether a system of N homogeneous polynomial equations in N variables admits a non-zero solution, or equivalently if the N polynomials defining the N equations have a shared non-zero root. Thus, the system of I polynomial equations $\mathcal{A}\mathbf{x}^{P-1} = 0$ admits a non-zero solution if and only if $\det(\mathcal{A}) = 0$, with $\det(\mathcal{A})$ as defined in [4.7]. The resultant is a homogeneous polynomial in the coefficients a_{i_p} of $\mathcal{A}$ of degree $I(P-1)^{I-1}$.

– In the case of a system of linear equations $\mathbf{A}\mathbf{x} = \mathbf{0}$, with $\mathbf{A} = [a_{ij}] \in \mathbb{R}^{I \times I}$, the resultant coincides exactly with the determinant of $\mathbf{A}$, which is a homogeneous polynomial of degree I in the coefficients a_{ij}.

– The theory of resultants is closely linked to the theories of discriminants and hyperdeterminants (Gelfand $et\ al.$ 1992; Morozov and Shakirov 2010). The discriminant of a polynomial Q is used to determine the existence of multiple roots. Thus, since the existence of a double root implies the existence of a root shared by Q and its derivative Q', the discriminant of a unitary polynomial Q of degree N, denoted $\mathrm{Dis}(Q)$, is given by $\mathrm{Dis}(Q) = (-1)^{N(N-1)/2}\mathrm{Res}(Q, Q')$, where $\mathrm{Res}(Q, Q')$ is the resultant of Q and Q'. See Coste (2001) for further details.

If the order P is even, the eigenvalues defined in [4.5] can be interpreted in terms of optimizing the following polynomial form (Lim 2005):

$$f(\mathbf{x}) = \mathcal{A}\mathbf{x}^P \text{ subject to the constraint } \|\mathbf{x}\|_P = 1, \tag{4.8}$$

where $\|.\|_P$ is the l_P norm defined by $\|\mathbf{x}\|_P = \left(\sum_{i=1}^{I} |x_i|^P\right)^{1/P}$, and $\mathcal{A}\mathbf{x}^P = \sum_{i_1,i_2,\cdots,i_P=1}^{I} a_{i_1,i_2,\cdots,i_P} \prod_{p=1}^{P} x_{i_p} = a_{\mathbf{i}_P} \prod_{p=1}^{P} x_{i_p}$, using the index convention.

PROOF.– This optimization with an equality constraint can be solved using the Lagrangian $L(\mathbf{x}, \lambda) = \mathcal{A}\mathbf{x}^P - \lambda(\|\mathbf{x}\|_P^P - 1)$, where λ is the Lagrange multiplier. For even P, since the gradient of the l_P norm is given by $\frac{\partial \|\mathbf{x}\|_P}{\partial \mathbf{x}} = \mathbf{x}^{[P-1]}/\|\mathbf{x}\|_P^{P-1}$, the Karush–Kuhn–Tucker (KKT) optimality conditions give:

$$\frac{\partial L(\mathbf{x}, \lambda)}{\partial \mathbf{x}} = P\mathcal{A}\mathbf{x}^{P-1} - \lambda P\mathbf{x}^{[P-1]} = 0 \ , \ \ \frac{\partial L(\mathbf{x}, \lambda)}{\partial \lambda} = 1 - \|\mathbf{x}\|_P^P = 0, \tag{4.9}$$

from which we deduce $\mathcal{A}\mathbf{x}^{P-1} = \lambda\mathbf{x}^{[P-1]}$, namely equation [4.5] defining the eigenvalues. The corresponding pair $(\lambda, \mathbf{x})$ is called an l_P-eigenpair by Lim (2005). We can therefore conclude that the H-eigenvalues are identical to the l_P-eigenvalues (see Table 4.1). □

REMARK 4.2.– For $P = 2$ (matrix case), we recover the interpretation of the eigenvalues of a matrix $\mathbf{A}$, defined as solutions of the equation $\mathbf{A}\mathbf{x} = \lambda\mathbf{x}$, in terms of optimizing the quadratic form $\mathbf{A}\mathbf{x}^2 \triangleq \mathbf{x}^T\mathbf{A}\mathbf{x}$ subject to the constraint $\|\mathbf{x}\|_2^2 = \mathbf{x}^T\mathbf{x} = 1$ (see Table 1.3).

Another definition of an eigenpair $(\lambda, \mathbf{x})$ was given by Qi (2005) as follows.

DEFINITION 4.2.– *A pair $(\lambda, \mathbf{x})$ is an eigenpair if it is a solution of the following system of non-homogeneous polynomial equations, subject to the constraint that $\mathbf{x}$ has unit Euclidean norm:*

$$\mathcal{A}\mathbf{x}^{P-1} = \lambda\mathbf{x} \text{ subject to the constraint } \|\mathbf{x}\|_2^2 = 1. \tag{4.10}$$

If λ and $\mathbf{x}$ are real, the pair $(\lambda, \mathbf{x})$ is called a Z-eigenpair[3]. In the case of a complex pair $(\lambda, \mathbf{x}) \in \mathbb{C} \times \{\mathbb{C}^I \setminus \{\mathbf{0}\}\}$, it is called an E-eigenpair by Qi (2005), where the letter E refers to the Euclidean norm in [4.10].

REMARK 4.3.– We can make the following remarks:

– In the matrix case, the definitions [4.5] and [4.10] are equivalent to $\mathbf{A}\mathbf{x} = \lambda\mathbf{x}$.

– Unlike the matrix case, where the eigenvalues and eigenvectors of a symmetric/Hermitian matrix are all real (see Table 1.4), the same is not true for a real symmetric tensor, which explains the importance of real H- and Z-eigenvalues when studying positive/negative definite tensors.

– Two E-eigenpairs $(\lambda, \mathbf{x})$ and $(\lambda', \mathbf{x}')$ are said to be equivalent if there exists a non-zero complex number $\alpha \in \mathbb{C}$ such that (Cartwright and Sturmfels 2013):

$$\lambda' = \alpha^{P-2}\lambda \text{ and } \mathbf{x}' = \alpha\mathbf{x}. \tag{4.11}$$

Indeed, we have $\mathcal{A}(\mathbf{x}')^{P-1} = \alpha^{P-1}\mathcal{A}\mathbf{x}^{P-1}$ and $\lambda'\mathbf{x}' = \alpha^{P-1}\lambda\mathbf{x}$. Hence, taking [4.10] into account, we deduce that $\mathcal{A}(\mathbf{x}')^{P-1} = \lambda'\mathbf{x}'$. This means that if $(\lambda, \mathbf{x})$ is an E-eigenvalue of $\mathcal{A}$, then $(\alpha^{P-2}\lambda, \alpha\mathbf{x})$ is also an E-eigenpair for any complex number $\alpha \neq 0$.

– Since the system of polynomial equations [4.5] is homogeneous of degree $P-1$, if $\mathbf{x}$ is an eigenvector associated with the eigenvalue λ, then $\alpha\mathbf{x}$ is also an eigenvector associated with λ, for any complex number $\alpha \neq 0$. As we saw in the previous remark, this is not the case for the system of polynomial equations [4.10] due to the presence of the constraint $\|\mathbf{x}\|_2 = 1$ instead of the constraint $\|\mathbf{x}\|_P = 1$. The normalization constraint on the eigenvectors based on the l_2 norm in [4.10] implies that there exists an E-eigenpair equivalent to $(\lambda, \mathbf{x})$ given by $((-1)^P\lambda, -\mathbf{x})$, which means that for odd P the eigenvalues are determined up to their sign, whereas for even P the sign indeterminacy disappears.

---

3 The name of Z-eigenvalue was suggested by Zhou (2004), as cited in Qi (2005).

– From the definitions [4.5] and [4.10], we can conclude that computing the eigenvalues of a tensor of order greater than two is a nonlinear problem in the sense that it requires us to solve a system of polynomial equations in the components x_i of the eigenvector $\mathbf{x}$. This problem has been addressed by various articles, in particular to compute the largest and smallest eigenvalues of a symmetric tensor (Hu *et al.* 2013; Hao *et al.* 2015).

Like for matrices, the spectral radius of a tensor $\mathcal{A}$ is the quantity defined as follows:

$$\rho(\mathcal{A}) = \max_i |\lambda_i(\mathcal{A})|, \qquad\qquad [4.12]$$

with $\lambda_i(\mathcal{A}) \in \text{sp}(\mathcal{A})$, where $\text{sp}(\mathcal{A})$ is the spectrum, i.e. the set of eigenvalues of $\mathcal{A}$. The geometric multiplicity m_i of an eigenvalue λ_i is defined as the maximum number of linearly independent eigenvectors associated with λ_i. Unlike the matrix case, the spectral radius of a tensor depends on the definition of its eigenvalues.

EXAMPLE 4.4.– For a third-order tensor $\mathcal{A} \in \mathbb{R}_S^{[3;2]}$ of dimensions two, equation [4.5] becomes:

$$\sum_{j,k=1}^{2} a_{ijk}x_j x_k = \lambda x_i^2 \quad \text{with} \quad i \in \{1,2\}, \qquad\qquad [4.13]$$

or in developed form:

$$a_{111}x_1^2 + (a_{112} + a_{121})x_1 x_2 + a_{122}x_2^2 = \lambda x_1^2$$
$$a_{211}x_1^2 + (a_{212} + a_{221})x_1 x_2 + a_{222}x_2^2 = \lambda x_2^2.$$

For a positive tensor $\mathcal{A}$ such that $a_{111} = a_{222} = 1, a_{122} = a_{211} = \epsilon$, with $0 < \epsilon < 1$, and $a_{ijk} = 0$ for the other triplets (i,j,k), the above system of equations simplifies as follows (Pearson 2010):

$$x_1^2 + \epsilon x_2^2 = \lambda x_1^2$$
$$\epsilon x_1^2 + x_2^2 = \lambda x_2^2.$$

By adding and subtracting both sides of these equations, we obtain the eigenvalues $\lambda_1 = 1 + \epsilon$ and $\lambda_2 = 1 - \epsilon$, with eigenvectors satisfying the equations $x_1^2 = x_2^2$ and $x_1^2 = -x_2^2$, respectively. For λ_1, we have the real eigenvectors $\mathbf{x} = (1,1)$ and $\mathbf{x} = (1,-1)$, whereas for λ_2 we have the complex eigenvectors $\mathbf{x} = (1,i)$ and $\mathbf{x} = (1,-i)$, where $i^2 = -1$. We can therefore conclude that the real geometric multiplicity of λ_1 and the complex geometric multiplicity of λ_2 are both equal to 2.

Qi (2005) proved the following properties for a real symmetric tensor $\mathcal{A} \in \mathbb{R}_S^{[P;I]}$:

– The number of eigenvalues is equal to $I(P - 1)^{I-1}$, which is the degree of the resultant defined in [4.7] (see remark 4.1). Their product is equal to $\det(\mathcal{A})$, the resultant of the I homogeneous polynomials $(\mathcal{A}\mathbf{x}^{P-1})_i$, with $i \in \langle I \rangle$. The sum of all eigenvalues is equal to $(P - 1)^{I-1} \, \mathrm{tr}(\mathcal{A})$.

– If P is even, then $\mathcal{A}$ is positive (semi-)definite if and only if all its H- and Z-eigenvalues are positive (non-negative); furthermore, a necessary condition for $\mathcal{A}$ to be positive semi-definite is $\det(\mathcal{A}) \geq 0$.

– The Z-eigenpair corresponding to the Z-eigenvalue with the largest absolute value gives the best rank-one approximation of $\mathcal{A}$ (Qi 2005; Qi et al. 2009) (see proposition 4.12).

We therefore have the following result (Qi 2005):

PROPOSITION 4.5.– Any real symmetric even-order tensor $\mathcal{A}$ is positive (semi-)definite (see the definition in Table 4.2) if and only if its smallest H-eigenvalue is positive (non-negative).

PROOF.– Let $(\lambda_{\mathrm{opt}}, \mathbf{x}_{\mathrm{opt}})$ be an H-eigenpair that minimizes the criterion [4.8] and is therefore a solution of the equation $\mathcal{A}\mathbf{x}_{\mathrm{opt}}^{P-1} = \lambda_{\mathrm{opt}}\mathbf{x}_{\mathrm{opt}}^{[P-1]}$. Taking the inner product of the two sides of this equation with the vector $\mathbf{x}_{\mathrm{opt}}$, we have:

$$\langle \mathcal{A}\mathbf{x}_{\mathrm{opt}}^{P-1}, \mathbf{x}_{\mathrm{opt}} \rangle = \mathcal{A}\mathbf{x}_{\mathrm{opt}}^{P} \quad \text{and} \quad \langle \mathbf{x}_{\mathrm{opt}}^{[P-1]}, \mathbf{x}_{\mathrm{opt}} \rangle = \sum_{i=1}^{I} x_{\mathrm{opt},i}^{P},$$

from which we deduce $\mathcal{A}\mathbf{x}_{\mathrm{opt}}^{P} = \lambda_{\mathrm{opt}} \sum_{i=1}^{I} x_{\mathrm{opt},i}^{P}$. Taking into account the hypothesis that P is even, which implies that $x_{\mathrm{opt},i}^{P} \geq 0$ for $i \in \langle I \rangle$, as well as the fact that $\mathcal{A}$ is positive definite, in the sense that $\mathcal{A}\mathbf{x}^{P} > 0, \forall \mathbf{x} \in \mathbb{R}^{I}$, we deduce from the above equality that $\lambda_{\mathrm{opt}} > 0$. Hence, the smallest eigenvalue λ_{opt} is positive (or non-negative for a positive semi-definite tensor $\mathcal{A}$). □

To illustrate the H- and Z-eigenvalues, let us consider the case of a symmetric third-order tensor $\mathcal{A} \in \mathbb{R}_S^{[3;I]}$ of dimensions I. In the case of an H-eigenpair, the eigenvalues are obtained by solving the system of homogeneous polynomial equations deduced from [4.6], which are of degree two in the components of $\mathbf{x} \in \mathbb{R}^I, \mathbf{x} \neq \mathbf{0}$:

$$\sum_{j,k=1}^{I} a_{ijk}x_j x_k = \lambda x_i^2 \ , \ i \in \langle I \rangle, \tag{4.14}$$

whereas in the case of a Z-eigenpair the system of non-homogeneous polynomial equations that must be solved is as follows:

$$\sum_{j,k=1}^{I} a_{ijk}x_j x_k = \lambda x_i \ , \ i \in \langle I \rangle. \tag{4.15}$$

Since equations [4.15] are not homogeneous, the eigenvectors are not invariant up to a scaling factor. This is due to the constraint $\|\mathbf{x}\|_2^2 = 1$. To satisfy this invariance property, we need to consider the l_3 norm for the constraint $\|\mathbf{x}\|_3 = (\sum_{i=1}^{I} |x_i|^3)^{1/3} = 1$, which then gives us equations [4.14].

The notion of generalized eigenpair introduced by Chang *et al.* (2009) is defined as follows. For more details, see the articles by Kolda and Mayo (2014) and Ding and Wei (2015).

DEFINITION 4.3.– *A pair* $(\lambda, \mathbf{x})$ *is said to be a* $\mathcal{B}$-*eigenpair of* $\mathcal{A} \in \mathbb{R}_S^{[P;I]}$ *if it is a solution of the following system of homogeneous polynomial equations:*

$$\mathcal{A}\mathbf{x}^{P-1} = \lambda \mathcal{B}\mathbf{x}^{P-1}.$$ [4.16]

It is also called a $\mathcal{B}$-*eigenvalue and a* $\mathcal{B}$-*eigenvector. This generalized eigenpair can be interpreted in terms of a constrained optimization (Kolda and Mayo 2014):*

$$min/max \, f(\mathbf{x}) = \frac{\mathcal{A}\mathbf{x}^P}{\mathcal{B}\mathbf{x}^P} \|\mathbf{x}\|_P^P \, subject \, to \, the \, constraint \, \|\mathbf{x}\|_P^P = 1,$$ [4.17]

where $\mathcal{A}, \mathcal{B} \in \mathbb{R}_S^{[P;I]}$ *are symmetric tensors of order* P *and dimensions* I, *the tensor* $\mathcal{A}$ *is positive semi-definite, the tensor* $\mathcal{B}$ *is positive definite (so that* $\mathcal{B}\mathbf{x}^P > 0$ *for* $\forall \mathbf{x} \neq \mathbf{0}$) *and* λ *is the Lagrange multiplier defined as:*

$$\lambda = \frac{\mathcal{A}\mathbf{x}^P}{\mathcal{B}\mathbf{x}^P}.$$ [4.18]

PROOF.– After defining the Lagrangian as $L(\mathbf{x}, \lambda) = f(\mathbf{x}) - \lambda(\|\mathbf{x}\|_P^P - 1)$, the KKT optimality conditions [4.9] become:

$$\frac{\partial L(\mathbf{x}, \lambda)}{\partial \mathbf{x}} = \frac{P}{\mathcal{B}\mathbf{x}^P} \left[(\mathcal{A}\mathbf{x}^P)\mathbf{x} + \mathcal{A}\mathbf{x}^{P-1} - (\frac{\mathcal{A}\mathbf{x}^P}{\mathcal{B}\mathbf{x}^P})\mathcal{B}\mathbf{x}^{P-1} \right] - P\lambda\mathbf{x} = 0$$

$$\frac{\partial L(\mathbf{x}, \lambda)}{\partial \lambda} = 1 - \|\mathbf{x}\|_P^P = 0,$$ [4.19]

from which we deduce equations [4.16] and [4.18]. For more details about the proof, see Kolda and Mayo (2014), who also suggest a method for computing generalized eigenpairs.   □

REMARK 4.6.– The definition [4.5] can be deduced as a special case of [4.16] by choosing $\mathcal{B} = \mathcal{I}_P$, the identity tensor of order P. Indeed, we then have, for $\mathbf{x} \in \mathbb{R}^I$:

$$\mathcal{B}\mathbf{x}^P = \langle \mathcal{I}_P, \mathbf{x}^{\circ P} \rangle = \sum_{i=1}^{I} x_i^p = \|\mathbf{x}\|_P^P$$ [4.20]

$$\mathcal{B}\mathbf{x}^{P-1} = \mathcal{I}_P\mathbf{x}^{P-1} = \mathbf{x}^{[P-1]}.$$ [4.21]

Equation [4.16] with $\mathcal{B} = \mathcal{I}_P$ then becomes $\mathcal{A}\mathbf{x}^{P-1} = \lambda\mathbf{x}^{[P-1]}$, i.e. equation [4.5].

As we mentioned earlier, the notion of eigenvalue can also be defined using the Einstein product (Liang *et al.* 2018).

DEFINITION 4.4.– *Given a complex square tensor* $\mathcal{A} \in \mathbb{C}^{\underline{I}_P \times \underline{I}_P}$ *of order* $2P$, *a pair* $(\lambda, \mathcal{X}) \in \mathbb{C} \times \mathbb{C}^{\underline{I}_P \times \underline{I}_P}$, *with* $\mathcal{X} \neq \mathcal{O}$, *is said to be an eigenvalue–eigentensor pair if it satisfies the following equation:*

$$\mathcal{A} \star_P \mathcal{X} = \lambda\mathcal{X}, \tag{4.22}$$

which amounts to solving the following system of tensor equations:

$$(\lambda\,\mathcal{J}_{2P} - \mathcal{A}) \star_P \mathcal{X} = \mathcal{O}. \tag{4.23}$$

Taking the property [3.24] into account, equation [4.23] can also be written as:

$$(\lambda\,\mathbf{I}_I - \mathbf{A}_{I\times I})\mathbf{X}_{I\times I} = \mathbf{0}\ ,\ I = \prod_{p=1}^{P} I_p. \tag{4.24}$$

The eigenvalues are then obtained by solving the characteristic equation of the square tensor $\mathcal{A}$ *expressed in terms of its matrix unfolding* $\mathbf{A}_{I\times I}$:

$$det(\lambda\,\mathbf{I}_I - \mathbf{A}_{I\times I}) = 0, \tag{4.25}$$

and the eigentensor associated with the eigenvalue λ *is determined by solving the tensor equation [4.23] or the matrix equation [4.24].*

DEFINITION 4.5.– *A variant of the eigenvalue–eigentensor problem was proposed by Cui et al. (2015) in the form of equation [4.22], with a tensor* $\mathcal{X} \in \mathbb{C}^{\underline{I}_P}$. *Equation [4.24] then becomes:*

$$(\lambda\,\mathbf{I}_I - \mathbf{A}_{I\times I})\mathbf{x}_I = \mathbf{0}, \tag{4.26}$$

where $\mathbf{x}_I \in \mathbb{C}^I$ *is a vectorized form of the tensor* $\mathcal{X}$ *obtained using a mode combination as defined in [3.16]. The eigenvalue–eigentensor problem is then reduced to a standard eigenvalue–eigenvector problem for the matrix* $\mathbf{A}_{I\times I}$, *and the eigentensor is computed in the vectorized form* $\mathbf{x}_I$ *by solving equation [4.26]. The eigenvalue problem for a Toeplitz tensor was also solved by Cui et al. (2015) and applied to image restoration.*

The various definitions of tensor eigenpairs introduced above are summarized in Table 4.1.

$(\lambda, \mathbf{x})$	Systems of polynomial equations	Eigenpairs
	$\mathcal{A} \in \mathbb{R}_S^{[P;I]}$	eigenvalue-eigenvector
$\in \mathbb{C} \times \{\mathbb{C}^I \backslash \{\mathbf{0}\}\}$	$\mathcal{A}\mathbf{x}^{P-1} = \lambda \mathbf{x}^{[P-1]}$	eigenpair
$\in \mathbb{C} \times \{\mathbb{C}^I \backslash \{\mathbf{0}\}\}$	$\mathcal{A}\mathbf{x}^{P-1} = \lambda \mathbf{x}$, $\mathbf{x}^T \mathbf{x} = 1$	E-eigenpair
$\in \mathbb{R} \times \{\mathbb{R}^I \backslash \{\mathbf{0}\}\}$	$\mathcal{A}\mathbf{x}^{P-1} = \lambda \mathbf{x}^{[P-1]}$	H-eigenpair
$\in \mathbb{R} \times \{\mathbb{R}^I \backslash \{\mathbf{0}\}\}$	$\mathcal{A}\mathbf{x}^{P-1} = \lambda \mathbf{x}$, $\mathbf{x}^T \mathbf{x} = 1$	Z-eigenpair
$(\lambda, \mathbf{x})$	Systems of polynomial equations	Generalized eigenpair
	$\mathcal{A}, \mathcal{B} \in \mathbb{R}_S^{[P;I]}$ with $\mathcal{A} \geq 0, \mathcal{B} > 0$	
$\in \mathbb{C} \times \{\mathbb{C}^I \backslash \{\mathbf{0}\}\}$	$\mathcal{A}\mathbf{x}^{P-1} = \lambda \mathcal{B}\mathbf{x}^{P-1}$	generalized eigenpair
$(\lambda, \mathcal{X}), \mathcal{X} \neq \mathcal{O}$	$\mathcal{A} \in \mathbb{C}^{I_P \times I_P}$	
$\in \mathbb{C} \times \{\mathbb{C}^{I_P} \backslash \{\mathcal{O}\}\}$	$\mathcal{A} \star_P \mathcal{X} = \lambda \mathcal{X}$	eigenvalue-eigentensor

Table 4.1. *Eigenpairs and generalized eigenpairs for tensors*

4.2.2. *Positive/negative (semi-)definite tensors*

The multiple tensor-vector product [3.274] and the notion of eigenvalue of a tensor have been used to define various classes of tensors, such as positive definite tensors, $\mathcal{M}$-tensors and P-tensors (Ding *et al.* 2013; Zhang *et al.* 2014; Song and Qi 2015).

In the same way that a positive (respectively, negative) definite matrix $\mathbf{A} \in \mathbb{R}_S^{I \times I}$ satisfies the property that the quadratic form $\mathbf{x}^T \mathbf{A} \mathbf{x} = \mathbf{A} \times_1 \mathbf{x}^T \times_2 \mathbf{x}^T$, denoted $\mathbf{A}\mathbf{x}^2$, is positive (respectively, negative), a real symmetric hypercubic tensor $\mathcal{A} \in \mathbb{R}_S^{[P;I]}$ is positive definite (respectively, negative) if the P-linear form $\mathcal{A}\mathbf{x}^P$ is itself positive (respectively, negative) definite, as summarized in Table 4.2. These definitions generalize those of Table 1.5 to tensors of order greater than two (see Chen and Qi 2015).

Properties	Conditions
$\mathcal{A}$ positive semi-definite $(\mathcal{A} \geq 0)$	$\mathcal{A}\mathbf{x}^P \geq 0, \forall \mathbf{x} \in \mathbb{R}^I$
$\mathcal{A}$ positive definite $(\mathcal{A} > 0)$	$\mathcal{A}\mathbf{x}^P > 0, \forall \mathbf{x} \in \mathbb{R}^I, \mathbf{x} \neq \mathbf{0}$
$\mathcal{A}$ negative semi-definite $(\mathcal{A} \leq 0)$	$\mathcal{A}\mathbf{x}^P \leq 0, \forall \mathbf{x} \in \mathbb{R}^I$
$\mathcal{A}$ negative definite $(\mathcal{A} < 0)$	$\mathcal{A}\mathbf{x}^P < 0, \forall \mathbf{x} \in \mathbb{R}^I, \mathbf{x} \neq \mathbf{0}$

Table 4.2. *Properties of hypercubic symmetric tensors* $\mathcal{A} \in \mathbb{R}_S^{[P;I]}$

4.2.3. *Orthogonally/unitarily similar tensors*

Analogously to the definition [1.21] of orthogonally similar matrices, we say that the tensors $\mathcal{A}$ and $\mathcal{B} \in \mathbb{R}^{[P;I]}$ are orthogonally similar if there exists an orthogonal matrix $\mathbf{Q} \in \mathbb{R}^{I \times I}$ such that (Qi 2012):

$$\mathcal{B} = \mathcal{A} \times_1 \mathbf{Q} \times_2 \mathbf{Q} \cdots \times_P \mathbf{Q} \triangleq \mathcal{A} \overset{P}{\underset{p=1}{\times}} \mathbf{Q}, \qquad [4.27]$$

or alternatively, in scalar form and using the index convention:

$$b_{i_1 i_2 \cdots i_P} = \sum_{k_1, k_2, \cdots, k_P = 1}^{I} q_{i_1 k_1} q_{i_2 k_2} \cdots q_{i_P k_P} a_{k_1 k_2 \cdots k_P} \qquad [4.28]$$

$$b_{\mathbf{i}_P} = a_{\mathbf{k}_P} \prod_{p=1}^{P} q_{i_p k_p}. \qquad [4.29]$$

Similarly, in the complex case, analogously to the definition [1.22] in the matrix case, we say that the tensors $\mathcal{A}$ and $\mathcal{B} \in \mathbb{C}^{[2P;I]}$ are unitarily similar if there exists a unitary matrix $\mathbf{Q} \in \mathbb{C}^{I \times I}$ such that:

$$\mathcal{B} = \mathcal{A} \times_1 \mathbf{Q} \times_2 \cdots \times_P \mathbf{Q} \times_{P+1} \mathbf{Q}^* \times_{P+2} \cdots \times_{2P} \mathbf{Q}^* \triangleq \mathcal{A} \overset{P}{\underset{p=1}{\times}} \mathbf{Q} \overset{2P}{\underset{p=P+1}{\times}} \mathbf{Q}^*. $$
$$[4.30]$$

The relations [4.28] and [4.29] then become:

$$b_{\mathbf{i}_P, \mathbf{j}_P} = \sum_{k_1, \cdots, k_P, n_1, \cdots, n_P = 1}^{I} q_{i_1 k_1} \cdots q_{i_P k_P} q^*_{j_1 n_1} \cdots q^*_{j_P n_P} a_{k_1 \cdots k_P n_1 \cdots n_P}$$
$$[4.31]$$

$$= a_{\mathbf{k}_P, \mathbf{n}_P} \prod_{p=1}^{P} q_{i_p k_p} q^*_{j_p n_p}. \qquad [4.32]$$

REMARK 4.7.– We have the following results:

– If $\mathcal{A}$ is symmetric, then $\mathcal{B}$ defined in [4.27] is also symmetric.

– Using the orthogonality property of $\mathbf{Q}$, we deduce from [4.27] that:

$$\mathcal{A} = \mathcal{B} \overset{P}{\underset{p=1}{\times}} \mathbf{Q}^{-1} = \mathcal{B} \overset{P}{\underset{p=1}{\times}} \mathbf{Q}^T. \qquad [4.33]$$

Similarly, from [4.30], with a unitary matrix $\mathbf{Q}$, we deduce:

$$\mathcal{A} = \mathcal{B} \overset{P}{\underset{p=1}{\times}} \mathbf{Q}^H \overset{2P}{\underset{p=P+1}{\times}} \mathbf{Q}^T. \qquad [4.34]$$

PROPOSITION 4.8.– Any two orthogonally similar symmetric tensors have the same eigenvalues in the sense of the definition [4.10]. Furthermore[4], if $(\lambda, \mathbf{x})$ is an eigenpair of $\mathcal{A}$, then $(\lambda, \mathbf{Qx})$ is an eigenpair for $\mathcal{B}$. This property generalizes the corresponding property of orthogonally similar matrices to the case of tensors of order greater than two (see Table 1.6).

PROOF.– Let $\mathcal{A}$ and $\mathcal{B}$ be symmetric and orthogonally similar tensors, according to the relation [4.27], and let $(\lambda, \mathbf{x})$ be an eigenpair of $\mathcal{A}$ in the sense of the definition $\mathcal{A}\mathbf{x}^{P-1} = \lambda\mathbf{x}$. Defining the vector $\mathbf{y} = \mathbf{Qx}$, or equivalently $\mathbf{x}^T = \mathbf{y}^T\mathbf{Q}$, we have:

$$\mathcal{A}\mathbf{x}^{P-1} = \mathcal{A} \underset{p=2}{\overset{P}{\times}} \mathbf{x}^T = \mathcal{A} \underset{p=2}{\overset{P}{\times}} \mathbf{y}^T\mathbf{Q} = (\mathcal{A} \underset{p=2}{\overset{P}{\times}} \mathbf{Q}) \underset{p=2}{\overset{P}{\times}} \mathbf{y}^T \qquad [4.35]$$

$$\lambda\mathbf{x} = \lambda\mathbf{Q}^{-1}\mathbf{Qx} = \lambda\mathbf{Q}^{-1}\mathbf{y}. \qquad [4.36]$$

From these two equations, we deduce:

$$(\mathcal{A} \underset{p=2}{\overset{P}{\times}} \mathbf{Q}) \underset{p=2}{\overset{P}{\times}} \mathbf{y}^T = \lambda\mathbf{Q}^{-1}\mathbf{y}. \qquad [4.37]$$

By pre-multiplying the two sides of this equation by $\mathbf{Q}$, we obtain:

$$(\mathcal{A} \underset{p=1}{\overset{P}{\times}} \mathbf{Q}) \underset{p=2}{\overset{P}{\times}} \mathbf{y}^T = \mathcal{B} \underset{p=2}{\overset{P}{\times}} \mathbf{y}^T = \lambda\mathbf{y}, \qquad [4.38]$$

which proves that $(\lambda, \mathbf{Qx})$ is an eigenpair of $\mathcal{B}$.   □

REMARK 4.9.– In the case of a symmetric tensor $\mathcal{A} \in \mathbb{R}_S^{[P;I]}$, the HOSVD decomposition [5.15] can be written as:

$$\mathcal{A} = \mathcal{G} \underset{p=1}{\overset{P}{\times}} \mathbf{U}, \qquad [4.39]$$

where $\mathcal{G}$ is symmetric and $\mathbf{U}$ is orthogonal. The tensors $\mathcal{A}$ and $\mathcal{G}$ are therefore orthogonally similar, and, by the previous proposition, we conclude that the eigenpair $(\lambda, \mathbf{x})$ of $\mathcal{G}$ corresponds to the eigenpair $(\lambda, \mathbf{Ux})$ of $\mathcal{A}$. Equivalently, each eigenpair $(\lambda, \mathbf{y})$ of $\mathcal{A}$ is associated with the eigenpair $(\lambda, \mathbf{U}^T\mathbf{y})$ of $\mathcal{G}$.

Note that, in the general (non-symmetric) case, the HOSVD of the tensor $\mathcal{A}$ is not determined by computing its eigenvalues but by computing the singular values of its P modal unfoldings. In the symmetric case, $\mathbf{U}$ is composed of the left singular vectors of one of the P modal unfoldings, which are all identical due to the symmetry of $\mathcal{A}$.

---

4 Our proof of this result is different to the one given in Qi (2005).

4.3. Best rank-one approximation

The best rank-one approximation of a tensor plays a very important role in applications involving the simplification of the representation of a data tensor (de Lathauwer *et al.* 2000b; Zhang and Golub 2001; Kofidis and Regalia 2002). Recall that, in the matrix case, the best rank-one approximation, denoted $\hat{\mathbf{A}}$, in the sense of the Frobenius norm, of a matrix $\mathbf{A} \in \mathbb{K}^{I \times J}$ of rank R, is directly obtained from the compact SVD, namely $\mathbf{A} = \sum_{r=1}^{R} \sigma_r \mathbf{u}_r \mathbf{v}_r^H$, as $\hat{\mathbf{A}} = \sigma_1 \mathbf{u}_1 \mathbf{v}_1^H$, where σ_1 is the largest singular value of $\mathbf{A}$, and $(\mathbf{u}_1, \mathbf{v}_1)$ are the left and right singular vectors associated with σ_1. The approximation error is given by $\|\mathbf{A} - \hat{\mathbf{A}}\|_F^2 = \sum_{r=2}^{R} \sigma_r^2$, with $\sigma_1 \geq \sigma_2 \geq \cdots \geq \sigma_R > 0$ (see section 1.5.7).

In this section, we present the link between the best rank-one approximation of a given tensor and the maximization of a homogeneous polynomial.

PROPOSITION 4.10.– Let $\mathcal{A} \in \mathbb{R}^{\underline{I}_P}$ be a tensor of order P and size $\underline{I}_P$. The best rank-one approximation of $\mathcal{A}$, in the sense of the Frobenius norm, is obtained by minimizing the following criterion with respect to the scalar λ and the vectors $\mathbf{x}^{(p)} \in \mathbb{R}^{I_p}$:

$$f(\lambda, \mathbf{x}^{(1)} \cdots, \mathbf{x}^{(P)}) = \|\mathcal{A} - \lambda \underset{p=1}{\overset{P}{\circ}} \mathbf{x}^{(p)}\|_F^2 \qquad [4.40]$$

$$= \sum_{i_1=1}^{I_1} \cdots \sum_{i_P=1}^{I_P} \left(a_{\mathbf{i}_P} - \lambda \prod_{p=1}^{P} x_{i_p}^{(p)} \right)^2 \qquad [4.41]$$

subject to the constraint $\|\mathbf{x}^{(p)}\|_2 = 1$ for every $p \in \langle P \rangle$. A sub-optimal solution is obtained by minimizing [4.40] in two steps. First, the criterion is minimized with respect to λ while assuming that the vectors $\mathbf{x}^{(p)}$ are fixed, then the criterion is minimized with respect to $\mathbf{x}^{(p)}$ using the optimized value of λ found during the first step. This second step is the object of proposition 4.13.

The first minimization leads us to maximize the following criterion with respect to the vectors $\mathbf{x}^{(p)}$:

$$\langle \mathcal{A}, \underset{p=1}{\overset{P}{\circ}} \mathbf{x}^{(p)} \rangle^2 = \left(\mathcal{A} \underset{p=1}{\overset{P}{\times}} \mathbf{x}^{(p)T} \right)^2 = \left(\sum_{i_1=1}^{I_1} \cdots \sum_{i_P=1}^{I_P} a_{\mathbf{i}_P} \prod_{p=1}^{P} x_{i_p}^{(p)} \right)^2 \qquad [4.42]$$

$$= \left(a_{\mathbf{i}_P} \prod_{p=1}^{P} x_{i_p}^{(p)} \right)^2, \qquad [4.43]$$

where the last equality follows from the index convention.

PROOF.– The criterion [4.40] can be developed as follows:

$$\|\mathcal{A} - \lambda \overset{P}{\underset{p=1}{\circ}} \mathbf{x}^{(p)}\|_F^2 = \langle \mathcal{A} - \lambda \overset{P}{\underset{p=1}{\circ}} \mathbf{x}^{(p)}, \mathcal{A} - \lambda \overset{P}{\underset{p=1}{\circ}} \mathbf{x}^{(p)} \rangle$$

$$= \|\mathcal{A}\|_F^2 - 2\lambda \langle \mathcal{A}, \overset{P}{\underset{p=1}{\circ}} \mathbf{x}^{(p)} \rangle + \lambda^2 \| \overset{P}{\underset{p=1}{\circ}} \mathbf{x}^{(p)}\|_F^2. \qquad [4.44]$$

Since the vectors $\mathbf{x}^{(p)}$ are assumed to be fixed, the minimization of [4.44] with respect to λ is obtained by setting the gradient with respect to λ equal to zero, which gives the following solution:

$$\lambda_{\text{opt}}(\mathbf{x}^{(p)}) = \frac{\langle \mathcal{A}, \overset{P}{\underset{p=1}{\circ}} \mathbf{x}^{(p)} \rangle}{\| \overset{P}{\underset{p=1}{\circ}} \mathbf{x}^{(p)}\|_F^2}. \qquad [4.45]$$

This expression can be interpreted as a high-order generalized Rayleigh quotient. Taking into account the constraint $\|\mathbf{x}^{(p)}\|_2 = 1$ for every $p \in \langle P \rangle$, as well as the expression [3.246] for the Frobenius norm of a rank-one tensor of order P, we have:

$$\| \overset{P}{\underset{p=1}{\circ}} \mathbf{x}^{(p)}\|_F^2 = \prod_{p=1}^{P} \|\mathbf{x}^{(p)}\|_2^2 = 1. \qquad [4.46]$$

By proposition 3.81, the solution [4.45] can also be written as:

$$\lambda_{\text{opt}}(\mathbf{x}^{(p)}) = \langle \mathcal{A}, \overset{P}{\underset{p=1}{\circ}} \mathbf{x}^{(p)} \rangle = \mathcal{A} \overset{P}{\underset{p=1}{\times}} \mathbf{x}^{(p)^T} = a_{\mathbf{i}_P} \prod_{p=1}^{P} x_{i_p}^{(p)}. \qquad [4.47]$$

After replacing λ with this value λ_{opt} in [4.44] and taking the relation [4.46] into account, the value of the minimized criterion is given by:

$$\|\mathcal{A} - \lambda_{\text{opt}} \overset{P}{\underset{p=1}{\circ}} \mathbf{x}^{(p)}\|_F^2 = \|\mathcal{A}\|_F^2 - 2\lambda_{\text{opt}} \langle \mathcal{A}, \overset{P}{\underset{p=1}{\circ}} \mathbf{x}^{(p)} \rangle + \lambda_{\text{opt}}^2$$

$$= \|\mathcal{A}\|_F^2 - \lambda_{\text{opt}}^2$$

$$= \|\mathcal{A}\|_F^2 - \left(a_{\mathbf{i}_P} \prod_{p=1}^{P} x_{i_p}^{(p)} \right)^2. \qquad [4.48]$$

Since the Frobenius norm of $\mathcal{A}$ is constant, minimizing the above criterion is equivalent to maximizing λ_{opt}^2 and hence to maximizing the criterion [4.43]. $\qquad \square$

REMARK 4.11.– In the case of a symmetric tensor $\mathcal{A} \in \mathbb{R}_S^{[P,I]}$ of order P and dimensions I, the criterion [4.40] is written as $f(\lambda, \mathbf{x}) = \|\mathcal{A} - \lambda \mathcal{I}_P \mathbf{x}^P\|_F^2$, where $\mathcal{I}_P$ is the identity tensor of order P. By the above proposition, minimizing this criterion is equivalent to maximizing the criterion $\mathcal{A} \overset{P}{\underset{p=1}{\times}} \mathbf{x}^T = \mathcal{A}\mathbf{x}^P$ subject to the constraint of unit Euclidean norm. Hence, we can conclude that maximizing the criterion [4.43] gives the largest eigenvalue of $\mathcal{A}$, in the sense of the definition [4.10]. This result is summarized in the next proposition.

PROPOSITION 4.12.– The best rank-one approximation of a symmetric tensor $\mathcal{A} \in \mathbb{R}_S^{[P,I]}$ is given by the eigenvalue according to the definition [4.10], with the largest absolute value, which can be obtained by solving the system of polynomial equations $\mathcal{A}\mathbf{x}^{P-1} = \lambda\mathbf{x}$.

In the general case of a tensor $\mathcal{A} \in \mathbb{R}^{I_P}$, maximizing the criterion [4.43] with respect to the vectors $\mathbf{x}^{(p)}$ subject to the constraint $\|\mathbf{x}^{(p)}\|_2 = 1$ for every $p \in \langle P \rangle$ constitutes the second step of proposition 4.10. This maximization leads to the solution described in the next proposition.

PROPOSITION 4.13.– The best rank-one approximation of a tensor $\mathcal{A} \in \mathbb{R}^{I_P}$ is given by $\lambda \overset{P}{\underset{p=1}{\circ}} \mathbf{x}^{(p)}$, where the unit-norm vectors $\mathbf{x}^{(p)}$ and the scalar λ are solutions of the following polynomial equations:

$$\mathcal{A} \overset{P}{\underset{\substack{q=1 \\ q \neq p}}{\times}} \mathbf{x}^{(q)^T} = \lambda\mathbf{x}^{(p)} \tag{4.49}$$

$$\mathcal{A} \overset{P}{\underset{p=1}{\times}} \mathbf{x}^{(p)^T} = \lambda. \tag{4.50}$$

PROOF.– The criterion [4.43] can be maximized subject to the constraint $\|\mathbf{x}^{(p)}\|_2 = 1$ for every $p \in \langle P \rangle$ using the Lagrangian:

$$L\big(\lambda_1, \cdots, \lambda_P, \mathbf{x}^{(1)}, \cdots, \mathbf{x}^{(P)}\big) = \mathcal{A} \overset{P}{\underset{p=1}{\times}} \mathbf{x}^{(p)^T} - \sum_{p=1}^{P} \frac{\lambda_p}{2}\big(\|\mathbf{x}^{(p)}\|_2^2 - 1\big). \tag{4.51}$$

By setting the partial derivatives of the Lagrangian [4.51] with respect to the optimization variables to zero, we obtain the following equations, for every $p \in \langle P \rangle$, as the KKT optimality conditions:

$$\frac{\partial L}{\partial \lambda_p} = \frac{1}{2}(1 - \|\mathbf{x}^{(p)}\|_2^2) = 0 \tag{4.52}$$

$$\frac{\partial L}{\partial \mathbf{x}^{(p)}} = \mathcal{A} \overset{P}{\underset{\substack{q=1 \\ q \neq p}}{\times}} \mathbf{x}^{(q)^T} - \lambda_p \mathbf{x}^{(p)} = \mathbf{0}. \tag{4.53}$$

By considering the inner product of the vector $\frac{\partial L}{\partial \mathbf{x}^{(p)}}$ with the vector $\mathbf{x}^{(p)}$ and taking equation [4.52] into account, we obtain:

$$\langle \frac{\partial L}{\partial \mathbf{x}^{(p)}}, \mathbf{x}^{(p)} \rangle = \mathcal{A} \underset{p=1}{\overset{P}{\times}} \mathbf{x}^{(p)^T} - \lambda_p \|\mathbf{x}^{(p)}\|_2^2 \qquad [4.54]$$

$$= \mathcal{A} \underset{p=1}{\overset{P}{\times}} \mathbf{x}^{(p)^T} - \lambda_p = 0, \qquad [4.55]$$

from which we deduce that the λ_p are all identical and equal to $\lambda = \mathcal{A} \underset{p=1}{\overset{P}{\times}} \mathbf{x}^{(p)^T}$. The optimality equations [4.53] therefore give the best rank-one approximation, as defined in [4.49], where λ is given by the expression [4.50]. $\qquad \square$

EXAMPLE 4.14.– To illustrate the above results, consider the case of a third-order tensor $\mathcal{A} \in \mathbb{R}^{I \times J \times K}$ (Zhang and Golub 2001). Writing the vectors $\mathbf{x}^{(p)}$, for $p \in \{1, 2, 3\}$ as $\mathbf{x} \in \mathbb{R}^I, \mathbf{y} \in \mathbb{R}^J, \mathbf{z} \in \mathbb{R}^K$, the optimality equations [4.49]–[4.50] become:

$$\mathcal{A} \times_2 \mathbf{y} \times_3 \mathbf{z} = \lambda \mathbf{x} \iff \sum_{j,k} a_{ijk} y_j z_k = \lambda x_i \ , \ i \in \langle I \rangle \qquad [4.56]$$

$$\mathcal{A} \times_1 \mathbf{x} \times_3 \mathbf{z} = \lambda \mathbf{y} \iff \sum_{i,k} a_{ijk} x_i z_k = \lambda y_j \ , \ j \in \langle J \rangle \qquad [4.57]$$

$$\mathcal{A} \times_1 \mathbf{x} \times_2 \mathbf{y} = \lambda \mathbf{z} \iff \sum_{i,j} a_{ijk} x_i y_j = \lambda z_k \ , \ k \in \langle K \rangle \qquad [4.58]$$

$$\mathcal{A} \times_1 \mathbf{x} \times_2 \mathbf{y} \times_3 \mathbf{z} = \sum_{i,j,k} a_{ijk} x_i y_j z_k = \lambda. \qquad [4.59]$$

In Zhang and Golub (2001), the Newton and alternating least squares (ALS) algorithms are suggested to solve these polynomial equations. See section 5.2.5.8 for the ALS algorithm.

REMARK 4.15.– Unlike the matrix case, the low-rank approximation problem is ill-posed for tensors. De Silva and Lim (2008) showed that a tensor $\mathcal{A} \in \mathbb{R}^{I_P}$ of order P does not admit a best approximation of rank R with $R \leq \min(I_1, \cdots, I_P)$ in general.

4.4. Orthogonal decompositions

In Chapter 1, we saw that any real symmetric matrix $\mathbf{A} \in \mathbb{R}_S^{I \times I}$ of rank R always admits an orthogonal decomposition, namely the eigendecomposition, which can be written as:

$$\mathbf{A} = \mathbf{U \Lambda U}^T = \sum_{r=1}^{R} \lambda_r \mathbf{u}_r \mathbf{u}_r^T = \sum_{r=1}^{R} \lambda_r \mathbf{u}_r \circ \mathbf{u}_r, \qquad [4.60]$$

where $\mathbf{\Lambda} = \mathrm{diag}(\lambda_1, \cdots, \lambda_R) \in \mathbb{R}^{R \times R}$ is a diagonal matrix, and $\mathbf{U} = [\mathbf{u}_1 \cdots \mathbf{u}_R] \in \mathbb{R}^{I \times R}$ is a column orthonormal matrix ($\mathbf{u}_i^T \mathbf{u}_j = \delta_{ij}$). The eigenvalues are determined by solving the equation $\mathbf{A}\mathbf{u}_i = \lambda_i \mathbf{u}_i$, which is equivalent to determining the extrema of the Rayleigh quotient $\frac{\mathbf{u}^T \mathbf{A} \mathbf{u}}{\|\mathbf{u}\|^2}$.

In the case of a real symmetric tensor $\mathcal{A} \in \mathbb{R}_S^{I_P}$ of order P, an orthogonal decomposition is defined as a decomposition of the form:

$$\mathcal{A} = \sum_{r=1}^{R} \lambda_r \underbrace{(\mathbf{u}_r \circ \cdots \circ \mathbf{u}_r)}_{P \text{ terms}} \triangleq \sum_{r=1}^{R} \lambda_r \mathbf{u}_r^{\circ P}, \qquad [4.61]$$

where $\mathbf{u}_r$ is the eigenvector of unit Euclidean norm associated with the eigenvalue λ_r, as defined in [4.10]. It can be obtained by solving the system of polynomial equations [4.10], written in scalar form as:

$$\sum_{i_2=1}^{I} \cdots \sum_{i_P=1}^{I} a_{i,i_2,\cdots,i_P} u_{i_2} \cdots u_{i_P} = \lambda u_i \ , \ i \in \langle I \rangle, \qquad [4.62]$$

with $\sum_{i=1}^{I} u_i^2 = 1$. It can also be obtained using a variational approach that maximizes the generalized Rayleigh quotient (Lim 2005). Note that, unlike the matrix case, not every real symmetric tensor admits such an orthogonal decomposition.

4.5. Singular values of a tensor

We define the singular values and singular vectors $\mathbf{u} \in \mathbb{R}^I$, $\mathbf{v} \in \mathbb{R}^J$, and $\mathbf{w} \in \mathbb{R}^K$ of a third-order tensor $\mathcal{A} \in \mathbb{R}^{I \times J \times K}$ as the values and critical points of the Rayleigh quotient:

$$\frac{\mathcal{A} \times_1 \mathbf{u} \times_2 \mathbf{v} \times_3 \mathbf{w}}{\|\mathbf{u}\|_3 \|\mathbf{v}\|_3 \|\mathbf{w}\|_3}. \qquad [4.63]$$

The solution is obtained from the Lagrangian $L(\mathbf{u}, \mathbf{v}, \mathbf{w}, \lambda) = \sum_{i,j,k=1}^{I,J,K} a_{ijk} u_i v_j w_k - \sigma(\|\mathbf{u}\|_3 \|\mathbf{v}\|_3 \|\mathbf{w}\|_3 - 1)$. Thus, we obtain the following system of homogeneous polynomial equations of degree two (Lim 2005):

$$\sum_{j,k=1}^{J,K} a_{ijk} v_j w_k = \sigma u_i^2 \ , \ i \in \langle I \rangle \qquad [4.64a]$$

$$\sum_{i,k=1}^{I,K} a_{ijk} u_i w_k = \sigma v_j^2 \ , \ j \in \langle J \rangle \qquad [4.64b]$$

$$\sum_{i,j=1}^{I,J} a_{ijk} u_i v_j = \sigma w_k^2 \ , \ k \in \langle K \rangle, \qquad [4.64c]$$

where σ is a singular value of $\mathcal{A}$, and $\mathbf{u}$, $\mathbf{v}$ and $\mathbf{w}$ are mode-1, -2 and -3 singular vectors, respectively. These equations generalize the equations in Table 1.9, which allow us to compute singular values and singular vectors of a matrix by solving a system of linear equations.

This definition can be generalized to a tensor $\mathcal{A} \in \mathbb{R}^{I_1 \times \cdots \times I_N}$ of order N. The mode-n singular vectors, denoted $\mathbf{u}^{(n)}$, with $n \in \langle N \rangle$, associated with the singular value σ, are then solutions of the following homogeneous polynomial equations:

$$\sum_{i_1,\cdots,i_{n-1},i_{n+1}\cdots,i_N=1}^{I_1,\cdots,I_{n-1},I_{n+1},\cdots,I_N} a_{i_1,\cdots,i_n,\cdots,i_N} u_{i_1}^{(1)} \cdots u_{i_{n-1}}^{(n-1)} u_{i_{n+1}}^{(n+1)} \cdots u_{i_N}^{(N)} = \sigma (u_{i_n}^{(n)})^{N-1},$$

where each equation is defined by the factor $u_{i_n}^{(n)}$, with $n \in \langle N \rangle$ and $i_n \in \langle I_n \rangle$.

For $N = 3$, we recover equations [4.64a]–[4.64c], with $(\mathbf{u}^{(1)}, \mathbf{u}^{(2)}, \mathbf{u}^{(3)}) = (\mathbf{u}, \mathbf{v}, \mathbf{w})$ and $(I_1, I_2, I_3) = (I, J, K)$.

Tensor Decompositions

5.1. Introduction

As we saw in the introductory chapter, tensor-based approaches have several advantages over traditional matrix-based ones, offering in particular the possibility to process multimodal and incomplete large-dimensional data efficiently. The exponential increase of the number of elements in a data tensor with the number of modes, i.e. the tensor order, is called the *curse of dimensionality* (Oseledets and Tyrtyshnikov 2009). Tensor decompositions, also called tensor models, allow us to represent data tensors by means of matrix factors and lower order tensors, called core tensors. Such representations can be exploited to reduce the storage memory while facilitating the data processing and leading to new solutions for various applications such as source separation, array processing, wireless communications, biomedical signal processing, data analysis and fusion, recommender systems, and estimation of missing data (tensor completion problem), to mention only a few.

Many different tensor models exist. Some of them have been developed within the framework of certain applications, as will be described in the next two volumes through the design of new wireless communication systems. Other models like the hierarchical Tucker (HT) and tensor train (TT) decompositions (Oseledets and Tyrtyshnikov 2010; Grasedyck and Hakbush 2011; Oseledets 2011; Ballani *et al.* 2013), which can be viewed as special cases of tensor networks (TNs) (Cichocki 2014), will also be presented in the next volume.

The aim of this chapter is to present the basic Tucker and PARAFAC decompositions, as well as some of their variants such as the Tucker-(N_1, N)decomposition, block Tucker and PARAFAC decompositions, constrained decompositions (PARALIND and CONFAC) and block term decomposition (BTD). The third-order Tucker and PARAFAC models will be described in further detail.

For the tensor models considered in this chapter, we will give different forms of representation (scalar forms, with mode-n products, with outer products, vectorized forms and matrix unfoldings). The uniqueness properties of the Tucker and PARAFAC models will be detailed. We will describe the HOSVD algorithm, which is based on computing the SVD of the modal matrix unfoldings, leading to a Tucker model with orthogonal matrix factors in the real case and unitary matrix factors in the complex case. For the parameter estimation of the PARAFAC models, we will present the alternating least squares (ALS) algorithm, which is the most widely used algorithm for estimating the parameters of tensor models.

In the last section, four examples of tensor models will be presented to illustrate the use of tensor decompositions:

– multidimensional harmonic model, based on the observation of a sum of complex exponentials;

– instantaneous linear mixture using the received signals, then using the fourth-order cumulants of the received signals, in the context of source separation;

– representation of a linear finite impulse response (FIR) system using the fourth-order cumulants of the output, with the perspective of system identification.

5.2. Tensor models

The Tucker model will first be presented in section 5.2.1, with the HOSVD algorithm, whereas the Tucker-(N_1, N) model will be defined in section 5.2.2. The PARAFAC decomposition will then be considered in section 5.2.5, with the presentation of the ALS algorithm. Some generalizations, including block tensor models (section 5.2.6) and constrained tensor models (section 5.2.7), will also be described.

5.2.1. *Tucker model*

5.2.1.1. *Definition in scalar form*

For a tensor $\mathcal{X} \in \mathbb{K}^{I_N}$ of order N, a Tucker model can be written in scalar form as follows (Tucker 1966):

$$x_{i_1,\cdots,i_N} = \sum_{r_1=1}^{R_1} \cdots \sum_{r_N=1}^{R_N} g_{r_1,\cdots,r_N} \prod_{n=1}^{N} a_{i_n,r_n}^{(n)}, \qquad [5.1]$$

with $i_n \in \langle I_n \rangle$ for $n \in \langle N \rangle$, where $g_{r_1,\cdots,r_N}$ is an element of the core tensor $\mathcal{G} \in \mathbb{K}^{R_N}$, and $a_{i_n,r_n}^{(n)}$ is an element of the factor matrix $\mathbf{A}^{(n)} \in \mathbb{K}^{I_n \times R_n}$, $n \in \langle N \rangle$.

Using the index convention, the Tucker model can be rewritten concisely as follows:

$$x_{i_1,\cdots,i_N} = g_{r_1,\cdots,r_N} \prod_{n=1}^{N} a_{i_n,r_n}^{(n)}. \qquad [5.2]$$

5.2.1.2. Expression in terms of mode-n products

Noting that the sum $\sum_{r_n=1}^{R_n} g_{r_1,\cdots,r_N} a_{i_n,r_n}^{(n)}$ in [5.1] represents the mode-n product of the core tensor with the factor matrix $\mathbf{A}^{(n)}$, i.e. $\mathcal{G} \times_n \mathbf{A}^{(n)}$, the Tucker model can also be written in terms of the mode-n products, $n \in \langle N \rangle$, as follows:

$$\mathcal{X} = \mathcal{G} \times_1 \mathbf{A}^{(1)} \times_2 \mathbf{A}^{(2)} \times_3 \cdots \times_N \mathbf{A}^{(N)}$$

$$\triangleq \mathcal{G} \overset{N}{\underset{n=1}{\times}} \mathbf{A}^{(n)}. \qquad [5.3]$$

This decomposition, abbreviated as $[\![\mathcal{G}; \mathbf{A}^{(1)}, \cdots, \mathbf{A}^{(N)}]\!]$, can be interpreted as N linear transformations of the core tensor $\mathcal{G}$, each associated with a matrix $\mathbf{A}^{(n)} \in \mathbb{K}^{I_n \times R_n}$ applied to the mode-n vector space of $\mathcal{G} \in \mathbb{K}^{\underline{R}_N}$. Indeed, for $\mathcal{X} = \mathcal{G} \times_n \mathbf{A}^{(n)}$, we have $\mathbf{X}_n = \mathbf{A}^{(n)} \mathbf{G}_n$, where $\mathbf{X}_n$ and $\mathbf{G}_n$ are the mode-n flat unfoldings of $\mathcal{X}$ and $\mathcal{G}$ as defined in [3.41].

5.2.1.3. Expression in terms of outer products

Noting that the factor $\prod_{n=1}^{N} a_{i_n,r_n}^{(n)}$ in [5.1] can be interpreted as the $(i_1, \cdots, i_N)$th element of the outer product of the column vectors $\mathbf{A}_{.r_n}^{(n)}$ of the matrix factors $\mathbf{A}^{(n)}$, the Tucker model can also be written as a linear combination of these outer products:

$$\mathcal{X} = \sum_{r_1=1}^{R_1} \cdots \sum_{r_N=1}^{R_N} g_{r_1,\cdots,r_N} \overset{N}{\underset{n=1}{\circ}} \mathbf{A}_{.r_n}^{(n)}. \qquad [5.4]$$

This expression gives $\mathcal{X}$ as a weighted sum of $\prod_{n=1}^{N} R_n$ rank-one tensors, where the elements $g_{r_1,\cdots,r_N}$ of the core tensor are the weights of the linear combination. Each element $g_{r_1,\cdots,r_N}$ can also be interpreted as a weight of the interactions between the columns $\mathbf{A}_{.r_n}^{(n)}, n \in \langle N \rangle$, of the factor matrices.

5.2.1.4. Matricization

The matrix unfolding [3.38] of the Tucker model along the mode subsets $\mathbb{S}_1$ and $\mathbb{S}_2$ is given by the following equation:

$$\mathbf{X}_{\mathbb{S}_1;\mathbb{S}_2} = \left(\underset{n \in \mathbb{S}_1}{\otimes} \mathbf{A}^{(n)} \right) \mathbf{G}_{\mathbb{S}_1;\mathbb{S}_2} \left(\underset{n \in \mathbb{S}_2}{\otimes} \mathbf{A}^{(n)} \right)^T \in \mathbb{K}^{M_1 \times M_2}, \qquad [5.5]$$

with $\mathbf{G}_{\mathbb{S}_1;\mathbb{S}_2} \in \mathbb{K}^{K_1 \times K_2}$, $K_i = \prod_{n \in \mathbb{S}_i} R_n$, and $M_i = \prod_{n \in \mathbb{S}_i} I_n$, for $i = 1$ and 2.

PROOF.– Let us define $(\mathbb{I}_1, \mathbb{I}_2)$ and $(\mathbb{R}_1, \mathbb{R}_2)$ as the sets of indices i_n and r_n associated with the mode subsets $(\mathbb{S}_1, \mathbb{S}_2)$ combined to form the rows and columns of $\mathbf{X}_{\mathbb{S}_1;\mathbb{S}_2}$, respectively. Applying the formula [3.40] to the core tensor allows us to write the element $g_{r_1,\cdots,r_N}$ as:

$$g_{r_1,\cdots,r_N} = \mathbf{e}^{\mathbb{R}_1} \mathbf{G}_{\mathbb{S}_1;\mathbb{S}_2} \mathbf{e}_{\mathbb{R}_2}. \qquad [5.6]$$

Using the expressions [5.2] and [5.6] of $x_{i_1,\cdots,i_N}$ and $g_{r_1,\cdots,r_N}$, we obtain:

$$\mathbf{X}_{\mathbb{S}_1;\mathbb{S}_2} = x_{i_1,\cdots,i_N} \mathbf{e}_{\mathbb{I}_1}^{\mathbb{I}_2} = \mathbf{e}_{\mathbb{I}_1} x_{i_1,\cdots,i_N} \mathbf{e}^{\mathbb{I}_2}$$

$$= \mathbf{e}_{\mathbb{I}_1} g_{r_1,\cdots,r_N} \Big(\prod_{n=1}^{N} a_{i_n,r_n}^{(n)} \Big) \mathbf{e}^{\mathbb{I}_2}$$

$$= \Big(\prod_{n=1}^{N} a_{i_n,r_n}^{(n)} \Big) \mathbf{e}_{\mathbb{I}_1} \mathbf{e}^{\mathbb{R}_1} \mathbf{G}_{\mathbb{S}_1;\mathbb{S}_2} \mathbf{e}_{\mathbb{R}_2} \mathbf{e}^{\mathbb{I}_2}$$

$$= \Big(\prod_{n \in \mathbb{S}_1} a_{i_n,r_n}^{(n)} \Big) \mathbf{e}_{\mathbb{I}_1}^{\mathbb{R}_1} \mathbf{G}_{\mathbb{S}_1;\mathbb{S}_2} \Big(\prod_{n \in \mathbb{S}_2} a_{i_n,r_n}^{(n)} \Big) \mathbf{e}_{\mathbb{R}_2}^{\mathbb{I}_2}.$$

Taking into account the expression [2.102] of the multiple Kronecker product, $\mathbf{X}_{\mathbb{S}_1;\mathbb{S}_2}$ can now be written as:

$$\mathbf{X}_{\mathbb{S}_1;\mathbb{S}_2} = \Big(\underset{n \in \mathbb{S}_1}{\otimes} \mathbf{A}^{(n)} \Big) \mathbf{G}_{\mathbb{S}_1;\mathbb{S}_2} \Big(\underset{n \in \mathbb{S}_2}{\otimes} \mathbf{A}^{(n)} \Big)^T,$$

which completes the proof of [5.5]. □

5.2.1.5. *Vectorization*

Applying the vectorization formula [2.113] to equation [5.5] gives the following expression for the vectorized form of the matrix unfolding $\mathbf{X}_{\mathbb{S}_1;\mathbb{S}_2}$:

$$\mathrm{vec}(\mathbf{X}_{\mathbb{S}_1;\mathbb{S}_2}) = \Big(\underset{n \in \mathbb{S}_2}{\otimes} \mathbf{A}^{(n)} \Big) \otimes \Big(\underset{n \in \mathbb{S}_1}{\otimes} \mathbf{A}^{(n)} \Big) \mathrm{vec}(\mathbf{G}_{\mathbb{S}_1;\mathbb{S}_2}). \qquad [5.7]$$

In the case where $\mathbb{S}_1$ is the singleton $\{n\}$ and $\mathbb{S}_2 = \{n+1, \cdots, N, 1, \cdots, n-1\}$, equation [5.5] becomes the mode-n flat matrix unfolding defined in Table 3.5:

$$\mathbf{X}_n = \mathbf{A}^{(n)} \mathbf{G}_n (\mathbf{A}^{(n+1)} \otimes \cdots \otimes \mathbf{A}^{(N)} \otimes \mathbf{A}^{(1)} \otimes \cdots \otimes \mathbf{A}^{(n-1)})^T \qquad [5.8]$$

$$\in \mathbb{K}^{I_n \times I_{n+1} \cdots I_N I_1 \cdots I_{n-1}}.$$

The corresponding vectorized form is obtained by applying the identity [2.113] to [5.8], which gives:

$$\text{vec}(\mathbf{X}_n) = (\mathbf{A}^{(n+1)} \otimes \cdots \otimes \mathbf{A}^{(N)} \otimes \mathbf{A}^{(1)} \otimes \cdots \otimes \mathbf{A}^{(n)})\text{vec}(\mathbf{G}_n). \qquad [5.9]$$

This expression can also be deduced from the general vectorized form [5.7] by choosing $\$_1 = \{n\}$ and $\$_2 = \{n+1, \cdots, N, 1, \cdots, n-1\}$.

Similarly, when $\$_2$ is the singleton $\{n\}$ and $\$_1 = \{n+1, \cdots, N, 1, \cdots, n-1\}$, we obtain the mode-$n$ tall matrix unfolding equal to the transpose of $\mathbf{X}_n$, defined in [5.8], of size $I_{n+1} \cdots I_N I_1 \cdots I_{n-1} \times I_n$.

5.2.1.6. *Case of a third-order tensor* $\mathcal{X} \in \mathbb{K}^{I \times J \times K}$

Setting $\mathbf{A}^{(1)} = \mathbf{A}, \mathbf{A}^{(2)} = \mathbf{B}, \mathbf{A}^{(3)} = \mathbf{C}$, and $(R_1, R_2, R_3) = (P, Q, S)$ and using the index convention, equations [5.1], [5.3] and [5.4] then become:

$$x_{ijk} = \sum_{p=1}^{P} \sum_{q=1}^{Q} \sum_{s=1}^{S} g_{pqs} a_{ip} b_{jq} c_{ks} = g_{pqs} a_{ip} b_{jq} c_{ks} \qquad [5.10]$$

$$\mathcal{X} = \mathcal{G} \times_1 \mathbf{A} \times_2 \mathbf{B} \times_3 \mathbf{C} \qquad [5.11]$$

$$= \sum_{p=1}^{P} \sum_{q=1}^{Q} \sum_{s=1}^{S} g_{pqs} \, \mathbf{A}_{.p} \circ \mathbf{B}_{.q} \circ \mathbf{C}_{.r} = g_{pqs} \, \mathbf{A}_{.p} \circ \mathbf{B}_{.q} \circ \mathbf{C}_{.s}, \qquad [5.12]$$

with $\mathcal{G} \in \mathbb{K}^{P \times Q \times S}$, and $\mathbf{A} \in \mathbb{K}^{I \times P}, \mathbf{B} \in \mathbb{K}^{J \times Q}, \mathbf{C} \in \mathbb{K}^{K \times S}$. Equation [5.12] translates the fact that the third-order Tucker model, denoted Tucker3, expresses the tensor $\mathcal{X}$ as the weighted sum of PQS first-order tensors, where the weights of these first-order tensors are given by the coefficients g_{pqs} of the core tensor $\mathcal{G}$. This Tucker model is denoted $[\![\mathcal{G}; \mathbf{A}, \mathbf{B}, \mathbf{C}]\!]$. See Figure I.2 for a graphical representation of this Tucker model.

The various representations of a third-order Tucker model are summarized in Table 5.1, where the flat matrix unfoldings and the vectorized form were deduced from the general formulae [5.8] and [5.9], with $\mathbf{X}_1 = \mathbf{X}_{I \times JK}, \mathbf{X}_2 = \mathbf{X}_{J \times KI}, \mathbf{X}_3 = \mathbf{X}_{K \times IJ}$, and $\mathbf{G}_1 = \mathbf{G}_{P \times QS}, \mathbf{G}_2 = \mathbf{G}_{Q \times SP}, \mathbf{G}_3 = \mathbf{G}_{S \times PQ}$.

5.2.1.7. *Uniqueness*

The Tucker model is not unique in general, since the factors $\mathbf{A}^{(n)}$ are invariant up to non-singular matrices $\mathbf{\Lambda}^{(n)}$ whose effect is compensated by the core tensor, i.e. $\mathcal{X} = [\![\mathcal{G}; \mathbf{A}^{(1)}, \cdots, \mathbf{A}^{(N)}]\!]$ is left unchanged if we replace the factors $\mathbf{A}^{(n)}$ with $\mathbf{A}^{(n)} \mathbf{\Lambda}^{(n)}$ and the core tensor with $\mathcal{C} = [\![\mathcal{G}; (\mathbf{\Lambda}^{(1)})^{-1}, \cdots, (\mathbf{\Lambda}^{(N)})^{-1}]\!]$. Indeed, using the property [3.125] of the mode-n product, we have:

$$[\![\mathcal{C}; \mathbf{A}^{(1)} \mathbf{\Lambda}^{(1)}, \cdots, \mathbf{A}^{(N)} \mathbf{\Lambda}^{(N)}]\!] = [\![\mathcal{G}; \mathbf{A}^{(1)} \mathbf{\Lambda}^{(1)} (\mathbf{\Lambda}^{(1)})^{-1}, \cdots, \mathbf{A}^{(N)} \mathbf{\Lambda}^{(N)} (\mathbf{\Lambda}^{(N)})^{-1}]\!]$$

$$= [\![\mathcal{G}; \mathbf{A}^{(1)}, \cdots, \mathbf{A}^{(N)}]\!] = \mathcal{X}.$$

This means that the Tucker model is not changed if we apply a linear transformation with matrix $\boldsymbol{\Lambda}^{(n)}$ to the column space of each factor $\mathbf{A}^{(n)}$ and the inverse transformation with matrix $\left(\boldsymbol{\Lambda}^{(n)}\right)^{-1}$ to the mode-n subspace of the tensor $\mathcal{G}$, with $n \in \langle N \rangle$, which is equivalent to replacing $\mathbf{G}_n$ with $\left(\boldsymbol{\Lambda}^{(n)}\right)^{-1}\mathbf{G}_n$. Indeed, under this double mode-n transformation, the matrix unfolding [5.8] becomes:

$$\mathbf{X}_n = \mathbf{A}^{(n)}\boldsymbol{\Lambda}^{(n)}\left(\boldsymbol{\Lambda}^{(n)}\right)^{-1}\mathbf{G}_n(\mathbf{A}^{(n+1)} \otimes \cdots \otimes \mathbf{A}^{(N)} \otimes \mathbf{A}^{(1)} \otimes \cdots \otimes \mathbf{A}^{(n-1)})^T$$
$$= \mathbf{A}^{(n)}\mathbf{G}_n(\mathbf{A}^{(n+1)} \otimes \cdots \otimes \mathbf{A}^{(N)} \otimes \mathbf{A}^{(1)} \otimes \cdots \otimes \mathbf{A}^{(n-1)})^T.$$

$\mathcal{X} \in \mathbb{K}^{I \times J \times K}$ $\mathcal{G} \in \mathbb{K}^{P \times Q \times S}, \mathbf{A} \in \mathbb{K}^{I \times P}, \mathbf{B} \in \mathbb{K}^{J \times Q}, \mathbf{C} \in \mathbb{K}^{K \times S}$
Scalar expression $x_{ijk} = \sum_{p=1}^{P} \sum_{q=1}^{Q} \sum_{s=1}^{S} g_{pqs} a_{ip} b_{jq} c_{ks}$
Expression with mode-n products $\mathcal{X} = \mathcal{G} \times_1 \mathbf{A} \times_2 \mathbf{B} \times_3 \mathbf{C}$
Expression with outer products $\mathcal{X} = \sum_{p=1}^{P} \sum_{q=1}^{Q} \sum_{s=1}^{S} g_{pqs} \mathbf{A}_{.p} \circ \mathbf{B}_{.q} \circ \mathbf{C}_{.s}$
Matrix unfoldings $\mathbf{X}_{I \times JK} = \mathbf{A}\mathbf{G}_{P \times QS}(\mathbf{B} \otimes \mathbf{C})^T$ $\mathbf{X}_{J \times KI} = \mathbf{B}\mathbf{G}_{Q \times SP}(\mathbf{C} \otimes \mathbf{A})^T$ $\mathbf{X}_{K \times IJ} = \mathbf{C}\mathbf{G}_{S \times PQ}(\mathbf{A} \otimes \mathbf{B})^T$
Vectorization $\text{vec}(\mathbf{X}_3) = \mathbf{x}_{IJK} = (\mathbf{A} \otimes \mathbf{B} \otimes \mathbf{C})\text{vec}(\mathbf{G}_3)$ $\text{vec}(\mathbf{G}_3) = \mathbf{g}_{PQS}$

Table 5.1. *Third-order Tucker model*

The matrix unfolding $\mathbf{X}_n$ is therefore left unchanged. The same reasoning holds for any mode-$n \in \langle N \rangle$, which implies that the Tucker model is unique up to these linear transformations, characterized by the non-singular matrices $\boldsymbol{\Lambda}^{(n)}$, for $n \in \langle N \rangle$.

Nevertheless, it is important to note that the Tucker model is guaranteed to be unique if certain conditions are satisfied, such as:

– the core tensor is perfectly known;

– there are certain zeroes in the core tensor, i.e. under certain sparseness constraints on the core tensor (ten Berge and Smilde 2002);

– certain structural constraints are imposed on the core tensor (Favier *et al.* 2012c), where the core tensor is characterized by Hankel and Vandermonde matrix slices.

See Smilde *et al.* (2004) for a review of uniqueness results for Tucker models.

5.2.1.8. *HOSVD algorithm*

Since the Tucker model of an Nth-order tensor is nonlinear with respect to the parameters $(\mathcal{G}; \mathbf{A}^{(1)}, \cdots, \mathbf{A}^{(N)})$, the ALS method can be used for parameter estimation through an alternating minimization of least squares criteria constructed from the matrix unfoldings [5.8] for $n \in \langle N \rangle$, and the vectorized form [5.9] for $n = 1$, for example. This type of algorithm will be presented in section 5.2.5.7 to estimate the parameters of a third-order PARAFAC model.

With regard to the identifiability of the Tucker model [5.1], the number of data points contained in the tensor $\mathcal{X} \in \mathbb{K}^{I_N}$, which is equal to $\prod_{n=1}^{N} I_n$, must be greater than or equal to the number of parameters that must be estimated in the core tensor $\mathcal{G} \in \mathbb{K}^{R_N}$ and the factor matrices $\mathbf{A}^{(n)} \in \mathbb{K}^{I_n \times R_n}$, $n \in \langle n \rangle$, which is $\prod_{n=1}^{N} R_n + \sum_{n=1}^{N} I_n R_n$, giving us the condition:

$$\prod_{n=1}^{N} I_n \geq \prod_{n=1}^{N} R_n + \sum_{n=1}^{N} I_n R_n. \tag{5.13}$$

We now present a non-iterative solution based on the orthogonality constraint on the factor matrices. This method, called HSOVD (higher order singular value decomposition) or MLSVD (multilinear SVD), has the advantage of being simple to implement, since it only involves computing N matrix SVDs for a tensor of order N (de Lathauwer *et al.* 2000a).

To simplify the presentation, we will consider the case of a complex third-order tensor $\mathcal{X} \in \mathbb{C}^{I \times J \times K}$. From the expressions of the matrix unfoldings $\mathbf{X}_{I \times JK}$, $\mathbf{X}_{J \times KI}$ and $\mathbf{X}_{K \times IJ}$ given in Table 5.1, we can conclude that the column spaces of these unfoldings are identical to the column spaces of the matrices $\mathbf{A}$, $\mathbf{B}$ and $\mathbf{C}$, respectively. The HOSVD method computes the SVD of these three unfoldings and takes the unitary matrices $(\mathbf{U}^{(1)}, \mathbf{U}^{(2)}, \mathbf{U}^{(3)})$ of the left singular vectors of these three SVDs as an estimate for $\mathbf{A} \in \mathbb{C}^{I \times I}$, $\mathbf{B} \in \mathbb{C}^{J \times J}$ and $\mathbf{C} \in \mathbb{C}^{K \times K}$, respectively. The core tensor $\mathcal{G} \in \mathbb{C}^{I \times J \times K}$ is then deduced from $\mathbf{X}_{I \times JK} = \mathbf{U}^{(1)} \mathbf{G}_{I \times JK} (\mathbf{U}^{(2)} \otimes \mathbf{U}^{(3)})^T$ in the following unfolded form:

$$\mathbf{G}_{I \times JK} = (\mathbf{U}^{(1)})^H \mathbf{X}_{I \times JK} (\mathbf{U}^{(2)} \otimes \mathbf{U}^{(3)})^*, \tag{5.14}$$

which is obtained by pre-multiplying $\mathbf{X}_{I \times JK}$ with $(\mathbf{U}^{(1)})^H$ and post-multiplying it with $(\mathbf{U}^{(2)} \otimes \mathbf{U}^{(3)})^*$.

PROOF.– The orthogonality property of the factor matrices implies the following identities: $(\mathbf{U}^{(1)})^H \mathbf{U}^{(1)} = \mathbf{U}^{(1)} (\mathbf{U}^{(1)})^H = \mathbf{I}_I$, $(\mathbf{U}^{(2)})^* (\mathbf{U}^{(2)})^T = \mathbf{I}_J$, $(\mathbf{U}^{(3)})^* (\mathbf{U}^{(3)})^T = \mathbf{I}_K$, from which we deduce $(\mathbf{U}^{(2)})^{-T} = (\mathbf{U}^{(2)})^*$ and $(\mathbf{U}^{(3)})^{-T} = (\mathbf{U}^{(3)})^*$.

The inverses of $\mathbf{U}^{(1)}$ and $(\mathbf{U}^{(2)} \otimes \mathbf{U}^{(3)})^T$ are therefore given by $(\mathbf{U}^{(1)})^{-1} = (\mathbf{U}^{(1)})^H$ and $(\mathbf{U}^{(2)} \otimes \mathbf{U}^{(3)})^{-T} = (\mathbf{U}^{(2)})^{-T} \otimes (\mathbf{U}^{(3)})^{-T} = (\mathbf{U}^{(2)})^* \otimes (\mathbf{U}^{(3)})^* = (\mathbf{U}^{(2)} \otimes \mathbf{U}^{(3)})^*$, respectively. The formula [5.14] is then deduced using these inverses. $\qquad \square$

The resulting estimation algorithm is summarized below.

1) Compute the SVD of $\mathbf{X}_{I \times JK}$, and choose $\mathbf{A}$ as the matrix $\mathbf{U}^{(1)}$ of left singular vectors.

2) Compute the SVD of $\mathbf{X}_{J \times KI}$, and choose $\mathbf{B}$ as the matrix $\mathbf{U}^{(2)}$ of left singular vectors.

3) Compute the SVD of $\mathbf{X}_{K \times IJ}$, and choose $\mathbf{C}$ as the matrix $\mathbf{U}^{(3)}$ of left singular vectors.

4) Compute the core tensor in the unfolded form [5.14].

In the case of an Nth-order tensor $\mathcal{X} \in \mathbb{K}^{I_N}$, its HOSVD can be written as:

$$\mathcal{X} = \mathcal{G} \overset{N}{\underset{n=1}{\times}} \mathbf{U}^{(n)}, \qquad [5.15]$$

where $\mathbf{U}^{(n)} \in \mathbb{K}^{I_n \times I_n}$ is unitary in the complex case ($\mathbb{K} = \mathbb{C}$) and orthogonal in the real case ($\mathbb{K} = \mathbb{R}$), given by the SVD of $\mathbf{X}_n = \mathbf{U}^{(n)} \mathbf{\Sigma}^{(n)} (\mathbf{V}^{(n)})^H$. The columns of $\mathbf{U}^{(n)}$ can be viewed as the mode-n singular vectors of $\mathcal{X}$. In the complex case, the core tensor $\mathcal{G} \in \mathbb{C}^{I_N}$, which has the same size as $\mathcal{X}$, can be obtained by applying the property [3.130] of the mode-n product, with $(\mathbf{U}^{(n)})^\dagger = (\mathbf{U}^{(n)})^H$:

$$\mathcal{G} = \mathcal{X} \overset{N}{\underset{n=1}{\times}} (\mathbf{U}^{(n)})^H. \qquad [5.16]$$

It satisfies the all-orthogonality and ordering constraints, which means that:

– all the slices along a fixed mode-n are mutually orthogonal:

$$\langle \mathcal{G}_{i_n=i}, \mathcal{G}_{i_n=j} \rangle = 0 \text{ for } i \neq j \text{ and } \forall n \in \langle N \rangle \qquad [5.17]$$

where $\mathcal{G}_{i_n=i}$ denotes the ith slice along the mode-n;

– the slices along each mode-n are arranged in such a way that their norms are in decreasing order, i.e. such that:

$$\|\mathcal{G}_{i_n=1}\|_F \geq \|\mathcal{G}_{i_n=2}\|_F \geq \cdots \geq \|\mathcal{G}_{i_n=I_n}\|_F \ \forall n \in \langle N \rangle. \qquad [5.18]$$

Furthermore, the Frobenius norm of the slice $\mathcal{G}_{i_n}$ of the core tensor is the mode-n singular value $\|\mathcal{G}_{i_n}\|_F = \sigma_{i_n}^{(n)} = (\mathbf{\Sigma}^{(n)})_{i_n,i_n}$ of $\mathcal{X}$.

REMARK 5.1.– We can make the following remarks:

– In the case of a symmetric tensor $\mathcal{X}$, all the factors $\mathbf{U}^{(n)}$ are identical, and the core tensor itself is symmetric. The HOSVD then coincides with the HOEVD (higher order eigenvalue decomposition).

– The THOSVD (truncated higher order singular value decomposition) method provides a low multilinear rank approximation of $\mathcal{X}$. Choosing $(R_1, \cdots, R_N)$ as an approximated multilinear rank, with $R_n \leq I_n$, implies that only the R_n principal mode-n left singular vectors associated with the R_n largest mode-n singular values are kept. The approximation of $\mathcal{X}$ is then given by:

$$\hat{\mathcal{X}} = \hat{\mathcal{G}} \overset{N}{\underset{n=1}{\times}} \hat{\mathbf{U}}^{(n)} \quad , \quad \hat{\mathcal{G}} = \mathcal{X} \overset{N}{\underset{n=1}{\times}} (\hat{\mathbf{U}}^{(n)})^H, \tag{5.19}$$

with $\hat{\mathbf{U}}^{(n)} \in \mathbb{K}^{I_n \times R_n}$, where the dimension R_n is often chosen to be much smaller than I_n. Note that, with the THOSVD, $\hat{\mathbf{U}}^{(n)}$ is a column orthonormal matrix, which implies $(\hat{\mathbf{U}}^{(n)})^\dagger = ((\hat{\mathbf{U}}^{(n)})^H \hat{\mathbf{U}}^{(n)})^{-1} (\hat{\mathbf{U}}^{(n)})^H = (\hat{\mathbf{U}}^{(n)})^H$.

This THOSVD method does not give the best multilinear rank-$(R_1, \cdots, R_N)$ approximation because it does not minimize the criterion $\|\mathcal{X} - \hat{\mathcal{X}}\|_F$ globally, since the optimization is performed separately for each mode-n via the best rank-R_n approximation for $\mathbf{X}_n$. Nevertheless, it is widely used in practice to reduce the dimensionality of data tensors (aspect compression) because it is very straightforward to implement and generally offers good performance if the approximated multilinear rank is suitably chosen. The THOSVD can be viewed as a compressed form of the Tucker decomposition. For more details, see the article by de Lathauwer *et al.* (2000b).

– To improve the performance of the THOSVD method, de Lathauwer *et al.* (2000b) proposed the HOOI (higher order orthogonal iteration) algorithm. The idea is to begin estimating the Tucker model with the HOSVD method, then refine the estimates of the factor matrices $\hat{\mathbf{U}}^{(n)}$ iteratively and alternately using the R_n principal left singular vectors obtained by computing the SVD of the mode-n matrix unfolding of the tensor reconstructed at each iteration. After convergence is achieved, the core tensor is computed using equation [5.19], after replacing the matrices $\hat{\mathbf{U}}^{(n)} \in \mathbb{K}^{I_n \times R_n}$ with the factors determined from this iterative procedure.

5.2.2. *Tucker-(N_1, N) model*

A Tucker-(N_1, N) model for an Nth-order tensor $\mathcal{X} \in \mathbb{K}^{I_N}$, with $N \geq N_1$, corresponds to the case where $N - N_1$ factor matrices are equal to identity matrices (Favier and de Almeida 2014a).

For example, if we assume that $\mathbf{A}^{(n)} = \mathbf{I}_{I_n}$, which implies $R_n = I_n$, for $n = N_1 + 1, \cdots, N$, and hence $\mathcal{G} \in \mathbb{K}^{R_1 \times \cdots \times R_{N_1} \times I_{N_1+1} \times \cdots \times I_N}$, then equations [5.1] and [5.3] become:

$$x_{i_1,\cdots,i_N} = \sum_{r_1=1}^{R_1} \cdots \sum_{r_{N_1}=1}^{R_{N_1}} g_{r_1,\cdots,r_{N_1},i_{N_1+1},\cdots,i_N} \prod_{n=1}^{N_1} a_{i_n,r_n}^{(n)} \tag{5.20}$$

$$\mathcal{X} = \mathcal{G} \times_1 \mathbf{A}^{(1)} \times_2 \cdots \times_{N_1} \mathbf{A}^{(N_1)} \times_{N_1+1} \mathbf{I}_{I_{N_1+1}} \cdots \times_N \mathbf{I}_{I_N} \tag{5.21}$$

$$= \mathcal{G} \mathop{\times}_{n=1}^{N_1} \mathbf{A}^{(n)}. \tag{5.22}$$

For a third-order tensor $\mathcal{X} \in \mathbb{K}^{I \times J \times K}$, two special cases are given by the Tucker-(2,3) and Tucker-(1,3) models, often called Tucker2 and Tucker1, respectively. These models are obtained by fixing one or two of the matrix factors equal to identity matrices.

Table 5.2 summarizes the equations of the Tucker-(2,3) and Tucker-(1,3) models, in the case where $\mathbf{C} = \mathbf{I}_K$ for the Tucker-(2,3) model, and $(\mathbf{B} = \mathbf{I}_J, \mathbf{C} = \mathbf{I}_K)$ for the Tucker-(1,3) model.

Tucker-(2,3) model		Tucker-(1,3) model
	$\mathcal{X} \in \mathbb{K}^{I \times J \times K}$	
$\mathcal{G} \in \mathbb{K}^{P \times Q \times K}$ $\mathbf{A} \in \mathbb{K}^{I \times P}, \mathbf{B} \in \mathbb{K}^{J \times Q}, \mathbf{C} = \mathbf{I}_K$		$\mathcal{G} \in \mathbb{K}^{P \times J \times K}$ $\mathbf{A} \in \mathbb{K}^{I \times P}, \mathbf{B} = \mathbf{I}_J, \mathbf{C} = \mathbf{I}_K$
	Scalar expression	
$x_{ijk} = \sum\limits_{p=1}^{P} \sum\limits_{q=1}^{Q} g_{pqk} a_{ip} b_{jq}$		$x_{ijk} = \sum\limits_{p=1}^{P} g_{pjk} a_{ip}$
	With mode-n products	
$\mathcal{X} = \mathcal{G} \times_1 \mathbf{A} \times_2 \mathbf{B}$		$\mathcal{X} = \mathcal{G} \times_1 \mathbf{A}$
	Matrix unfoldings	
$\mathbf{X}_{IJ \times K} = (\mathbf{A} \otimes \mathbf{B}) \mathbf{G}_{PQ \times K}$ $\mathbf{X}_{JK \times I} = (\mathbf{B} \otimes \mathbf{I}_K) \mathbf{G}_{QK \times P} \mathbf{A}^T$ $\mathbf{X}_{KI \times J} = (\mathbf{I}_K \otimes \mathbf{A}) \mathbf{G}_{KP \times Q} \mathbf{B}^T$		$\mathbf{X}_{IJ \times K} = (\mathbf{A} \otimes \mathbf{I}_J) \mathbf{G}_{PJ \times K}$ $\mathbf{X}_{JK \times I} = \mathbf{G}_{JK \times P} \mathbf{A}^T$ $\mathbf{X}_{KI \times J} = (\mathbf{I}_K \otimes \mathbf{A}) \mathbf{G}_{KP \times J}$
	Vectorization	
$\mathbf{x}_{IJK} = (\mathbf{A} \otimes \mathbf{B} \otimes \mathbf{I}_K) \mathbf{g}_{PQK}$		$\mathbf{x}_{IJK} = (\mathbf{A} \otimes \mathbf{I}_{JK}) \mathbf{g}_{PJK}$

Table 5.2. *Tucker-(2,3) and Tucker-(1,3) models*

5.2.3. *Tucker model of a transpose tensor*

Given the Tucker model $[\![\mathcal{G}; \mathbf{A}^{(1)}, \cdots, \mathbf{A}^{(N)}]\!]$ of an Nth-order tensor $\mathcal{X}$, the transpose tensor $\mathcal{X}^{T,\pi}$ associated with the permutation π admits the Tucker model $[\![\mathcal{G}^{T,\pi}; \mathbf{A}^{\pi(1)}, \cdots, \mathbf{A}^{\pi(N)}]\!]$, or alternatively:

$$\mathcal{X}^{T,\pi} = (\mathcal{G} \times_1 \mathbf{A}^{(1)} \times_2 \cdots \times_N \mathbf{A}^{(N)})^{T,\pi}$$

$$= \mathcal{G}^{T,\pi} \times_1 \mathbf{A}^{\pi(1)} \times_2 \cdots \times_N \mathbf{A}^{\pi(N)}, \qquad [5.23]$$

where $\mathbf{A}^{\pi(n)}$ is the matrix factor of $\mathcal{X}$ associated with the mode $\pi(n)$, i.e. the permuted mode of n. The core tensor of the Tucker model of $\mathcal{X}^{T,\pi}$ is therefore the transpose $\mathcal{G}^{T,\pi}$ of the core tensor $\mathcal{G}$, with the matrix factor $\mathbf{A}^{\pi(n)}$ for the mode n.

Thus, for the third-order tensor $\mathcal{X} \in \mathbb{K}^{I \times J \times K}$ admitting the Tucker model $[\![\mathcal{G}; \mathbf{A}, \mathbf{B}, \mathbf{C}]\!]$, with $\mathcal{G} \in \mathbb{K}^{P \times Q \times S}$, and the permutation $\pi(1,2,3) = (2,3,1)$, the transpose tensor $\mathcal{X}^{T,\pi} \in \mathbb{K}^{J \times K \times I}$ admits the Tucker model $[\![\mathcal{G}^{T,\pi}; \mathbf{B}, \mathbf{C}, \mathbf{A}]\!]$, with $\mathcal{G}^{T,\pi} \in \mathbb{K}^{Q \times S \times P}$.

5.2.4. *Tucker decomposition and multidimensional Fourier transform*

Given the tensor $\mathcal{X} \in \mathbb{K}^{\underline{I}_N}$ of order N, its discrete N-dimensional Fourier transform is the tensor $\mathcal{Y} \in \mathbb{C}^{\underline{I}_N}$ defined as:

$$\mathcal{Y} = \mathcal{X} \times_1 \mathbf{F}_{I_1} \times_2 \mathbf{F}_{I_2} \times_3 \cdots \times_N \mathbf{F}_{I_N} \triangleq \mathcal{X} \overset{N}{\underset{n=1}{\times}} \mathbf{F}_{I_n}. \qquad [5.24]$$

The tensor $\mathcal{Y}$ is obtained by applying the Fourier transform with matrix $\mathbf{F}_{I_n}$ to the mode-n subspace of $\mathcal{X}$, for $n \in \langle N \rangle$. It can be written in scalar form as:

$$y_{k_1, \cdots, k_N} = \left(\prod_{n=1}^{N} \frac{1}{\sqrt{I_n}} \right) \sum_{i_1=1}^{I_1} \cdots \sum_{i_N=1}^{I_N} x_{i_1, \cdots, i_N} \omega_{I_1}^{(i_1-1)(k_1-1)} \cdots \omega_{I_N}^{(i_N-1)(k_N-1)},$$

for $k_n \in \langle I_n \rangle, n \in \langle N \rangle$, with $\omega_{I_n} = \exp(-2\pi i/I_n)$, $i^2 = -1$, and where $\mathbf{F}_{I_n}$ is the Fourier matrix of order I_n defined as:

$$\mathbf{F}_{I_n} = \frac{1}{\sqrt{I_n}} \begin{bmatrix} 1 & 1 & 1 & \cdots & 1 \\ 1 & \omega_{I_n} & \omega_{I_n}^2 & \cdots & \omega_{I_n}^{I_n-1} \\ 1 & \omega_{I_n}^2 & \omega_{I_n}^4 & \cdots & \omega_{I_n}^{I_n-2} \\ \vdots & \vdots & \vdots & \ddots & \vdots \\ 1 & \omega_{I_n}^{I_n-1} & \omega_{I_n}^{I_n-2} & \cdots & \omega_{I_n} \end{bmatrix}. \qquad [5.25]$$

The Fourier transform [5.24] of $\mathcal{X}$ can be interpreted as the Tucker decomposition $[\![\mathcal{X}; \mathbf{F}_{I_1}, \cdots, \mathbf{F}_{I_N}]\!]$ of $\mathcal{Y}$, where $\mathcal{X}$ is the core tensor and the Fourier matrices are the matrix factors. Goulart and Favier (2014) applied such a multidimensional Fourier transform to a circulant-constraint canonical polyadic decomposition (CPD), i.e. a CPD of a hypercubic tensor whose factors are circulant matrices, to derive a specialized estimation algorithm that exploits the Fourier transform of such structured tensors.

If the tensor $\mathcal{X}$ satisfies a Tucker model $[\![\mathcal{G}; \mathbf{A}^{(1)}, \cdots, \mathbf{A}^{(N)}]\!]$, its multidimensional Fourier transform leads to the Tucker model $[\![\mathcal{G}; \mathbf{F}_{I_1}\mathbf{A}^{(1)}, \cdots, \mathbf{F}_{I_N}\mathbf{A}^{(N)}]\!]$, which amounts to pre-multiplying each factor matrix $\mathbf{A}^{(n)}$ by the Fourier matrix $\mathbf{F}_{I_n}$.

5.2.5. *PARAFAC model*

5.2.5.1. *Scalar expression*

The PARAFAC (parallel factors) decomposition of a tensor, also called CANDECOMP (canonical decomposition), was introduced independently by Harshman (1970) and Carroll and Chang (1970) for applications in psychometrics and phonetics, respectively. It is also called CP for CANDECOMP/PARAFAC in Kiers (2000) and CPD following work by Hitchcock (1927), who defined a tensor as a sum of polyads.

Thus, a third-order tensor can be expressed as a sum of triads, i.e. a sum of products of three factors:

$$x_{ijk} = \sum_{r=1}^{R} a_{ir}b_{jr}c_{kr}. \tag{5.26}$$

A PARAFAC model of an Nth-order tensor corresponds to the special case of a Tucker model whose core tensor is the identity tensor of order N and size $R \times \cdots \times R$, i.e.:

$$\mathcal{G} = \mathcal{I}_{N,R} \quad (\text{also denoted } \mathcal{I}_R) \quad \Leftrightarrow \quad g_{r_1, \cdots, r_N} = \delta_{r_1, \cdots, r_N},$$

where $\delta_{r_1, \cdots, r_N}$ is the generalized Kronecker delta, which takes the value 1 if $r_1 = \cdots = r_N = r$, with $r \in \langle R \rangle$, and 0 otherwise. Equations [5.1]–[5.2] then become:

$$x_{i_1, \cdots, i_N} = \sum_{r=1}^{R} \prod_{n=1}^{N} a_{i_n,r}^{(n)} = \prod_{n=1}^{N} a_{i_n,r}^{(n)} \tag{5.27}$$

where $\mathbf{A}^{(n)} = \left[a_{i_n,r}^{(n)}\right] \in \mathbb{C}^{I_n \times R}, n \in \langle N \rangle$, are the factor matrices, and the last equality follows from using the index convention, since the repetition of the index r implies summation over r. We will write this decomposition as $[\![\mathbf{A}^{(1)}, \cdots, \mathbf{A}^{(N)}; R]\!]$.

5.2.5.2. *Other expressions*

Equations [5.3] and [5.4] in terms of mode-n products and outer products now become:

$$\mathcal{X} = \mathcal{I}_R \overset{N}{\underset{n=1}{\times}} \mathbf{A}^{(n)}, \tag{5.28}$$

$$\mathcal{X} = \sum_{r=1}^{R} \overset{N}{\underset{n=1}{\circ}} \mathbf{A}_{.r}^{(n)}. \tag{5.29}$$

We can make the following remarks:

– the expression [5.27] in the form of a sum of polyads was called a polyadic form of $\mathcal{X}$ by Hitchcock (1927);

– the PARAFAC model [5.27]–[5.29] is equivalent to decomposing the tensor $\mathcal{X}$ into a sum of R components, with each component being a rank-one tensor, equal to the outer product of the rth columns of the N factor matrices. When R is minimal, it is called the tensor rank or canonical rank of $\mathcal{X}$ (Kruskal 1977). This explains why the PARAFAC model has also been called the rank-revealing decomposition.

This marks a distinction between PARAFAC and Tucker models, which are instead a sum of $\prod_{n=1}^{N} R_n$ components, where each component is a rank-one tensor obtained from the outer product of one column from each factor matrix. Unlike PARAFAC models, which only involve interactions between the same columns ($r \in \langle R \rangle$) of the factor matrices, Tucker models take into account all interactions between distinct columns ($r_n \in \langle R_n \rangle, n \in \langle N \rangle$) of the factor matrices;

– unlike matrices, for which the rank is always at most equal to the smallest dimension, the rank of a tensor of order greater than two can be larger than its largest dimension $I_n, n \in \langle N \rangle$; in other words, it is possible to have $R > \max_{n}(I_n)$. This property allows us to solve the source separation problem when the number of sources is greater than the number of sensors, which corresponds to an underdetermined system.

The maximum rank is defined as the largest attainable rank for a given set of tensors. In Kruskal (1989), it was shown that the maximum rank R_{max} for the set of third-order tensors $\mathcal{X} \in \mathbb{K}^{I \times J \times K}$ satisfies the following inequalities:

$$\max(I, J, K) \le R_{max}(I, J, K) \le \min(IJ, JK, KI). \tag{5.30}$$

There are other definitions of rank for tensors, like the typical and generic ranks, as well as the symmetric rank for a symmetric tensor. For more details, see Comon *et al.* (2008, 2009a);

– every symmetric tensor $\mathcal{A}$ of dimensions I can be written as a linear combination of symmetric rank-one tensors:

$$\mathcal{A} = \mathbf{x}_1^{\circ N} + \cdots + \mathbf{x}_{R_S}^{\circ N} \triangleq \sum_{r=1}^{R_S} \mathbf{x}_r^{\circ N} \, , \ \mathbf{x}_r \in \mathbb{K}^I \, , \ r \in \langle R_S \rangle. \qquad [5.31]$$

The smallest integer R_S such that $\mathcal{A}$ can be written as a linear combination of R_S symmetric rank-one tensors is called the symmetric rank of $\mathcal{A}$. The symmetric rank of a tensor is difficult to determine. A general formula was established by Alexander and Hirschowitz (1995) for a symmetric tensor of order greater than two. Comon *et al.* (2008) formulated the conjecture that the symmetric rank is equal to the rank. See Comon *et al.* (2008), Landsberg (2012), and Nie (2017) for more details about symmetric tensors and their applications, as well as the computation of symmetric tensor decompositions.

– In telecommunications applications, the structure parameters (rank, dimensions of the modes and of the core tensor) of a tensor model are design parameters that are chosen as a function of the desired system performance. However, in most applications, like in data analysis, the structure parameters are generally unknown and need to be determined.

Several approaches have been proposed to determine these parameters (see Bro and Kiers 2003; da Costa *et al.* 2008, 2010, 2011, and the references given in these articles). In practice, the structure parameters are often determined by a trial-and-error approach. It is important to mention that the determination of the tensor rank is an NP-hard problem (Håstad 1990; Hillar and Lim 2009).

5.2.5.3. *Matricization*

The matrix unfolding [3.38] of the PARAFAC model is given by:

$$\mathbf{X}_{\mathbb{S}_1;\mathbb{S}_2} = \left(\underset{n \in \mathbb{S}_1}{\diamond} \mathbf{A}^{(n)} \right) \left(\underset{n \in \mathbb{S}_2}{\diamond} \mathbf{A}^{(n)} \right)^T . \qquad [5.32]$$

PROOF.– Using the expression [5.27] of $x_{i_1,\cdots,i_N}$ and the identity [2.206] with $\mathbf{A}^{(n)}_{.r}$ instead of $\mathbf{u}^{(n)}$, we obtain:

$$\mathbf{X}_{\mathbb{S}_1;\mathbb{S}_2} = x_{i_1,\cdots,i_N} \mathbf{e}_{\mathbb{I}_1}^{\mathbb{I}_2} = \left(\prod_{n \in \mathbb{S}_1} a_{i_n,r}^{(n)} \mathbf{e}_{\mathbb{I}_1} \right) \left(\prod_{n \in \mathbb{S}_2} a_{i_n,r}^{(n)} \mathbf{e}^{\mathbb{I}_2} \right) \qquad [5.33]$$

$$= \left(\underset{n \in \mathbb{S}_1}{\diamond} \mathbf{A}^{(n)}_{.r} \right) \left(\underset{n \in \mathbb{S}_2}{\diamond} \mathbf{A}^{(n)}_{.r} \right)^T = \left(\underset{n \in \mathbb{S}_1}{\diamond} \mathbf{A}^{(n)} \right) \left(\underset{n \in \mathbb{S}_2}{\diamond} \mathbf{A}^{(n)} \right)^T . \qquad [5.34]$$

The last equality follows from the sum over r implicit in the expression [5.34], using the index convention. This completes the proof of [5.32]. □

In particular, the mode-n flat unfolding corresponding to $\mathbb{S}_1 = \{n\}$ and $\mathbb{S}_2 = \{n+1, \cdots, N, 1, \cdots, n-1\}$ is given by:

$$\mathbf{X}_n = \mathbf{A}^{(n)} (\mathbf{A}^{(n+1)} \diamond \cdots \diamond \mathbf{A}^{(N)} \diamond \mathbf{A}^{(1)} \diamond \cdots \diamond \mathbf{A}^{(n-1)})^T. \qquad [5.35]$$

PROPOSITION 5.2.– Any matrix unfolding $\mathbf{X}_{\mathbb{S}_1;\mathbb{S}_2}$ has rank at most equal to R, and the rank of $\mathcal{X}$ satisfies $R \geq \max_n (R_n)$, with $R_n \leq I_n$ for $n \in \langle N \rangle$.

PROOF.– From [5.32], we deduce:

$$\left. \begin{array}{l} \dim(\underset{n \in \mathbb{S}_1}{\diamond} \mathbf{A}^{(n)}) = \underset{n \in \mathbb{S}_1}{\prod} I_n \times R \\[2mm] \dim(\underset{n \in \mathbb{S}_2}{\diamond} \mathbf{A}^{(n)}) = \underset{n \in \mathbb{S}_2}{\prod} I_n \times R \end{array} \right\} \Rightarrow \mathrm{r}(\mathbf{X}_{\mathbb{S}_1;\mathbb{S}_2}) \leq R. \qquad [5.36]$$

In particular, we have $\mathrm{r}(\mathbf{X}_n) \leq R$, i.e. the mode-$n$ rank satisfies $R_n \leq R$, for every $n \in \langle N \rangle$, which implies:

$$R \geq \max(R_1, R_2, \cdots, R_N). \qquad [5.37]$$

From [5.35], we can also deduce that:

$$R_n \leq I_n , \ n \in \langle N \rangle. \qquad [5.38]$$

This completes the proof of the proposition. □

5.2.5.4. *Vectorization*

Applying the property [2.257] to [5.32], with $\mathbf{A} = \underset{n \in \mathbb{S}_1}{\diamond} \mathbf{A}^{(n)}$, $\mathbf{C} = (\underset{n \in \mathbb{S}_2}{\diamond} \mathbf{A}^{(n)})^T$ and $\lambda_1 = \cdots = \lambda_R = 1$, gives the following vectorization formula, associated with the unfolding $\mathbf{X}_{\mathbb{S}_1;\mathbb{S}_2}$:

$$\mathrm{vec}(\mathbf{X}_{\mathbb{S}_1;\mathbb{S}_2}) = \left((\underset{n \in \mathbb{S}_2}{\diamond} \mathbf{A}^{(n)}) \diamond (\underset{n \in \mathbb{S}_1}{\diamond} \mathbf{A}^{(n)}) \right) \mathbf{1}_R. \qquad [5.39]$$

In particular, the vectorization associated with a mode combination according to lexicographical order is given by:

$$\mathbf{x}_{I_1 \cdots I_N} = (\mathbf{A}^{(1)} \diamond \mathbf{A}^{(2)} \diamond \cdots \diamond \mathbf{A}^{(N)}) \mathbf{1}_R. \qquad [5.40]$$

5.2.5.5. *Normalized form*

The PARAFAC model is often defined in the following normalized form:

$$x_{i_1,\cdots,i_N} = \sum_{r=1}^{R} g_r \prod_{n=1}^{N} b_{i_n,r}^{(n)} \quad \text{with} \quad g_r > 0, \tag{5.41}$$

where the column vectors $\mathbf{B}_{.r}^{(n)}$ are obtained by normalizing the columns $\mathbf{A}_{.r}^{(n)}$:

$$\mathbf{B}_{.r}^{(n)} = \frac{1}{\|\mathbf{A}_{.r}^{(n)}\|_F}\mathbf{A}_{.r}^{(n)} \ , \ r \in \langle R \rangle \, , \, n \in \langle N \rangle \tag{5.42}$$

$$g_r = \prod_{n=1}^{N} \|\mathbf{A}_{.r}^{(n)}\|_F. \tag{5.43}$$

In this case, the identity tensor $\mathcal{I}_R$ in [5.28] is replaced by the diagonal tensor $\mathcal{G} \in \mathbb{R}^{R \times \cdots \times R}$ whose diagonal elements are equal to the scaling factors g_r:

$$g_{r_1,\cdots,r_N} = \begin{cases} g_r & \text{if } r_1 = \cdots = r_N = r \, , \, r \in \langle R \rangle \\ 0 & \text{otherwise} \end{cases}.$$

The PARAFAC model in normalized form [5.41] can be viewed as a Tucker model $[\![\mathcal{G}, \mathbf{B}^{(1)}, \cdots, \mathbf{B}^{(N)}]\!]$ with a diagonal core tensor.

We can also write $[\![\mathbf{g}, \mathbf{B}^{(1)}, \cdots, \mathbf{B}^{(N)}]\!]$, where $\mathbf{g} = [g_1, \cdots, g_R]^T$ is the vector whose components are the weights $g_r, r \in \langle R \rangle$. The matricization [5.32] and vectorization [5.40] formulae then become:

$$\mathbf{X}_{\mathbb{S}_1;\mathbb{S}_2} = \left(\underset{n \in \mathbb{S}_1}{\diamond} \mathbf{B}^{(n)} \right) \text{diag}(\mathbf{g}) \left(\underset{n \in \mathbb{S}_2}{\diamond} \mathbf{B}^{(n)} \right)^T \tag{5.44}$$

$$\mathbf{x}_{I_1 \cdots I_N} = (\mathbf{B}^{(1)} \diamond \mathbf{B}^{(2)} \diamond \cdots \diamond \mathbf{B}^{(N)})\mathbf{g}, \tag{5.45}$$

where $\text{diag}(\mathbf{g})$ is a diagonal matrix whose diagonal elements are the coefficients g_r.

Note that, with this normalized form, the PARAFAC model is unique up to sign ambiguities $\delta_r^{(n)}$ for each column of the factors $\mathbf{B}^{(n)}$, satisfying $\prod_{n=1}^{N} \delta_r^{(n)} = 1$ for every $r \in \langle R \rangle$. To eliminate the column permutation ambiguities, we can arrange them in such a way that they are associated with the factors g_r in decreasing order $(g_1 \geq g_2 \geq \cdots \geq g_R)$.

5.2.5.6. *Case of a third-order tensor*

In the case of a third-order tensor $\mathcal{X} \in \mathbb{K}^{I \times J \times K}$, equations [5.27]–[5.29] become:

$$x_{ijk} = \sum_{r=1}^{R} a_{ir} b_{jr} c_{kr} \tag{5.46}$$

$$\mathcal{X} = \mathcal{I}_R \times_1 \mathbf{A} \times_2 \mathbf{B} \times_3 \mathbf{C} \tag{5.47}$$

$$= \sum_{r=1}^{R} \mathbf{A}_{.r} \circ \mathbf{B}_{.r} \circ \mathbf{C}_{.r}, \tag{5.48}$$

where $\mathbf{A} \in \mathbb{K}^{I \times R}, \mathbf{B} \in \mathbb{K}^{J \times R}, \mathbf{C} \in \mathbb{K}^{K \times R}$ are the factor matrices.

This PARAFAC model of rank R is denoted $[\![\mathbf{A}, \mathbf{B}, \mathbf{C}; R]\!]$. Figure I.1 in the introductory chapter illustrates this model in the form of a sum of R rank-one tensors, each tensor being the outer product of the rth column vectors of the three factor matrices, denoted $(\mathbf{a}_r^{(1)}, \mathbf{a}_r^{(2)}, \mathbf{a}_r^{(3)})$ in Figure I.1, with $r \in \langle R \rangle$.

REMARK 5.3.– In the case of a rank-one tensor $\mathcal{X} \in \mathbb{K}^{I \times J \times K}$, PARAFAC models will be denoted $[\![\mathbf{a}, \mathbf{b}, \mathbf{c}]\!]$, with $\mathbf{a} \in \mathbb{K}^I, \mathbf{b} \in \mathbb{K}^J, \mathbf{c} \in \mathbb{K}^K$ and $x_{ijk} = a_i b_j c_k$, $i \in \langle I \rangle, j \in \langle J \rangle, k \in \langle K \rangle$.

We can also decompose the tensor into matrix slices. For example, using index notation and the identity [2.84], the frontal slices obtained by fixing the index k in the third-order PARAFAC model [5.46] are given by:

$$\mathbf{X}_{..k} = c_{kr} a_{ir} b_{jr} \mathbf{e}_i^j = c_{kr} (a_{ir} \mathbf{e}_i)(b_{jr} \mathbf{e}^j) = c_{kr} \mathbf{A}_{.r} \mathbf{B}_{.r}^T \tag{5.49}$$

$$= \mathbf{A} \operatorname{diag}(c_{k1}, \cdots, c_{kR}) \mathbf{B}^T$$

$$= \mathbf{A} \operatorname{diag}(\mathbf{C}_{k.}) \mathbf{B}^T = \mathbf{A} D_k(\mathbf{C}) \mathbf{B}^T \in \mathbb{K}^{I \times J}, \tag{5.50}$$

where $D_k(\mathbf{C}) \triangleq \operatorname{diag}(\mathbf{C}_{k.}) \triangleq \operatorname{diag}(c_{k1}, \cdots, c_{kR})$, i.e. the diagonal matrix whose diagonal entries are the elements of the kth row of $\mathbf{C}$.

Similarly, it is easy to deduce the following expressions for the horizontal and lateral matrix slices:

$$\mathbf{X}_{i..} = a_{ir} b_{jr} c_{kr} \mathbf{e}_j^k = \mathbf{B} D_i(\mathbf{A}) \mathbf{C}^T \in \mathbb{K}^{J \times K} \tag{5.51}$$

$$\mathbf{X}_{.j.} = b_{jr} c_{kr} a_{ir} \mathbf{e}_k^i = \mathbf{C} D_j(\mathbf{B}) \mathbf{A}^T \in \mathbb{K}^{K \times I}. \tag{5.52}$$

The matricized and vectorized forms can be deduced from equations [5.32] and [5.39]. The various representations of a third-order PARAFAC model are summarized in Table 5.3.

$$\mathcal{X} \in \mathbb{K}^{I \times J \times K}$$
$$\mathbf{A} \in \mathbb{K}^{I \times R}, \mathbf{B} \in \mathbb{K}^{J \times R}, \mathbf{C} \in \mathbb{K}^{K \times R}$$

Scalar expression
$$x_{ijk} = \sum_{r=1}^{R} a_{ir} b_{jr} c_{kr}$$

Expression with mode-n products
$$\mathcal{X} = \mathcal{I}_R \times_1 \mathbf{A} \times_2 \mathbf{B} \times_3 \mathbf{C}$$

Expression with outer products
$$\mathcal{X} = \sum_{r=1}^{R} \mathbf{A}_{.r} \circ \mathbf{B}_{.r} \circ \mathbf{C}_{.r}$$

Matrix unfoldings
$$\mathbf{X}_{IJ \times K} = (\mathbf{A} \diamond \mathbf{B}) \mathbf{C}^T$$
$$\mathbf{X}_{JK \times I} = (\mathbf{B} \diamond \mathbf{C}) \mathbf{A}^T$$
$$\mathbf{X}_{KI \times J} = (\mathbf{C} \diamond \mathbf{A}) \mathbf{B}^T$$

Vectorization
$$\mathbf{x}_{IJK} = (\mathbf{A} \diamond \mathbf{B} \diamond \mathbf{C}) \mathbf{1}_R$$

Matrix slices
$$\mathbf{X}_{..k} = \mathbf{A} \, D_k(\mathbf{C}) \, \mathbf{B}^T \in \mathbb{K}^{I \times J}$$
$$\mathbf{X}_{i..} = \mathbf{B} \, D_i(\mathbf{A}) \, \mathbf{C}^T \in \mathbb{K}^{J \times K}$$
$$\mathbf{X}_{.j.} = \mathbf{C} \, D_j(\mathbf{B}) \, \mathbf{A}^T \in \mathbb{K}^{K \times I}$$

Table 5.3. *Third-order PARAFAC model*

5.2.5.7. *ALS algorithm for estimating a PARAFAC model*

The PARAFAC model [5.27] is multilinear (more precisely N-linear) in its parameters in the sense that it is linear with respect to each of its N factors. ALS-based estimation of the parameters $\mathbf{A}^{(n)}, n \in \langle N \rangle$, relies on the minimization of the quadratic error between the data tensor $\mathcal{X}$ and its PARAFAC model, which amounts to minimizing the following cost function:

$$\min_{\mathbf{A}^{(n)}, n \in \langle N \rangle} \left\| \mathcal{X} - \sum_{r=1}^{R} \overset{N}{\underset{n=1}{\circ}} \mathbf{A}_{.r}^{(n)} \right\|_F^2 = \min_{\mathbf{A}^{(1)}, \cdots, \mathbf{A}^{(N)}} \left\| \mathcal{X} - [\![\mathbf{A}^{(1)}, \cdots, \mathbf{A}^{(N)}]\!] \right\|_F^2,$$

$$[5.53]$$

where the rank R was fixed beforehand or estimated using a trial-and-error method. This non-linear optimization problem can be solved using the ALS algorithm (Harshman 1970; Carroll and Chang 1970).

This algorithm estimates each factor matrix separately and alternately by minimizing a quadratic error criterion conditional on fixing the other factor matrices with their previously estimated values.

Thus, using the mode-n matrix unfolding [5.35], we can estimate the factor $\mathbf{A}^{(n)}$ at the iteration t by minimizing the LS criterion conditional on the estimated values of the other factors, i.e.:

$$\min_{\mathbf{A}^{(n)}} \|\mathbf{X}_n^T - (\mathbf{A}_{t-1}^{(n+1)} \diamond \cdots \diamond \mathbf{A}_{t-1}^{(N)} \diamond \mathbf{A}_t^{(1)} \diamond \cdots \diamond \mathbf{A}_t^{(n-1)})(\mathbf{A}^{(n)})^T\|_F^2, \quad [5.54]$$

where $\mathbf{A}_t^{(n)}$ denotes the estimate of $\mathbf{A}^{(n)}$ at the iteration t. The details of this ALS algorithm are presented below for a third-order tensor.

To guarantee the uniqueness of the LS solution resulting from the minimization of [5.54], the matrix to be pseudo-inverted must have full column rank, i.e. $R \leq \prod_{p=1,p\neq n}^{N} I_p$. Hence, to guarantee the identifiability of the N factors, the rank R must satisfy the following necessary, but not sufficient, condition:

$$R \leq \min_n \left(\prod_{p=1,p\neq n}^{N} I_p \right). \quad [5.55]$$

In the case of the third-order PARAFAC model, the factors $(\mathbf{A}, \mathbf{B}, \mathbf{C})$ are estimated by minimizing the following LS criterion:

$$\min_{\mathbf{A},\mathbf{B},\mathbf{C}} \|\mathcal{X} - \sum_{r=1}^{R} \mathbf{A}_{.r} \circ \mathbf{B}_{.r} \circ \mathbf{C}_{.r}\|_F^2. \quad [5.56]$$

Since the PARAFAC model is trilinear with respect to its parameters, the ALS algorithm replaces the criterion [5.56] with three quadratic criteria obtained using the three matrix unfoldings given in Table 5.3, while fixing two of these matrices using the values estimated at the previous iterations:

$$\min_{\mathbf{A}} \|\mathbf{X}_{JK \times I} - (\mathbf{B}_{t-1} \diamond \mathbf{C}_{t-1})\mathbf{A}^T\|_F^2 \;\Rightarrow\; \mathbf{A}_t \quad [5.57]$$

$$\min_{\mathbf{B}} \|\mathbf{X}_{KI \times J} - (\mathbf{C}_{t-1} \diamond \mathbf{A_t})\mathbf{B}^T\|_F^2 \;\Rightarrow\; \mathbf{B}_t \quad [5.58]$$

$$\min_{\mathbf{B}} \|\mathbf{X}_{IJ \times K} - (\mathbf{A_t} \diamond \mathbf{B_t})\mathbf{C}^T\|_F^2 \;\Rightarrow\; \mathbf{C}_t. \quad [5.59]$$

The ALS algorithm for a third-order PARAFAC model is summarized as follows:

1) Initialization: $t = 0$; $\mathbf{B}_0$, $\mathbf{C}_0$.

2) Increment t and compute:

$$(\mathbf{A}_t)^T = (\mathbf{B}_{t-1} \diamond \mathbf{C}_{t-1})^\dagger \mathbf{X}_{JK \times I}$$

$$(\mathbf{B}_t)^T = (\mathbf{C}_{t-1} \diamond \mathbf{A}_t)^\dagger \mathbf{X}_{KI \times J}$$

$$(\mathbf{C}_t)^T = (\mathbf{A}_t \diamond \mathbf{B}_t)^\dagger \mathbf{X}_{IJ \times K}.$$

3) Return to step 2 until convergence.

Stopping criterion and identifiability condition: the stopping criterion is often based on the variation (between two consecutive iterations $t - 1$ and t) in the quadratic error between the measured data and the data reconstructed using the estimated model, i.e.:

$$|e_t - e_{t-1}| \leq \epsilon \text{ with } e_t = \|\mathbf{X}_{JK \times I} - (\mathbf{B}_t \diamond \mathbf{C}_t)\mathbf{A}_t^T\|_F^2, \qquad [5.60]$$

where the threshold, for example, is set to the value $\epsilon = 10^{-6}$. We can also use a stopping criterion based on the variation in the estimated parameters between two consecutive iterations.

For a third-order PARAFAC model, the necessary condition [5.55] for identifiability, corresponding to the constraint that $\mathbf{A} \diamond \mathbf{B}$, $\mathbf{B} \diamond \mathbf{C}$ and $\mathbf{C} \diamond \mathbf{A}$ have full column rank, becomes:

$$R \leq \min(IJ, JK, KI). \qquad [5.61]$$

Hence, taking into account the lower bound [5.37] for $N = 3$, the rank R must satisfy:

$$\max(R_1, R_2, R_3) \leq R \leq \min(IJ, JK, KI). \qquad [5.62]$$

We can make the following remarks:

– the ALS algorithm was originally proposed by Harshman (1970) and Carroll and Chang (1970) for third-order PARAFAC models. However, these two articles do not explicitly mention Khatri–Rao products of the factor matrices;

– taking into account the property [2.255] of the Khatri–Rao product, the computation of the pseudo-inverses in the ALS algorithm can be simplified as follows:

$$(\mathbf{B}_{t-1} \diamond \mathbf{C}_{t-1})^\dagger = [(\mathbf{B}_{t-1} \diamond \mathbf{C}_{t-1})^H (\mathbf{B}_{t-1} \diamond \mathbf{C}_{t-1})]^{-1}(\mathbf{B}_{t-1} \diamond \mathbf{C}_{t-1})^H$$

$$= (\mathbf{B}_{t-1}^H \mathbf{B}_{t-1} \odot \mathbf{C}_{t-1}^H \mathbf{C}_{t-1})^{-1}(\mathbf{B}_{t-1} \diamond \mathbf{C}_{t-1})^H \qquad [5.63]$$

$$(\mathbf{C}_{t-1} \diamond \mathbf{A}_t)^\dagger = (\mathbf{C}_{t-1}^H \mathbf{C}_{t-1} \odot \mathbf{A}_t^H \mathbf{A}_t)^{-1}(\mathbf{C}_{t-1} \diamond \mathbf{A}_t)^H \qquad [5.64]$$

$$(\mathbf{A}_t \diamond \mathbf{B}_t)^\dagger = (\mathbf{A}_t^H \mathbf{A}_t \odot \mathbf{B}_t^H \mathbf{B}_t)^{-1}(\mathbf{A}_t \diamond \mathbf{B}_t)^H. \qquad [5.65]$$

This amounts to replacing the computation of the pseudo-inverses of matrices of size $JK \times R$, $KI \times R$, and $IJ \times R$ by the computation of the inverses of three matrices of size $R \times R$.

If a matrix factor is also column orthonormal (e.g. $\mathbf{C}^H \mathbf{C} = \mathbf{I}_R$), then the formulae [5.63] and [5.64] simplify as follows:

$$(\mathbf{B}_{t-1} \diamond \mathbf{C}_{t-1})^\dagger = \left[\mathrm{diag}\left(\|(\mathbf{B}_{.1})_{t-1}\|_2^2, \cdots, \|(\mathbf{B}_{.R})_{t-1}\|_2^2\right)\right]^{-1} (\mathbf{B}_{t-1} \diamond \mathbf{C}_{t-1})^H$$

$$(\mathbf{C}_{t-1} \diamond \mathbf{A}_t)^\dagger = \left[\mathrm{diag}\left(\|(\mathbf{A}_{.1})_t\|_2^2, \cdots, \|(\mathbf{A}_{.R})_t\|_2^2\right)\right]^{-1} (\mathbf{C}_{t-1} \diamond \mathbf{A}_t)^H.$$

– Many algorithms have been proposed for the parametric estimation of PARAFAC models. See, for example, Faber *et al.* (2003) for a review of various estimation algorithms for third-order PARAFAC models and the associated identifiability conditions. Similarly, in Tomasi and Bro (2006), different algorithms are compared for third-order models, with a presentation of two possible approaches for accelerating convergence (line search and compression). Other optimization-based estimation methods like the Gauss–Newton algorithm can also be used (Phan *et al.* 2013). For an overview of various parameter estimation algorithms for PARAFAC, see the article by Comon *et al.* (2009b).

– The main advantages of the ALS algorithm are its simplicity to implement and the fact that it can be straightforwardly generalized to an arbitrary order N. On the other hand, it has the drawbacks of being slow to converge and potentially converging to a local minimum, depending on the choice of initialization. This slow convergence occurs especially in the case of degenerate PARAFAC models, i.e. when there is a linear dependency between the columns of the factor matrices that can induce a model order modification (Kruskal *et al.* 1989).

In Paatero (2000), the degeneracy phenomenon is illustrated for third-order PARAFAC models of size $2 \times 2 \times 2$, as in the example presented below. The case of $p \times p \times 2$ models is considered in Stegeman (2006).

EXAMPLE 5.4.– Consider the sequence of rank-two tensors $\mathcal{X}_n \in \mathbb{K}^{I \times J \times K}$ satisfying the PARAFAC model $[[\mathbf{A}_n, \mathbf{B}_n, \mathbf{C}_n; 2]]$ defined as:

$$\mathbf{A}_n = \begin{bmatrix} n\mathbf{a} & | & -n\mathbf{a} + \frac{1}{n^2}\mathbf{v} \end{bmatrix} ; \mathbf{B}_n = \begin{bmatrix} n\mathbf{b} & | & -n\mathbf{b} + \frac{1}{n^2}\mathbf{w} \end{bmatrix}$$

$$\mathbf{C}_n = \begin{bmatrix} n\mathbf{c} + \frac{1}{n^2}\mathbf{u} & | & -n\mathbf{c} \end{bmatrix},$$

i.e.:

$$\mathcal{X}_n = (n\mathbf{a}) \circ (n\mathbf{b}) \circ (n\mathbf{c} + \frac{1}{n^2}\mathbf{u}) + (-n\mathbf{a} + \frac{1}{n^2}\mathbf{v}) \circ (-n\mathbf{b} + \frac{1}{n^2}\mathbf{w}) \circ (-n\mathbf{c})$$

$$= \mathcal{X} - \frac{1}{n^3}(\mathbf{v} \circ \mathbf{w} \circ \mathbf{c}) \qquad [5.66]$$

$$\mathcal{X} = \mathbf{a} \circ \mathbf{b} \circ \mathbf{u} + \mathbf{v} \circ \mathbf{b} \circ \mathbf{c} + \mathbf{a} \circ \mathbf{w} \circ \mathbf{c}, \qquad [5.67]$$

where $\mathbf{a}, \mathbf{v} \in \mathbb{K}^I$, $\mathbf{b}, \mathbf{w} \in \mathbb{K}^J$ and $\mathbf{c}, \mathbf{u} \in \mathbb{K}^K$ are the generating vectors of the model, assumed to be such that $(\mathbf{a}, \mathbf{v})$ are independent, as well as $(\mathbf{b}, \mathbf{w})$ and $(\mathbf{c}, \mathbf{u})$.

As $n \to \infty$, each of the columns of the three factors $(\mathbf{A}_n, \mathbf{B}_n, \mathbf{C}_n)$ tends to infinity proportionally to n, becoming increasingly dependent on the fact that the contributions of the vectors $(\mathbf{u}, \mathbf{v}, \mathbf{w})$ decrease proportionally to $\frac{1}{n^2}$ and tend to zero, giving:

$$\mathbf{A}_n \simeq [\ n\mathbf{a} \quad | \quad -n\mathbf{a}\ ] \quad \mathbf{B}_n \simeq [\ n\mathbf{b} \quad | \quad -n\mathbf{b}\ ] \quad \mathbf{C}_n \simeq [\ n\mathbf{c} \quad | \quad -n\mathbf{c}\ ].$$

However, equations [5.66]–[5.67] show that the tensors $\mathcal{X}_n$ remain finite. The original second-order PARAFAC model is constructed in such a way that the sequence $\mathcal{X}_n$ tends to the third-order tensor $\mathcal{X}$ defined in [5.67], while the difference $\mathcal{X}_n - \mathcal{X}$ tends to zero proportionally to $\frac{1}{n^3}$.

In general, PARAFAC degeneracy occurs when a tensor is approximated by a lower rank tensor. As in the above example, some factors tending to infinity cancel each other, which induces the convergence toward a higher rank tensor. In practice, the degeneracy problem can be avoided by imposing orthogonality constraints as with the HOSVD algorithm or non-negativity ones. Thus, Lim and Comon (2009) showed that, for any non-negative tensor, a best non-negative rank-R approximation always exists in the sense of minimizing the LS criterion [5.53].

5.2.5.8. *Estimation of a third-order rank-one tensor*

Consider a rank-one tensor $\mathcal{X} \in \mathbb{R}^{I \times J \times K}$ admitting a PARAFAC model of the form [5.41], with factors $\mathbf{u} \in \mathbb{R}^I$, $\mathbf{v} \in \mathbb{R}^J$ and $\mathbf{w} \in \mathbb{R}^K$, of unit norm, i.e.:

$$\mathcal{X} = g\,\mathbf{u} \circ \mathbf{v} \circ \mathbf{w} \quad \Leftrightarrow \quad x_{ijk} = g\,u_i v_j w_k \ , \ g > 0. \qquad [5.68]$$

The LS parameter estimation problem is solved by minimizing the following cost function $f(g, \mathbf{u}, \mathbf{v}, \mathbf{w}) = \|\mathcal{X} - g(\mathbf{u} \circ \mathbf{v} \circ \mathbf{w})\|_F^2$ with respect to $(g, \mathbf{u}, \mathbf{v}, \mathbf{w})$ subject to the constraints $\|\mathbf{u}\|_2 = \|\mathbf{v}\|_2 = \|\mathbf{w}\|_2 = 1$.

Using the relations [3.132]–[3.135], and noting that the vectors $\mathbf{u}, \mathbf{v}, \mathbf{w}$ have unit norm, we can deduce the following equations:

$$\mathcal{X} \times_1 \mathbf{u}^T \times_2 \mathbf{v}^T = \sum_{i=1}^{I} \sum_{j=1}^{J} u_i v_j \mathbf{X}_{ij.} = g \sum_{i=1}^{I} \sum_{j=1}^{J} u_i^2 v_j^2 \mathbf{w} = g\mathbf{w} \quad [5.69]$$

$$\mathcal{X} \times_1 \mathbf{u}^T \times_3 \mathbf{w}^T = \sum_{i=1}^{I} \sum_{k=1}^{K} u_i w_k \mathbf{X}_{i.k} = g \sum_{i=1}^{I} \sum_{k=1}^{K} u_i^2 w_k^2 \mathbf{v} = g\mathbf{v} \quad [5.70]$$

$$\mathcal{X} \times_2 \mathbf{v}^T \times_3 \mathbf{w}^T = \sum_{j=1}^{J} \sum_{k=1}^{K} v_j w_k \mathbf{X}_{.jk} = g \sum_{j=1}^{J} \sum_{k=1}^{K} v_j^2 w_k^2 \mathbf{u} = g\mathbf{u} \quad [5.71]$$

$$\mathcal{X} \times_1 \mathbf{u}^T \times_2 \mathbf{v}^T \times_3 \mathbf{w}^T = \sum_{i=1}^{I} \sum_{j=1}^{J} \sum_{k=1}^{K} x_{ijk} u_i v_j w_k = g \sum_{i=1}^{I} \sum_{j=1}^{J} \sum_{k=1}^{K} u_i^2 v_j^2 w_k^2 = g.$$

From these equations, and taking into account the constraints of unit norm for the vectors $(\mathbf{u}, \mathbf{v}, \mathbf{w})$, we deduce the ALS estimation algorithm described below.

Initialization: $it = 0 : \mathbf{u}_0, \mathbf{v}_0$.

(i) $\mathbf{w}_{it+1} = \mathcal{X} \times_1 \mathbf{u}_{it}^T \times_2 \mathbf{v}_{it}^T / \|\mathcal{X} \times_1 \mathbf{u}_{it}^T \times_2 \mathbf{v}_{it}^T\|_2$

(ii) $\mathbf{v}_{it+1} = \mathcal{X} \times_1 \mathbf{u}_{it}^T \times_3 \mathbf{w}_{it+1}^T / \|\mathcal{X} \times_1 \mathbf{u}_{it}^T \times_3 \mathbf{w}_{it+1}^T\|_2$

(iii) $\mathbf{u}_{it+1} = \mathcal{X} \times_2 \mathbf{v}_{it+1}^T \times_3 \mathbf{w}_{it+1}^T / \|\mathcal{X} \times_2 \mathbf{v}_{it+1}^T \times_3 \mathbf{w}_{it+1}^T\|_2$

Return to (i) until convergence.

$$g_\infty = \mathcal{X} \times_1 \mathbf{u}_\infty^T \times_2 \mathbf{v}_\infty^T \times_3 \mathbf{w}_\infty^T,$$

where $(\mathbf{u}_\infty, \mathbf{v}_\infty, \mathbf{w}_\infty)$ are the estimated values of $(\mathbf{u}, \mathbf{v}, \mathbf{w})$ at convergence.

REMARK 5.5.– Note that, for any iteration it, we have:

$$f(\mathbf{u}_{it}, \mathbf{v}_{it}, \mathbf{w}_{it}) \le f(\mathbf{u}_{it}, \mathbf{v}_{it}, \mathbf{w}_{it-1}) \le f(\mathbf{u}_{it}, \mathbf{v}_{it-1}, \mathbf{w}_{it-1})$$

$$\le f(\mathbf{u}_{it-1}, \mathbf{v}_{it-1}, \mathbf{w}_{it-1}),$$

which means that the cost function $f(\mathbf{u}_{it}, \mathbf{v}_{it}, \mathbf{w}_{it})$ decreases monotonely.

5.2.5.9. *Case of a fourth-order tensor*

The various possible representations of the PARAFAC model $(\mathbf{A}, \mathbf{B}, \mathbf{C}, \mathbf{D}; R)$ of a fourth-order tensor $\mathcal{X} \in \mathbb{K}^{I \times J \times K \times L}$ are summarized in Table 5.4.

$$\mathcal{X} \in \mathbb{K}^{I \times J \times K \times L}$$
$$\mathbf{A} \in \mathbb{K}^{I \times R}, \mathbf{B} \in \mathbb{K}^{J \times R}, \mathbf{C} \in \mathbb{K}^{K \times R}, \mathbf{D} \in \mathbb{K}^{L \times R}$$

Scalar expression
$$x_{ijkl} = \sum_{r=1}^{R} a_{ir} b_{jr} c_{kr} d_{lr}$$

Expression with mode-n products
$$\mathcal{X} = \mathcal{I}_R \times_1 \mathbf{A} \times_2 \mathbf{B} \times_3 \mathbf{C} \times_4 \mathbf{D}$$

Expression with outer products
$$\mathcal{X} = \sum_{r=1}^{R} \mathbf{A}_{.r} \circ \mathbf{B}_{.r} \circ \mathbf{C}_{.r} \circ \mathbf{D}_{.r}$$

Matrix unfoldings
$$\mathbf{X}_{IJ \times KL} = (\mathbf{A} \diamond \mathbf{B})(\mathbf{C} \diamond \mathbf{D})^T$$
$$\mathbf{X}_{IJK \times L} = (\mathbf{A} \diamond \mathbf{B} \diamond \mathbf{C})\mathbf{D}^T$$

Vectorization
$$\mathbf{x}_{IJKL} = (\mathbf{A} \diamond \mathbf{B} \diamond \mathbf{C} \diamond \mathbf{D})\mathbf{1}_R$$

Matrix slices
$$\mathbf{X}_{ij..} = \mathbf{C}\mathrm{diag}(\mathbf{B}_{j.})\mathrm{diag}(\mathbf{A}_{i.})\mathbf{D}^T \in \mathbb{K}^{K \times L}$$

Table 5.4. *Fourth-order PARAFAC model*

In particular, applying the formula [5.32] with $\mathbb{S}_1 = \{1, 2, 3\}$ and $\mathbb{S}_2 = \{4\}$, as well as the formula [2.202] defining the Khatri–Rao product, we have:

$$\mathbf{X}_{IJK \times L} = (\mathbf{A} \diamond \mathbf{B} \diamond \mathbf{C})\mathbf{D}^T$$

$$= \begin{bmatrix} (\mathbf{B} \diamond \mathbf{C})D_1(\mathbf{A}) \\ \vdots \\ (\mathbf{B} \diamond \mathbf{C})D_I(\mathbf{A}) \end{bmatrix} \mathbf{D}^T = \begin{bmatrix} \mathbf{C}D_1(\mathbf{B})D_1(\mathbf{A}) \\ \vdots \\ \mathbf{C}D_J(\mathbf{B})D_1(\mathbf{A}) \\ \vdots \\ \mathbf{C}D_1(\mathbf{B})D_I(\mathbf{A}) \\ \vdots \\ \mathbf{C}D_J(\mathbf{B})D_I(\mathbf{A}) \end{bmatrix} \mathbf{D}^T \in \mathbb{K}^{IJK \times L}.$$

Since the matrix unfolding $\mathbf{X}_{IJK \times L}$ is obtained by stacking the IJ matrix slices $\mathbf{X}_{ij..} \in \mathbb{K}^{K \times L}$, we deduce the following expression for these slices:

$$\mathbf{X}_{ij..} = \mathbf{C}D_j(\mathbf{B})D_i(\mathbf{A})\mathbf{D}^T \in \mathbb{K}^{K \times L}. \tag{5.72}$$

Noting that:

$$D_j(\mathbf{B})D_i(\mathbf{A}) = D_{(j-1)I+i}(\mathbf{B} \diamond \mathbf{A}) = D_{(i-1)J+j}(\mathbf{A} \diamond \mathbf{B}), \tag{5.73}$$

where $D_{(j-1)I+i}(\mathbf{B} \diamond \mathbf{A})$ and $D_{(i-1)J+j}(\mathbf{A} \diamond \mathbf{B})$ are diagonal matrices whose diagonal entries are the elements of the rows $(j-1)I + i$ and $(i-1)J + j$ of the Khatri–Rao products $\mathbf{B} \diamond \mathbf{A}$ and $\mathbf{A} \diamond \mathbf{B}$, respectively, the matrix unfolding [5.72] can be rewritten as:

$$\mathbf{X}_{ij..} = CD_{(j-1)I+i}(\mathbf{B} \diamond \mathbf{A})\mathbf{D}^T = CD_{(i-1)J+j}(\mathbf{A} \diamond \mathbf{B})\mathbf{D}^T. \qquad [5.74]$$

Similarly, we have:

$$\mathbf{X}_{.jk.} = AD_{(j-1)K+k}(\mathbf{B} \diamond \mathbf{C})\mathbf{D}^T = AD_{(k-1)J+j}(\mathbf{C} \diamond \mathbf{B})\mathbf{D}^T \in \mathbb{K}^{I \times L} \qquad [5.75]$$

$$\mathbf{X}_{..kl} = AD_{(k-1)L+l}(\mathbf{C} \diamond \mathbf{D})\mathbf{B}^T = AD_{(l-1)K+k}(\mathbf{D} \diamond \mathbf{C})\mathbf{B}^T \in \mathbb{K}^{I \times J}. \qquad [5.76]$$

Other slices and matrix unfoldings can be deduced from the expressions presented above using simple permutations of the factor matrices. Thus, for example, we have:

$$\mathbf{X}_{IL \times JK} = (\mathbf{A} \diamond \mathbf{D})(\mathbf{B} \diamond \mathbf{C})^T \quad ; \quad \mathbf{X}_{ILJ \times K} = (\mathbf{A} \diamond \mathbf{D} \diamond \mathbf{B})\mathbf{C}^T$$

$$\mathbf{X}_{i..l} = BD_l(\mathbf{D})D_i(\mathbf{A})\mathbf{C}^T \in \mathbb{K}^{J \times K}$$

$$= BD_{(l-1)I+i}(\mathbf{D} \diamond \mathbf{A})\mathbf{C}^T = BD_{(i-1)L+l}(\mathbf{A} \diamond \mathbf{D})\mathbf{C}^T.$$

By combining two modes (for example the modes 2 and 3), we can reshape the tensor $\mathcal{X}$ into a third-order tensor $\mathcal{Y} \in \mathbb{K}^{I \times M \times L}$, where $M = JK$ and the combined mode is defined as $m = k + (j-1)K$. Setting $f_{mr} = b_{jr}c_{kr}$, or equivalently $\mathbf{F} = \mathbf{B} \diamond \mathbf{C} \in \mathbb{K}^{M \times R}$, the scalar equation of the tensor $\mathcal{Y}$ can be written as:

$$y_{iml} = \sum_{r=1}^{R} a_{ir}f_{mr}d_{lr}, \qquad [5.77]$$

which is a PARAFAC model $[\![\mathbf{A}, \mathbf{F}, \mathbf{D}; R]\!]$. From Table 5.3, we can deduce the following vectorized and matricized forms of $\mathcal{Y}$, as well as the matrix slice $\mathbf{Y}_{i..}$:

$$\mathbf{Y}_{IM \times L} = (\mathbf{A} \diamond \mathbf{F})\mathbf{D}^T \quad ; \quad \mathbf{y}_{IML} = (\mathbf{A} \diamond \mathbf{F} \diamond \mathbf{D})\mathbf{1}_R \quad ;$$

$$\mathbf{Y}_{i..} = \mathbf{F}\text{diag}(\mathbf{A}_{i.})\mathbf{D}^T \in \mathbb{K}^{M \times L}.$$

REMARK 5.6.– In general, by combining P modes of an Nth-order tensor $\mathcal{X}$ that admits a PARAFAC model of rank R, we obtain a tensor $\mathcal{Y}$ of reduced order $N - P + 1$ that admits a PARAFAC model of rank R characterized by $N - P + 1$ matrix factors, one of which is given by the Khatri–Rao products of the P factor matrices associated with the P combined modes.

5.2.5.10. *Variants of the PARAFAC model*

Table 5.5 presents a few variants of the third-order PARAFAC model. The acronyms mean "individual differences scaling" (INDSCAL), "symmetric CP"

(Sym CP), "doubly symmetric CP" (DSym CP), "shifted CP" (ShiftCP) and "convolutive CP" (ConvCP). The symmetric CP model was used to model Volterra kernels in Favier and Bouilloc (2010) and Favier *et al.* (2012a), whereas the doubly symmetric model was used to model nonlinear communication channels (Bouilloc and Favier 2012; Crespo-Cadenas *et al.* 2014).

Models	Ref	$x_{i,j,k}$	Applications
CP	(Harshman 1970)	$\sum\limits_{r=1}^{R} a_{ir}b_{jr}c_{kr}$	Various fields
INDSCAL	(Carroll and Chang 1970)	$\sum\limits_{r=1}^{R} a_{ir}a_{jr}c_{kr}$	Psychometrics
Sym CP	(Comon *et al.* 2008)	$\sum\limits_{r=1}^{R} a_{ir}a_{jr}a_{kr}$	Volterra models
DSym CP	(Bouilloc and Favier 2012; Crespo-Cadenas *et al.* 2014)	$\sum\limits_{r=1}^{R} a_{ir}a_{jr}a_{kr}^*$	Nonlinear channels
ShiftCP	(Harshman *et al.* 2003; Morup 2011)	$\sum\limits_{r=1}^{R} a_{ir}b_{j-t_k,r}c_{kr}$	Neuro-imaging
ConvCP	(Morup 2011; Morup *et al.* 2011)	$\sum\limits_{r=1}^{R}\sum\limits_{t=1}^{T} a_{ir}b_{j-t,r}c_{k,r,t}$	Neuro-imaging

Table 5.5. *Variants of the third-order PARAFAC model*

The convolutive CP model described by the equation

$$x_{i,j,k} = \sum_{r=1}^{R}\sum_{t=1}^{T} a_{ir}b_{j-t,r}c_{k,r,t} \qquad [5.78]$$

was used to model three-dimensional neuro-imaging data whose first two modes are space–time and whose third mode is the experiment number. The dependency of the components of the time factor $b_r = \mathbf{B}_{.r}$ with respect to the time shift t reflects variability between experiments by taking into account, for the experiment k, an amplitude that itself depends on t via the coefficient $c_{k,r,t}$. The sum $\sum\limits_{t=1}^{T} b_{j-t,r}c_{k,r,t}$ can be interpreted as a time convolution, which explains the name of the convolutive CP model. The ShiftCP model corresponds to the special case of the ConvCP model where each time factor b_r only depends on a shift t_k for the experiment k. The ConvCP model can therefore be viewed as a generalization of the ShiftCP model that introduces a convolution on the time interval T. For more details, see Morup *et al.* (2011).

5.2.5.11. *PARAFAC model of a transpose tensor*

Given the PARAFAC model $[\![\mathbf{A}^{(1)}, \cdots, \mathbf{A}^{(N)}; R]\!]$ of an Nth-order tensor $\mathcal{X}$, the transpose tensor $\mathcal{X}^{T,\pi}$ associated with the permutation π admits the PARAFAC model $[\![\mathbf{A}^{\pi(1)}, \cdots, \mathbf{A}^{\pi(N)}; R]\!]$, or alternatively:

$$\mathcal{X}^{T,\pi} = (\mathcal{I}_R \times_1 \mathbf{A}^{(1)} \times_2 \cdots \times_N \mathbf{A}^{(N)})^{T,\pi}$$

$$= \mathcal{I}_R \times_1 \mathbf{A}^{\pi(1)} \times_2 \cdots \times_N \mathbf{A}^{\pi(N)}, \qquad [5.79]$$

where $\mathbf{A}^{\pi(n)}$ is the matrix factor of $\mathcal{X}$ associated with the mode $\pi(n)$, i.e. the permuted mode of n. The PARAFAC model of $\mathcal{X}^{T,\pi}$ therefore has $\mathbf{A}^{\pi(n)}$ as its mode-n factor matrix. Note that the rank is not changed by a permutation of the modes. Thus, for example, for the third-order tensor $\mathcal{X} \in \mathbb{K}^{I \times J \times K}$ admitting the PARAFAC model $[\![\mathbf{A}, \mathbf{B}, \mathbf{C}; R]\!]$ and the permutation $\pi(1, 2, 3) = (2, 3, 1)$, the transpose tensor $\mathcal{X}^{T,\pi}$ admits the PARAFAC model $[\![\mathbf{B}, \mathbf{C}, \mathbf{A}; R]\!]$.

5.2.5.12. *Uniqueness and identifiability*

Since the pioneering work of Harshman (1970) and Kruskal (1977), various articles have been written about the problem of essential uniqueness of PARAFAC models, i.e. uniqueness up to trivial indeterminacies corresponding to permutation and scaling ambiguities among the columns of the factor matrices. The permutation ambiguity arises from the fact that the order of the rank-one tensors in [5.29] can be changed without changing the result of the sum, due to the commutativity of addition. This is equivalent to permuting the columns of the factors $\mathbf{A}^{(n)}$ using a permutation matrix $\mathbf{\Pi}$. Similarly, it is possible to multiply each column $\mathbf{A}_{.r}^{(n)}$ with a scalar factor $\lambda_r^{(n)}$ without changing equation [5.29] of the model, provided that $\prod_{n=1}^{N} \lambda_r^{(n)} = 1$. In other words, the factors $\mathbf{A}^{(n)}$ can be replaced with $\mathbf{A}^{(n)} \mathbf{\Pi} \mathbf{\Lambda}^{(n)}$ without changing the model if $\prod_{n=1}^{N} \mathbf{\Lambda}^{(n)} = \mathbf{I}_R$, where $\mathbf{\Lambda}^{(n)} = \mathrm{diag}(\lambda_1^{(n)}, \cdots, \lambda_R^{(n)})$, for $n \in \langle N \rangle$.

This result can also be recovered from a matrix unfolding using the following properties of the Khatri–Rao product:

$$(\mathbf{A}\mathbf{\Pi}) \diamond (\mathbf{B}\mathbf{\Pi}) = (\mathbf{A} \diamond \mathbf{B})\mathbf{\Pi} \qquad [5.80]$$

$$(\mathbf{A}\mathbf{\Lambda}_1) \diamond (\mathbf{B}\mathbf{\Lambda}_2) = (\mathbf{A} \diamond \mathbf{B})\mathbf{\Lambda}_1\mathbf{\Lambda}_2 \qquad [5.81]$$

for any permutation matrix $\mathbf{\Pi}$ and any diagonal matrices $\mathbf{\Lambda}_1$ and $\mathbf{\Lambda}_2$.

For example, consider the mode-1 flat unfolding defined in [5.35] for the PARAFAC model $[\![\mathbf{A}, \mathbf{B}, \mathbf{C}; R]\!]$ and the triplet $(\mathbf{A}\mathbf{\Pi}\mathbf{\Lambda}^{(A)}, \mathbf{B}\mathbf{\Pi}\mathbf{\Lambda}^{(B)}, \mathbf{C}\mathbf{\Pi}\mathbf{\Lambda}^{(C)})$. Using the properties [5.80] and [5.81] gives:

$$\mathbf{A}\mathbf{\Pi}\mathbf{\Lambda}^{(A)}(\mathbf{B}\mathbf{\Pi}\mathbf{\Lambda}^{(B)} \diamond \mathbf{C}\mathbf{\Pi}\mathbf{\Lambda}^{(C)})^T = \mathbf{A}\mathbf{\Pi}\mathbf{\Lambda}^{(A)}\mathbf{\Lambda}^{(B)}\mathbf{\Lambda}^{(C)}\mathbf{\Pi}^T(\mathbf{B} \diamond \mathbf{C})^T.$$

$$[5.82]$$

Taking into account the properties $\mathbf{\Lambda}^{(A)}\mathbf{\Lambda}^{(B)}\mathbf{\Lambda}^{(C)} = \mathbf{\Pi\Pi}^T = \mathbf{I}_R$, we conclude that the model $[\![\mathbf{A\Pi\Lambda}^{(A)}, \mathbf{B\Pi\Lambda}^{(B)}, \mathbf{C\Pi\Lambda}^{(C)}; R]\!]$ is equivalent to the model $[\![\mathbf{A}, \mathbf{B}, \mathbf{C}; R]\!]$.

An overview of the main uniqueness conditions of PARAFAC models can be found in Domanov and de Lathauwer (2013) for the deterministic case, i.e. for a particular PARAFAC model, and in Domanov and de Lathauwer (2014) for the generic case, i.e. when the elements of the factor matrices are random and generated from a continuous distribution. Below, we briefly recall a few results about the uniqueness of PARAFAC models, after defining the notion of k-rank.

The k-rank (also known as the Kruskal rank) of a matrix $\mathbf{A}$, denoted $k_\mathbf{A}$, is the largest integer such that every set of $k_\mathbf{A}$ columns of $\mathbf{A}$ is linearly independent.

From this definition, we can deduce the following results:

$- k_\mathbf{A} \leq r_\mathbf{A};$

$-$ if $\mathbf{A} \in \mathbb{K}^{I \times J}$ is a matrix whose elements are randomly generated from an absolutely continuous distribution, then $\mathbf{A}$ has full rank, which implies that $k_\mathbf{A} = r_\mathbf{A} = \min(I, J);$

$-$ if $\mathbf{A}$ contains a zero column, then $k_\mathbf{A} = 0;$

$-$ if $\mathbf{A}$ contains two proportional columns, then $k_\mathbf{A} = 1.$

EXAMPLE 5.7.– Below are three examples of square matrices of order three with the same rank but different column k-ranks:

$$\mathbf{A}_1 = \begin{pmatrix} 1 & 0 & 0 \\ 0 & 1 & 0 \\ 0 & 0 & 0 \end{pmatrix} \Rightarrow r_{\mathbf{A}_1} = 2, \ k_{\mathbf{A}_1} = 0$$

$$\mathbf{A}_2 = \begin{pmatrix} 1 & 1 & 1 \\ 1 & 1 & 0 \\ 1 & 1 & 0 \end{pmatrix} \Rightarrow r_{\mathbf{A}_2} = 2, \ k_{\mathbf{A}_2} = 1$$

$$\mathbf{A}_3 = \begin{pmatrix} 1 & 0 & 1 \\ 0 & 1 & 1 \\ 0 & 0 & 0 \end{pmatrix} \Rightarrow r_{\mathbf{A}_3} = k_{\mathbf{A}_3} = 2.$$

The next example shows that the column and row k-ranks are not necessarily the same.

$$\mathbf{A} = \begin{pmatrix} 1 & 0 & 0 & 1 \\ 0 & 1 & 0 & 0 \\ 0 & 0 & 1 & 0 \end{pmatrix}.$$

We have $r_\mathbf{A} = 3$, with $k_\mathbf{A} = 1$ for the column k-rank, and $k_\mathbf{A} = 3$ for the row k-rank.

PROPOSITION 5.8.– The PARAFAC model of order N described by equations [5.27]–[5.29] is essentially unique if the following condition is satisfied (Sidiropoulos and Bro 2000):

$$\sum_{n=1}^{N} k_{\mathbf{A}^{(n)}} \geq 2R + N - 1. \tag{5.83}$$

In the generic case, the factor matrices have full rank with probability one, which implies that $k_{\mathbf{A}^{(n)}} = \min(I_n, R)$, and the condition [5.83] becomes:

$$\sum_{n=1}^{N} \min(I_n, R) \geq 2R + N - 1. \tag{5.84}$$

For a third-order PARAFAC model $[\![\mathbf{A}, \mathbf{B}, \mathbf{C}; R]\!]$, the Kruskal condition can be written as follows:

$$k_{\mathbf{A}} + k_{\mathbf{B}} + k_{\mathbf{C}} \geq 2R + 2. \tag{5.85}$$

REMARK 5.9.– We can make the following remarks.

– The condition [5.83] is sufficient but not necessary to guarantee essential uniqueness. This condition does not hold for $R = 1$. It is necessary for $R = 2$ and $R = 3$ but not for $R > 3$ (ten Berge and Sidiropoulos 2002).

If all the factor matrices have full column rank, then every PARAFAC model of a tensor of order $N > 3$ and rank strictly greater than one is essentially unique.

– The first sufficient condition for the essential uniqueness of a third-order PARAFAC model was established by Harshman (1972), then generalized by (Kruskal 1977) using the notion of k-rank. A more accessible proof of the Kruskal condition is given by Stegeman and Sidiropoulos (2007). This condition was then extended to complex-valued tensors by Sidiropoulos et al. (2000) and to tensors of order $N > 3$ by Sidiropoulos and Bro (2000).

– From the condition [5.85], we can conclude that if two factor matrices $(\mathbf{A}, \mathbf{B})$ have full column rank ($k_{\mathbf{A}} = k_{\mathbf{B}} = R$), then the PARAFAC model is essentially unique if the third factor does not have any proportional columns ($k_{\mathbf{C}} > 1$). Hence, we can deduce the following condition for the essential uniqueness of a third-order PARAFAC model:

$$\min(k_{\mathbf{A}}, k_{\mathbf{B}}, k_{\mathbf{C}}) \geq 2. \tag{5.86}$$

– If a factor matrix (e.g. $\mathbf{C}$) has full column rank, then the condition [5.85] becomes:

$$k_{\mathbf{A}} + k_{\mathbf{B}} \geq R + 2. \tag{5.87}$$

– In Jiang and Sidiropoulos (2004) and de Lathauwer (2006) necessary and sufficient uniqueness conditions, less strict than the Kruskal condition, are established for third- and fourth-order tensors subject to the hypothesis that at least one factor matrix has full column rank. However, these conditions are difficult to exploit in practice.

Other less strict conditions were independently derived by Stegeman (2008) and Guo et al. (2011) for third-order tensors with a factor matrix (e.g. $\mathbf{C}$) of full column rank. It has been shown that the PARAFAC model is essentially unique if the factors $(\mathbf{A}, \mathbf{B})$ satisfy the following conditions:

1) $k_\mathbf{A}, k_\mathbf{B} \geq 2$

2) $r_\mathbf{A} + k_\mathbf{B} \geq R + 2$ or $r_\mathbf{B} + k_\mathbf{A} \geq R + 2.$ [5.88]

These conditions are less restrictive than [5.87]. Indeed, if, for example, $k_\mathbf{A} = 2$ and $r_\mathbf{A} = k_\mathbf{A} + \delta$ with $\delta > 0$, then applying [5.87] implies that $k_\mathbf{B} = R$, i.e. $\mathbf{B}$ must have full column rank, whereas [5.88] gives $k_\mathbf{B} \geq R - \delta$, which does not require $\mathbf{B}$ to have full column rank.

– When a factor matrix (e.g. $\mathbf{C}$) is known and the Kruskal condition [5.85] is satisfied, as is often the case in telecommunications applications, essential uniqueness is guaranteed without any permutation ambiguity, leaving only scaling ambiguities $(\mathbf{\Lambda_A}, \mathbf{\Lambda_B})$ such that $\mathbf{\Lambda_A}\mathbf{\Lambda_B} = \mathbf{I}_R$.

Table 5.6 summarizes a few important results about the rank and k-rank of certain matrices frequently encountered in signal processing applications, namely random matrices with i.i.d. (independent and identically distributed) elements with a uniformly continuous law, Vandermonde matrices and Khatri–Rao products, as well as the sufficient conditions guaranteeing the essential uniqueness of a third-order PARAFAC model presented above.

With regard to the identifiability of a PARAFAC model $[\![\mathbf{A}^{(1)}, \cdots, \mathbf{A}^{(N)}; R]\!]$ of order N, with $\mathbf{A}^{(n)} \in \mathbb{K}^{I_n \times R}$, $n \in \langle N \rangle$, the number of data points contained in the tensor $\mathcal{X} \in \mathbb{K}^{I_N}$, which is equal to $\prod_{n=1}^{N} I_n$, must be greater than or equal to the number of unknown parameters to be estimated[1], which is equal to $(\sum_{n=1}^{N} I_n - N + 1)R$, giving us the following upper bound for the rank of the tensor:

$$R \leq \frac{\prod_{n=1}^{N} I_n}{(\sum_{n=1}^{N} I_n - N + 1)R}.$$ [5.89]

---

1 The term $(N-1)R$ subtracted from the total number of coefficients of the factor matrices corresponds to the number of scaling ambiguity factors contained in the diagonal matrices $\mathbf{\Lambda}^{(n)} \in \mathbb{K}^{R \times R}$, $n \in \langle N \rangle$, minus R due to the relation $\prod_{n=1}^{N} \mathbf{\Lambda}^{(n)} = \mathbf{I}_R$, which allows us to compute an ambiguity matrix if the other $N-1$ are known. The number $(N-1)R$ therefore represents the amount of a priori information needed to eliminate the scaling ambiguities.

References	Conditions	Properties
	Random matrix $A \in K^{I \times J}$ with i.i.d. elements from a uniformly continuous law	$k_A = r_A = \min(I, J)$
(Sidiropoulos and Liu 2001)	Vandermonde matrix $A = \begin{bmatrix} 1 & 1 & \cdots & 1 \\ \alpha_1 & \alpha_2 & \cdots & \alpha_J \\ \vdots & \vdots & & \vdots \\ \alpha_1^{I-1} & \alpha_2^{I-1} & \cdots & \alpha_J^{I-1} \end{bmatrix} \in K^{I \times J}$ $\alpha_j \neq 0 \ \forall j \in \langle J \rangle , \ \alpha_k \neq \alpha_j \ \forall k \neq j$	$k_A = r_A = \min(I, J)$
	Khatri-Rao product $A \diamond B ; A \in K^{I \times R}, B \in K^{J \times R}$	
(Sidiropoulos et al. 2000a)	$k_A + k_B \geq R + 1$	$r_{A \diamond B} = R$
(Sidiropoulos and Liu 2001)	$k_A \geq 1, \ k_B \geq 1$	$k_{A \diamond B} \geq \min(k_A + k_B - 1, R)$
(Jiang et al. 2001)	A, B random or Vandermonde with random generators	$r_{A \diamond B} = k_{A \diamond B} = \min(IJ, R)$
	Essential uniqueness of a PARAFAC model $[\![A, B, C; R]\!]$	
(Kruskal 1977)	$k_A + k_B + k_C \geq 2R + 2$	$\begin{cases} \overline{A} = A\Pi\Lambda_A \\ \overline{B} = B\Pi\Lambda_B \\ \overline{C} = C\Pi\Lambda_C \\ \Lambda_A\Lambda_B\Lambda_C = I_R \end{cases}$
$\begin{cases} \text{(Stegeman 2008)} \\ \text{(Guo et al. 2011)} \end{cases}$	$\begin{cases} k_C = R \\ k_A, k_B \geq 2 \\ r_A + k_B \geq R + 2 \ \text{or} \ r_B + k_A \geq R + 2 \end{cases}$	$\begin{cases} \overline{A} = A\Pi\Lambda_A \\ \overline{B} = B\Pi\Lambda_B \\ \overline{C} = C\Pi\Lambda_C \\ \Lambda_A\Lambda_B\Lambda_C = I_R \end{cases}$
	$k_A + k_B + k_C \geq 2R + 2$ with C known	$\begin{cases} \overline{A} = A\Lambda_A \\ \overline{B} = B\Lambda_B \\ \Lambda_A\Lambda_B = I_R \end{cases}$

Table 5.6. k-rank of certain matrices and essential uniqueness of a third-order PARAFAC model

5.2.6. Block tensor models

In some applications, the data tensor $\mathcal{X} \in K^{I_N}$ of order N is decomposed into a sum of P tensor components, each with a differently structured tensor model. For example, in the case of a tensor $\mathcal{X}$ decomposed into the sum of P tensors $\mathcal{X}^{(p)} \in K^{I_N}$ admitting a Tucker model, we have:

$$\mathcal{X} = \sum_{p=1}^{P} \mathcal{X}^{(p)} , \quad \mathcal{X}^{(p)} = \mathcal{G}^{(p)} \overset{N}{\underset{n=1}{\times}} A^{(p,n)}, \qquad [5.90]$$

where $\mathcal{G}^{(p)} \in \mathbb{K}^{R^{(p,1)} \times \cdots \times R^{(p,N)}}$ and $\mathbf{A}^{(p,n)} \in \mathbb{K}^{I_n \times R^{(p,n)}}, n \in \langle N \rangle$, are the core tensor and the factor matrices of $\mathcal{X}^{(p)}$. The matrix unfolding [5.5] then becomes:

$$\mathbf{X}_{\mathbb{S}_1;\mathbb{S}_2} = \sum_{p=1}^{P} \mathbf{X}_{\mathbb{S}_1;\mathbb{S}_2}^{(p)} = \sum_{p=1}^{P} \left(\bigotimes_{n \in \mathbb{S}_1} \mathbf{A}^{(p,n)} \right) \mathbf{G}_{\mathbb{S}_1;\mathbb{S}_2}^{(p)} \left(\bigotimes_{n \in \mathbb{S}_2} \mathbf{A}^{(p,n)} \right)^T . \quad [5.91]$$

Let us define the partitioned matrices:

$$\bigotimes_{n \in \mathbb{S}_1}{}_b \mathbf{A}^{(n)} = \left[\begin{array}{cccc} \bigotimes_{n \in \mathbb{S}_1} \mathbf{A}^{(1,n)}, & \bigotimes_{n \in \mathbb{S}_1} \mathbf{A}^{(2,n)}, & \cdots, & \bigotimes_{n \in \mathbb{S}_1} \mathbf{A}^{(P,n)} \end{array} \right] \quad [5.92]$$

$$\mathbf{G}_{\mathbb{S}_1;\mathbb{S}_2} = \mathrm{bdiag}\left(\mathbf{G}_{\mathbb{S}_1;\mathbb{S}_2}^{(1)}, \mathbf{G}_{\mathbb{S}_1;\mathbb{S}_2}^{(2)}, \cdots, \mathbf{G}_{\mathbb{S}_1;\mathbb{S}_2}^{(P)} \right), \quad [5.93]$$

where $\otimes_b$ denotes a block-wise Kronecker product, and *bdiag* is the block-wise diagonalization operator. The matrix unfolding [5.91] can then be rewritten in the following compact form, corresponding to a block Tucker model:

$$\mathbf{X}_{\mathbb{S}_1;\mathbb{S}_2} = \left(\bigotimes_{n \in \mathbb{S}_1}{}_b \mathbf{A}^{(n)} \right) \mathbf{G}_{\mathbb{S}_1;\mathbb{S}_2} \left(\bigotimes_{n \in \mathbb{S}_2}{}_b \mathbf{A}^{(n)} \right)^T . \quad [5.94]$$

Similarly, for a block PARAFAC model, we have:

$$\mathcal{X} = \sum_{p=1}^{P} \mathcal{X}^{(p)} , \quad \mathcal{X}^{(p)} = \mathcal{I}_{N,R^{(p)}} \mathop{\times}_{n=1}^{N} \mathbf{A}^{(p,n)}, \quad [5.95]$$

where $\mathbf{A}^{(p,n)} \in \mathbb{K}^{I_n \times R^{(p)}}, n \in \langle N \rangle$, are the factor matrices of the component $\mathcal{X}^{(p)}$.

The matrix unfolding [5.32] then becomes:

$$\mathbf{X}_{\mathbb{S}_1;\mathbb{S}_2} = \sum_{p=1}^{P} \mathbf{X}_{\mathbb{S}_1;\mathbb{S}_2}^{(p)} = \sum_{p=1}^{P} \left(\diamond_{n \in \mathbb{S}_1} \mathbf{A}^{(p,n)} \right) \left(\diamond_{n \in \mathbb{S}_2} \mathbf{A}^{(p,n)} \right)^T$$

$$= \left(\diamond_{n \in \mathbb{S}_1}{}_b \mathbf{A}^{(n)} \right) \left(\diamond_{n \in \mathbb{S}_2}{}_b \mathbf{A}^{(n)} \right)^T , \quad [5.96]$$

where $\diamond_b$ denotes the block Khatri–Rao product, defined as:

$$\diamond_{n \in \mathbb{S}_1}{}_b \mathbf{A}^{(n)} = \left[\begin{array}{cccc} \diamond_{n \in \mathbb{S}_1} \mathbf{A}^{(1,n)}, & \diamond_{n \in \mathbb{S}_1} \mathbf{A}^{(2,n)}, & \cdots, & \diamond_{n \in \mathbb{S}_1} \mathbf{A}^{(P,n)} \end{array} \right]. \quad [5.97]$$

Constrained block PARAFAC decompositions were introduced by de Almeida *et al.* (2007) to represent three different multiuser wireless communication systems in a unified way, whereas constrained block Tucker models were used to design MIMO OFDM systems with tensor-based space–time multiplexing codes (de Almeida *et al.* 2006) and for blind beamforming (de Almeida *et al.* 2009b).

5.2.7. *Constrained tensor models*

5.2.7.1. *Interpretation and use of constraints*

Constraints in tensor models can be interpreted as interactions or linear dependencies between the matrix factors. Such dependencies are encountered in applications in psychometrics and chemometrics for example, which are at the root of the PARATUCK-2 (Harshman and Lundy 1996) and PARALIND (Bro *et al.* 2009) models, respectively. The PARALIND model was introduced by Carroll *et al.* (1980) under the name CANDELINC (canonical decomposition with linear constraints). The earliest applications of the PARATUCK-2 model in signal processing concerned the identification and blind equalization of Wiener–Hammerstein communication channels (Kibangou and Favier 2007), followed by the design of point-to-point (de Almeida *et al.* 2009) and cooperative (Ximenes *et al.* 2014) MIMO communication systems. The PARALIND model was used to estimate propagation parameters in the context of antenna processing (Xu *et al.* 2012).

Constraints can also be used for resource allocation. For example, this is the case with the CONFAC (constrained factors) (de Almeida *et al.* 2008; Favier and de Almeida 2014a) and PARATUCK-(N_1, N) (Favier *et al.* 2012b) models. For a unified presentation of constrained PARAFAC models, see Favier and de Almeida (2014a). In the following two sections, we present two examples of constrained tensor models: the CONFAC model and an extension of the PARAFAC model, called BTD (de Lathauwer 2008), while giving an interpretation as a CONFAC model.

5.2.7.2. *Constrained PARAFAC models*

Various constrained PARAFAC models can be defined using a Tucker model $[\![\mathcal{G}, \mathbf{A}^{(1)}, \cdots, \mathbf{A}^{(N)}]\!]$, with $\mathbf{A}^{(n)} \in \mathbb{K}^{I_n \times R_n}$. If we assume that the core tensor $\mathcal{G} \in \mathbb{R}^{R_1 \times \cdots \times R_N}$ satisfies a PARAFAC model $[\![\boldsymbol{\Phi}^{(1)}, \cdots, \boldsymbol{\Phi}^{(N)}; R]\!]$, then its factors $\boldsymbol{\Phi}^{(n)} \in \mathbb{R}^{R_n \times R}$ allow us to introduce constraints on the matrices $\mathbf{A}^{(n)}$. Indeed, using the property [3.125] of the mode-n product, we have:

$$\mathcal{X} = \mathcal{G} \overset{N}{\underset{n=1}{\times}} \mathbf{A}^{(n)} \;,\; \mathcal{G} = \mathcal{I}_{N,R} \overset{N}{\underset{n=1}{\times}} \boldsymbol{\Phi}^{(n)} \quad \Rightarrow \quad \mathcal{X} = \mathcal{I}_{N,R} \overset{N}{\underset{n=1}{\times}} \mathbf{A}^{(n)} \boldsymbol{\Phi}^{(n)}.$$

$$[5.98]$$

The tensor $\mathcal{X}$ therefore satisfies a constrained PARAFAC model $[\![\overline{\mathbf{A}}^{(1)}, \cdots, \overline{\mathbf{A}}^{(N)}; R]\!]$ whose factors are given by $\overline{\mathbf{A}}^{(n)} = \mathbf{A}^{(n)} \boldsymbol{\Phi}^{(n)}$, $n \in \langle N \rangle$. The matrices $\boldsymbol{\Phi}^{(n)}$ can be interpreted in two different ways:

– either in terms of linear dependency between the columns of $\mathbf{A}^{(n)}$, as is the case for the PARALIND model;

– or in terms of resource allocation, as is the case for the CONFAC model, in the context of digital communications, where the columns of $\mathbf{\Phi}^{(n)}$ are chosen to be canonical vectors of the space $\mathbb{R}^{R_n}$.

Equations [5.98] can be written in scalar form as follows:

$$g_{r_1,\cdots,r_N} = \sum_{r=1}^{R} \prod_{n=1}^{N} \varphi_{r_n,r}^{(n)} \tag{5.99}$$

$$x_{i_1,\cdots,i_N} = \sum_{r_1=1}^{R_1} \cdots \sum_{r_N=1}^{R_N} g_{r_1,\cdots,r_N} \prod_{n=1}^{N} a_{i_n,r_n}^{(n)} \tag{5.100}$$

$$= \sum_{r=1}^{R} \prod_{n=1}^{N} \overline{a}_{i_n,r}^{(n)} \text{ with } \overline{a}_{i_n,r}^{(n)} = \sum_{r_n=1}^{R_N} a_{i_n,r_n}^{(n)} \varphi_{r_n,r}^{(n)}. \tag{5.101}$$

Using the property [2.256], the matrix unfolding [5.32] can be developed in the following different ways:

$$\mathbf{X}_{\mathbb{S}_1;\mathbb{S}_2} = \left(\underset{n \in \mathbb{S}_1}{\diamond} \overline{\mathbf{A}}^{(n)} \right) \left(\underset{n \in \mathbb{S}_2}{\diamond} \overline{\mathbf{A}}^{(n)} \right)^T, \tag{5.102}$$

$$= \left(\underset{n \in \mathbb{S}_1}{\diamond} \mathbf{A}^{(n)} \mathbf{\Phi}^{(n)} \right) \left(\underset{n \in \mathbb{S}_2}{\diamond} \mathbf{A}^{(n)} \mathbf{\Phi}^{(n)} \right)^T \tag{5.103}$$

$$= \left(\underset{n \in \mathbb{S}_1}{\otimes} \mathbf{A}^{(n)} \right) \left(\underset{n \in \mathbb{S}_1}{\diamond} \mathbf{\Phi}^{(n)} \right) \left(\underset{n \in \mathbb{S}_2}{\diamond} \mathbf{\Phi}^{(n)} \right)^T \left(\underset{n \in \mathbb{S}_2}{\otimes} \mathbf{A}^{(n)} \right)^T. \tag{5.104}$$

After defining the matrix unfolding $\mathbf{G}_{\mathbb{S}_1;\mathbb{S}_2} = \left(\underset{n \in \mathbb{S}_1}{\diamond} \mathbf{\Phi}^{(n)} \right) \left(\underset{n \in \mathbb{S}_2}{\diamond} \mathbf{\Phi}^{(n)} \right)^T$ of the PARAFAC model $[\![\mathbf{\Phi}^{(1)}, \cdots, \mathbf{\Phi}^{(N)}; R]\!]$ satisfied by the core tensor $\mathcal{G}$, we also have:

$$\mathbf{X}_{\mathbb{S}_1;\mathbb{S}_2} = \left(\underset{n \in \mathbb{S}_1}{\otimes} \mathbf{A}^{(n)} \right) \mathbf{G}_{\mathbb{S}_1;\mathbb{S}_2} \left(\underset{n \in \mathbb{S}_2}{\otimes} \mathbf{A}^{(n)} \right)^T. \tag{5.105}$$

5.2.7.3. BTD model

The BTD in blocks of multilinear rank-$(R_p, R_p, 1)$ for a third-order tensor $\mathcal{X} \in \mathbb{K}^{I \times J \times K}$ can be viewed as a constrained PARAFAC decomposition. Indeed, it is defined by de Lathauwer (2008) as:

$$\mathcal{X} = \sum_{p=1}^{P} \left(\mathbf{A}^{(p)} (\mathbf{B}^{(p)})^T \right) \circ \mathbf{c}^{(p)}, \tag{5.106}$$

where the matrices $\mathbf{A}^{(p)} = [a_{i,r_p}^{(p)}] \in \mathbb{K}^{I \times R_p}$ and $\mathbf{B}^{(p)} = [b_{j,r_p}^{(p)}] \in \mathbb{K}^{J \times R_p}$ have full column rank, and $\mathbf{c}^{(p)} = [c_1^{(p)}, \cdots, c_K^{(p)}]^T \in \mathbb{K}^K$ is a column vector whose components are assumed to be non-zero, for $p \in \langle P \rangle$.

Defining $d_{k,r_p}^{(p)} = c_k^{(p)}$, for $r_p \in \langle R_p \rangle$, the BTD model can be rewritten in scalar form as:

$$x_{i,j,k} = \sum_{p=1}^{P} \left(\sum_{r_p=1}^{R_p} a_{i,r_p}^{(p)} b_{j,r_p}^{(p)} \right) c_k^{(p)} = \sum_{p=1}^{P} \left(\sum_{r_p=1}^{R_p} a_{i,r_p}^{(p)} b_{j,r_p}^{(p)} d_{k,r_p}^{(p)} \right). \qquad [5.107]$$

We can therefore interpret the BTD model as the sum of P third-order PARAFAC models $[\![\mathbf{A}^{(p)}, \mathbf{B}^{(p)}, \mathbf{D}^{(p)}; R_p]\!]$, with $p \in \langle P \rangle$, and $\mathbf{D}^{(p)} = \mathbf{c}^{(p)} \mathbf{1}_{R_p}^T \in \mathbb{K}^{K \times R_p}$, i.e. a matrix where the R_p columns are equal to $\mathbf{c}^{(p)}$. Each PARAFAC model is associated with a tensor $\mathcal{X}^{(p)} \in \mathbb{K}^{I \times J \times K}$, and we have $\mathcal{X} = \sum_{p=1}^{P} \mathcal{X}^{(p)}$.

After defining matrices partitioned into P column blocks $\mathbf{A} = [\mathbf{A}^{(1)} \cdots \mathbf{A}^{(P)}] \in \mathbb{K}^{I \times R}$, $\mathbf{B} = [\mathbf{B}^{(1)} \cdots \mathbf{B}^{(P)}] \in \mathbb{K}^{J \times R}$, $\mathbf{D} = [\mathbf{D}^{(1)} \cdots \mathbf{D}^{(P)}] \in \mathbb{K}^{K \times R}$ and $\mathbf{C} = [\mathbf{c}^{(1)} \cdots \mathbf{c}^{(P)}] \in \mathbb{K}^{K \times P}$, with $R = \sum_{p=1}^{P} R_p$, the BTD model [5.106] can be rewritten as the following block PARAFAC model:

$$\mathcal{X} = \mathcal{I}_{3,R} \times_1 \mathbf{A} \times_2 \mathbf{B} \times_3 \mathbf{D}. \qquad [5.108]$$

The factor matrix $\mathbf{D}$ can also be written as:

$$\mathbf{D} = [\underbrace{\mathbf{c}^{(1)} \cdots \mathbf{c}^{(1)}}_{R_1}, \cdots, \underbrace{\mathbf{c}^{(P)} \cdots \mathbf{c}^{(P)}}_{R_P}] = \mathbf{C}\boldsymbol{\Phi}, \qquad [5.109]$$

where $\boldsymbol{\Phi} = \begin{bmatrix} \mathbf{1}_{R_1}^T & & \\ & \ddots & \\ & & \mathbf{1}_{R_P}^T \end{bmatrix} \in \mathbb{K}^{P \times R}$ can be viewed as a constraint matrix.

We can therefore conclude that the BTD model is equivalent to the constrained block PARAFAC model $[\![\mathbf{A}, \mathbf{B}, \mathbf{C}\boldsymbol{\Phi}; R]\!]$ (Favier and de Almeida 2014a).

Using the formulae in Table 5.3, the matrix unfoldings of $\mathcal{X}$ are given by:

$$\mathbf{X}_{JK \times I} = (\mathbf{B} \diamond \mathbf{C}\boldsymbol{\Phi})\mathbf{A}^T = \sum_{p=1}^{P} (\mathbf{B}^{(p)} \diamond \mathbf{c}^{(p)} \mathbf{1}_{R_p}^T)(\mathbf{A}^{(p)})^T \qquad [5.110]$$

$$\mathbf{X}_{KI \times J} = (\mathbf{C}\boldsymbol{\Phi} \diamond \mathbf{A})\mathbf{B}^T \qquad [5.111]$$

$$\mathbf{X}_{IJ \times K} = (\mathbf{A} \diamond \mathbf{B})\boldsymbol{\Phi}^T \mathbf{C}^T. \qquad [5.112]$$

Noting that, by the hypotheses, $\mathbf{A}^{(p)}$ and $\mathbf{B}^{(p)}$ have full column rank equal to R_p and $\mathbf{D}^{(p)} = \mathbf{c}_p \mathbf{1}_{R_p}^T$ has rank one, we can conclude that the BTD model amounts to decompose a tensor $\mathcal{X}$ into a sum of P tensors $\mathcal{X}^{(p)}$ of multilinear rank $(R_p, R_p, 1)$.

The BTD model is unique up to permutations of terms of same rank R_p, scaling ambiguities $(\alpha_p, \beta_p, \gamma_p)$ in the factors of each term, with $\alpha_p \beta_p \gamma_p = 1, \forall p \in \langle P \rangle$, and a non-singular multiplicative ambiguity matrix $\boldsymbol{\Lambda}^{(p)} \in \mathbb{C}^{R_p \times R_p}$, since any triplet $(\alpha_p \mathbf{A}^{(p)} \boldsymbol{\Lambda}^{(p)}, \beta_p \mathbf{B}^{(p)} (\boldsymbol{\Lambda}^{(p)})^{-T}, \gamma_p \mathbf{c}^{(p)})$ gives the same tensor as defined in [5.106].

In the special case where $R_p = 1, \forall p \in \langle P \rangle$, the BTD model [5.106] reduces to a standard CPD model of rank P:

$$\mathcal{X} = \sum_{p=1}^{P} (\mathbf{a}_p \mathbf{b}_p^T) \circ \mathbf{c}^{(p)} = \sum_{p=1}^{P} (\mathbf{a}_p \circ \mathbf{b}_p) \circ \mathbf{c}^{(p)} = \sum_{p=1}^{P} \mathbf{a}_p \circ \mathbf{b}_p \circ \mathbf{c}_p, \quad [5.113]$$

$$x_{ijk} = \sum_{p=1}^{P} a_{ip} b_{jp} c_{kp}, \quad [5.114]$$

which corresponds to a CPD model whose matrix factors are $\mathbf{A} = [\mathbf{a}_1 \cdots \mathbf{a}_P] \in \mathbb{K}^{I \times P}, \mathbf{B} = [\mathbf{b}_1 \cdots \mathbf{b}_P] \in \mathbb{K}^{J \times P}, \mathbf{C} = [\mathbf{c}_1 \cdots \mathbf{c}_P] \in \mathbb{K}^{K \times P}$, with $\mathbf{c}_p = \mathbf{c}^{(p)}$.

By comparing [5.113] with [5.106], we can conclude that the BTD decomposition into terms of multilinear rank-$(R_p, R_p, 1)$ replaces the rank-one factors $\mathbf{a}_p \mathbf{b}_p^T$ with the factors $\mathbf{A}^{(p)} (\mathbf{B}^{(p)})^T$ of rank R_p. Using the formula [5.50] and the column-block-partitioned forms of $(\mathbf{A}, \mathbf{B}, \mathbf{D})$, the kth frontal slice for the BTD model is given by:

$$\mathbf{X}_{..k} = \mathbf{A}\mathrm{diag}(\mathbf{D}_{k.})\mathbf{B}^T$$

$$= [\mathbf{A}^{(1)} \cdots \mathbf{A}^{(P)}] \mathrm{diag}\big([\underbrace{c_k^{(1)}, \cdots, c_k^{(1)}}_{R_1}, \cdots, \underbrace{c_k^{(P)}, \cdots, c_k^{(P)}}_{R_P}] \big) [\mathbf{B}^{(1)} \cdots \mathbf{B}^{(P)}]^T$$

$$= \sum_{p=1}^{P} c_k^{(p)} \mathbf{A}^{(p)} (\mathbf{B}^{(p)})^T \in \mathbb{K}^{I \times J}. \quad [5.115]$$

This frontal slice is therefore equal to the sum of P products of two matrices of rank R_p, each weighted by the kth component $c_k^{(p)}$ of the vector $\mathbf{c}^{(p)}$.

This expression can be deduced directly from [5.106] as follows:

$$\mathcal{X} = \sum_{p=1}^{P} \left(\mathbf{A}^{(p)} (\mathbf{B}^{(p)})^T \right) \circ \left(\sum_{k=1}^{K} c_k^{(p)} \mathbf{e}_k^{(K)} \right)$$

$$= \sum_{k=1}^{K} \left(\sum_{p=1}^{P} c_k^{(p)} \mathbf{A}^{(p)} (\mathbf{B}^{(p)})^T \right) \circ \mathbf{e}_k^{(K)})$$

$$= \sum_{k=1}^{K} \mathbf{X}_{..k} \circ \mathbf{e}_k^{(K)}, \qquad\qquad [5.116]$$

with $\mathbf{X}_{..k}$ defined in [5.115].

5.3. Examples of tensor models

In this section, we present four examples of tensor models for multidimensional harmonics and an instantaneous linear mixture, using both a BTD model of the received signals and a PARAFAC model of the tensor of fourth-order cumulants of the received signals. The fourth example concerns the representation of a linear FIR system using a tensor of fourth-order cumulants of the system output.

5.3.1. *Model of multidimensional harmonics*

Multidimensional harmonics retrieval (MHR) from the observation of a sum of complex exponentials is an old and classical problem in the field of signal processing (Pisarenko 1973). The signal model corresponding to the superposition of M undamped exponentials on a P-dimensional grid can be written as:

$$x_{i_1,\cdots,i_P} = \sum_{m=1}^{M} \alpha_m \prod_{p=1}^{P} z_{m,p}^{i_p-1} \ , \ \ i_p \in \langle I_p \rangle, \qquad\qquad [5.117]$$

where α_m is the complex amplitude of the mth exponential $z_{m,p} = e^{j\omega_{m,p}}$, and $\omega_{m,p}$ is the angular frequency of the pth dimension. The tensor of received signals $\mathcal{X} \in \mathbb{C}^{I_P}$ satisfies a structured PARAFAC model $[\![\mathbf{g}, \mathbf{V}^{(1)}, \cdots, \mathbf{V}^{(P)}; M]\!]$ of rank M, where $\mathbf{g} \triangleq [\alpha_1, \cdots, \alpha_M]^T$ is the amplitude vector, and the factors $\mathbf{V}^{(p)}$ are Vandermonde matrices defined as:

$$\mathbf{V}^{(p)} = \left[\mathbf{V}_{.1}^{(p)}, \cdots, \mathbf{V}_{.M}^{(p)} \right] \in \mathbb{C}^{I_p \times M} \ , \ \ p \in \langle P \rangle, \qquad\qquad [5.118]$$

with:

$$\mathbf{V}_{.m}^{(p)} = \left[1, z_{m,p}, \cdots, z_{m,p}^{I_p-1} \right]^T \in \mathbb{C}^{I_p} \ , \ \ m \in \langle M \rangle. \qquad\qquad [5.119]$$

This PARAFAC model is called a higher order Vandermonde decomposition (HOVDMD) by Papy *et al.* (2005). Its uniqueness properties were studied by Sidiropoulos (2001). Using a correspondence between PARAFAC models and TT models (Zniyed *et al.* 2019a), the model [5.117] can be expressed as a train of

Vandermonde tensors (Zniyed *et al.* 2019b). In the case of a high-order tensor, this approach allows us to reduce the computational complexity of parameter estimation, since the high-order PARAFAC model is transformed into a train of coupled third-order tensors, each containing a Vandermonde factor.

5.3.2. *Source separation*

5.3.2.1. *Instantaneous mixture modeled using the received signals*

One of the earliest applications of the BTD model was the blind source separation (BSS) problem (de Lathauwer 2011), which we will briefly discuss here. Consider the following matrix equation, which models an instantaneous linear mixture, (also called a memoryless or non-convolutive mixture) of P sources, measured using K sensors, over N sampling periods:

$$\mathbf{Y} = \mathbf{MS} = \sum_{p=1}^{P} \mathbf{M}_{.p} \mathbf{S}_{p.} = \sum_{p=1}^{P} \mathbf{Y}^{(p)}, \qquad [5.120]$$

or in scalar form:

$$y_{k,n} = \sum_{p=1}^{P} m_{k,p} s_{p;n} = \sum_{p=1}^{P} y_{k,n}^{(p)} \; , \; k \in \langle K \rangle, n \in \langle N \rangle. \qquad [5.121]$$

The matrix $\mathbf{M} \in \mathbb{K}^{K \times P}$ is called the instantaneous mixture, whereas $\mathbf{S} \in \mathbb{K}^{P \times N}$ and $\mathbf{Y} \in \mathbb{K}^{K \times N}$ are the matrices containing the source signals and the observation signals over N sampling periods, respectively. The BSS problem is to estimate the matrices $\mathbf{M}$ and $\mathbf{S}$ from the observed signals in $\mathbf{Y}$ only. This problem can be reformulated using a BTD model. We assume that N is odd and set $N = I + J - 1$ with $I = J$, and we construct a Hankel matrix $\mathbf{H}^{(p)} \in \mathbb{K}^{I \times I}$ from the row vector $\mathbf{S}_{p.} \in \mathbb{K}^{1 \times (2I-1)}$, as in [3.305]:

$$\mathbf{H}^{(p)} = \begin{bmatrix} s_{p1} & s_{p2} & \cdots & s_{pI} \\ s_{p2} & s_{p3} & \cdots & s_{p,I+1} \\ \vdots & & & \vdots \\ s_{pI} & s_{p,I+1} & \cdots & s_{p,2I-1} \end{bmatrix} = [h_{ij}^{(p)}] \qquad [5.122]$$

$$h_{ij}^{(p)} = s_{p,i+j-1}. \qquad [5.123]$$

The matrix $\mathbf{H}^{(p)}$ corresponds to a Hankelization of the row vector $\mathbf{S}_{p.}$. The matrix $\mathbf{Y}^{(p)} = \mathbf{M}_{.p} \mathbf{S}_{p.} \in \mathbb{K}^{K \times N}$, which is the component of $\mathbf{Y}$ associated with the source p, is transformed into a third-order tensor $\mathcal{X}^{(p)} \in \mathbb{K}^{I \times J \times K}$ by defining the frontal slice $\mathbf{X}_{..k}^{(p)}$ as:

$$x_{ijk}^{(p)} = m_{kp} h_{ij}^{(p)} = m_{kp} s_{p,i+j-1} \; , \; i, j \in \langle I \rangle, k \in \langle K \rangle,$$

or equivalently $\mathbf{X}_{..k}^{(p)} = m_{kp} \mathbf{H}^{(p)}$.

Using the expression [5.116], the tensor $\mathcal{X}^{(p)}$ can be written as:

$$\mathcal{X}^{(p)} = \sum_{k=1}^{K} \mathbf{X}_{..k}^{(p)} \circ \mathbf{e}_k^{(K)} = \sum_{k=1}^{K} m_{kp} \mathbf{H}^{(p)} \circ \mathbf{e}_k^{(K)}$$

$$= \mathbf{H}^{(p)} \circ \left(\sum_{k=1}^{K} m_{kp} \mathbf{e}_k^{(K)} \right) = \mathbf{H}^{(p)} \circ \mathbf{m}_p, \qquad [5.124]$$

where $\mathbf{m}_p = \mathbf{M}_{.p} \in \mathbb{K}^K$ is the pth column vector of the mixture matrix $\mathbf{M}$.

It is important to note that rewriting the row vector $\mathbf{S}_{p.}$ as the Hankel matrix $\mathbf{H}^{(p)}$ defined in [5.122] introduces redundancy into the model by repeating the source signals. The observation matrix $\mathbf{Y}$ defined in [5.120] is transformed into the third-order tensor:

$$\mathcal{X} = \sum_{p=1}^{P} \mathcal{X}^{(p)} = \sum_{p=1}^{P} \mathbf{H}^{(p)} \circ \mathbf{m}_p \in \mathbb{K}^{I \times I \times K}, \qquad [5.125]$$

which satisfies a BTD model. Since the measurement matrix $\mathbf{Y}$ is replaced with the redundant measurement tensor $\mathcal{X}$ after replacing each row $\mathbf{Y}_{k.}^{(p)} \in \mathbb{K}^{1 \times (2I-1)}$ of $\mathbf{Y}^{(p)}$ with a frontal slice $\mathbf{X}_{..k}^{(p)} \in \mathbb{K}^{I \times I}$ of the tensor $\mathcal{X}^{(p)}$, we conclude that this transformation of [5.120] to [5.125] increases the number of measurements from $KN = K(2I - 1)$ to $IJK = KI^2$.

Writing $\mathcal{H}$ for the Hankelization operator, the transformation of the matrix model [5.120] to the tensor model [5.125] is summarized by the following equations:

$$\mathbb{K}^{1 \times (2I-1)} \ni \mathbf{S}_{p.} \xrightarrow{\mathcal{H}} \mathbf{H}^{(p)} \in \mathbb{K}^{I \times I} \mapsto \mathbf{X}_{..k} = m_{kp} \mathbf{H}^{(p)}$$

$$\Downarrow$$

$$\mathbb{K}^{K \times (2I-1)} \ni \mathbf{Y}^{(p)} \mapsto \mathcal{X}^{(p)} \in \mathbb{K}^{I \times I \times K} , \quad p \in \langle P \rangle. \qquad [5.126]$$

De Lathauwer (2011) used the BTD model for mixtures corresponding to linear combinations of exponentials observed during $N = 2I - 1$ sampling periods, i.e.

$$s_{p,n+1} = \sum_{r_p=1}^{R_p} c_{r_p,p} z_{r_p,p}^n , \quad p \in \langle P \rangle , \quad n = 0, 1, \cdots, N - 1. \qquad [5.127]$$

The Hankel matrix $\mathbf{H}^{(p)}$ defined in [5.122] then admits the following Vandermonde factorization:

$$\mathbf{H}^{(p)} = \mathbf{V}^{(p)} \mathbf{C}^{(p)} (\mathbf{V}^{(p)})^T, \qquad [5.128]$$

with:

$$\mathbf{V}^{(p)} = \begin{bmatrix} 1 & 1 & \cdots & 1 \\ z_{1,p} & z_{2,p} & \cdots & z_{R_p,p} \\ \vdots & & & \vdots \\ z_{1,p}^{I-1} & z_{2,p}^{I-1} & \cdots & z_{R_p,p}^{I-1} \end{bmatrix} \in \mathbb{K}^{I \times R_p}$$

$$\mathbf{C}^{(p)} = \begin{bmatrix} c_{1,p} & 0 & \cdots & 0 \\ 0 & c_{2,p} & \cdots & 0 \\ \vdots & & & \vdots \\ 0 & 0 & \cdots & c_{R_p,p} \end{bmatrix} = \mathrm{diag}\Big(c_{1,p}, \quad \cdots, c_{R_p,p}\Big) \in \mathbb{K}^{R_p \times R_p}.$$

If we assume that $I \geq \max(R_1, R_2, \cdots, R_p)$ and the exponentials are distinct, i.e. the generators $z_{r_p,p}$ of the Vandermonde matrix $\mathbf{V}^{(p)}$ are distinct for $r_p \in \langle R_p \rangle$, then $\mathbf{V}^{(p)}$ has full column rank. Therefore, assuming that all coefficients $c_{r_p,p}$ are non-zero, we can deduce that $\mathbf{H}^{(p)}$ has full rank equal to R_p, which implies that the tensor slice $\mathcal{X}^{(p)}$ has trilinear rank $(R_p, R_p, 1)$, by [5.124] and [5.128].

5.3.2.2. *Instantaneous mixture modeled using cumulants of the received signals*

Consider the output of a sensor network in the blind source separation problem in the case of an instantaneous and noisy linear mixture, analogous to equation [1.137], with the correspondences $(\mathbf{A}, \mathbf{s}_n) \Leftrightarrow (\mathbf{H}, \mathbf{x}(n))$:

$$\mathbf{y}(n) = \mathbf{H}\mathbf{x}(n) + \mathbf{e}(n), \tag{5.129}$$

where $\mathbf{x}(n) \in \mathbb{C}^P$, $\mathbf{y}(n)$ and $\mathbf{e}(n) \in \mathbb{C}^K$ represent the complex-valued vector of P sources that must be separated, the output vector of K sensors and an additive white Gaussian noise (AWGN) sequence that is independent of the input signal $\mathbf{x}(n)$ at time n, respectively, and $\mathbf{H} \in \mathbb{C}^{K \times P}$ is the mixture matrix. The sources are assumed to be statistically independent and stationary in the strict sense, and circular to fourth order, which implies that their fourth-order cumulant is a diagonal tensor:

$$\mathrm{cum}(x_{p_1}(n), x_{p_2}(n), x_{p_3}^*(n), x_{p_4}^*(n)) = \kappa_{4x}(p_1)\delta_{p_1 p_2 p_3 p_4}, \tag{5.130}$$

where $\delta_{p_1 p_2 p_3 p_4}$ is the generalized Kronecker delta, which is equal to 1 for $p_1 = p_2 = p_3 = p_4$ and 0 otherwise, with $p_1, p_2, p_3, p_4 \in \langle P \rangle$.

Consider the fourth-order cumulants of the output signal[2]:

$$c_{\mathbf{y},2,2} \triangleq c_{\mathbf{y},\mathbf{y},\mathbf{y}^*,\mathbf{y}^*}(\tau_1, \tau_2, \tau_3) = \mathrm{cum}\big(\mathbf{y}(n), \mathbf{y}(n - \tau_1), \mathbf{y}^*(n - \tau_2), \mathbf{y}^*(n - \tau_3)\big).$$

---

[2] For complex-valued signals, we can define different cumulants of order P according to the numbers P_1 and P_2 of non-conjugated and conjugated terms, with $P_1 + P_2 = P$. In the case of circular signals in the strict sense, the cumulants of order $P = P_1 + P_2$ are zero if $P_1 \neq P_2$, i.e. if the number of non-conjugated and conjugated terms is different (Picinbono 1994). See the reminders on the cumulants of random signals in section A1.3.3 of the Appendix.

Due to the stationarity hypothesis, these cumulants only depend on the time shifts (τ_1, τ_2, τ_3). They form a seventh-order tensor with four spatial dimensions and three time dimensions. In the case where the time shifts are zero $(\tau_1 = \tau_2 = \tau_3 = 0)$, the cumulants $\mathrm{cum}(\mathbf{y}, \mathbf{y}, \mathbf{y}^*, \mathbf{y}^*)$ form a fourth-order tensor $\mathcal{C}_{4\mathbf{y}} \in \mathbb{C}^{K \times K \times K \times K}$ with only four spatial modes.

PROPOSITION 5.10.– The tensor $\mathcal{C}_{4\mathbf{y}}$ of output cumulants satisfies a PARAFAC model $[\![\mathbf{g}, \mathbf{H}, \mathbf{H}, \mathbf{H}^*, \mathbf{H}^*; P]\!]$ of rank P with a Hermitian symmetry, factor matrices equal to the mixture matrix and its conjugate, and $\mathbf{g} = [\kappa_{4x}(1), \cdots, \kappa_{4x}(P)]^T$.

PROOF.– Using the index convention, the output of the sensor k at time n can be written as:

$$y_k(n) = \sum_{p=1}^{P} h_{kp} x_p(n) + e_k(n) = h_{kp} x_p(n) + e_k(n) , \quad k \in \langle K \rangle.$$

Since the additive white noise is assumed to be Gaussian, its fourth-order cumulant is zero, which implies that the fourth-order cumulant of the output $\mathbf{y}(n)$ is independent of the noise. Omitting the time index, this cumulant is given by:

$$\mathrm{cum}(y_{k_1}, y_{k_2}, y_{k_3}^*, y_{k_4}^*) = \mathrm{cum}(h_{k_1 p_1} x_{p_1}, h_{k_2 p_2} x_{p_2}, h_{k_3 p_3}^* x_{p_3}^*, h_{k_4 p_4}^* x_{p_4}^*)$$

with $k_i \in \langle K \rangle$ for $i \in \langle 4 \rangle$.

Taking into account the multilinearity property of the cumulant, and using the expression [5.130] of the source cumulants, the element $c_{k_1, k_2, k_3, k_4} \triangleq \mathrm{cum}(y_{k_1}, y_{k_2}, y_{k_3}^*, y_{k_4}^*)$ of the tensor $\mathcal{C}_{4\mathbf{y}}$ is given by:

$$\begin{aligned}
c_{k_1, k_2, k_3, k_4} &= h_{k_1 p_1} h_{k_2 p_2} h_{k_3 p_3}^* h_{k_4 p_4}^* \mathrm{cum}(x_{p_1}, x_{p_2}, x_{p_3}^*, x_{p_4}^*) \\
&= \kappa_{4x}(p_1) \delta_{p_1 p_2 p_3 p_4} h_{k_1 p_1} h_{k_2 p_2} h_{k_3 p_3}^* h_{k_4 p_4}^* && [5.131] \\
&= \kappa_{4x}(p) h_{k_1 p} h_{k_2 p} h_{k_3 p}^* h_{k_4 p}^* \\
&= \sum_{p=1}^{P} \kappa_{4x}(p) h_{k_1 p} h_{k_2 p} h_{k_3 p}^* h_{k_4 p}^*. && [5.132]
\end{aligned}$$

The fourth-order cumulants of the output therefore form a fourth-order tensor characterized by a Hermitian symmetry. Equation [5.132] describes a PARAFAC model $[\![\mathbf{g}, \mathbf{H}, \mathbf{H}, \mathbf{H}^*, \mathbf{H}^*; P]\!]$ of rank P whose four matrix factors are equal to the mixture matrix or its conjugate, with a scaling factor $\kappa_{4x}(p)$ for each column p equal to the fourth-order cumulant of the source P, which defines $\mathbf{g} = [\kappa_{4x}(1), \cdots, \kappa_{4x}(P)]^T$. $\qquad \square$

This model generalizes the DSym CP model presented in Table 5.5 to order four. A matrix unfolding $\mathbf{C}_{4y} \in \mathbb{C}^{K^2 \times K^2}$ of the cumulants tensor $\mathcal{C}_{4y} \in \mathbb{C}^{K \times K \times K \times K}$ can be deduced directly from the general matricization formula [5.44], namely:

$$\mathbf{C}_{4y} = (\mathbf{H} \diamond \mathbf{H})\mathrm{diag}\Big(\kappa_{4x}(1), \cdots, \kappa_{4x}(P)\Big)(\mathbf{H} \diamond \mathbf{H})^H. \tag{5.133}$$

Thus, we obtain a Hermitian matrix due to the Hermitian symmetry of the cumulants tensor $\mathcal{C}_{4y}$. This equation can be viewed as a diagonalization of the tensor $\mathcal{C}_{4y}$ of fourth-order cumulants. It provides the basis for developing source separation methods (Comon and Cardoso 1990).

Applying the formula [2.257] also allows us to vectorize the cumulants tensor $\mathcal{C}_{4y}$ in the following form:

$$\mathbf{c}_{4y} = \mathrm{vec}(\mathbf{C}_{4y}) = (\mathbf{H}^* \diamond \mathbf{H}^* \diamond \mathbf{H} \diamond \mathbf{H}) \begin{bmatrix} \kappa_{4x}(1) \\ \vdots \\ \kappa_{4x}(P) \end{bmatrix}. \tag{5.134}$$

5.3.3. *Model of a FIR system using fourth-order output cumulants*

Consider a stationary FIR system of order N described by the following equation:

$$y(t) = \sum_{n=0}^{N} b_n x(t-n) + e(t), \tag{5.135}$$

where $\{x(t)\}$ is a sequence of independent, centered, stationary, complex-valued random variables admitting the following fourth-order cumulant:

$$\mathrm{cum}\Big(x(t), x(t - \tau_1), x^*(t - \tau_2), x^*(t - \tau_3)\Big) = \kappa_{4x}\delta_{0,\tau_1,\tau_2,\tau_3}. \tag{5.136}$$

Here, note that we are assuming the samples of the input signal to be time independent, whereas in the source separation problem above the sources were space independent. The model [5.135] can also be viewed as a convolutive mixture of sources, or a noisy moving average (MA) model.

PROPOSITION 5.11.– The fourth-order output cumulants $c_{4y}(i_1, i_2, i_3) \triangleq c_{y,2,2} \triangleq \mathrm{cum}\Big(y(t), y(t - i_1), y^*(t - i_2), y^*(t - i_3)\Big)$ satisfy a third-order PARAFAC model $[\![\mathbf{g}, \mathbf{H}, \mathbf{H}^*, \mathbf{H}^*; N + 1]\!]$ of rank $N + 1$, with:

$$\mathbf{H} = \begin{bmatrix} b_0 & b_1 & \cdots & b_N \\ 0 & b_0 & \ddots & \vdots \\ \vdots & \ddots & \ddots & b_1 \\ 0 & \cdots & 0 & b_0 \end{bmatrix} \in \mathbb{C}^{(N+1)\times(N+1)} \qquad [5.137]$$

$$\mathbf{g} = \kappa_{4x}[b_0, b_1, \cdots, b_N]^T = \kappa_{4x}\mathbf{H}_{1.}^T, \qquad [5.138]$$

where $\mathbf{H}_{1.}$ represents the first row of $\mathbf{H}$, which is a Toeplitz matrix.

PROOF.– As in the previous example, the hypothesis of additive white Gaussian noise implies that the fourth-order cumulant of the output does not depend on the noise. Using the index convention, the equation of the model, and the multilinearity property of the cumulant, we obtain:

$$c_{4y}(i_1, i_2, i_3) = \mathrm{cum}(y(t), y(t - i_1), y^*(t - i_2), y^*(t - i_3))$$

$$= b_n b_{\tau_1} b_{\tau_2}^* b_{\tau_3}^* \mathrm{cum}\Big(x(t - n), x(t - i_1 - \tau_1),$$

$$x^*(t - i_2 - \tau_2), x^*(t - i_3 - \tau_3)\Big).$$

The stationarity and independence hypotheses on the input imply that:

$$c_{4y}(i_1, i_2, i_3) = \kappa_{4x} b_n b_{\tau_1} b_{\tau_2}^* b_{\tau_3}^* \delta_{n, i_1 + \tau_1, i_2 + \tau_2, i_3 + \tau_3}$$

$$= \kappa_{4x} \sum_{n=0}^{N} b_n b_{n-i_1} b_{n-i_2}^* b_{n-i_3}^*. \qquad [5.139]$$

Since $b_n = 0, \forall n \notin [0, N]$, we deduce that $c_{4y}(i_1, i_2, i_3) = 0, \forall |i_1|, |i_2|, |i_3| > N$. These cumulants form a third-order tensor $\mathcal{C}_{4y} \in \mathbb{C}^{(N+1)\times(N+1)\times(N+1)}$.

Defining $h_{i_m, n} = b_{n-i_m}$, for $m = 1, 2, 3$, with $i_m = 0, 1, \cdots, N$, the cumulant of the output can also be written as:

$$c_{4y}(i_1, i_2, i_3) = \kappa_{4x} \sum_{n=0}^{N} b_n h_{i_1, n} h_{i_2, n}^* h_{i_3, n}^*, \qquad [5.140]$$

which is the equation of a third-order PARAFAC model $[\![\mathbf{g}, \mathbf{H}, \mathbf{H}^*, \mathbf{H}^*; N + 1]\!]$ of rank $N + 1$, where $\mathbf{H}$ and $\mathbf{g}$ are as defined in [5.137] and [5.138]. $\qquad \square$

Using the general matricization formula [5.32] for a PARAFAC model, a matrix unfolding $[\mathbf{C}_{4y}]_{I_1 I_2 \times I_3}$ of the cumulants tensor $\mathcal{C}_{4y}$ is given by:

$$\mathbf{C}_{4y} = (\mathbf{H} \diamond \mathbf{H}^*)\mathbf{D}\mathbf{H}^H \in \mathbb{C}^{(N+1)^2 \times (N+1)}. \qquad [5.141]$$

with $\mathbf{D} = \kappa_{4x}\mathrm{diag}(b_0, b_1, \cdots, b_N)$. Applying the identity [2.257], with $\mathbf{A} = \mathbf{H} \diamond \mathbf{H}^*$, $\mathbf{C} = \mathbf{H}^H$, and $\lambda_n = \kappa_{4x}b_{n-1}$, for $n = 1, \cdots, N + 1$, allows us to deduce the following vectorized form for the output cumulants tensor:

$$\mathbf{c}_{4y} = \mathrm{vec}(\mathbf{C}_{4y}) = (\mathbf{H}^* \diamond \mathbf{H} \diamond \mathbf{H}^*)\mathbf{g}. \qquad [5.142]$$

Note that the matricized and vectorized forms [5.141] and [5.142] do not take the cumulants $c_{4y}(i_1, i_2, i_3)$ into account for $i_1, i_2, i_3 \in [-N, -1]$. One such matricization with $i_1, i_2, i_3 \in [-N, N]$ is considered in Fernandes $et\ al.$ (2008), as well as the case of a MIMO FIR system, i.e. with $\mathbf{y}(t) \in \mathbb{C}^K$ and $\mathbf{x}(t) \in \mathbb{C}^P$, corresponding to the case of K sensors and P sources.

Appendix

Random Variables and Stochastic Processes

A1.1. Introduction

The goal of this appendix is to give a brief overview of a few fundamental results about the higher order statistics (HOS) of random signals (i.e. of order greater than two). HOS play an important role in digital signal processing (SP) for the representation, detection, analysis, classification, equalization and filtering of signals such as radar, sonar, seismic, biomedical and communication signals. In many applications, second-order statistics are not sufficient to completely characterize the signals to be processed. It is then beneficial to use HOS (higher order cumulants and their associated Fourier transforms, known as polyspectra).

HOS-based SP methods, originally developed to solve the blind source separation (BSS) and blind deconvolution problems, offer the following advantages:

– they are robust to additive white Gaussian noise (AWGN), since the cumulants of order greater than two of Gaussian signals are zero, and the cumulants of the sum of independent stationary random signals are equal to the sum of the cumulants of each random signal; consequently, in the presence of an additive Gaussian measurement noise, independent of the noiseless output signal, as commonly assumed in most applications, HOS-based SP methods are blind to this additive noise;

– they allow us to identify non-minimum phase linear systems (i.e. with unstable inverses); this is not possible with second-order statistics (autocorrelation and power spectrum) that do not preserve non-minimum phase information, making phase reconstruction only achievable for minimum phase systems;

– they are the basis of certain blind channel equalization techniques that compensate the distortion undergone by a signal transmitted through a channel. In the context of digital communications, the function of an equalizer is to invert the channel in such a way that the cascade of the channel and the equalizer is as close as possible to the identity transfer function in order to recover the transmitted symbols. Since a blind equalizer does not use a learning sequence, we also speak of unsupervised equalization, with the goal of improving the transmission rate of the communication system;

– they are useful for analyzing non-Gaussian signals, detecting and characterizing non-linearities, and identifying nonlinear systems.

Higher-order cumulants can be viewed as tensors (McCullagh 1987), so it is natural for methods based on cumulants to take advantage of tensor decompositions, as illustrated in Chapter 5.

In the first section of this appendix, we will recall a few definitions and properties relating to random variables (r.v.s), while distinguishing between real and complex r.v.s, as well as between scalar and multidimensional r.v.s. The definition and properties of cumulants will be presented for real r.v.s, then for complex r.v.s, and the notion of circular complex r.v.s will be introduced. The Gaussian distribution, one of the most widely used distributions in SP applications, will be considered in particular detail.

In the second section, HOS of discrete-time random signals will be presented for real random signals, then for complex random signals. Here, it should be noted that complex-valued random signals play an important role in SP, especially in array processing and digital communications. This was the motivation for writing this appendix, which should also prove useful for the examples given in Chapter 5 involving the application of cumulant tensors to model signals and systems. The notions of polyspectra, also called higher order spectra, and circular signals will be defined. While the power spectrum allows a second-order spectral analysis, polyspectra such as the bispectrum or the trispectrum allow us to perform a so-called higher order spectral (or polyspectral) analysis.

In the third section, the use of HOS will be illustrated with the supervised identification of linear time-invariant (LTI) systems and homogeneous quadratic systems using higher order spectra and cross-spectra of input and output signals. This will also allow us to introduce the link between Volterra models and tensors via Volterra kernels. In Volume 4 of this set of books, this link will be used to reduce the parametric complexity of Volterra models via symmetrization and tensor decomposition of the kernels, as well as for their identification (Favier and Bouilloc 2010; Favier et al. 2012a).

For a more detailed presentation of HOS and their applications in SP, see the books by Nikias and Petropulu (1993), Picinbono (1993), Amblard et al. (1996b), Lacoume *et al.* (1997) and Nandi (1999), as well as the articles by Brillinger (1965), Nikias and Raghuveer (1987), Mendel (1991), and Nikias and Mendel (1993).

A1.2. Random variables

First, let us recall a few results about second-order statistics of real scalar r.v.s (section A1.2.1) and multidimensional r.v.s (section A1.2.2). The first and second characteristic functions are defined, also called moment-generating and cumulant-generating functions, respectively, since they allow us to generate moments and cumulants. Some relationships between cumulants and moments are presented, as well as some properties of cumulants.

A1.2.1. *Real scalar random variables*

A1.2.1.1. *Distribution function and probability density*

Let $(\Omega, \mathcal{F}, P)$ be a probability space, where Ω is the set of all possible events for a random experiment, $\mathcal{F}$ is a set of events, called a σ-algebra on Ω, and P denotes a probability measure. For more details, see Picinbono (1993) and Favier (2019).

Given a real-valued scalar r.v. $x(\omega)$, with $\omega \in \Omega$, its distribution function, denoted F_x, is a mapping from $\mathbb{R}$ into $[0, 1]$ that defines the probability that x takes a value less than or equal to u:

$$F_x(u) = \text{Prob}\big[x(\omega) \leq u\big].$$

If $F_x(u)$ is assumed to be continuous and differentiable, its derivative $p_x(u) \triangleq \frac{dF_x(u)}{du}$ is called the probability density function (pdf) of x. It satisfies:

$$\text{Prob}\big[a < x \leq b\big] = F_x(b) - F_x(a) = \int_a^b p_x(u)du \ \ \text{with} \ \ \int_{-\infty}^{\infty} p_x(u)du = 1.$$

A1.2.1.2. *Moments and central moments*

DEFINITION.– *The ith-order moment $\mu_{x,i}$ of the real, continuous, scalar r.v. $x(\omega)$ with pdf $p_x(u)$ is defined as[1]:*

$$\mu_{x,i} \triangleq E[x^i] = \int_{-\infty}^{\infty} u^i p_x(u)du, \qquad\qquad [A1.1]$$

where E is the mathematical expectation. The first-order moment $\mu_{x,1} \triangleq E[x] = \mu_x$ is called the mean of x. We say that the r.v. x is centered if its mean is zero. If not, x

---

1 In the rest of the appendix, $x(\omega)$ will be written x to alleviate the notation.

can be centered by subtracting its mean, giving $E[x - \mu_x] = 0$. The ith-order moment will also be denoted $m_{x,i}$.

In the case of a discrete r.v. x taking the values x_n with the probabilities p_n, the mean value of x is defined by:

$$E[x] = \sum_n x_n p_n, \qquad [A1.2]$$

which corresponds to the average of all possible values x_n weighted by their probabilities p_n.

If the distribution of x is symmetric, i.e. if its pdf is an even function ($p_x(-u) = p_x(u)$), we deduce from the definition [A1.1] that the odd-order moments are zero:

$$\mu_{x,2k+1} = 0 \; , \; \forall k \geq 0. \qquad [A1.3]$$

DEFINITION.– *The ith-order central moment of x, denoted $\nu_{x,i}$, is the ith-order moment about the mean, defined as:*

$$\nu_{x,i} \triangleq E[(x - \mu_x)^i] = \int_{-\infty}^{\infty} (u - \mu_x)^i p_x(u) du. \qquad [A1.4]$$

The second-order central moment $\nu_{x,2} \triangleq \sigma_x^2$ is called the variance of x, and its square root σ_x is called the standard deviation. We have the relation:

$$\sigma_x^2 = E[(x - \mu_x)^2] = E[x^2] - [E(x)]^2 = E[x^2] - \mu_x^2, \qquad [A1.5]$$

i.e. the variance is equal to the difference between the mean of the square and the square of the mean.

To characterize the shape of a distribution, we define two dimensionless quantities in terms of the third- and fourth-order central moments, namely the skewness coefficient $\gamma_{x,3}$, and the kurtosis $\gamma_{x,4}$, in the following normalized forms:

$$\gamma_{x,3} = \frac{\nu_{x,3}}{\sigma_x^3} \; , \; \gamma_{x,4} = \frac{\nu_{x,4} - 3\sigma_x^4}{\sigma_x^4}. \qquad [A1.6]$$

Note that the kurtosis is zero for a zero-mean Gaussian r.v.

Table A1.1 recalls a few definitions and results regarding the second-order statistics of jointly distributed scalar r.v.s, with the joint pdf $p_{x,y}(u, v)$, denoted $p(u, v)$.

Quantities	Definitions
Joint moment	$E[x^i y^j] = \int_{-\infty}^{\infty} \int_{-\infty}^{\infty} u^i v^j p(u, v) du dv$
Central joint moment	$E[(x - \mu_x)^i (y - \mu_y)^j] = \int_{-\infty}^{\infty} \int_{-\infty}^{\infty} (u - \mu_x)^i (v - \mu_y)^j p(u, v) du dv$
	Special cases
Cross-correlation	$\varphi_{xy} \triangleq E[xy]$
Cross-covariance	$\sigma_{xy} \triangleq Cov[x, y] = E[(x - \mu_x)(y - \mu_y)]$
Correlation coefficient	$\rho_{xy} \triangleq \frac{Cov[x,y]}{\sigma_x \sigma_y} = \frac{\sigma_{xy}}{\sigma_x \sigma_y}$
	Relation and property
	$Cov[x, y] = E[xy] - \mu_x \mu_y$ or $\sigma_{xy} = \varphi_{xy} - \mu_x \mu_y$
	$|\rho_{xy}| \leq 1 \Leftrightarrow |Cov[x, y]| \leq \sigma_x \sigma_y$

Table A1.1. *Some definitions for jointly distributed r.v.s x and y*

A1.2.1.3. *Independence, non-correlation and orthogonality*

Table A1.2 gives definitions of uncorrelated r.v.s and orthogonal r.v.s. The notion of orthogonality stems from the fact that $E[xy]$ can be interpreted as the inner product of x and y in the Hilbert space of scalar r.v.s. These two properties are equivalent for centered random variables.

Properties	Definitions
x and y uncorrelated	$\rho_{xy} = 0 \Leftrightarrow Cov[x, y] = 0 \Leftrightarrow \varphi_{xy} = \mu_x \mu_y$
x and y orthogonal	$\varphi_{xy} = E[xy] = 0$

Table A1.2. *Definitions of uncorrelated and orthogonal r.v.s x and y*

The r.v.s x and y are said to be independent if their joint pdf is separable[2]:

$$p(u, v) = p_x(u) p_y(v) \Rightarrow E[x^i y^j] = E[x^i] E[y^j] \; \forall i, j \in \mathbb{N}^*. \qquad [A1.7]$$

---

2 When two continuous r.v.s are independent, their marginal pdfs $p_x(u)$ and $p_y(v)$ are equal to the conditional probability densities $p_x(u/v)$ of x given $y = v$ and $p_y(v/u)$ of y given $x = u$, respectively:

$$p_x(u/v) \triangleq \frac{p(u, v)}{p_y(v)} = \frac{p_x(u) p_y(v)}{p_y(v)} = p_x(u) \triangleq \int p(u, v) dv$$

$$p_y(v/u) \triangleq \frac{p(u, v)}{p_x(u)} = \frac{p_x(u) p_y(v)}{p_x(u)} = p_y(v) \triangleq \int p(u, v) du.$$

In the case of two independent discrete r.v.s, we have: $p_x(i/j) = \frac{p(i,j)}{p_y(j)} = \frac{p_x(i) p_y(j)}{p_y(j)} =$
$p_x(i) \triangleq P[x = x_i] = \sum_j p(i, j)$, where $p_x(i/j) \triangleq P[x = x_i / y = y_j]$ is the conditional

In particular, for $i = j = 1$, we then have $\varphi_{xy} = \mu_x \mu_y$. This means that, if the r.v.s x and y are statistically independent, then they are uncorrelated, but they are not orthogonal in general. Orthogonality ($\varphi_{xy} = 0$) is satisfied if at least one of the r.v.s is zero-mean. The converse is not true, i.e. two uncorrelated r.v.s are not independent in general. Independence is therefore a more restrictive concept that non-correlation. An exception is the case of two jointly Gaussian r.v.s, where non-correlation implies statistical independence. The proof of this result is given in section A1.2.3.3. These properties are summarized in Table A1.3.

Hypotheses	Properties
x and y independent	$\Rightarrow$ x and y uncorrelated
$\begin{bmatrix} x \text{ and } y \text{ uncorrelated} \\ \text{and} \\ x \text{ or/and } y \text{ is/are zero-mean} \end{bmatrix}$	$\Rightarrow$ x and y orthogonal
$\begin{bmatrix} x \text{ and } y \text{ jointly Gaussian} \\ \text{and} \\ \text{uncorrelated} \end{bmatrix}$	$\Rightarrow$ x and y independent

Table A1.3. *Properties of r.v.s*

REMARK A1.1.– In SP, the notions of non-correlation and statistical independence are the basis of two classes of fundamental methods for solving the BSS problem in an instantaneous, i.e. non-convolutive, mixture: principal component analysis (PCA) and independent component analysis (ICA).

PCA methods attempt to decorrelate the signals of the mixture to obtain sources that are spatially decorrelated but not independent in general. These methods either diagonalize the covariance matrix of the received signals using the EVD decomposition or exploit the SVD decomposition of a data matrix directly (see Chapter 1).

ICA methods, on the other hand, seek to maximize the statistical independence of the sources, either by using adaptive algorithms or HOS-based processings of data blocks. Here, the separation principle corresponds to maximizing an independence criterion, called a contrast function (Comon 1994), and independence of the sources to p-th order means that all cross-cumulants of the sources are zero up to order p.

A1.2.1.4. *Ensemble averages and empirical averages*

In practice, the statistics of r.v.s are estimated as averages computed from finitely many measurements performed with identical systems or by repeating the same

---

probability, $p_x(i)$ and $p_y(j)$ are the marginal probabilities, and $p(i,j) \triangleq P[(x = x_i) \text{ and } (y = y_j)]$. Similarly, we have: $p_y(j/i) = p_y(j) \triangleq P[y = y_j] = \sum_i p(i,j)$.

experiment multiple times under the same conditions. This is equivalent to replacing ensemble averages (in the sense of the mathematical expectation) with empirical averages. Thus, if we assume that N samples $\{x_n, y_n; n \in \langle N \rangle\}$ of independent realizations of the variables x and y are known, we can estimate the empirical moments, cross-correlation and cross-covariance as:

$$\widehat{\mu}_{x,i} = \frac{1}{N} \sum_{n=1}^{N} x_n^i \ , \quad \widehat{\mu}_x = \frac{1}{N} \sum_{n=1}^{N} x_n \ , \quad \widehat{\mu}_y = \frac{1}{N} \sum_{n=1}^{N} y_n$$

$$\widehat{\sigma}_{xy} = \frac{1}{N} \sum_{n=1}^{N} (x_n - \widehat{\mu}_x)(y_n - \widehat{\mu}_y) \ , \quad \widehat{\varphi}_{xy} = \frac{1}{N} \sum_{n=1}^{N} x_n y_n.$$

In the case of discrete-time stationary random signals, the statistics are estimated using time averages computed from signals measured over a sufficiently long window. Replacing the ensemble averages by time averages corresponds to the hypothesis that the signals are ergodic, which is very commonly assumed in SP to estimate the HOS. Note that ergodicity requires stationarity, whereas a stationary random signal might not be ergodic (see section A1.3.2).

A1.2.1.5. *Characteristic functions, moments and cumulants*

The Pth-order moment of the scalar r.v. x, defined in [A1.1], is generated by the first characteristic function using the following formula:

$$m_{x,P} = \mathrm{E}[x^P] = \frac{1}{j^P} \frac{d^P}{du^P} \Xi_x(u) \Big|_{u=0} \triangleq \frac{1}{j^P} \Xi_x^{(P)}(0), \qquad \text{[A1.8]}$$

where $j^2 = -1$ and $\Xi_x^{(P)}(0)$ is the P-th derivative of $\Xi_x(u)$ calculated at $u = 0$. In the continuous case, the characteristic function is defined as the mean of e^{jux}:

$$\Xi_x(u) = \mathrm{E}[e^{jux}] = \int_{-\infty}^{+\infty} e^{juv} p_x(v) dv,$$

and in the discrete case:

$$\Xi_x(u) = \sum_n e^{jux_n} p_n.$$

PROOF.– Using the Taylor–MacLaurin expansion of e^{jux}, we obtain:

$$\Xi_x(u) = \mathrm{E}[e^{jux}] = \sum_{k=0}^{\infty} \frac{(ju)^k}{k!} \mathrm{E}[x^k] = \sum_{i=0}^{\infty} \frac{j^k m_{x,k}}{k!} u^k.$$

By differentiating P times and taking the derivative at $u = 0$, we deduce the expression [A1.8] for the Pth-order moment. $\qquad \square$

The second characteristic function is defined as the natural logarithm of $\Xi_x(u)$ in Papoulis (1984):

$$\Psi_x(u) = \text{Log}[\Xi_x(u)]. \tag{A1.9}$$

In the case of a non-centered r.v., we have:

$$\text{cum}(x, x) = E[x^2] - \left[E[x]\right]^2 = \nu_{x,2} = \sigma_x^2$$

$$\text{cum}(x, x, x) = E[x^3] - 3E[x]E[x^2] + 2\left[E[x]\right]^3 = \nu_{x,3}$$

$$\text{cum}(x, x, x, x) = E[x^4] - 4E[x]E[x^3] - 3\left[E[x^2]\right]^2 + 12\left[E[x]\right]^2 E[x^2] - 6\left[E[x]\right]^4$$

$$= \nu_{x,4} - 3\sigma_x^4,$$

where $\nu_{x,3}$ and $\nu_{x,4}$ are the third- and fourth-order central moments as defined in [A1.4]. Note that the cumulants are identical to the central moments up to third order. In the case of a zero-mean r.v., the second-, third-, and fourth-order cumulants satisfy the following relations with the moments:

$$\text{cum}(x, x) = E[x^2] \; ; \; \text{cum}(x, x, x) = E[x^3] \tag{A1.10}$$

$$\text{cum}(x, x, x, x) = E[x^4] - 3\left[E[x^2]\right]^2. \tag{A1.11}$$

A1.2.2. *Real multidimensional random variables*

A1.2.2.1. *Second-order statistics*

In the case of a real-valued N-dimensional r.v. $(x_1, \cdots, x_N)$, we define the vector $\mathbf{x} \in \mathbb{R}^N$ whose components are the r.v.s x_n. We say that $\mathbf{x}$ is a (real) random vector of size N, and the second-order statistics (cross-correlation and cross-covariance) of the r.v.s x_n define the autocorrelation and covariance matrices of the random vector $\mathbf{x}$.

Table A1.4 gives the definitions of the autocorrelation, covariance, cross-correlation and cross-covariance matrices of real random vectors.

REMARK A1.2.– We can make the following remarks:

– The element (i, j) of the autocorrelation matrix $\boldsymbol{\Phi}_\mathbf{x}$ is the correlation $\varphi_{x_i x_j} = E[x_i x_j]$ between the r.v.s x_i and x_j. The autocorrelation matrix $\boldsymbol{\Phi}_\mathbf{x}$ is symmetric and non-negative definite, i.e. $\mathbf{u}^T \boldsymbol{\Phi}_\mathbf{x} \mathbf{u} \geq 0$ for every non-zero real vector $\mathbf{u}$. Indeed, we have $\mathbf{u}^T E[\mathbf{x}\mathbf{x}^T] \mathbf{u} = E[\mathbf{u}^T \mathbf{x}\mathbf{x}^T \mathbf{u}] = E[y^2] \geq 0$, where $y \triangleq \mathbf{x}^T \mathbf{u} = \mathbf{u}^T \mathbf{x}$.

– When the r.v.s $x_i, i \in \langle n \rangle$, are mutually orthogonal ($\varphi_{x_i x_j} = 0 \; \forall i, j \in \langle n \rangle, i \neq j$), the autocorrelation matrix is diagonal, and the ith element of the diagonal is equal to $\varphi_{x_i} = E[x_i^2]$.

Quantities	Definitions
Autocorrelation matrix	$\Phi_\mathbf{x} \triangleq E[\mathbf{x}\mathbf{x}^T]$
Covariance matrix	$\Sigma_\mathbf{x} \triangleq E[(\mathbf{x} - \mu_\mathbf{x})(\mathbf{x} - \mu_\mathbf{x})^T]$
Cross-correlation matrix	$\Phi_\mathbf{xy} \triangleq E[\mathbf{x}\mathbf{y}^T]$
Cross-covariance matrix	$\Sigma_\mathbf{xy} \triangleq E[(\mathbf{x} - \mu_\mathbf{x})(\mathbf{y} - \mu_\mathbf{y})^T]$
Hypotheses	Relations and properties
	$\Sigma_\mathbf{x} = \Phi_\mathbf{x} - \mu_\mathbf{x}\mu_\mathbf{x}^T$
x centered	$\Sigma_\mathbf{x} = \Phi_\mathbf{x}$
	$\Phi_\mathbf{xy} = \Phi_\mathbf{yx}^T$
	$\Sigma_\mathbf{xy} = \Sigma_\mathbf{yx}^T = \Phi_\mathbf{xy} - \mu_\mathbf{x}\mu_\mathbf{y}^T$
x and y uncorrelated	$\Sigma_\mathbf{xy} = 0 \Leftrightarrow \Phi_\mathbf{xy} = \mu_\mathbf{x}\mu_\mathbf{y}^T$
x and y orthogonal	$\Phi_\mathbf{xy} = 0$

Table A1.4. *Definitions and properties of the second-order statistics of real random vectors* x *and* y

– The element (i, j) of $\Sigma_\mathbf{x}$ is the cross-covariance $\sigma_{x_i x_j} = E[(x_i - \mu_{x_i})(x_j - \mu_{x_j})]$ between the r.v.s x_i and x_j, where $\sigma_{x_i}^2 = E[(x_i - \mu_{x_i})^2]$ is the variance of x_i. The matrix $\Sigma_\mathbf{x}$ is also symmetric.

– When the r.v.s x_i, $i \in \langle n \rangle$, are mutually uncorrelated, we have $\sigma_{x_i x_j} = \sigma_{x_i}^2 \delta_{ij}$, the matrix $\Sigma_\mathbf{x}$ is diagonal, and the ith element $\sigma_{x_i}^2$ of the diagonal is equal to the variance of x_i.

– From the relation $\Sigma_\mathbf{xy} = \Phi_\mathbf{xy} - \mu_\mathbf{x}\mu_\mathbf{y}^T$, we can conclude that if at least one of the two random vectors x and y is zero-mean, then the cross-correlation and cross-covariance matrices are identical: $\Phi_\mathbf{xy} = \Sigma_\mathbf{xy}$. If so, non-correlation of the random vectors ($\Sigma_\mathbf{xy} = 0$) is equivalent to orthogonality ($\Phi_\mathbf{xy} = 0$).

– Like for scalar r.v.s, if the random vectors are independent, then they are uncorrelated, since independence of these vectors implies:

$$\Sigma_\mathbf{xy} = E[(\mathbf{x} - \mu_\mathbf{x})(\mathbf{y} - \mu_\mathbf{y})^T] = E[\mathbf{x} - \mu_\mathbf{x}]E[(\mathbf{y} - \mu_\mathbf{y})^T] = 0. \quad \text{[A1.12]}$$

The converse is not true in general, with one exception being the case of jointly Gaussian vectors, for which non-correlation implies statistical independence. This result is proven in section A1.2.3.3.

In the case of complex-valued random vectors, the definitions and relations recalled in Table A1.4 become:

$$\mathbf{\Phi_x} \triangleq E[\mathbf{x}\mathbf{x}^H] = \mathbf{\Phi}_\mathbf{x}^H \tag{A1.13}$$

$$\mathbf{\Sigma_x} \triangleq E[(\mathbf{x} - \boldsymbol{\mu_x})(\mathbf{x} - \boldsymbol{\mu_x})^H] = \mathbf{\Phi_x} - \boldsymbol{\mu_x}\boldsymbol{\mu}_\mathbf{x}^H = \mathbf{\Sigma}_\mathbf{x}^H \tag{A1.14}$$

$$\mathbf{\Phi_{xy}} \triangleq E[\mathbf{x}\mathbf{y}^H] = \mathbf{\Phi}_\mathbf{yx}^H \tag{A1.15}$$

$$\mathbf{\Sigma_{xy}} \triangleq E[(\mathbf{x} - \boldsymbol{\mu_x})(\mathbf{y} - \boldsymbol{\mu_y})^H] = \mathbf{\Phi_{xy}} - \boldsymbol{\mu_x}\boldsymbol{\mu}_\mathbf{y}^H = \mathbf{\Sigma}_\mathbf{yx}^H. \tag{A1.16}$$

The autocorrelation matrix $\mathbf{\Phi_x}$ is then Hermitian and non-negative definite, in the sense that $\mathbf{u}^H \mathbf{\Phi_x}\,\mathbf{u} \geq 0$ for every non-zero complex vector $\mathbf{u}$. The same holds for the covariance matrix $\mathbf{\Sigma_x}$.

A1.2.2.2. *Characteristic functions, moments and cumulants*

Given the real random vector $\mathbf{x} \in \mathbb{R}^N$, the Pth-order moments and cumulants of its components define the Pth-order tensors $\mathcal{M}_{\mathbf{x},P} \in \mathbb{R}^{[P;N]}$ and $\mathcal{C}_{\mathbf{x},P} \in \mathbb{R}^{[P;N]}$ such that, for $i_p \in \langle N \rangle$:

$$\left(m_{\mathbf{x},P}\right)_{i_1,i_2,\cdots,i_P} = E[x_{i_1} x_{i_2} \cdots x_{i_P}] \tag{A1.17}$$

$$\left(c_{\mathbf{x},P}\right)_{i_1,i_2,\cdots,i_P} = \mathrm{cum}(x_{i_1}, x_{i_2}, \cdots, x_{i_P}). \tag{A1.18}$$

The order P of the moment and the cumulant corresponds to the number of indices $\{i_p\}$. For $P = 2$, the second-order moment $\mathbf{M}_{\mathbf{x},2}$ corresponds to the autocorrelation matrix $\mathbf{\Phi_x}$.

In the case of a real random vector $\mathbf{x} \in \mathbb{R}^N$, the characteristic function, denoted $\Xi_\mathbf{x}(\mathbf{u})$, is defined as the mean of $h(\mathbf{x}) = e^{j\mathbf{u}^T\mathbf{x}}$, with $\mathbf{u} \in \mathbb{R}^N$:

$$\Xi_\mathbf{x}(\mathbf{u}) = E[e^{j\mathbf{u}^T\mathbf{x}}] = \int_{-\infty}^{\infty} e^{j\mathbf{u}^T\mathbf{v}} p_\mathbf{x}(\mathbf{v})d\mathbf{v}. \tag{A1.19}$$

Using the geometric series expansion $e^{j\mathbf{u}^T\mathbf{x}} = \sum_{k=0}^{\infty} \frac{j^k}{k!}[\mathbf{u}^T\mathbf{x}]^k$, we deduce that:

$$\Xi_\mathbf{x}(\mathbf{u}) = \sum_{k=0}^{\infty} \frac{j^k}{k!} E[(\mathbf{u}^T\mathbf{x})^k]. \tag{A1.20}$$

The characteristic function allows us to generate the moments. Thus, the cross-moment of order $P = \sum_{n=1}^{N} p_n$ of the components of the random vector $\mathbf{x}$ can be expressed as follows in terms of the partial derivatives of $\Xi_\mathbf{x}(\mathbf{u})$ at the point $\mathbf{u} = \mathbf{0}$:

$$E\Big[\prod_{n=1}^{N} x_n^{p_n}\Big] = j^{-P}\Big(\frac{\partial}{\partial u_1}\Big)^{p_1} \cdots \Big(\frac{\partial}{\partial u_N}\Big)^{p_N} \Xi_\mathbf{x}(\mathbf{u})\Big|_{\mathbf{u}=\mathbf{0}}. \tag{A1.21}$$

The second characteristic function is defined as the natural logarithm of $\Xi_{\mathbf{x}}(\mathbf{u})$ in Papoulis (1984):

$$\Psi_{\mathbf{x}}(\mathbf{u}) = \mathrm{Log}[\Xi_{\mathbf{x}}(\mathbf{u})]. \tag{A1.22}$$

This function allows us to generate the cumulants. For example, we define the cross-cumulant of order $P = \sum_{n=1}^{N} p_n$ as:

$$\mathrm{cum}\Big(\underbrace{x_1,\cdots,x_1}_{p_1 \text{ terms}},\cdots,\underbrace{x_N,\cdots,x_N}_{p_N \text{ terms}}\Big) = j^{-P}\Big(\frac{\partial}{\partial u_1}\Big)^{p_1}\cdots\Big(\frac{\partial}{\partial u_N}\Big)^{p_N}\Psi_{\mathbf{x}}(\mathbf{u})\Big|_{\mathbf{u}=0}.$$

$$\tag{A1.23}$$

Here, the order of the cumulant corresponds to the sum of the repetitions p_n of each r.v. x_n, with $n \in \langle N \rangle$.

A1.2.2.3. *Relationship between cumulants and moments*

A formula established by Leonov and Shiryaev (1959) allows us to pass from moments to cumulants. Thus, the joint cumulant of P random variables can be expressed in terms of the moments of order smaller than or equal to P as:

$$\mathrm{cum}(x_1, x_2, \cdots, x_P) = \sum_{q=1}^{P} (-1)^{q-1}(q-1)! \sum_{P_{q_j}\in\mathcal{P}_q} \prod_{\mathcal{I}_k\in P_{q_j}} E\Big[\prod_{i_m\in\mathcal{I}_k} x_{i_m}\Big],$$

$$\tag{A1.24}$$

where $\mathcal{P}_q$ represents the set of partitions of order q of the index set $\mathcal{I} = \{1, \cdots, P\}$, i.e. the set of partitions with q disjoint non-empty subsets $\mathcal{I}_k$ of $\mathcal{I}$ whose union is the set $\mathcal{I}$, and P_{q_j} is an element of the set $\mathcal{P}_q$.

Using this formula for centered r.v.s, we can check that the cumulants of order less than or equal to four satisfy the following relations with the moments:

$$\mathrm{cum}(x_i) = E[x_i] = 0 \tag{A1.25}$$

$$\mathrm{cum}(x_{i_1}, x_{i_2}) = E[x_{i_1}x_{i_2}] = \varphi_{x_1 x_2} \tag{A1.26}$$

$$\mathrm{cum}(x_{i_1}, x_{i_2}, x_{i_3}) = E[x_{i_1}x_{i_2}x_{i_3}] \tag{A1.27}$$

$$\mathrm{cum}(x_{i_1}, x_{i_2}, x_{i_3}, x_{i_4}) = E[x_{i_1}x_{i_2}x_{i_3}x_{i_4}] - E[x_{i_1}x_{i_2}]E[x_{i_3}x_{i_4}]$$
$$- E[x_{i_1}x_{i_3}]E[x_{i_2}x_{i_4}] - E[x_{i_1}x_{i_4}]E[x_{i_2}x_{i_3}]. \tag{A1.28}$$

REMARK A1.3.– If we choose the variables $x_{i_n} = x$ for $n \in \langle 4 \rangle$, the formulae [A1.26]–[A1.28] give the second-, third- and fourth-order cumulants [A1.10]–[A1.11] of a centered scalar r.v.

In the case of non-centered r.v.s, we have:

$$\text{cum}(x_{i_1}, x_{i_2}) = E[x_{i_1} x_{i_2}] - E[x_{i_1}]E[x_{i_2}] = \sigma_{x_{i_1} x_{i_2}} \qquad [\text{A1.29}]$$

$$\text{cum}(x_{i_1}, x_{i_2}, x_{i_3}) = E[x_{i_1} x_{i_2} x_{i_3}] - E[x_{i_1}]E[x_{i_2} x_{i_3}]$$

$$-E[x_{i_2}]E[x_{i_1} x_{i_3}] - E[x_{i_3}]E[x_{i_1} x_{i_2}] + 2E[x_{i_1}]E[x_{i_2}]E[x_{i_3}]. \qquad [\text{A1.30}]$$

A1.2.2.4. *Properties of cumulants*

The cumulants satisfy several important properties, which are described below (Brillinger 1965; Mendel 1991; Nikias and Mendel 1993; Nikias and Petropulu 1993; Picinbono 1993):

– **P1**: The higher order cumulants (i.e. of order greater than two) of any set of jointly Gaussian r.v.s are zero, which explains why higher order processing methods are robust to additive Gaussian noise. This property, which will be proven in section A1.2.3.3, allows us to define Gaussianity tests for r.v.s and linearity tests for systems (Hinich 1982).

– **P2**: The odd-order cumulants of a r.v. with a symmetric distribution are zero. This is, for example, the case for the uniform, Gaussian and Laplacian distributions. This property also holds for the odd-order moments, as indicated in [A1.3].

– **P3**: Changing the order of the partial derivatives in [A1.23] does not change the result, from which we can conclude that the cumulants are symmetric functions with respect to their arguments, i.e.:

$$\text{cum}(x_1, \cdots, x_N) = \text{cum}(x_{\pi(1)}, \cdots, x_{\pi(N)}) \qquad [\text{A1.31}]$$

for any permutation π of the set $\langle N \rangle$. This property means that the tensor $\mathcal{C}_{\mathbf{x}, P} \in \mathbb{R}_S^{[P; N]}$ of Pth-order cumulants of a set of N r.v.s is symmetric. This property also holds for moments: $\mathcal{M}_{\mathbf{x}, P} \in \mathbb{R}_S^{[P; N]}$.

– **P4**: The cumulants (and moments) satisfy the following multilinearity property: for a set of N random vectors $\mathbf{u}^{(n)} \in \mathbb{R}^{I_n}$ linearly transformed to $\mathbf{y}^{(n)} = \mathbf{B}^{(n)} \mathbf{u}^{(n)}$, with $\mathbf{B}^{(n)} \in \mathbb{R}^{J_n \times I_n}$, the tensor of Nth-order cumulants of the vectors $\mathbf{y}^{(n)} \in \mathbb{R}^{J_n}$ is given by:

$$\mathcal{C}_{\mathbf{y}, N} = \text{cum}(\mathbf{y}^{(1)}, \cdots, \mathbf{y}^{(N)}) = \mathcal{C}_{\mathbf{u}, N} \overset{N}{\underset{n=1}{\times}} \mathbf{B}^{(n)} \in \mathbb{R}^{\underline{J}_N} \qquad [\text{A1.32}]$$

$$\mathcal{C}_{\mathbf{u}, N} = \text{cum}(\mathbf{u}^{(1)}, \cdots, \mathbf{u}^{(N)}) \in \mathbb{R}^{\underline{I}_N}. \qquad [\text{A1.33}]$$

This key property establishes a close link between cumulants and tensors.

Developing the above, the cross-cumulant of the components $y_{j_n}^{(n)}$ of the vectors $\mathbf{y}^{(n)}$, with $n \in \langle N \rangle$, can be written as:

$$\text{cum}\big(y_{j_1}^{(1)}, \cdots, y_{j_N}^{(N)}\big) = \sum_{i_1} \cdots \sum_{i_N} b_{j_1, i_1}^{(1)} \cdots b_{j_N, i_N}^{(N)} \text{cum}\big(u_{i_1}^{(1)}, \cdots, u_{i_N}^{(N)}\big)$$

$$= \prod_{n=1}^{N} b_{j_n, i_n}^{(n)} \text{cum}\big(u_{i_1}^{(1)}, \cdots, u_{i_N}^{(N)}\big), \qquad \text{[A1.34]}$$

where the last equality follows from the use of the index convention.

This multilinearity property implies the following two properties:

– **P5**: The cumulants (respectively, moments) of the r.v.s x_n, $n \in \langle N \rangle$, multiplied by constant scaling factors λ_n are equal to the cumulants (respectively, moments) of the r.v.s multiplied by the product of all scaling factors:

$$\text{cum}(\lambda_1 x_1, \cdots, \lambda_N x_N) = \Big(\prod_{n=1}^{N} \lambda_n \Big) \text{cum}(x_1, \cdots, x_N). \qquad \text{[A1.35]}$$

In particular, we have $\text{cum}(\lambda x_1, x_2, \cdots, x_N) = \lambda \, \text{cum}(x_1, \cdots, x_N)$.

– **P6**: The cumulants are additive with respect to their arguments:

$$\text{cum}(x_1 + y, x_2 \cdots, x_N) = \text{cum}(x_1, x_2 \cdots, x_N) + \text{cum}(y, x_2 \cdots, x_N).$$

This property holds for both real and complex r.v.s. It is also satisfied by moments.

– **P7**: The cumulants of the sum of two random vectors whose components are statistically independent are equal to the sum of the cumulants of each random vector, considered separately:

$$\text{cum}(x_1 + y_1, \cdots, x_N + y_N) = \text{cum}(x_1, \cdots, x_N) + \text{cum}(y_1, \cdots, y_N).$$

$$\text{[A1.36]}$$

Indeed, since the r.v.s x_n are independent of the r.v.s y_n, for $n \in \langle N \rangle$, we have:

$$\Psi_{\mathbf{x}+\mathbf{y}}(\mathbf{u}) = \text{Log}\big[E[e^{j\mathbf{u}^T(\mathbf{x}+\mathbf{y})}]\big] = \text{Log}\big[E[e^{j\mathbf{u}^T\mathbf{x}}]\big] + \text{Log}\big[E[e^{j\mathbf{u}^T\mathbf{y}}]\big]$$

$$= \Psi_{\mathbf{x}}(\mathbf{u}) + \Psi_{\mathbf{y}}(\mathbf{u}).$$

Since the second characteristic function can be written as a sum of functions, we can deduce the property [A1.36]. This property justifies the name of cumulant, since the cumulant of a sum of two sets of independent r.v.s is equal to the sum of the cumulants of each set of r.v.s, considered separately. It should be noted that this property is not satisfied by moments.

This property is very often used in SP in the case of an additive Gaussian noise $e(k)$ that is assumed to be independent of the noiseless component $y(k)$ of the measured signal $s(k) = y(k) + e(k)$. Since the cumulants of order greater than two of Gaussian noise are zero, using these cumulants of the measured signal makes it possible to eliminate the effect of additive Gaussian noise. This property will be used in section A1.4.

– **P8**: The cumulants of a set of statistically independent r.v.s are zero. Indeed, the independence property implies that the pdf $p(\mathbf{x})$ can be factorized into $\prod_{n=1}^{N} p(x_n)$, and the second characteristic function defined in [A1.22] can then be written as a sum $\Psi_{\mathbf{x}}(\mathbf{u}) = \sum_{n=1}^{N} \Psi_{x_n}(u_n)$. Therefore, the partial derivatives in [A1.23] of $\Psi_{x_n}(u_n)$ with respect to u_m, with $m \neq n$, are zero. This property is also satisfied in the case where only one subset of the N r.v.s is statistically independent of the other r.v.s. It is not satisfied by the moments in general, since $\Phi_{\mathbf{x}}(\mathbf{u}) = \prod_{n=1}^{N} \Phi_{x_n}(u_n)$.

In conclusion, the properties P1, P7 and P8, which distinguish cumulants from moments, are the main arguments in favor of using the former rather than the latter in SP applications. Moreover, the fourth-order cumulants are often used because the third-order cumulants of symmetrically distributed random signals are zero. This is in particular the case when solving blind communication channel identification/deconvolution problems.

A1.2.2.5. *Cumulants of complex random variables*

Complex-valued random variables and complex-valued random signals were studied by Amblard *et al.* (1996a, 1996b), extending the notion of circularity to non-Gaussian r.v.s (Picinbono 1994).

A complex r.v. z can be defined as $z = x_1 + jx_2$, with $j^2 = -1$ and $x_1, x_2 \in \mathbb{R}$. It can be viewed as a real two-dimensional r.v. whose real and imaginary parts, x_1 and x_2, respectively, have a joint pdf. The pdf of z is also a function of the conjugated variable $z^* = x_1 - jx_2$, i.e. $p(z, z^*)$. The first characteristic function is then defined as:

$$\Xi_{z,z^*}(w, w^*) = E\left[e^{\frac{j}{2}(wz^* + w^* z)}\right], \tag{A1.37}$$

with $w = u_1 + ju_2$, which implies $\frac{1}{2}(wz^* + w^* z) = u_1 x_1 + u_2 x_2 = \mathbf{u}^T \mathbf{x}$, with $\mathbf{u}^T = [u_1 \ u_2]$ and $\mathbf{x}^T = [x_1 \ x_2]$. The first characteristic function can therefore be viewed as the characteristic function [A1.19] of a real two-dimensional r.v. with components (x_1, x_2) corresponding to the real and imaginary parts of z, i.e.:

$$\Xi_{z,z^*}(\mathbf{u}) = E\left[e^{j\mathbf{u}^T \mathbf{x}}\right]. \tag{A1.38}$$

In the same way as for real r.v.s, the second characteristic function is defined as:

$$\Psi_{z,z^*}(w, w^*) = \text{Log}\left[\Xi_{z,z^*}(w, w^*)\right].$$ [A1.39]

Using the Taylor–McLaurin expansion of the exponential in [A1.37] and the binomial formula, we obtain:

$$\Xi_{z,z^*}(w, w^*) = \sum_{k=0}^{\infty} \frac{j^k}{2^k k!} \sum_{m=1}^{k} C_k^m w^{k-m}(w^*)^m E\left[(z^*)^{k-m} z^m\right],$$ [A1.40]

where $C_k^m = \frac{k!}{m!(k-m)!}$ are the binomial coefficients.

This expression of the characteristic function shows the introduction of complex moments of the form $E\left[(z^*)^{k-m} z^m\right]$, i.e. which depend on both z and z^*. For the Pth-order moment, there are therefore $P + 1$ different moments, depending on how many conjugated terms are considered. Thus, to second order, there are three different moments: $E[z^2], E[zz^*]$, and $E[z^{*2}]$. From equation [A1.40], we can deduce the following expression for the moment $E\left[(z^*)^{k-m} z^m\right]$ involving $k - m$ partial derivatives of $\Xi_{z,z^*}(w, w^*)$ with respect to w, and m partial derivatives with respect to w^*, calculated at the point $(w, w^*) = (0, 0)$:

$$E\left[(z^*)^{k-m} z^m\right] = \frac{2^k}{j^k}\left(\frac{\partial}{\partial w}\right)^{k-m}\left(\frac{\partial}{\partial w^*}\right)^m \Xi_{z,z^*}(w, w^*)\Big|_{w=0, w^*=0}.$$ [A1.41]

Similarly, from the second characteristic function, we deduce:

$$\text{cum}[\underbrace{z, \cdots, z}_{m}, \underbrace{z^*, \cdots, z^*}_{k-m}] = \frac{2^k}{j^k}\left(\frac{\partial}{\partial w}\right)^{k-m}\left(\frac{\partial}{\partial w^*}\right)^m \Psi_{z,z^*}(w, w^*)\Big|_{w=0, w^*=0}.$$

[A1.42]

Using the formula [A1.30] with $(x_{i_1}, x_{i_2}, x_{i_3}) = (z, z^*, z^*)$, we deduce the following expression for the cumulant $\text{cum}(z, z^*, z^*)$:

$$\text{cum}(z, z^*, z^*) = E[zz^{*2}] - E[z]E[z^{*2}] - 2E[z^*]E[zz^*] + 2E[z]\left(E[z^*]\right)^2.$$

A1.2.2.6. *Circular complex random variables*

The notion of circularity, introduced by Goodman (1963) in the Gaussian case, was generalized to the non-Gaussian case by Amblard *et al.* (1996a).

DEFINITION.– *We say that a complex r.v. z is circular to order n if and only if its statistics of order less than or equal to n involving a number (p) of non-conjugated terms that is not equal to the number (q) of conjugated terms are zero, i.e. for every (p, q) such that $p + q \leq n$, with $p \neq q$, we have:*

$$m_{z,p,q} \triangleq E\left[z^p (z^*)^q\right] = 0 \ , \ c_{z,p,q} \triangleq \text{cum}[\underbrace{z, \cdots, z}_{p}, \underbrace{z^*, \cdots, z^*}_{q}] = 0.$$ [A1.43]

PROPERTY.– Let $z = x + jy$, with $x, y \in \mathbb{R}$, be a circular complex r.v. By the definition of circularity, we have:

$$E[z] = 0 \;\Rightarrow\; E[x] + jE[y] = 0 \;\Rightarrow\; E[x] = E[y] = 0$$

$$E[z^2] = 0 \;\Rightarrow\; E[x^2] - E[y^2] + 2jE[xy] = 0$$

$$\Rightarrow\; E[x^2] = E[y^2] \text{ and } E[xy] = 0.$$

We can therefore conclude that the real (x) and imaginary (y) parts of z are centered, with the same standard deviation $(\sigma_x = \sigma_y)$, and uncorrelated $(\varphi_{xy} = E[xy] = 0)$.

Furthermore, if z is a complex Gaussian r.v., which implies that x and y are real Gaussian r.v.s, then x and y are independent, since non-correlation implies independence for Gaussian r.v.s (see section A1.2.3.3).

EXAMPLE A1.4.– Let $\mathbf{x}$ be a complex random vector of size N whose components x_n, with $n \in \langle N \rangle$, are circular complex r.v.s. Only one type of fourth-order cumulant is non-zero, namely $c_{\mathbf{x},2,2} = \mathrm{cum}(x_i, x_j, x_k^*, x_l^*)$, obtained by considering two conjugated and two non-conjugated terms. This tensor of fourth-order cumulants is called the quadricovariance by Cardoso (1990) in the context of antenna processing to localize and identify sources.

If we write $c_{i,j,k,l} \triangleq \mathrm{cum}(x_i, x_j, x_k^*, x_l^*)$, where the indices (i, j) are associated with the non-conjugated terms and the indices (k, l) are associated with the conjugated terms, the quadricovariance tensor satisfies the following symmetries:

$$c_{i,j,k,l} = c_{j,i,l,k} = c_{k,l,i,j}^* = c_{l,k,j,i}^*. \tag{A1.44}$$

A1.2.3. *Gaussian distribution*

In SP, and more generally in statistics, the Gaussian distribution, also called the normal distribution, is very widely used to model various physical phenomena and certain signals, such as measurement noise that results from the addition of a large number of random perturbations. The importance of the Gaussian distribution is mainly due to the central limit theorem, which states that a sum of N independent identically distributed (i.i.d.) r.v.s tends toward a Gaussian distribution as N tends to infinity, even if the added r.v.s are not themselves Gaussian.

A1.2.3.1. *Case of a scalar Gaussian variable*

DEFINITION.– *The pdf of a real scalar Gaussian r.v. x is given by:*

$$p(u) = \frac{1}{\sigma\sqrt{2\pi}} exp\left(- \frac{(u - \mu)^2}{2\sigma^2} \right), \tag{A1.45}$$

where μ is a real number and σ is a positive real number representing the mean and the standard deviation, respectively, i.e.:

$$E[x] = \int_{-\infty}^{+\infty} u\, p(u)du = \mu \ , \ E[(x-\mu)^2] = \int_{-\infty}^{+\infty} (u-\mu)^2 p(u)du = \sigma^2.$$

The Gaussian pdf is completely determined by the two parameters μ and σ, and it is often written as $x \sim \mathcal{N}(\mu, \sigma^2)$.

A1.2.3.2. Characteristic functions and HOS

The first and second characteristic functions are given by:

$$\Xi_x(u) = E[e^{jux}] = e^{j\mu u}\, e^{-(1/2)\sigma^2 u^2} \tag{A1.46}$$

$$\Psi_x(u) = \text{Log}[\Xi_x(u)] = j\mu u - (1/2)\sigma^2 u^2. \tag{A1.47}$$

By [A1.8] and [A1.23], the Pth-order moments and cumulants are given by:

$$m_{x,P} = E[x^P] = \frac{1}{j^P}\Xi_x^{(P)}(0) \tag{A1.48}$$

$$c_{x,P} = \text{cum}(\underbrace{x, \cdots, x}_{P \text{ terms}}) = \frac{1}{j^P}\Psi_x^{(P)}(0), \tag{A1.49}$$

where $\Xi_x^{(P)}(0)$ and $\Psi_x^{(P)}(0)$ are the P-th derivatives of $\Xi_x(u)$ and $\Psi_x(u)$ at $u = 0$.

In the case of a zero-mean Gaussian r.v. $x \sim \mathcal{N}(0, \sigma^2)$, the series expansion of [A1.46] with $\mu = 0$ can be written as:

$$\Xi_x(u) = e^{-(1/2)\sigma^2 u^2} = \sum_{k=0}^{\infty} \frac{(-1)^k \sigma^{2k}}{2^k k!} u^{2k}.$$

Since $\Xi_x(u)$ only depends on the even powers of u, using the formula [A1.48] for $P = 2k + 1$ and $P = 2k$ allows us to deduce the following expressions for the odd- and even-order moments:

$$m_{x,2k+1} = E(x^{2k+1}) = 0 \ , \ m_{x,2k} = E(x^{2k}) = \frac{(2k)!\,\sigma^{2k}}{2^k k!}. \tag{A1.50}$$

It is worth highlighting that all odd-order moments of a centered Gaussian variable are zero. Furthermore, we have:

$$m_{x,2} = \sigma^2 \ ; \ m_{x,4} = 3\sigma^4, \tag{A1.51}$$

and consequently the kurtosis is zero.

In the case of a non-zero-mean Gaussian r.v., we have:

$$m_{x,1} = \mu \ ; \ m_{x,2} = \sigma^2 + \mu^2 \ ; \ m_{x,3} = 3\sigma^2\mu + \mu^3 \ ; \ m_{x,4} = 3\sigma^4 + 6\mu^2\sigma^2 + \mu^4.$$

These expressions can be proven by computing the kth-order central moments of x. For example:

$$E[(x - \mu)^3] = E[x^3] - 3\mu E[x^2] + 3\mu^2 E[x] - \mu^3$$
$$= m_{x,3} - 3\mu(\sigma^2 + \mu^2) + 2\mu^3 = 0,$$

from which we deduce the expression of $m_{x,3} = 3\sigma^2\mu + \mu^3$.

From the formulae [A1.47] and [A1.49], we can conclude that all the cumulants $c_{x,P}$ of order greater than two of a Gaussian r.v. are zero, i.e. $c_{x,P} = 0$, $\forall P > 2$. This proves the property P1 stated in section A1.2.2.4. Furthermore, we have $c_{x,1} = \mu$, and $c_{x,2} = \sigma^2$.

This nullity property of the cumulants of order greater than two of a Gaussian r.v. provides the basis for developing blind identification methods for systems excited by a non-Gaussian input and corrupted by additive Gaussian noise, as will be illustrated in section A1.4.1. These identification methods based on the use of cumulants of order greater than two of the output signal are said to be robust with respect to additive Gaussian noise, as the output cumulants do not depend on the cumulants of the Gaussian noise, which are zero for orders greater than two.

A1.2.3.3. *Case of a Gaussian random vector*

DEFINITION.– *A real random vector* $\mathbf{x} \in \mathbb{R}^N$ *is Gaussian if every linear combination* $\mathbf{a}^T\mathbf{x}$ *of its components follows a one-dimensional Gaussian distribution. We also say that the components* x_n *of* $\mathbf{x}$ *are jointly Gaussian.*

The pdf of a real Gaussian random vector of size N, mean $\boldsymbol{\mu}$, and covariance matrix $\boldsymbol{\Sigma}$, is given by:

$$p(\mathbf{u}) = \frac{1}{(2\pi)^{N/2}[\det(\boldsymbol{\Sigma})]^{1/2}} e^{-\frac{1}{2}(\mathbf{u}-\boldsymbol{\mu})^T\boldsymbol{\Sigma}^{-1}(\mathbf{u}-\boldsymbol{\mu})}. \tag{A1.52}$$

Like in the scalar case, the pdf $p(\mathbf{u})$ is fully defined by the first- and second-order statistics, which explains the notation $\mathcal{N}(\boldsymbol{\mu}, \boldsymbol{\Sigma})$. The first and second characteristic functions [A1.46] and [A1.47] become:

$$\boldsymbol{\Xi}_{\mathbf{x}}(\mathbf{u}) = E[e^{j\mathbf{u}^T\mathbf{x}}] = e^{j\mathbf{u}^T\boldsymbol{\mu}} e^{-(1/2)\mathbf{u}^T\boldsymbol{\Sigma}\mathbf{u}} \tag{A1.53}$$

$$\Psi_x(\mathbf{u}) = \mathrm{Log}[\Xi_x(\mathbf{u})] = j\mathbf{u}^T\boldsymbol{\mu} - (1/2)\mathbf{u}^T\boldsymbol{\Sigma}\mathbf{u}. \tag{A1.54}$$

EXAMPLE A1.5.– Case of a real two-dimensional Gaussian vector: The components x_1 and x_2 of $\mathbf{x}$ are jointly Gaussian if their joint pdf is of the form:

$$p(\mathbf{u}) = \frac{1}{2\pi\sigma_1\sigma_2\sqrt{1-\rho^2}}\exp\Big(-\frac{1}{2(1-\rho^2)}\Big[(\frac{u_1-\mu_1}{\sigma_1})^2$$

$$-2\rho\frac{(u_1-\mu_1)(u_2-\mu_2)}{\sigma_1\sigma_2} + (\frac{u_2-\mu_2}{\sigma_2})^2\Big]\Big), \qquad \text{[A1.55]}$$

where

$$\boldsymbol{\mu} = \begin{bmatrix} \mu_1 \\ \mu_2 \end{bmatrix}, \quad \boldsymbol{\Sigma} = \begin{bmatrix} \sigma_1^2 & \rho\sigma_1\sigma_2 \\ \rho\sigma_1\sigma_2 & \sigma_2^2 \end{bmatrix}, \qquad \text{[A1.56]}$$

and ρ is the correlation coefficient between x_1 and x_2, with $\det(\boldsymbol{\Sigma}) = \sigma_1^2\sigma_2^2(1-\rho^2)$.

PROPERTIES.–

– If the r.v.s x_1 and x_2 are jointly Gaussian, then they are marginally Gaussian, with the marginal pdfs:

$$p_{x_i}(u_i) = \frac{1}{\sigma_i\sqrt{2\pi}}\exp\Big(-\frac{(u_i-\mu_i)^2}{2\sigma_i^2}\Big), \ i \in \{1,2\}. \qquad \text{[A1.57]}$$

PROOF.– The marginal pdf of x_i is given by:

$$p_{x_i}(u_i) = \int_{-\infty}^{\infty} p(u_1,u_2)du_j, \ i,j \in \{1,2\}, \ i \neq j,$$

where $p(u_1,u_2) = p(\mathbf{u})$ is defined in [A1.55].

For example, if we choose $j = 1, i = 2$, and we rewrite the bracket of the exponential in [A1.55] in the following form:

$$\Big(\frac{u_1-\mu_1}{\sigma_1} - \rho\frac{u_2-\mu_2}{\sigma_2}\Big)^2 + (1-\rho^2)\frac{(u_2-\mu_2)^2}{\sigma_2^2}, \qquad \text{[A1.58]}$$

we obtain:

$$p_{x_2}(u_2) = \int_{-\infty}^{\infty} p(u_1,u_2)du_1$$

$$= A\exp\Big(-\frac{(u_2-\mu_2)^2}{2\sigma_2^2}\Big) \int_{-\infty}^{\infty} \exp\Big(-\frac{1}{2(1-\rho^2)}\Big[\frac{u_1-\mu_1}{\sigma_1} - \rho\frac{u_2-\mu_2}{\sigma_2}\Big]^2\Big)du_1$$

with $A = \frac{1}{2\pi\sigma_1\sigma_2\sqrt{1-\rho^2}}$. By performing the change of variables: $u = \frac{u_1-\mu_1}{\sigma_1} - \rho\frac{u_2-\mu_2}{\sigma_2}$, which gives $du = \frac{du_1}{\sigma_1}$, and using the result $\int_{-\infty}^{\infty} e^{-au^2}du = \sqrt{\frac{\pi}{a}}$, with $a = \frac{1}{2(1-\rho^2)}$, the integral term in $p_{x_2}(u_2)$ simplifies as follows:

$$\int_{-\infty}^{\infty} \exp\Big(-\frac{1}{2(1-\rho^2)}\Big[\frac{u_1-\mu_1}{\sigma_1} - \rho\frac{u_2-\mu_2}{\sigma_2}\Big]^2\Big)du_1$$

$$= \sigma_1 \int_{-\infty}^{\infty} \exp\Big(-\frac{1}{2(1-\rho^2)}u^2\Big)du = \sigma_1\sqrt{2\pi(1-\rho^2)}.$$

Hence, the marginal pdf $p_{x_2}(u_2)$ can be written as:

$$p_{x_2}(u_2) = \frac{1}{\sigma_2\sqrt{2\pi}}\exp\left(-\frac{(u_2-\mu_2)^2}{2\sigma_2^2}\right),$$

which proves that x_2 is marginally Gaussian $\mathcal{N}(\mu_2, \sigma_2^2)$. We can prove that x_1 is marginally Gaussian $\mathcal{N}(\mu_1, \sigma_1^2)$ in the same way. $\square$

– More generally, the components of a Gaussian vector of size $N \geq 2$ are marginally Gaussian.

– The converse of the above property is not true. Gaussianity of each component x_i of $\mathbf{x}$ is not sufficient to guarantee that $\mathbf{x}$ is a Gaussian vector.

– If jointly Gaussian r.v.s are uncorrelated, then they are independent.

PROOF.– Consider a Gaussian vector $\mathbf{x} \sim \mathcal{N}(\boldsymbol{\mu}, \boldsymbol{\Sigma})$ of size N whose components $x_n \sim \mathcal{N}(\mu_n, \sigma_n^2)$ are assumed to be uncorrelated. The covariance matrix is then diagonal $\boldsymbol{\Sigma} = \mathrm{diag}(\sigma_n^2)$, so $\boldsymbol{\Sigma}^{-1} = \mathrm{diag}(\sigma_n^{-2})$. Hence, the pdf of $\mathbf{x}$ can be written as:

$$p(u) = \frac{1}{(2\pi)^{N/2}\prod_{n=1}^N \sigma_n}\exp\left(-\sum_{n=1}^N \frac{(u_n-\mu_n)^2}{2\sigma_n^2}\right)$$

$$= \prod_{n=1}^N p(u_n) \text{ with } p(u_n) = \frac{1}{\sqrt{2\pi}\sigma_n}\exp\left(-\frac{(u_n-\mu_n)^2}{2\sigma_n^2}\right).$$

We can therefore conclude that non-correlation of the Gaussian r.v.s x_n implies their independence. This proves the property stated in Table A1.3. $\square$

– Let $\mathbf{x} \sim \mathcal{N}(\mathbf{0}, \boldsymbol{\Sigma_x})$ be a centered Gaussian vector. The cross-moments $E\left[\prod_{n=1}^N x_{i_n}^{p_n}\right]$ of odd order $P = \sum_{n=1}^N p_n = 2k + 1$ of its components are zero. The cross-cumulants $\left(\underbrace{x_1, \cdots, x_1}_{p_1 \text{ terms}}, \cdots, \underbrace{x_N, \cdots, x_N}_{p_N \text{ terms}}\right)$ of order greater than two $(P = \sum_{n=1}^N p_n > 2)$ are zero.

– Given a random vector $\mathbf{y}$ of size M obtained by applying an affine transformation to a Gaussian random vector $\mathbf{x} \sim \mathcal{N}(\boldsymbol{\mu_x}, \boldsymbol{\Sigma_x})$ of size N, i.e. $\mathbf{y} = \mathbf{Ax} + \mathbf{b}$, with $\mathbf{A} \in \mathbb{R}^{M \times N}$, the vector $\mathbf{y}$ is itself Gaussian $\mathbf{y} \sim \mathcal{N}(\boldsymbol{\mu_y}, \boldsymbol{\Sigma_y})$ with:

$$\boldsymbol{\mu_y} = \mathbf{A}\boldsymbol{\mu_x} + \mathbf{b} , \quad \boldsymbol{\Sigma_y} = \mathbf{A}\boldsymbol{\Sigma_x}\mathbf{A}^T.$$

This Gaussianity preservation property under any linear transformation of a Gaussian random vector plays a fundamental role in SP applications. In particular, it can be exploited to test the nonlinearity of a system. Indeed, it is sufficient for any cumulant of order greater than two of the output of a system excited by a Gaussian input to be non-zero in order to conclude that the system is nonlinear.

A1.3. Discrete-time random signals

A discrete-time random signal (also called a stochastic process) can be viewed as a sequence of random variables indexed by time, denoted $x(k)$, with $k \in \mathbb{N}$ if the signal is assumed to be causal. Therefore, like for random variables, we can define the moments and cumulants of a random signal. Below, we first recall the second-order statistics, then the notion of stationary and ergodic signals, before introducing higher-order statistics, consisting of cumulants and polyspectra.

A1.3.1. *Second-order statistics*

Let $x(k)$ and $y(k)$ be two discrete-time, real, stationary scalar random signals. Table A1.5 gives the definitions of the mean, variance, autocorrelation and covariance of the signal $x(k)$, as well as the cross-correlation and cross-covariance of the signals $x(k)$ and $y(k)$.

Quantities	Definitions
Mean	$\mu_x(k) = E[x(k)]$
Variance	$\sigma_x^2(k) = E[(x(k) - \mu_x(k))^2] = E[x^2(k)] - \mu_x^2(k)$
Autocorrelation	$\varphi_x(k, t) = E[x(k)x(t)]$
Covariance	$\sigma_x(k, t) = E\big[(x(k) - \mu_x(k))(x(t) - \mu_x(t))\big] = \varphi_x(k, t) - \mu_x(k)\mu_x(t)$
Cross-correlation	$\varphi_{x,y}(k, t) = E[x(k)y(t)]$
Cross-covariance	$\sigma_{x,y}(k, t) = E\big[(x(k) - \mu_x(k))(y(t) - \mu_y(t))\big] = \varphi_{x,y}(k, t) - \mu_x(k)\mu_y(t)$
	Properties
$x(k)$ and $y(k)$ uncorrelated	$\sigma_{x,y}(k, t) = 0 , \forall k, t \quad \Leftrightarrow \quad \varphi_{x,y}(k, t) = \mu_x(k)\mu_y(t)$
$x(k)$ and $y(k)$ orthogonal	$\varphi_{x,y}(k, t) = 0 , \forall k, t$

Table A1.5. *Definitions and properties for real random signals $x(k)$ and $y(k)$*

REMARK A1.6.– Note that:

– if the processes are uncorrelated and at least one of them is zero-mean, then they are orthogonal;

– the covariance gives information about the fluctuations of the signal around its mean.

A1.3.2. *Stationary and ergodic random signals*

We say that a random signal is strictly stationary (or stationary in the strict sense) if its statistics are independent of the time origin, or equivalently if they are invariant under any translation in time. This means, for example, that the autocorrelation function $\varphi_x(k,t) = E[x(k)x(t)]$ only depends on the time interval $\tau = k - t$, in which case it is defined as $\varphi_x(\tau) = E[x(k)x(k - \tau)]$.

The hypothesis of stationarity is often used in SP, together with the hypothesis of ergodicity, since this allows us to substitute ensemble averages with time averages. This amounts to replacing averages calculated using an ensemble of realizations of a random signal with averages computed using samples measured over a time window of finite duration[3], for only a single realization of the signal.

In practice, the hypothesis of stationarity is impossible to verify, since the measurements are performed over a finite period of time. Nevertheless, it is very often assumed in order to allow us to estimate the statistics used in processing algorithms. Thus, for second-order methods, we assume the hypothesis of stationarity to second order (also called weak stationarity or stationarity in the wide sense), which means that the signals are assumed to have constant mean, and the autocorrelation function $\varphi_x(k,t)$ only depends on the time interval $\tau = k - t$. Similarly, for methods based on fourth-order cumulants, we would assume the stationarity to fourth order.

Under the hypotheses of causality ($x(k) = 0, \forall k < 0$), stationarity to second order, and ergodicity, we can estimate the mean and the autocorrelation function as follows:

$$\widehat{\mu}_x = \frac{1}{T+1} \sum_{k=0}^{T} x(k) \; ; \; \widehat{\varphi}_x(\tau) = \frac{1}{T+1} \sum_{k=\tau}^{T+\tau} x(k)x(k - \tau).$$

These estimators tend asymptotically to μ_x and $\varphi_x(\tau)$ as T tends to infinity. In the case of stationarity to order $P > 2$, we can estimate the moments $m_{x,p} = E[x(k)x(k - \tau_1) \cdots x(k - \tau_{p-1})]$ of order $p \leq P$ using time averages of products of p shifted signals:

$$\widehat{m}_{x,p}(\tau_1, \cdots, \tau_{p-1}) = \frac{1}{T+1} \sum_{k=\tau}^{T+\tau} x(k)x(k - \tau_1) \cdots x(k - \tau_{p-1}), \quad [A1.59]$$

where $\tau = \max(\tau_q)$, with $q \in \langle p - 1 \rangle$.

---

3 The time window considered to estimate the statistics of a signal needs to be sufficiently large to guarantee high-quality estimates. The higher the order of the statistics being estimated, the longer the window needs to be. For some signals, like periodic signals, the stationarity hypothesis is replaced by the cyclo-stationarity hypothesis, which states that the statistics of these signals vary periodically (Gardner 1991).

Two processes $x(k)$ and $y(k)$ are said to be jointly wide-sense stationary if each of them is wide-sense stationary, and their cross-correlation function $\varphi_{x,y}(k,t)$ only depends on the interval $\tau = k - t$. If so, we have:

$$\varphi_{x,y}(k-t) = E[x(k)y(t)] \quad \Leftrightarrow \quad \varphi_{x,y}(\tau) = E[x(k)y(k-\tau)].$$

Similarly, their cross-covariance is given by:

$$\sigma_{x,y}(\tau) = E\big[(x(k) - m_x)(y(k-\tau) - m_y)\big] = \varphi_{x,y}(\tau) - m_x m_y.$$

The random signals $x(k)$ and $y(k)$ are said to be orthogonal (respectively, uncorrelated) if their cross-correlation (respectively, cross-covariance) function is zero.

PROPERTY.– Given the real-valued random signals $x(k)$ and $y(k)$, assumed to be wide-sense stationary, their autocorrelation and cross-correlation functions have the following symmetry properties:

$$\varphi_x(\tau) = \varphi_x(-\tau) \; , \; \varphi_{yx}(\tau) = \varphi_{xy}(-\tau). \qquad [A1.60]$$

For a stationary discrete-time random signal, we define the power spectrum, also called the power spectral density (PSD) or simply the spectrum, as the (one-dimensional) discrete-time Fourier transform (DTFT) of the autocorrelation function[4]:

$$\Sigma_x(\omega) = \sum_{\tau=-\infty}^{\infty} \varphi_x(\tau) e^{-j\omega\tau} \; , \; |\omega| \le \pi. \qquad [A1.61]$$

The autocorrelation function is given by the inverse Fourier transform of the spectrum:

$$\varphi_x(\tau) = \frac{1}{2\pi} \int_{-\pi}^{\pi} \Sigma_x(\omega) e^{j\omega\tau} d\omega. \qquad [A1.62]$$

In particular, the average power is given by:

$$E[|x|^2] = \varphi_x(0) = \frac{1}{2\pi} \int_{-\pi}^{\pi} \Sigma_x(\omega) d\omega. \qquad [A1.63]$$

The power spectrum is an even function of ω, taking real non-negative values:

$$\Sigma_x(-\omega) = \Sigma_x(\omega) \in \mathbb{R}^+. \qquad [A1.64]$$

---

4 This result, known as the Wiener–Khintchine identity or theorem, was published by Norber Wiener in 1930 and independently by Aleksandr Khintchine in 1934. It states that a stationary ergodic random signal admits a spectral decomposition given by the DTFT of its autocorrelation function.

Note that $\Sigma_x(\omega)$ is periodic[5], with period 2π.

With normal frequencies ($f = \omega/2\pi$), the spectrum is given by:

$$S_x(f) = \sum_{\tau=-\infty}^{\infty} \varphi_x(\tau)\, e^{-j2\pi f \tau}. \qquad [\text{A1.65}]$$

For two jointly stationary signals $x(k)$ and $y(k)$, we define the cross-spectrum, also called the cross PSD, as the Fourier transform of the cross-correlation function:

$$\Sigma_{xy}(\omega) = \sum_{\tau=-\infty}^{\infty} \varphi_{xy}(\tau) e^{-j\omega\tau}. \qquad [\text{A1.66}]$$

For real signals, the cross-correlation $\varphi_{xy}(\tau)$ is real, and the cross-spectrum is such that:

$$\Sigma_{xy}(\omega) = \Sigma_{xy}^*(-\omega), \qquad [\text{A1.67}]$$

which implies that its magnitude is even, whereas its phase is odd.

The cross-correlation can be obtained from the inverse Fourier transform of $\Sigma_{xy}(\omega)$ as:

$$\varphi_{xy}(\tau) = \frac{1}{2\pi} \int_{-\pi}^{\pi} \Sigma_{xy}(\omega) e^{j\omega\tau} d\omega. \qquad [\text{A1.68}]$$

In Table A1.6, we summarize the definitions and the properties of the second-order statistics for two stationary complex random signals $x(k)$ and $y(k)$.

REMARK A1.7.– We can make the following remarks:

– For a stationary complex-valued random signal, we have:

$$\varphi_x(\tau) = E[x(k)x^*(k-\tau)] = E[x(k+\tau)x^*(k)] = \varphi_x^*(-\tau)\,, \quad -\infty < \tau < \infty.$$

This relation corresponds to the Hermitian symmetry property of the autocorrelation function of a complex random signal that is stationary to second order.

– The cross-spectrum $\Sigma_{x,y}(\omega)$ of two stationary complex random signals is complex-valued, satisfies $\Sigma_{xy}(\omega) = \Sigma_{yx}^*(\omega)$, and is periodic, with period 2π.

---

[5] Some authors use the notation $\Sigma_x(e^{j\omega})$ instead of $\Sigma_x(\omega)$ to better highlight the periodicity with period 2π of the spectrum. Indeed, we have:

$$\Sigma_x(e^{j(\omega+2k\pi)}) = \Sigma_x(e^{j\omega})\,, \forall k \in \mathbb{Z}.$$

Correlations	Definitions / Properties
Autocorrelation	$\varphi_x(\tau) = E[x(k)x^*(k-\tau)] = \varphi_x^*(-\tau)$
Covariance	$\sigma_x(\tau) = E\big[(x(k)-\mu_x)(x(k-\tau)-\mu_x)^*\big] = \varphi_x(\tau) - |\mu_x|^2 = \sigma_x^*(-\tau)$
Cross-correlation	$\varphi_{x,y}(\tau) = E[x(k)y^*(k-\tau)] = \varphi_{y,x}^*(-\tau)$
Cross-covariance	$\sigma_{x,y}(\tau) = E\big[(x(k)-\mu_x)(y(k-\tau)-\mu_y)^*\big] = \sigma_{y,x}^*(-\tau)$

Spectra	Properties
Spectrum	$\Sigma_x(\omega) = \Sigma_x^*(\omega)$
Cross-spectrum	$\Sigma_{x,y}(\omega) = \Sigma_{y,x}^*(\omega)$

Table A1.6. *Definitions and properties of second-order statistics for stationary complex random signals $x(k)$ and $y(k)$*

We have the following inequality: $|\Sigma_{xy}(\omega)|^2 \leq \Sigma_x(\omega)\Sigma_y(\omega)$, from which we define the coherence function:

$$C_{xy}(\omega) = \frac{|\Sigma_{xy}(\omega)|^2}{\Sigma_x(\omega)\Sigma_y(\omega)} \quad \text{such that } 0 \leq C_{xy}(\omega) \leq 1. \qquad [A1.69]$$

Although the correlation functions give information about the correlation, and hence the similarity, of two samples of the same signal (autocorrelation) or of two different signals (cross-correlation) separated by a time interval τ, the power spectrum gives information about the frequency content of a signal (in terms of the frequency f or the angular frequency $\omega = 2\pi f$). This leads to spectral analysis or frequency analysis of signals and is generally performed using a fast Fourier transform (FFT) algorithm.

A1.3.3. *Higher order statistics of random signals*

A1.3.3.1. *Cumulants of real random signals*

Let $x(k)$ be a real random signal. The Pth-order cumulant, denoted $c_{x,P}(k_1, \cdots, k_P)$, is the joint Pth-order cumulant of the r.v.s $x(k_p)$, for $p \in \langle P \rangle$:

$$c_{x,P}(k_1, \cdots, k_P) \triangleq \text{cum}\big(x(k_1), x(k_2), \cdots, x(k_P)\big). \qquad [A1.70]$$

The stationarity hypothesis in the strict sense for the signal $x(k)$ implies that the Pth-order cumulant only depends on $P-1$ time lags $\tau_p \in \mathbb{Z}$ for $p \in \langle P-1 \rangle$.

If we select the time instant $k_1 = k$ as reference time and define $\tau_p = k - k_{p+1}$, the cumulant defined in [A1.70] can be written as:

$$c_{x,P}(\tau_1, \cdots, \tau_{P-1}) = \text{cum}\big(x(k), x(k-\tau_1), \cdots, x(k-\tau_{P-1})\big). \qquad [A1.71]$$

The Pth-order cumulants of a stationary random signal therefore form a tensor of order $P - 1$. The cumulants are also called multicorrelations. In particular, the third- and fourth-order cumulants are called the bicorrelation and the tricorrelation, respectively (Lacoume *et al.* 1997).

Recall that, for a stationary random signal $x(k)$, the Pth-order moment is defined as:

$$m_{x,P}(\tau_1, \cdots, \tau_{P-1}) \triangleq E\big[x(k)x(k - \tau_1) \cdots x(k - \tau_{P-1})\big]. \qquad [A1.72]$$

In the case of a non-zero-mean stationary signal $x(k)$, using the relations [A1.25], [A1.29] and [A1.30] with $x_{i_1} = x(k), x_{i_2} = x(k - \tau_1), x_{i_3} = x(k - \tau_2)$ gives us the following expressions for the first-, second- and third-order cumulants:

$$c_{x,1} = m_{x,1} \; ; \; c_{x,2}(\tau) = m_{x,2}(\tau) - [m_{x,1}]^2 \qquad [A1.73]$$

$$c_{x,3}(\tau_1, \tau_2) = m_{x,3}(\tau_1, \tau_2) - m_{x,1}\big[m_{x,2}(\tau_2 - \tau_1) + m_{x,2}(\tau_1) + m_{x,2}(\tau_2)\big]$$

$$+ 2[m_{x,1}]^3. \qquad [A1.74]$$

Analogously to the formulae [A1.26]–[A1.28], the second-, third- and fourth-order cumulants of the random signal $x(k)$, assumed to be stationary and centered, are given by:

$$c_{x,2}(\tau) = E[x(k)x(k - \tau)] = m_{x,2}(\tau) \qquad [A1.75]$$

$$= \varphi_x(\tau) = \varphi_x(-\tau) = \text{cum}(x(k), x(k - \tau)) \qquad [A1.76]$$

$$c_{x,3}(\tau_1, \tau_2) = E[x(k)x(k - \tau_1)x(k - \tau_2)] = m_{x,3}(\tau_1, \tau_2) \qquad [A1.77]$$

$$c_{x,4}(\tau_1, \tau_2, \tau_3) = E[x(k)x(k - \tau_1)x(k - \tau_2)x(k - \tau_3)]$$

$$- c_{x,2}(\tau_1)c_{x,2}(\tau_2 - \tau_3) - c_{x,2}(\tau_2)c_{x,2}(\tau_3 - \tau_1)$$

$$- c_{x,2}(\tau_3)c_{x,2}(\tau_1 - \tau_2). \qquad [A1.78]$$

Note that, for a zero-mean random signal, the second- and third-order cumulants are identical to the second- and third-order moments, respectively, whereas the fourth-order cumulant depends on both the second- and fourth-order moments.

REMARK A1.8.– The relations [A1.77] and [A1.78] can be used to estimate the third- and fourth-order cumulants of a stationary, centered random signal $x(k)$, after replacing the moments with values estimated using time averages, such as:

$$\hat{c}_{x,3}(\tau_1, \tau_2) = \frac{1}{T + 1} \sum_{k=0}^{T} x(k)x(k - \tau_1)x(k - \tau_2).$$

By fixing $\tau = \tau_1 = \tau_2 = \tau_3 = 0$ in the relations [A1.76]–[A1.78], we obtain the second-, third- and fourth-order cumulants with zero time lags, which correspond to the variance, the non-normalized forms of skewness ($\gamma_{x,3}$) and kurtosis ($\gamma_{x,4}$) of the signal $x(k)$, expressed in terms of the cumulants:

$$c_{x,2}(0) = E[x^2(k)] = m_{x,2}(0) \ , \ c_{x,3}(0,0) = E[x^3(k)] = \gamma_{x,3}$$

$$c_{x,4}(0,0,0) = E[x^4(k)] - 3c_{x,2}^2(0) = \gamma_{x,4}.$$

The symmetry property [A1.31] of the cumulants implies that, for the Pth-order cumulant defined in [A1.70], there are $P!$ possible ways to choose the order of the time instants $(k_1, \cdots, k_P)$ without changing the cumulant.

In the case of a stationary signal, there are $(P-1)!$ ways to choose the order of the time lags τ_p, $p \in \langle P-1 \rangle$, for each choice of reference time. For example, for the bicorrelation, choosing $k_1 = k$ as the reference time leads to the following symmetry relation:

$$c_{x,3}(\tau_1, \tau_2) = c_{x,3}(\tau_2, \tau_1). \tag{A1.79}$$

Similarly, by choosing $k' = k - \tau_1$ as the reference time, we deduce the following symmetry relations:

$$\text{cum}\big(x(k), x(k-\tau_1), x(k-\tau_2)\big) = \text{cum}\big(x(k'), x(k'+\tau_1), x(k'+\tau_1-\tau_2)\big)$$

$$\Downarrow$$

$$c_{x,3}(\tau_1, \tau_2) = c_{x,3}(-\tau_1, \tau_2 - \tau_1) = c_{x,3}(\tau_2 - \tau_1, -\tau_1). \tag{A1.80}$$

Likewise, if we choose $k' = k - \tau_2$ as the reference time, permuting the time lags τ_1 and τ_2 in the above relations gives:

$$c_{x,3}(\tau_1, \tau_2) = c_{x,3}(-\tau_2, \tau_1 - \tau_2) = c_{x,3}(\tau_1 - \tau_2, -\tau_2). \tag{A1.81}$$

These symmetry relations define six regions in the (τ_1, τ_2)-plane, as illustrated in Figure A1.1. Knowing the third-order cumulants in one of these six regions is sufficient to deduce the values of $c_{x,3}(\tau_1, \tau_2)$ in the other five regions. Thus, we can restrict the third-order cumulant estimation to the region defined by: $0 < \tau_2 \leq \tau_1$, with $\tau_1 \geq 0$ and $\tau_2 \geq 0$ (region I in Figure A.1).

A1.3.3.2. *Polyspectra*

The multidimensional DTFT[6] of the multicorrelations of a stationary discrete-time random signal gives us the polyspectra of the signal, also called cumulant spectra, a notion that was introduced by Brillinger (1965).

---

6 This is a transformation in the reduced frequency f associated with the frequency F of the continuous-time Fourier transform via the relation $f = F/T$, where T is the sampling period. Recall that the discrete-time Fourier transform (DTFT) of a sampled signal $x(k)$, assumed to be absolutely summable, is defined as the function $G(f) = \sum_{k=-\infty}^{\infty} x(k)e^{-j2\pi fk}$, with $f \in [-1/2, 1/2]$, since G is a periodic function with period 1, whereas the continuous Fourier transform (CFT) of a continuous-time signal $x(t)$ is defined as $G(F) = \int_{-\infty}^{\infty} x(t)e^{-j2\pi Ft}dt$, with $F \in (-\infty, \infty)$, since the continuous time implies that G is not periodic.

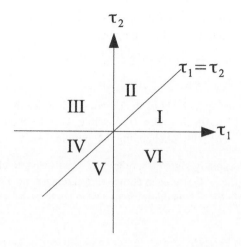

Figure A1.1. *Symmetry regions of the third-order cumulant*

If we assume that $c_{x,P}(\tau_1, \cdots, \tau_{P-1})$ is absolutely summable, i.e.:

$$\sum_{\tau_1=-\infty}^{+\infty} \cdots \sum_{\tau_{P-1}=-\infty}^{+\infty} \left| c_{x,P}(\tau_1, \cdots, \tau_{P-1}) \right| < \infty,$$

the Pth-order cumulant spectrum of the signal $x(k)$ is defined as the $(P-1)$-dimensional Fourier transform of $c_{x,P}(\tau_1, \cdots, \tau_{P-1})$:

$$S_{x,P}(f_1, \cdots, f_{P-1}) \triangleq \mathrm{DTFT}\left[c_{x,P}(\tau_1, \cdots, \tau_{P-1}) \right] \qquad [\text{A1.82}]$$

$$= \sum_{\tau_1=-\infty}^{+\infty} \cdots \sum_{\tau_{P-1}=-\infty}^{+\infty} c_{x,P}(\tau_1, \cdots, \tau_{P-1}) \, e^{-j2\pi \sum_{p=1}^{P-1} f_p \tau_p}. \qquad [\text{A1.83}]$$

The cumulant spectrum is periodic with period 1, i.e. $S_{x,P}(f_1 + 1, \cdots, f_{P-1} + 1) = S_{x,P}(f_1, \cdots, f_{P-1})$, which enables us to restrict consideration of the cumulant spectrum to only one period, i.e., $|f_p| \leq \frac{1}{2}, p \in \langle P - 1 \rangle$ and $\left| \sum_{p=1}^{P-1} f_p \right| \leq \frac{1}{2}$.

The cumulant spectrum can also be defined in terms of the angular frequency $\omega = 2\pi f$ as:

$$\Sigma_{x,P}(\omega_1, \cdots, \omega_{P-1}) = \sum_{\tau_1=-\infty}^{+\infty} \cdots \sum_{\tau_{P-1}=-\infty}^{+\infty} c_{x,P}(\tau_1, \cdots, \tau_{P-1}) \, e^{-j \sum_{p=1}^{P-1} \omega_p \tau_p}.$$

$$[\text{A1.84}]$$

The cumulant spectrum is then periodic with period 2π, i.e. $\Sigma_{x,P}(\omega_1, \cdots, \omega_{P-1}) = \Sigma_{x,P}(\omega_1 + 2\pi, \cdots, \omega_{P-1} + 2\pi)$, with $|\omega_p| \leq \pi$, $p \in \langle P - 1 \rangle$, and $\left| \sum_{p=1}^{P-1} \omega_p \right| \leq \pi$.

We also define the Pth-order z-cumulant spectrum, denoted $C_{x,P}(z_1, \cdots, z_{P-1})$, as the bilateral multidimensional z-transform of the Pth-order cumulant:

$$C_{x,P}(z_1, \cdots, z_{P-1}) = \sum_{\tau_1=-\infty}^{+\infty} \cdots \sum_{\tau_{P-1}=-\infty}^{+\infty} c_{x,P}(\tau_1, \cdots, \tau_{P-1}) \, z_1^{-\tau_1} \cdots z_{P-1}^{-\tau_{P-1}}.$$

[A1.85]

The expressions [A1.83] and [A1.84] can be deduced from [A1.85] using the transformations:

$$z_p = e^{j2\pi f_p} = e^{j\omega_p} \;,\; p \in \langle P-1 \rangle,$$

[A1.86]

which give the relations:

$$S_{x,P}(f_1, \cdots, f_{P-1}) = C_{x,P}(e^{j2\pi f_1}, \cdots, e^{j2\pi f_{P-1}})$$

[A1.87]

$$\Sigma_{x,P}(\omega_1, \cdots, \omega_{P-1}) = C_{x,P}(e^{j\omega_1}, \cdots, e^{j\omega_{P-1}}).$$

[A1.88]

Polyspectra allow us to study random signals in the frequency domain to identify relationships between frequencies, whereas cumulants are used in the time domain to obtain information relating to the temporal (multi)correlation of signals.

Taking the inverse Fourier transform of [A1.83] and [A1.84], we have:

$$c_{x,P}(\tau_1, \cdots, \tau_{P-1}) = \int_{-\frac{1}{2}}^{+\frac{1}{2}} \cdots \int_{-\frac{1}{2}}^{+\frac{1}{2}} S_{x,P}(f_1, \cdots, f_{P-1}) \, e^{j2\pi \sum_{p=1}^{P-1} f_p \tau_p}$$

$$df_1 \cdots df_{P-1}$$

$$= \frac{1}{(2\pi)^{P-1}} \int_{-\pi}^{+\pi} \cdots \int_{-\pi}^{+\pi} \Sigma_{x,P}(\omega_1, \cdots, \omega_{P-1}) \, e^{j \sum_{p=1}^{P-1} \omega_p \tau_p}$$

$$d\omega_1 \cdots d\omega_{P-1}.$$

It is a well-known fact that the PSD $S_{x,2}(f) = \sum_{\tau=-\infty}^{+\infty} c_{x,2}(\tau) e^{-j2\pi f \tau}$ corresponds to the DTFT of the autocorrelation function $\varphi_x(\tau) = c_{x,2}(\tau) = \text{cum}_{x,2}(\tau)$. In the cases of bicorrelation $c_{x,3}(\tau_1, \tau_2)$ and tricorrelation $c_{x,4}(\tau_1, \tau_2, \tau_3)$, the two- and three-dimensional DTFTs give the bispectrum $S_{x,3}(f_1, f_2)$ and the trispectrum $S_{x,4}(f_1, f_2, f_3)$, respectively.

Like cumulants, polyspectra are characterized by various symmetries. This makes it easier to estimate them using the DTFT in a reduced frequency domain. Thus, the spectrum satisfies $S_{x,2}(f) = S_{x,2}(-f)$. As a result of this parity property and the periodicity of period 1, we can restrict consideration to the frequencies $0 \leq f \leq \frac{1}{2}$ when estimating the spectrum.

In the case of third-order statistics, the bispectrum (corresponding to $P = 3$) is defined as the two-dimensional DTFT of the bicorrelation:

$$\Sigma_{x,3}(\omega_1, \omega_2) = \sum_{\tau_1=-\infty}^{+\infty} \sum_{\tau_2=-\infty}^{+\infty} c_{x,3}(\tau_1, \tau_2)\, e^{-j(\omega_1\tau_1 + \omega_2\tau_2)}. \qquad [A1.89]$$

Taking into account the symmetries [A1.79]–[A1.81] of the bicorrelation, we can deduce the following symmetries for the bispectrum (Therrien 1992; Nikias and Petropulu 1993):

$$S_{x,3}(f_1, f_2) = S_{x,3}(f_2, f_1) = S_{x,3}(f_1, -f_1 - f_2) = S_{x,3}(f_2, -f_1 - f_2)$$

$$= S_{x,3}(-f_1 - f_2, f_1) = S_{x,3}(-f_1 - f_2, f_2). \qquad [A1.90]$$

As a result of these symmetries, we can restrict the bispectrum estimation to the domain $0 \le f_2 \le f_1$, with $f_1 + f_2 \le \frac{1}{2}$. Similar symmetry properties exist for the trispectrum.

EXAMPLE A1.9.– Case of a stationary zero-mean non-Gaussian signal, white to order P[7]:

This type of signal is often considered in SP to model source signals. The Pth-order cumulant and cumulant spectrum of such a signal $e(k)$ are given by:

$$c_{e,P}(\tau_1, \cdots, \tau_{P-1}) = \gamma_{e,P}\, \delta(0, \tau_1, \cdots, \tau_{P-1}) \qquad [A1.91]$$

$$S_{e,P}(f_1, \cdots, f_{P-1}) = \gamma_{e,P} \qquad [A1.92]$$

where $\gamma_{e,P} = \mathrm{cum}(e(k), \cdots, e(k))$ is a constant and $\delta(0, \tau_1, \cdots, \tau_{P-1})$ is the generalized Kronecker delta, which is equal to 0 except if $\tau_p = 0$ for $p \in \langle P - 1 \rangle$. We can therefore conclude that the cumulant spectrum is constant for every frequency, and the cumulant of order P is non-zero only if all the time lags τ_p are zero. We say that $e(k)$ is a sequence of white noise to order P.

---

7 When we are only interested in second-order statistics, the notion of white noise corresponds to a sequence $e(k)$ of uncorrelated r.v.s (to second order) with autocorrelation function $\varphi_e(\tau) = E[e(k)e(k - \tau)] = \sigma^2\delta(\tau)$ and hence a constant power spectrum. Using HOS, the notion of white noise was extended to orders greater than two by (Bondon and Picinbono 1990), defining whiteness to orders $P > 2$ in terms of the cumulants. We say that a stationary signal $e(k)$ is white to order P if all its cumulants of orders less than or equal to P are zero for non-zero time lags. A white noise is said to be pure if it is a sequence of independent random variables, which implies whiteness to every order P. Furthermore, the stationarity hypothesis means that the r.v.s are identically distributed. The sequence is then described as i.i.d. (independent and identically distributed). Generally, the Gaussian distribution is chosen, as is the case in signal processing with additive white Gaussian noise (AWGN).

This result should be compared against the case of a sequence of i.i.d. zero-mean Gaussian noise, a hypothesis widely used to model measurement noise. As we saw earlier, all cumulants of order greater than two of a centered Gaussian r.v. are zero, the only non-zero cumulant being the second-order cumulant. Therefore, unlike the non-Gaussian case, the hypothesis of Gaussian white noise implies that every cumulant of order greater than two is zero ($\gamma_{e,P} = 0$ for $P > 2$ in [A1.91]). In particular, the fourth-order cumulants are zero, a property that is exploited by the applications considered in Chapter 5. Moreover, the spectrum is equal to $S_{e,2}(f) = \sum_{\tau=-\infty}^{\infty} \varphi_e(\tau) e^{-j2\pi f\tau} = \sigma_e^2$, with $\varphi_e(\tau) = \sigma_e^2 \delta_{\tau 0}$.

A1.3.3.3. Cumulants of complex random signals

The case of stationary complex-valued random signals leads to different definitions of the cumulants involving the conjugation of certain terms. Thus, we define the cumulant of order $p + q$ as follows:

$$c_{x,p,q}(\tau_1, \cdots, \tau_{p+q-1}) = \mathrm{cum}\big[x(k), x(k - \tau_1), \cdots, x(k - \tau_{p-1}), x^*(k - \tau_p), \cdots,$$
$$x^*(k - \tau_{p+q-1})\big],$$

with p non-conjugated terms and q conjugated terms. The stationarity hypothesis implies a dependency on $p + q - 1$ time lags τ_i with $i \in \langle p + q - 1 \rangle$. The cumulants of order $p + q$ therefore form a tensor of order $p + q - 1$. In general, there are 2^{p+q} different definitions of a complex cumulant of order $p + q$.

For example, for the third-order cumulants, there are eight possible definitions, including the following three:

$$c_{x,3,0}(\tau_1, \tau_2) = \mathrm{cum}\big[x(k), x(k - \tau_1), x(k - \tau_2)\big]$$
$$c_{x,2,1}(\tau_1, \tau_2) = \mathrm{cum}\big[x(k), x^*(k - \tau_1), x(k - \tau_2)\big]$$
$$c_{x,1,2}(\tau_1, \tau_2) = \mathrm{cum}\big[x(k), x^*(k - \tau_1), x^*(k - \tau_2)\big].$$

A1.3.3.4. Case of complex circular random signals

Complex random signals are characterized by an important property: circularity. Analogously to circular complex r.v.s (see [A1.43]), we say that a complex random signal is circular to order n if and only if its statistics of order less than or equal to n involving p non-conjugated terms and q conjugated terms are zero for $p \neq q$, and, for all (p, q) such that $p + q \leq n$:

$$m_{x,p,q} = E\big[x(k)x(k - \tau_1) \cdots x(k - \tau_{p-1})x^*(k - \tau_p) \cdots x^*(k - \tau_{p+q-1})\big] = 0$$
$$c_{x,p,q} = \mathrm{cum}[x(k), \cdots, x(k - \tau_{p-1}), x^*(k - \tau_p), \cdots, x^*(k - \tau_{p+q-1})] = 0.$$

Thus, for a complex signal circular to fourth order, we have $E[x^2(k)] = E[x^2(k)x^*(k)] = E[x^3(k)x^*(k)] = 0$. Exactly one type of fourth-order cumulant is non-zero, namely $c_{x,2,2} = \mathrm{cum}[x(k), x(k - \tau_1), x^*(k - \tau_2), x^*(k - \tau_3)]$, obtained by considering two conjugated terms and two non-conjugated terms. These cumulants, denoted $c_{i,j,m} \triangleq \mathrm{cum}[x(k), x(k - \tau_i), x^*(k - \tau_j), x^*(k - \tau_m)]$, form a third-order tensor, where the index i is associated with the non-conjugated term $x(k - \tau_i)$, and the indices (j, m) are associated with the conjugated terms $x^*(k - \tau_j)$ and $x^*(k - \tau_m)$. This tensor satisfies the following partial symmetry:

$$c_{i,j,m} = c_{i,m,j}. \hspace{4cm} \text{[A1.93]}$$

A1.4. Application to system identification

The goal of this section is to present methods for identifying linear systems and homogeneous quadratic systems using the spectra and cumulant spectra of input and output signals. If the input is not measurable, only the output statistics can be used for identification. This is called blind or unsupervised identification. In the case of a linear system represented by means of its transfer function, and hence its frequency response, we will show that the second-order statistics of the output do not contain any information about the phase and hence do not allow non-minimum phase systems to be identified. For this type of system, which is very widespread in practice, we need to use HOS like the bispectrum or trispectrum to estimate the gain and the phase of the system (Alshebeili and Cetin 1990; Nikias and Petropulu 1993; Li and Ding 1994).

REMARK A1.10.– In the case of autoregressive (AR), moving average (MA) and ARMA systems, the non-measurable input is assumed to be a non-Gaussian i.i.d. sequence. For these systems, very often considered in SP applications, various methods of blind identification based on HOS of the output have been proposed in the literature (Giannakis 1987; Nikias 1988; Giannakis and Mendel 1989; Tugnait 1990; Swami and Mendel 1990; Alshebeili et al. 1993; Favier et al. 1994; Na et al. 1995; Nandi 1999; Abderrahim et al. 2001; Favier 2004).

A1.4.1. *Case of linear systems*

Consider a stable, discrete-time linear system represented using the following input–output equation:

$$y(k) = \sum_{i=-\infty}^{\infty} h(i)u(k - i) = \sum_{i=-\infty}^{\infty} h(k - i)u(i) = h(k) * u(k)$$

$$s(k) = y(k) + e(k),$$

where the symbol $*$ denotes time convolution, and $h(k)$ is the impulse response (i.r.) of the system, whose bilateral z-transform corresponds to the discrete transfer function of the system: $H(z) = \sum_{k=-\infty}^{\infty} h(k)z^{-k}$. For a causal system, we have $h(k) = 0, \forall k < 0$. As the system is assumed to be stable, its i.r. is absolutely summable: $\sum_{k=-\infty}^{\infty} |h(k)| < \infty$.

The complex-valued signals $u(k)$, $y(k)$ and $s(k)$ denote the input, assumed to be non-Gaussian, centered and stationary, the noiseless output and the noisy measured output of the system. The sequence of additive white noise $e(k)$ is assumed to be Gaussian, centered, with variance σ^2, and independent of the input. Since the system is assumed to be stable and the input is assumed to be centered and stationary, the output is also centered and stationary. Furthermore, since the input $u(k)$ is assumed to be independent of the additive noise, the noiseless output $y(k)$ is also independent of the noise. This allows us to exploit the additivity property [A1.36] of the cumulants for the noisy output $s(k)$ to separate the contributions of the noise and the noiseless output signal in the statistics of the noisy measured output. Thus, the second- and third-order statistics of the output $s(k)$ are given by:

$$c_{s,2}(\tau) = c_{y,2}(\tau) + c_{e,2}(\tau) = c_{y,2}(\tau) + \sigma^2 \delta(\tau) \qquad \text{[A1.94]}$$

$$\Phi_s(z) = \Phi_y(z) + \Phi_e(z) = H(z)H^*(z^{-*})\Phi_u(z) + \sigma^2 \qquad \text{[A1.95]}$$

$$c_{s,3,0}(\tau_1, \tau_2) = c_{y,3,0}(\tau_1, \tau_2) \qquad \text{[A1.96]}$$

$$\Sigma_{s,3}(\omega_1, \omega_2) = \Sigma_{y,3}(\omega_1, \omega_2) = H(\omega_1)H(\omega_2)H(-\omega_1 - \omega_2)\Sigma_{u,3}(\omega_1, \omega_2). \qquad \text{[A1.97]}$$

As we mentioned earlier, using the bispectrum [A1.97] enables us to make the statistics of the noisy output independent of the additive Gaussian noise, which is not the case with the spectrum [A1.95].

Table A1.7 summarizes the expressions of the second-order statistics of the noiseless output $y(k)$. Below, we will prove some of the formulae stated in this table. The others are easy to prove using the same reasoning.

PROOF.– By the definition of the autocorrelation function of the output, we have:

$$c_{s,2}(\tau) \triangleq \varphi_s(\tau) = E[s(k)s^*(k - \tau)]$$

$$= E[(y(k) + e(k))(y^*(k - \tau) + e^*(k - \tau))].$$

Taking into account the hypotheses on the additive noise, we obtain:

$$\varphi_s(\tau) \triangleq \varphi_y(\tau) + \sigma^2 \delta(\tau) \qquad \text{[A1.98]}$$

with

$$\varphi_y(\tau) \triangleq E[y(k)y^*(k - \tau)] = E\left[\sum_i h(i)u(k - i)y^*(k - \tau)\right]$$

$$= \sum_i h(i)\varphi_{uy}(\tau - i) = h(\tau) * \varphi_{uy}(\tau). \qquad \text{[A1.99]}$$

Quantities	Expressions
	Noiseless output
$y(k)$	$y(k) = \sum_{i=-\infty}^{\infty} h(i)u(k-i) = \sum_{i=-\infty}^{\infty} h(k-i)u(i) = h(k) * u(k)$
	Cross-correlation
$\varphi_{yu}(\tau)$	$\varphi_{yu}(\tau) \triangleq E[y(k)u^*(k-\tau)] = h(\tau) * \varphi_u(\tau) = \varphi_{uy}^*(-\tau)$
$\varphi_{uy}(\tau)$	$\varphi_{uy}(\tau) = h^*(-\tau) * \varphi_u(\tau)$
	Autocorrelation
$\varphi_y(\tau)$	$\varphi_y(\tau) \triangleq E[y(k)y^*(k-\tau)] = h(\tau) * \varphi_{uy}(\tau)$ $= h(\tau) * h^*(-\tau) * \varphi_u(\tau) = \varphi_y^*(-\tau)$
	z-transform of the output $Y(z) = H(z)U(z)$
	Cross power spectrum in z
$\Phi_{yu}(z)$	$\Phi_{yu}(z) \triangleq \sum_{\tau=-\infty}^{\infty} \varphi_{yu}(\tau)z^{-\tau} = H(z)\Phi_u(z) = \Phi_{uy}^*(z^{-*})$
$\Phi_{uy}(z)$	$\Phi_{uy}(z) \triangleq \sum_{\tau=-\infty}^{\infty} \varphi_{uy}(\tau)z^{-\tau} = H^*(z^{-*})\Phi_u(z)$
	Cross power spectrum in ω
$\Sigma_{yu}(\omega)$	$\Sigma_{yu}(\omega) \triangleq \sum_{\tau=-\infty}^{\infty} \varphi_{yu}(\tau)e^{-j\omega\tau} = H(\omega)\Sigma_u(\omega) = \Sigma_{uy}^*(\omega)$
$\Sigma_{uy}(\omega)$	$\Sigma_{uy}(\omega) \triangleq \sum_{\tau=-\infty}^{\infty} \varphi_{uy}(\tau)e^{-j\omega\tau} = H^*(\omega)\Sigma_u(\omega)$
	Power spectrum of the output
$\Phi_y(z)$	$\Phi_y(z) \triangleq \sum_{\tau=-\infty}^{\infty} \varphi_y(\tau)z^{-\tau} = H(z)\Phi_{uy}(z)$ $= H(z)H^*(z^{-*})\Phi_u(z) = \Phi_y^*(z^{-*})$
$\Sigma_y(\omega)$	$\Sigma_y(\omega) \triangleq \sum_{\tau=-\infty}^{\infty} \varphi_y(\tau)e^{-j\omega\tau} = H(\omega)\Sigma_{uy}(\omega)$ $= |H(\omega)|^2\Sigma_u(\omega) = \Sigma_y^*(\omega)$

Table A1.7. *Second-order statistics of the output of a linear system*

Similarly, for the cross-correlation function, we have:

$$\varphi_{su}(\tau) \triangleq E[s(k)u^*(k-\tau)] = E[(y(k) + e(k))u^*(k-\tau)] = \varphi_{yu}(\tau)$$

[A1.100]

$$\varphi_{yu}(\tau) \triangleq E[y(k)u^*(k-\tau)] = \sum_i h(i)\varphi_u(\tau-i) = h(\tau) * \varphi_u(\tau).$$

[A1.101]

Taking into account the symmetry property of the autocorrelation function of the input $\varphi_u^*(-\tau) = \varphi_u(\tau)$ and therefore $\varphi_u^*(-\tau - i) = \varphi_u(\tau + i)$, we also obtain:

$$\varphi_{us}(\tau) = \varphi_{uy}(\tau) \triangleq E\big[u(k)y^*(k - \tau)\big] = E\Big[u(k) \sum_i h^*(i)u^*(k - \tau - i)\Big]$$

$$= \sum_i h^*(i)\varphi_u(\tau + i) = \varphi_{yu}^*(-\tau) = h^*(-\tau) * \varphi_u(\tau).$$

Hence, replacing $\varphi_{uy}(\tau)$ with this expression in [A1.99], we deduce that:

$$\varphi_y(\tau) = h(\tau) * h^*(-\tau) * \varphi_u(\tau) = \varphi_y^*(-\tau) \qquad \text{[A1.102]}$$

or equivalently:

$$\varphi_y(\tau) = \sum_{i=-\infty}^{\infty} \sum_{m=-\infty}^{\infty} h(i)\, h^*(m)\, \varphi_u(\tau - i + m). \qquad \text{[A1.103]}$$

The double convolution in [A1.102] corresponds to the double sum in [A1.103].

Consider now the spectrum in z of the noiseless output, defined as the bilateral z-transform of the autocorrelation function, expressed in the form [A1.103]:

$$\Phi_y(z) \triangleq \sum_{\tau=-\infty}^{\infty} \varphi_y(\tau)z^{-\tau} = \sum_{\tau=-\infty}^{\infty} \sum_{i=-\infty}^{\infty} \sum_{m=-\infty}^{\infty} h(i)\, h^*(m)\, \varphi_u(\tau - i + m)z^{-\tau}.$$

$$\text{[A1.104]}$$

By decomposing $z^{-\tau} = z^{-(\tau-i+m)}\, z^{-i}\, z^m$ and setting $t = \tau - i + m$, the triple sum can be expressed as:

$$\Phi_y(z) = \Big(\sum_{i=-\infty}^{\infty} h(i)z^{-i} \Big)\Big(\sum_{m=-\infty}^{\infty} h^*(m)z^m \Big)\Big(\sum_{t=-\infty}^{\infty} \varphi_u(t)z^{-t} \Big). \qquad \text{[A1.105]}$$

Noting that $\sum_{m=-\infty}^{\infty} h^*(m)z^m = \big[\sum_{m=-\infty}^{\infty} h(m)(z^*)^m\big]^* = H^*(z^{-*})$, equation [A1.105] gives us:

$$\Phi_y(z) = H(z)H^*(z^{-*})\Phi_u(z). \qquad \text{[A1.106]}$$

From this expression, we can deduce that the term $h^*(-\tau)$ of the double convolution [A1.102], which comes from the conjugated term $y^*(k - \tau)$ in the autocorrelation function $\varphi_y(\tau) = E[y(k)y^*(k - \tau)]$, is associated with the term $H^*(z^{-*})$ in the spectrum in z.

Proceeding in the same way for the spectrum in ω, we have[8]:

$$\Sigma_y(\omega) \triangleq \sum_{\tau=-\infty}^{\infty} \varphi_y(\tau)e^{-j\omega\tau}$$

$$= \Big(\sum_{i=-\infty}^{\infty} h(i)e^{-j\omega i} \Big)\Big(\sum_{m=-\infty}^{\infty} h^*(m)e^{j\omega m} \Big)\Big(\sum_{t=-\infty}^{\infty} \varphi_u(t)e^{-j\omega t} \Big).$$

Noting that $\sum_{m=-\infty}^{\infty} h^*(m)e^{j\omega m} = \big[\sum_{m=-\infty}^{\infty} h(m)e^{-j\omega m}\big]^* = H^*(\omega)$, we deduce the following expression for the spectrum in ω:

$$\Sigma_y(\omega) = H(\omega)H^*(\omega)\Sigma_u(\omega) = |H(\omega)|^2 \Sigma_u(\omega), \qquad [\text{A1.107}]$$

where $\Sigma_u(\omega)$ is the spectrum of the input signal. We therefore conclude that the term $H^*(w)$ is now associated with the term $h^*(-\tau)$ of the double convolution [A1.102].

$\square$

REMARK A1.11.– We can make the following remarks:

– The power spectrum in ω satisfies the Hermitian symmetry property $\Sigma_y(\omega) = \Sigma_y^*(\omega)$ and the non-negativity property $\Sigma_y(\omega) \geq 0$, $0 \leq \omega \leq 2\pi$.

– The expression [A1.107] clearly shows that the output spectrum does not contain any information about the phase of the system, unlike the bispectrum (see Table A1.9). Only the modulus of $H(\omega)$ can be determined using the spectra $\Sigma_u(\omega)$ and $\Sigma_y(\omega)$ of the input and output signals. This means that the same input signal filtered by various filters with the same magnitude but different phases provides different output signals that have the same second-order statistics (autocorrelation function and spectrum).

– The stationarity property in the wide sense of the zero-mean complex random signal $y(k)$ is reflected in the following relation satisfied by its power spectrum in z:

$$\varphi_y(\tau) = \varphi_y^*(-\tau) \, , \quad -\infty < \tau < \infty \qquad [\text{A1.108}]$$

$$\Downarrow$$

$$\Phi_y(z) = \Phi_y^*(z^{-*}). \qquad [\text{A1.109}]$$

In other words, the spectrum in z has the property of para-Hermitian symmetry. We say that $\Phi_y(z)$ is a para-Hermitian polynomial.

8 Some authors use the notation $\Sigma_y(e^{j\omega})$, $\Sigma_u(e^{j\omega})$, and $H(e^{j\omega})$ instead of $\Sigma_y(\omega)$, $\Sigma_u(\omega)$, and $H(\omega)$ to better highlight the periodicity with period 2π of the discrete-time Fourier transform.

PROOF.– Using the definition of the power spectrum in z and the property of Hermitian symmetry [A1.102] of the autocorrelation function, we have:

$$\Phi_y^*(z^{-*}) = \sum_{\tau=-\infty}^{\infty} \varphi_y^*(\tau)z^\tau = \sum_{\tau=-\infty}^{\infty} \varphi_y^*(-\tau)z^{-\tau} \text{ (by changing the sign of } \tau)$$

$$= \sum_{\tau=-\infty}^{\infty} \varphi_y(\tau)z^{-\tau} = \Phi_y(z) \quad \text{(by } [A1.102]) \quad\quad [A1.110]$$

which proves the para-Hermitian symmetry property [A1.109] of the spectrum.   □

– In the case of a system whose transfer function $H(z)$ has real coefficients, we have $H^*(z^{-*}) = \sum_{-\infty}^{\infty} h(\tau)z^\tau = H(z^{-1})$. Hence, the expression [A1.106] of the output spectrum in z can be written as:

$$\Phi_y(z) = H(z)H(z^{-1})\Phi_u(z) = \Phi_y(z^{-1}). \quad\quad [A1.111]$$

It should be noted that identifying a linear system requires knowledge of the second-order (or higher-order) statistics of the input. In the case of a Volterra system of order P, i.e. with input nonlinearities of order P, parameter estimation requires us to appeal to input statistics of order at least $2P$. Thus, as we will see in the next section, identifying a quadratic Volterra system requires the use of input statistics of order at least 4.

Consider now the cumulant spectrum of the output associated with the cumulant $c_{y,P,Q} = \text{cum}\big(y(k), y(k - \tau_1), \cdots, y(k - \tau_{P-1}), y^*(k - \tau_P, \cdots, y^*(k - \tau_{P+Q-1})\big)$ involving P non-conjugated terms and Q conjugated terms. From our remarks on the expressions [A1.106] and [A1.107] of the spectra in z and ω of the output signal $y(k)$, we can deduce the rules stated in Table A1.8 for determining the factors of the polyspectra in z, ω and f associated with each term $y(k - \tau_p)$, for $p \in \langle P - 1\rangle$, $y^*(k - \tau_q)$, for $q \in \{P, P + 1, \cdots, P + Q - 1\}$, and with $y(k)$.

Polyspectra	$y(k - \tau_p)$ term	$y(k - \tau_q)$ term	$y(k)$ term
$C_{y,P,Q}(z_1, \cdots, z_{P+Q-1})$	$H(z_p)$	$H^*(z_q^{-*})$	$H\big(\prod_{p=1}^{P} z_p^{-1} \prod_{q=1}^{Q} z_q\big)$
$\Sigma_{y,P,Q}(\omega_1, \cdots, \omega_{P+Q-1})$	$H(\omega_p)$	$H^*(\omega_q)$	$H^*\big(-\sum_{p=1}^{P} \omega_p + \sum_{q=1}^{Q} \omega_q\big)$
$S_{y,P,Q}(f_1, \cdots, f_{P+Q-1})$	$H(f_p)$	$H^*(f_q)$	$H^*\big(-\sum_{p=1}^{P} f_p + \sum_{q=1}^{Q} f_q\big)$

Table A1.8. *Factors of the polyspectra*

Applying these rules allows us to easily deduce the bispectra ($P + Q = 3$) and trispectra ($P + Q = 4$) in z and ω for different choices of P and Q, as stated in Table A1.9.

As mentioned earlier, the output spectrum does not allow us to estimate the phase of the transfer function. For an input signal that is white to third order, and therefore non-Gaussian, with a non-symmetric pdf, the bispectrum $\Sigma_{u,3,0}(\omega_1, \omega_2)$ is a non-zero constant $\gamma_{u,3}$. We can therefore use the bispectrum of the output $\Sigma_{y,3,0}(\omega_1, \omega_2)$ given in Table A1.9 to estimate the phase φ_H of the system from the following relation:

$$\varphi_{\Sigma_{y,3,0}}(\omega_1, \omega_2) = \varphi_H(\omega_1) + \varphi_H(\omega_2) - \varphi_H(\omega_1 + \omega_2), \qquad \text{[A1.112]}$$

where $\varphi_{\Sigma_{y,3,0}}(\omega_1, \omega_2)$ is the phase of the output bispectrum. Different methods for reconstructing the phase of the system from the relation [A1.112] are given by Nikias and Petropulu (1993).

Bispectra in z
$C_{y,3,0}(z_1, z_2) = H(\frac{1}{z_1 z_2})H(z_1)H(z_2)\, C_{u,3,0}(z_1, z_2)$
$C_{y,2,1}(z_1, z_2) = H(\frac{z_2}{z_1})H(z_1)H^*(z_2^{-*})\, C_{u,2,1}(z_1, z_2)$

Bispectra in ω
$\Sigma_{y,3,0}(\omega_1, \omega_2) = H(-\omega_1 - \omega_2)H(\omega_1)H(\omega_2)\, \Sigma_{u,3,0}(\omega_1, \omega_2)$
$\Sigma_{y,2,1}(\omega_1, \omega_2) = H^*(-\omega_1 + \omega_2)H(\omega_1)H^*(\omega_2)\, \Sigma_{u,2,1}(\omega_1, \omega_2)$

Trispectra in z and ω
$C_{y,2,2}(z_1, z_2, z_3) = H(\frac{z_2 z_3}{z_1})H(z_1)H^*(z_2^{-*})H^*(z_3^{-*})\, C_{u,2,2}(z_1, z_2, z_3)$
$\Sigma_{y,2,2}(\omega_1, \omega_2, \omega_3) = H^*(-\omega_1 + \omega_2 + \omega_3)H(\omega_1)H^*(\omega_2)H^*(\omega_3)\, \Sigma_{u,2,2}(\omega_1, \omega_2, \omega_3)$

Table A1.9. *Bispectra and trispectra of the output of a linear system*

Note that the equation of the bispectrum also gives us the following relation to compute the modulus of $H(\omega)$:

$$|\Sigma_{y,3,0}(\omega_1, \omega_2)| = |\gamma_{u,3}|\, |H(\omega_1)|\, |H(\omega_2)|\, |H(\omega_1 + \omega_2)|. \qquad \text{[A1.113]}$$

Relations similar to [A1.112] and [A1.113] can be deduced from the trispectrum of the output for an input signal that is white to fourth order. Similarly, we can use the cross-spectra of the input–output signals to identify the transfer function of a linear system. Thus, from the relations established in Table A1.7, we have:

$$H(\omega) = \frac{\Sigma_{yu}(\omega)}{\Sigma_u(\omega)} \quad \text{and} \quad \frac{H(\omega)}{H^*(\omega)} = \frac{\Sigma_{yu}(\omega)}{\Sigma_{uy}(\omega)} = e^{j2\varphi_H}, \qquad \text{[A1.114]}$$

where φ_H is the phase of $H(\omega)$.

Likewise, for the cross-bicorrelation, we have, for example:

$$c_{s,u,u}(\tau_1,\tau_2) = \mathrm{cum}\big(s(k), u(k-\tau_1), u(k-\tau_2)\big) = c_{y,u,u}(\tau_1,\tau_2)$$

$$c_{y,u,u}(\tau_1,\tau_2) = \sum_m h(m)\mathrm{cum}\big(u(k-m), u(k-\tau_1), u(k-\tau_2)\big)$$

$$= \sum_m h(m)\mathrm{cum}\big(u(k), u(k+m-\tau_1), u(k+m-\tau_2)\big).$$

Hence, setting $t_i = \tau_i - m$, with $i \in \{1,2\}$, the corresponding cross-trispectrum is given by:

$$\Sigma_{y,u,u}(\omega_1,\omega_2) = \sum_m h(m)e^{-jm(\omega_1+\omega_2)} \sum_{t_1,t_2} \mathrm{cum}\big(u(k), u(k-t_1), u(k-t_2)\big)$$

$$e^{-j(\omega_1 t_1 + \omega_2 t_2)}$$

$$= H(\omega_1+\omega_2)\Sigma_{u,u,u}(\omega_1,\omega_2).$$

From this example, we can conclude that different cross-polyspectra can be used according to the hypotheses formulated on the input.

A1.4.2. *Case of homogeneous quadratic systems*

Truncated Volterra models, also called discrete-time Volterra series expansions, are an extension of linear finite impulse response (FIR) models to account for nonlinearities in the input.

This extension provides a link to tensors via Volterra kernels. Recall the input–output equation of a Volterra model of order P:

$$y(k) = h_0 + \sum_{p=1}^{P} \sum_{m_1=0}^{M_p-1} \cdots \sum_{m_P=0}^{M_p-1} h_{m_1,\cdots,m_p}^{(p)} \prod_{i=1}^{p} u(k-m_i), \qquad [\text{A1.115}]$$

where M_p is the memory of the pth-order kernel $\mathcal{H}^{(p)} \in \mathbb{R}^{M_p \times \cdots \times M_P}$, for which $h_{m_1,\cdots,m_p}^{(p)}$ is a coefficient. This kernel can be interpreted as a pth-order tensor.

Volterra models are very widely used in many fields of application. They satisfy several important properties:

– linearity with respect to their parameters, i.e. the coefficients of the Volterra kernels, with the possibility of reducing the parametric complexity via symmetrization and decomposition of the kernels, viewed as tensors (Favier and Bouilloc 2010; Favier *et al.* 2012a);

– interpretation of the homogeneous term of order P as a P-dimensional convolution of the Pth-order kernel with the input defined as $u(k_1, \cdots, k_P) = u(k_1) \cdots u(k_P)$;

– sufficient condition for BIBO (bounded-input bounded-output) stability of a homogeneous system of order P in terms of absolute summability of the Pth-order kernel, generalizing the BIBO stability condition for FIR linear systems.

Below, we will consider a homogeneous quadratic system modeled using a second-order Volterra model described by the following equation:

$$y(k) = \sum_{m_1=0}^{M-1} \sum_{m_2=0}^{M-1} h(m_1, m_2) u(k - m_1) u(k - m_2), \qquad [\text{A1.116}]$$

where $h(m_1, m_2)$ is the second-order Volterra kernel with memory M, and $0 \leq m_1, m_2 \leq M - 1$, and $u(k)$ and $y(k)$ are the input and output signals, assumed to be real-valued. After defining the input vector $\mathbf{u}^T(k) = [u(k)\, u(k-1) \cdots u(k-M+1)] \in \mathbb{R}^M$, the output can be rewritten as a quadratic form in the input vector:

$$y(k) = \mathbf{u}^T(k)\mathbf{H}\mathbf{u}(k), \qquad [\text{A1.117}]$$

where $\mathbf{H}$ is a square matrix of order M containing the coefficients $h(m_1, m_2)$ of the kernel.

Unlike the case of linear systems previously considered, we assume that the input is a measured centered Gaussian signal. This is known as supervised identification. Our goal is to show that the quadratic kernel can be estimated using the cross-cumulant $c_{yuu}(\tau_1, \tau_2) \triangleq \text{cum}\Big(y(k), u(k-\tau_1), u(k-\tau_2)\Big)$ and the cumulant spectrum associated with it.

We will omit the additive Gaussian measurement noise, which is assumed to be independent of the input, since we will be considering third-order statistics, which are zero for a Gaussian signal. Taking into account the multilinearity property [A1.32] of the cumulants, the cross-cumulant $c_{yuu}(\tau_1, \tau_2)$ is given by:

$$c_{yuu}(\tau_1, \tau_2) = \text{cum}\big(y(k), u(k-\tau_1), u(k-\tau_2)\big) \qquad [\text{A1.118}]$$

$$= \sum_{m_1=0}^{M-1} \sum_{m_2=0}^{M-1} h(m_1, m_2) \text{cum}\big(u(k-m_1)u(k-m_2), u(k-\tau_1), u(k-\tau_2)\big).$$

$$[\text{A1.119}]$$

Using the relation [A1.30], with $x_{i_1} = u(k - m_1)u(k - m_2)$, $x_{i_2} = u(k - \tau_1)$, $x_{i_3} = u(k - \tau_2)$, we obtain:

$$\mathrm{cum}\big(u(k - m_1)u(k - m_2), u(k - \tau_1), u(k - \tau_2)\big) =$$

$$E\big[u(k - m_1)u(k - m_2)u(k - \tau_1)u(k - \tau_2)\big] - E\big[u(k - m_1)u(k - m_2)\big]$$

$$E\big[u(k - \tau_1)u(k - \tau_2)\big] = m_{u,4}(m_1, m_2, \tau_1, \tau_2) - c_{u,2}(m_2 - m_1)c_{u,2}(\tau_2 - \tau_1).$$

$$[\text{A1.120}]$$

By the hypotheses, since the input is a centered Gaussian signal, its Nth-order moment is given by Schetzen (1980) and Picinbono (1993):

$$E\left[\prod_{n=1}^{N} u(k - \tau_n)\right] = \begin{cases} 0 & \text{if } N = 2k + 1, \ k \in \mathbb{N} \\ \sum \prod c_{u,2}(\tau_i - \tau_j) & \text{if } N = 2k \end{cases} \qquad [\text{A1.121}]$$

where the sum ranges over all partitionings of the N indices $\tau_1, \cdots, \tau_N$ into pairs (τ_i, τ_j). There are $\frac{(2k)!}{k! \, 2^k} = (2k - 1)(2k - 3) \cdots (1)$ such partitionings in total. From this formula, we deduce that all odd-order moments of the input are zero (as we saw earlier for a Gaussian r.v.) and that every even-order moment can be expressed as a sum of products of second-order moments, i.e. the autocorrelation. Thus, for the fourth-order moment $m_{u,4}(m_1, m_2, \tau_1, \tau_2)$, we have:

$$m_{u,4}(m_1, m_2, \tau_1, \tau_2) = c_{u,2}(m_2 - m_1)c_{u,2}(\tau_2 - \tau_1)$$

$$+ c_{u,2}(m_1 - \tau_1) \, c_{u,2}(m_2 - \tau_2) + c_{u,2}(m_1 - \tau_2)c_{u,2}(m_2 - \tau_1).$$

$$[\text{A1.122}]$$

Replacing the fourth-order moment of the input with this expression [A1.122] in [A1.120], the cross-cumulant [A1.119] can be written as follows, for $\tau_1, \tau_2 \in \{0, 1, \cdots, M - 1\}$:

$$c_{yuu}(\tau_1, \tau_2) = \sum_{m_1=0}^{M-1} \sum_{m_2=0}^{M-1} \Big[c_{u,2}(m_1 - \tau_1)c_{u,2}(m_2 - \tau_2)$$

$$+ c_{u,2}(m_1 - \tau_2)c_{u,2}(m_2 - \tau_1)\Big] h(m_1, m_2).$$

$$[\text{A1.123}]$$

After writing the kernel in triangularized form using the equations:

$$h_{\mathrm{tri}}(m_1, m_2) = \begin{cases} h(m_1, m_2) + h(m_2, m_1) & \forall m_1 < m_2 \le M - 1 \\ h(m_1, m_2) & \forall m_1 = m_2 \\ 0 & \forall m_1 > m_2 \end{cases}$$

$$[\text{A1.124}]$$

equation [A1.116] of the Volterra model can also be written as:

$$y(k) = \sum_{m_1=0}^{M-1} \sum_{m_2=m_1}^{M-1} h_{\text{tri}}(m_1, m_2) u(k - m_1) u(k - m_2). \qquad [A1.125]$$

By defining the matrices $\mathbf{C}_{uu}$ and $\mathbf{C}_{yuu} \in \mathbb{R}^{M \times M}$ such that $[\mathbf{C}_{uu}]_{\tau_1, \tau_2} = c_{u,2}(\tau_1 - \tau_2)$ and $[\mathbf{C}_{yuu}]_{\tau_1, \tau_2} = c_{yuu}(\tau_1, \tau_2)$, with $\tau_1, \tau_2 \in \{0, 1, \cdots, M - 1\}$, equation [A1.123] of the cross-cumulant can be rewritten in matrix form as:

$$\mathbf{C}_{yuu} = \mathbf{C}_{uu}(\mathbf{H}_{\text{tri}} + \mathbf{H}_{\text{tri}}^T)\mathbf{C}_{uu}, \qquad [A1.126]$$

where $\mathbf{H}_{\text{tri}} \in \mathbb{R}^{M \times M}$ is the matrix of coefficients of the quadratic kernel, in upper triangular form, and $\mathbf{C}_{uu}$ and $\mathbf{C}_{yuu}$ are the autocorrelation matrix of the input signal and the matrix of cross-cumulants between the input and output formed by the cumulants $c_{yuu}(\tau_1, \tau_2)$. These two matrices are symmetric and of size $M \times M$. If we assume that the autocorrelation matrix is non-singular, the quadratic kernel can be estimated using the following equation:

$$(\mathbf{H}_{\text{tri}} + \mathbf{H}_{\text{tri}}^T) = \mathbf{C}_{uu}^{-1} \mathbf{C}_{yuu} \mathbf{C}_{uu}^{-1}. \qquad [A1.127]$$

EXAMPLE A1.12.– For $M = 2$, we have:

$$\mathbf{C}_{yuu} = \begin{pmatrix} c_{yuu}(0,0) & c_{yuu}(0,1) \\ c_{yuu}(1,0) & c_{yuu}(1,1) \end{pmatrix} \ , \ \mathbf{C}_{uu} = \begin{pmatrix} c_{u,2}(0) & c_{u,2}(1) \\ c_{u,2}(1) & c_{u,2}(0) \end{pmatrix}$$

$$\mathbf{H}_{\text{tri}} = \begin{pmatrix} h(0,0) & h(0,1) + h(1,0) \\ 0 & h(1,1) \end{pmatrix}.$$

Taking the two-dimensional Fourier transform of the two sides of equation [A1.123], we obtain the cross-bispectrum of the output and input:

$$\Sigma_{yuu}(\omega_1, \omega_2) \triangleq \sum_{\tau_1} \sum_{\tau_2} c_{yuu}(\tau_1, \tau_2) e^{-j(\omega_1 \tau_1 + \omega_2 \tau_2)}$$

$$= \sum_{\tau_1} \sum_{\tau_2} \sum_{m_1=0}^{M-1} \sum_{m_2=0}^{M-1} h(m_1, m_2) \Big[c_{u,2}(m_1 - \tau_1) c_{u,2}(m_2 - \tau_2)$$

$$+ c_{u,2}(m_1 - \tau_2) c_{u,2}(m_2 - \tau_1) \Big] e^{-j(\omega_1 \tau_1 + \omega_2 \tau_2)}. \qquad [A1.128]$$

By performing the change of variables $(\sigma_1, \sigma_2) = (m_1 - \tau_1, m_2 - \tau_2)$ in the first term of the bracket and $(\sigma_1, \sigma_2) = (m_1 - \tau_2, m_2 - \tau_1)$ in the second term, we can rewrite the cross-bispectrum as:

$$\Sigma_{yuu}(\omega_1, \omega_2) \triangleq \sum_{m_1=0}^{M-1} \sum_{m_2=0}^{M-1} h(m_1, m_2) \left[e^{-j(\omega_1 m_1 + \omega_2 m_2)} + e^{-j(\omega_1 m_2 + \omega_2 m_1)} \right]$$

$$\times \sum_{\sigma_1} c_{u,2}(\sigma_1) e^{j\omega_1 \sigma_1} \sum_{\sigma_2} c_{u,2}(\sigma_2) e^{j\omega_2 \sigma_2} \qquad [\text{A1.129}]$$

Under the hypothesis of real signals, with the symmetry $(\Sigma_u(-\omega) = \Sigma_u(\omega))$ of the input spectrum, we obtain:

$$\Sigma_{yuu}(\omega_1, \omega_2) = \left[H(\omega_1, \omega_2) + H(\omega_2, \omega_1) \right] \Sigma_u(\omega_1) \Sigma_u(\omega_2). \qquad [\text{A1.130}]$$

Considering the kernel in symmetrized form $h_{\text{sym}}(\tau_1, \tau_2)$, the corresponding transfer function in ω satisfies $H_{\text{sym}}(\omega_1, \omega_2) = \frac{1}{2}\left[H(\omega_1, \omega_2) + H(\omega_2, \omega_1) \right]$. We can therefore estimate the two-dimensional transform of the symmetrized quadratic kernel using the spectrum of the input and the cross-bispectrum of the input and output as follows:

$$H_{\text{sym}}(\omega_1, \omega_2) = \frac{\Sigma_{yuu}(\omega_1, \omega_2)}{2 \Sigma_u(\omega_1) \Sigma_u(\omega_2)}. \qquad [\text{A1.131}]$$

This formula was established by Tick (1961) for continuous-time quadratic Volterra systems. An extension to the case of discrete-time Volterra systems of arbitrary order was proposed by Koukoulas and Kalouptsidis (1995) for a Gaussian input (see also Nikias and Petropulu 1993; Mathews and Sicuranza 2000).

REMARK A1.13.– The cross-bispectrum $\Sigma_{yuu}(\omega_1, \omega_2)$ can be estimated by computing the two-dimensional DTFT of the cross-cumulant defined in [A1.118], which is in turn estimated using the expression [A1.120] with an empirical estimator of the second- and fourth-order moments of the input.

References

Abderrahim, K., Ben Abdennour, R., Favier, G., Ksouri, M., Faouzi, M. (2001). New results on FIR system identification using cumulants. *APII-JESA*, 35(5), 601–622.

Acar, E., Aykut-Bingol, C., Bingol, H., Bro, R., Yener, B. (2007). Multiway analysis of epilepsy tensors. *Bioinformatics*, 23, i10–i18.

Acar, E., Dunlavy, D.M., Kolda, T.G., Morup, M. (2011a). Scalable tensor factorizations for incomplete data. *Chemometrics and Intelligent Laboratory Systems*, 106(1), 41–56.

Acar, E., Kolda, T.G., Dunlavy, D.M. (2011b). All-at-once optimization for coupled matrix and tensor factorizations. *KDD Workshop on Mining and Learning with Graphs*. arXiv.org:1105.3422.

Acar, E., Levin-Schwartz, Y., Calhoun, V.D., Adal, T. (2017). Tensor-based fusion of EEG and fMRI to understand neurological changes in schizophrenia. *IEEE Int. Symp. on Circuits and Systems (ISCAS'2017)*, Baltimore, USA.

Alexander, J. and Hirschowitz, A. (1995). Polynomial interpolation in several variables. *J. Algebraic Geom.*, 4(4), 201–222.

de Almeida, A.L.F. and Favier, G. (2013). Double Khatri-Rao space-time-frequency coding using semi-blind PARAFAC based receiver. *IEEE Signal Processing Letters*, 20(5), 471–474.

de Almeida, A.L.F., Favier, G., Mota, J.C.M. (2006). Tensor-based space-time multiplexing codes for MIMO-OFDM systems with blind detection. *Proc. of 17th IEEE Symp. Pers. Ind. Mob. Radio Com. (PIMRC'2006)*, Helsinki, Finland.

de Almeida, A.L.F., Favier, G., Mota, J.C.M. (2007). PARAFAC-based unified tensor modeling for wireless communication systems with application to blind multiuser equalization. *Signal Processing*, 87(2), 337–351.

de Almeida, A.L.F., Favier, G., Mota, J.C.M. (2008). A constrained factor decomposition with application to MIMO antenna systems. *IEEE Trans. Signal Process.*, 56(6), 2429–2442.

de Almeida, A.L.F., Favier, G., Mota, J.C.M. (2009a). Space-time spreading-multiplexing for MIMO wireless communication systems using the PARATUCK-2 tensor model. *Signal Processing*, 89(11), 2103–2116.

de Almeida, A.L.F., Favier, G., Mota, J.C.M. (2009b). Constrained Tucker-3 model for blind beamforming. *Signal Process*, 89, 1240–1244.

de Almeida, A.L.F., Favier, G., da Costa, J.P.C.L., Mota, J.C. (2016). Overview of tensor decompositions with applications to communications. In *Signals and Images: Advances and Results in Speech, Estimation, Compression, Recognition, Filtering, and Processing*, Coelho, R.F., Nascimento, V.H., de Queiroz, R.L., Romano, J.M., Cavalcante, C.C. (eds). CRC Press, Boca Raton.

Alshebeili, S.A. and Cetin, A.E. (1990). A phase reconstruction algorithm from bispectrum. *IEEE Tr. on Geoscience and Remote Sensing*, 28, 166–170.

Alshebeili, S.A., Venetsanopoulos, A.N., Cetin, A.E. (1993). Cumulant based identification approaches for nonminimum phase FIR systems. *IEEE Tr. Signal Proc.*, 41(4), 1576–1588.

Amblard, P.O., Gaeta, M., Lacoume, J.L. (1996a). Statistics for complex variables and signals. Part I: Variables. *Signal Processing*, 53, 1–13.

Amblard, P.O., Gaeta, M., Lacoume, J.L. (1996b). Statistics for complex variables and signals. Part II: Signals. *Signal Processing*, 53, 15–25.

Anandkumar, A., Ge, R., Hsu, D., Kakade, S.M., Telgarsky, M. (2014). Tensor decompositions for learning latent variable models. *J. Machine Learning Res.*, 15(1), 2773–2832 [Online]. Available at: http://dl.acm.org/citation.cfm?id=2627435.2697055.

Arachchilage, S.W. and Izquierdo, E. (2020). Deep-learned faces: A survey. *EURASIP J. on Image and Video Processing*, 25 [Online]. Available at: https://doi.org/10.1186/ s13640-020-00510-w.

Ballani, J., Grasedyck, L., Kluge, M. (2013). Black box approximation of tensors in hierarchical Tucker format. *Linear Algebra and Its Applications*, 438, 639–657.

Banerjee, A., Char, A., Mondal, B. (2017). Spectra of general hypergraphs. *Linear Algebra and Its Applications*, 518, 14–30.

Bartels, R. and Stewart, G. (1972). Solution of the matrix equation AX + XB = C. *Comm. of the ACM*, 15(9), 820–826.

Becker, H., Albera, L., Comon, P., Haardt, M., Birot, G., Wendling, F., Gavaret, M., Bénar, C.-G., Merlet, I. (2014). EEG extended source localization: Tensor-based vs. conventional methods. *NeuroImage*, 96, 143–157.

Beckmann, C.F. and Smith, S.M. (2005). Tensorial extensions of independent component analysis for multisubject fMRI analysis. *NeuroImage*, 25(1), 294–311.

Behera, R. and Mishra, D. (2017). Further results on generalized inverses of tensors via the Einstein product. *Linear and Multilinear Algebra*, 65(8), 1662–1682.

Behera, R., Maji, S., Mohapatra, R.N. (2020). Weighted Moore–Penrose inverses of arbitrary-order tensors. *Computational and Applied Mathematics*, 39, 284.

Benetos, E. and Kotropoulos, C. (2008). A tensor-based approach for automatic music genre classification. *EUSIPCO*, Lausanne, Switzerland.

Bengua, J.A., Phien, H.N., Tuan, H.D., Do, M.N. (2017). Efficient tensor completion for color image and video recovery: Low-rank tensor train. *IEEE Tr. Image Processing*, 26(5), 2466–2479.

ten Berge, J.M.F. and Sidiropoulos, N.D. (2002). On uniqueness in CANDECOMP/ PARAFAC. *Psychometrika*, 67(3), 399–409.

ten Berge, J.M.F. and Smilde, A.K. (2002). Non-triviality and identification of a constrained Tucker3 analysis. *Journal of Chemometrics*, 16, 609–612.

Bobadilla, J., Ortega, F., Hernando, A., Gutiérrez, A. (2013). Recommender systems survey. *Knowledge-Based Systems*, 46, 109–132.

Bokde, D., Girase, S., Mukhopadhyay, D. (2015). Matrix factorization model in collaborative filtering algorithms: A survey. *Procedia Computer Science*, 49(1), 136–146.

Bondon, P. and Picinbono, B. (1990). De la blancheur et de ses transformations. *Traitement du signal*, 7(5), 385–395.

Bouilloc, T. and Favier, G. (2012). Nonlinear channel modeling and identification using bandpass Volterra-PARAFAC models. *Signal Processing*, 92(6), 1492–1498.

Boutsidis, C., Mahoney, M.W., Drineas, P. (2010). An improved approximation algorithm for the column subset selection problem. *Proc. 19th Annual ACM-SIAM Symp. on Discrete Algorithms (SODA)*, 968–977.

Boutsidis, C. and Woodruff, D.P. (2014). Optimal CUR matrix decompositions. *Proc. 46th ACM Symp. on Theory of Computing (STOC'14)*, 353–362.

Boyd, S., Parikh, N., Chu, E., Peleato, B., Eckstein, J. (2011). Distributed optimization and statistical learning via the alternating direction method of multipliers. *Foundations and Trends in Machine Learning*, 3(1), 1–122.

Brandoni, D. and Simoncini, V. (2020). Tensor-train decomposition for image recognition. *Calcolo*, 57(9), 1–24.

Brazell, M., Li, N., Navasca, C., Tamon, C. (2013). Solving multilinear systems via tensor inversion. *SIAM J. Matrix Analysis and Applications*, 34(2), 542–570.

Brewer, J.W. (1978). Kronecker products and matrix calculus in system theory. *IEEE Tr. on Circuits and Systems*, 25(9), 772–781.

Brillinger, D.R. (1965). An introduction to polyspectra. *Ann. Math. Statist.*, 36, 1351–1374.

Bro, R. (1997). PARAFAC. Tutorial and applications. *Chemometrics and Intelligent Laboratory Systems*, 38(2), 149–171.

Bro, R. (2006). Review on multiway analysis in chemisty – 2000–2005. *Critical Reviews in Analytical Chemistry*, 36, 279–293.

Bro, R. and Kiers, H.A.L. (2003). A new efficient method for determining the number of components in PARAFAC models. *J. Chemometrics*, 17(5), 274–286.

Bro, R., Harshman, R.A., Sidiropoulos, N.D., Lundy, M.E. (2009). Modeling multi-way data with linearly dependent loadings. *Chemometrics*, 23(7–8), 324–340.

Bu, C., Zhang, X., Zhou, J., Wang, W., Wei, Y. (2014). The inverse, rank and product of tensors. *Linear Algebra and Its Applications*, 446, 269–280.

Burns, F., Carlson, D., Haynsworth, E., Markham, T. (1974). Generalized inverse formulas using the Schur complement. *SIAM J. Appl. Math.*, 26(2), 254–259.

Caiafa, C.F. and Cichocki, A. (2010). Generalizing the column–row matrix decomposition to multi-way arrays. *Linear Algebra and Its Applications*, 433, 557–573.

Candès, E.J. and Plan, Y. (2010). Matrix completion with noise. *Proc. IEEE*, 98(6), 925–936.

Candès, E.J. and Recht, B. (2009). Exact matrix completion via convex optimization. *Found. Comput. Math.*, 9, 717–772.

Candès, E.J. and Wakin, M.B. (2008). An introduction to compressive sampling. *IEEE Signal Proc. Mag.*, 25(2), 21–30.

Cardoso, J.F. (1990). Localisation et identification par la quadricovariance. *Traitement du signal*, 7(5), 397–406.

Carroll, J.D. and Chang, J. (1970). Analysis of individual differences in multidimensional scaling via an N-way generalization of "Eckart-Young" decomposition. *Psychometrika*, 35(3), 283–319.

Carroll, J.D., Pruzansky, S., Kruskal, J.B. (1980). CANDELINC: A general approach to multidimensional analysis of many-way arrays with linear constraints on parameters. *Psychometrika*, 45(1), 3–24.

Cartwright, D. and Sturmfels, B. (2013). The number of eigenvalues of a tensor. *Linear Algebra and Its Applications*, 438, 942–952.

Chang, K.C. and Zhang, T. (2013). On the uniqueness and non-uniqueness of the positive Z-eigenvector for transition probability tensors. *J. Math. Anal. Appl.*, 408, 525–540.

Chang, K.C., Pearson, K., Zhang, T. (2009). On eigenvalue problems of real symmetric tensors. *J. Math. Anal. and Appl.*, 350, 416–422.

Chang, K.C., Qi, L., Zhang, T. (2013). A survey on the spectral theory of nonnegative tensors. *Numerical Linear Algebra Appl.*, 20(6), 891–912.

Che, M., Cichocki, A., Wei, Y. (2017). Neural networks for computing best rank-one approximations of tensors and its applications. *Neurocomputing*, 267(6), 114–133.

Chen, H. and Qi, L. (2015). Positive definiteness and semi-definiteness of even order symmetric Cauchy tensors. *American Inst. of Math. Sciences*, 11(4), 1263–1274.

Chen, C., Surana, A., Bloch, A., Rajapakse, I. (2019). Multilinear time invariant system theory. *Proc. of SIAM Conf. on Control and its Applications*, 118–125 [Online]. Available at: arXiv:1905.07427v1.

Chien, J.-T. and Bao, Y.-T. (2017). Tensor factorized neural networks. *IEEE Tr. on Neural Networks and Learning Systems*, 29(5), 1998–2011.

Cichocki, A. (2014). Tensor networks for big data analytics and large-scale optimization problems [Online]. Available at: arXiv:1407.3124.

Cline, A.K. and Dhillon, I.S. (2007). Computation of the singular value decomposition. In *Handbook of Linear Algebra*, Hogben, L. (ed.). Chapman & Hall/CRC Press, Boca Raton.

Comon, P. (1994). Independent component analysis, a new concept? *Signal Processing*, 36(3), 287–314.

Comon, P. and Cardoso, J.F. (1990). Eigenvalue decomposition of a cumulant tensor with applications. *SPIE Conf. on Advanced Signal Processing Algorithms, Architectures, and Implementations*, 361–372, San Diego, USA.

Comon, P., Golub, G., Lim, L.-H., Mourrain, B. (2008). Symmetric tensors and symmetric tensor rank. *SIAM J. Matrix Anal. Appl.*, 30(3), 1254–1279.

Comon, P., ten Berge, J.M.F., de Lathauwer, L., Castaing, J. (2009a). Generic and typical ranks of multi-way arrays. *Linear Algebra and Its Applications*, 430(11), 2997–3007.

Comon, P., Luciani, X., de Almeida, A.L.F. (2009b). Tensor decompositions, alternating least squares and other tales. *J. of Chemometrics*, 23(7–8), 393–405.

Cong, F., Lin, Q.-H., Kuang, L.-D., Gong, X.-F., Astikainen, P., Ristaniemi, T. (2015). Tensor decomposition of EEG signals: A brief review. *J. Neuroscience Methods*, 248, 59–69.

da Costa, J.P.C.L., Haardt, M., Roemer, F. (2008). Robust methods based on HOSVD for estimating the model order in PARAFAC models. *Proc. of 5th IEEE Sensor Array and Multich. Signal Proc. Workshop (SAM 2008), 510–514*, Darmstadt, Germany.

da Costa, J.P.C.L., Roemer, F., Weis, M., Haardt, M. (2010). Robust R-D parameter estimation via closed-form PARAFAC. *Proc. of ITG Workshop on Smart Antennas (WSA'2010), 99–106*, Bremen, Germany.

da Costa, J.P.C.L., Roemer, F., Haardt, M., de Sousa, R.T. (2011). Multi-dimensional model order selection. *EURASIP J. on Advances in Signal Processing*, 26 [Online]. Available at: http://asp.eurasipjournals.com/content/2011/1/26.

da Costa, M.N., Favier, G., Romano, J.-M. (2018). Tensor modelling of MIMO communication systems with performance analysis and Kronecker receivers. *Signal Processing*, 145, 304–316.

Coste, M. (2001). Elimination, résultant. Discriminant. Agrégation preparation. Université de Rennes 1, Rennes.

Crespo-Cadenas, C., Aguilera-Bonet, P., Becerra-Gonzalez, J.A., Cruces, S. (2014). On nonlinear amplifier modeling and identification using baseband Volterra-PARAFAC models. *Signal Processing*, 96, 401–405.

Cui, L.-B., Chen, C., Li, W., Ng, M.K. (2015). An eigenvalue problem for even order tensors with its applications. *Linear and Multilinear Algebra*, 64(4), 602–621.

Culp, J., Pearson, K.J., Zhang, T. (2017). On the uniqueness of the Z1-eigenvector of transition probability tensors. *Linear and Multilinear Algebra*, 65(5), 891–896.

Debals, O. and de Lathauwer, L. (2017). The concept of tensorization. Technical report. ESAT-STADIUS, KU Leuven, 1–34 [Online]. Available at: ftp://134.58.56.3/pub/sista/odebals/debals2017concept.pdf.

Ding, W. and Wei, Y. (2015). Generalized tensor eigenvalue problems. *SIAM J. Matrix Anal. Appl.*, 36(3), 1073–1099.

Ding, W., Qi, L., Wei, Y. (2013). $\mathcal{M}$-tensors and nonsingular $\mathcal{M}$-tensors. *Linear Algebra and Its Applications*, 439(10), 3264–3278.

Domanov, I. and de Lathauwer, L. (2013). On the uniqueness of the canonical polyadic decomposition of third-order tensors. Part I: Basic results and uniqueness of one factor matrix. *SIAM J. Matrix Anal. Appl.*, 34(3), 855–875.

Domanov, I. and de Lathauwer, L. (2014). Generic uniqueness conditions for the canonical polyadic decomposition and INDSCAL. *SIAM J. Matrix Anal. Appl.*, 36(4), 1567–1589.

Drineas, P., Kannan, R., Mahoney, M.W. (2006). Fast Monte Carlo algorithms for matrices III: Computing a compressed approximate matrix decomposition. *SIAM J. Comput.*, 36(1), 184–206.

Drineas, P., Mahoney, M.W., Muthukrishnan, S. (2008). Relative-error CUR matrix decompositions. *SIAM J. Matrix Anal. Appl.*, 30(2), 844–881.

Duarte, M.F. and Baraniuk, R.G. (2012). Kronecker compressive sensing. *IEEE Tr. on Image Processing*, 21(2), 494–504.

Eckart, C. and Young, G. (1936). The approximation of one matrix by another of lower rank. *Psychometrika*, 1(3), 211–218.

Elisei-Iliescu, C., Dogariu, L.-M., Paleologu, C., Bebnesty, J., Enescu, A.-A., Ciochina, S. (2020). A recursive least-squares algorithm for the identification of trilinear forms. *MDPI/Algorithms*, 13, 135.

Faber, N.M., Bro, R., Hopke, P.K. (2003). Recent developments in CANDECOMP/PARAFAC algorithms: A critical review. *Chemometrics and Intell. Lab. Syst.*, 65, 119–137.

Favier, G. (1982). *Filtrage, modélisation et identification de systèmes linéaires stochastiques à temps discret.* CNRS, Paris.

Favier, G. (2004). Estimation paramétrique de modèles entrée-sortie. In *Signaux aléatoires: modélisation, estimation, détection*, Guglielmi, M. (ed.). Hermes, Lavoisier, Cachan.

Favier, G. (2019). *From Algebraic Structures to Tensors*, ISTE Ltd, London and John Wiley & Sons, New York.

Favier, G. and de Almeida, A.L.F. (2014a). Overview of constrained PARAFAC models. *EURASIP J. Advances in Signal Processing*, 5, 41.

Favier, G. and de Almeida, A.L.F. (2014b). Tensor space-time-frequency coding with semi-blind receivers for MIMO wireless communication systems. *IEEE Tr. Signal Processing*, 62(22), 5987–6002.

Favier, G. and Bouilloc, T. (2009). Parametric complexity reduction of Volterra models using tensor decompositions. *17th European Signal Proc. Conf. (EUSIPCO)*, Glasgow.

Favier, G. and Bouilloc, T. (2010). Identification de modèles de Volterra basée sur la décomposition PARAFAC de leurs noyaux et le filtre de Kalman étendu. *Traitement du signal*, 27(1), 27–51.

Favier, G. and Kibangou, A.Y. (2009). Tensor-based methods for system identification. Part 2: Three examples of tensor-based system identification methods. *Int. Journal on Sciences and Techniques of Automatic Control (IJ-STA)*, 3(1), 870–889.

Favier, G., Dembélé, D., Peyre, J.L. (1994). ARMA identification using high-order statistics based linear methods: A unified presentation. *EUSIPCO'94*, 203–207, Edinburgh.

Favier, G., Kibangou, A., Bouilloc, T. (2012a). Nonlinear system modeling and identification using Volterra-PARAFAC models. *Int. J. of Adaptive Control and Signal Proc.*, 26(1), 30–53.

Favier, G., da Costa, M.N., de Almeida, A.L.F., Romano, J.M.T. (2012b). Tensor space–time (TST) coding for MIMO wireless communication systems. *Signal Processing*, 92(4), 1079–1092.

Favier, G., Bouilloc, T., de Almeida, A.L.F. (2012c). Blind constrained block-Tucker2 receiver for multiuser SIMO NL-CDMA communication systems. *Signal Processing*, 92(7), 1624–1636.

Favier, G., Fernandes, C.E.R, de Almeida, A.L.F. (2016). Nested Tucker tensor decomposition with application to MIMO relay systems using tensor space–time coding (TSTC). *Signal Processing*, 128, 318–331.

Filipovic, M. and Jukic, A. (2015). Tucker factorization with missing data with application to low-n-rank tensor completion. *Multidimensional Systems and Signal Processing*, 26, 677–692.

Freitas, W., Favier, G., de Almeida, A.L.F. (2018). Generalized Khatri-Rao and Kronecker space-time coding for MIMO relay systems with closed-form semi-blind receivers. *Signal Processing*, 151, 19–31.

Frolov, E. and Oseledets, I. (2017). Tensor methods and recommender systems. *WIREs Data Mining Knowl. Discov.*, 7(3).

Fu, T., Jiang, B., Li, Z. (2018). On decompositions and approximations of conjugate partial-symmetric complex tensors [Online]. Available at: arXiv:1802.09013v1.

Gardner, W.A. (1991). Exploitation of spectral redundancy in cyclostationary signals. *IEEE Signal Proc. Magazine*, 8(2), 14–36.

Gelfand, I.M., Kapranov, M.M., Zelevinsky, A.V. (1992). Hyperdeterminants. *Advances in Mathematics*, 96, 226–263.

Giannakis, G.B. (1987). Cumulants: A powerful tool in signal processing. *Proc. of the IEEE*, 75, 1333–1334.

Giannakis, G.B. and Mendel, J.M. (1989). Identification of nonminimum phase systems using higher order statistics. *IEEE Tr. on Acoustics, Speech and Signal Proc.*, 37(3), 360–377.

Golub, G.H. and Van Loan, C.F. (1983). *Matrix Computations*. Johns Hopkins University Press, Oxford.

Goodman, N.R. (1963). Statistical analysis based on certain multivariate complex Gaussian distribution. *Ann. Math. Statist.*, 34, 152–176.

Goulart, J.H.M. and Favier, G. (2014). An algebraic solution for the CANDECOMP/PARAFAC decomposition with circulant factors. *SIAM J. Matrix Anal. Appl.*, 35(4), 1543–1562.

Goulart, J.H.M., Kibangou, A., Favier, G. (2017). Traffic data imputation via tensor completion based on soft thresholding of Tucker core. *Transportation Research, Part C: Emerging Technologies*, 85, 348–362.

Grasedyck, L. and Hackbusch, W. (2011). An introduction to hierarchical (h-)rank and TT-rank of tensors with examples. *Comput. Meth. in Appl. Math.*, 11, 291–304.

Grasedyck, L., Kluge, M., Kramer, S. (2015). Variants of alternating least squares tensor completion in the tensor train format. *SIAM J. Sci. Comput.*, 37(5), A2424–A2450.

Guo, X., Miron, S., Brie, D., Zhu, S., Liao, X. (2011). A CANDECOMP/PARAFAC perspective on uniqueness of DOA estimation using a vector sensor array. *IEEE Trans. Signal Process.*, 59(7), 3475–3481.

Hao, N., Kilmer, M.E., Braman, K., Hoover, R.C. (2013). Facial recognition using tensor-tensor product decompositions. *SIAM J. Imaging Sci.*, 6, 437–463.

Hao, C., Cui, C., Dai, Y.-H. (2015). A sequential subspace projection method for extreme Z-eigenvalues of supersymmetric tensors. *Numerical Linear Algebra with Applications*, 22(2), 283–298.

Harshman, R.A. (1970). Foundations of the PARAFAC procedure: Model and conditions for an "explanatory" multimodal factor analysis. UCLA Working Papers in Phonetics, 16, 1–84.

Harshman, R.A. (1972). Determination and proof of minimum uniqueness conditions for PARAFAC1. UCLA Working Papers in Phonetics, 22, 111–117.

Harshman, R.A. and Lundy, M.E. (1996). Uniqueness proof for a family of models sharing features of Tucker's three-mode factor analysis and PARAFAC/CANDECOMP. *Psychometrika*, 61, 133–154.

Harshman, R.A., Hong, S., Lundy, M.E. (2003). Shifted factor analysis. Part I: Models and properties. *J. Chemometrics*, 17, 363–378.

Hastad, J. (1990). Tensor rank is NP-complete. *J. Algorithms*, 11(4), 644–654.

Henderson, H.V. and Searle, S.R. (1979). Vec and Vech operators for matrices, with some uses in Jacobians and multivariate statistics. *Canad. J. Statist*, 7(1), 65–81.

Henderson, H.V. and Searle, S.R. (1981). The vec-pemutation matrix, the vec operator and Kronecker products: A review. *Linear and Multilinear Algebra*, 9, 271–288.

Henderson, H.V., Pukelsheim, F., Searle, S.R. (1983). On the history of the Kronecker product. *Linear and Multilinear Algebra*, 14, 113–120.

Hillar, C.J. and Lim, L.H. (2013). Most tensor problems are NP-hard. *J. of the ACM*, 60(6), 45:1–45:39.

Hinich, M.J. (1982). Testing for Gaussianity and linearity of a stationary time series. *J. Time Series Anal.*, 3, 169–176.

Hitchcock, F.L. (1927). The expression of a tensor or a polyadic as a sum of products. *Journal of Mathematics and Physics*, 6(3), 164–189.

Horn, R.A. (1990). The Hadamard product. *Proc. Symp. Appl. Math.*, 40, 87–169.

Horn, R.A. and Johnson, C.A. (1985). *Matrix Analysis*. Cambridge University Press, Cambridge.

Horn, R.A. and Johnson, C.A. (1991). *Topics in Matrix Analysis*. Cambridge University Press, Cambridge.

Hu, S. and Qi, L. (2014). The eigenvectors associated with the zero eigenvalues of the Laplacian and signless Laplacian tensors of a uniform hypergraph. *Discrete Applied Mathematics*, 169, 140–151.

Hu, S., Huang, Z.H., Qi, L. (2013). Finding the extreme Z-eigenvalues of tensors via a sequential semidefinite programming method. *Numerical Linear Algebra with Applications*, 20, 972–984.

Huang, Z. and Qi, L. (2018). Positive definiteness of paired symmetric tensors and elasticity tensors. *Computational and Applied Mathematics*, 338, 22–43.

Hyland, D.C. and Collins, E.G. (1989). Block Kronecker products and block norm matrices in large-scale systems analysis. *SIAM J. Matrix Anal. Appl.*, 10(1), 18–29.

Jiang, T. and Sidiropoulos, N.D. (2004). Kruskal's permutation lemma and the identification of CANDECOMP/PARAFAC and bilinear models with constant modulus constraints. *IEEE Trans. Signal Process.*, 52(9), 2625–2636.

Jiang, T., Sidiropoulos, N.D., ten Berge, J.M.F. (2001). Almost-sure identifiability of multidimensional harmonic retrieval. *IEEE Trans. Signal Process.*, 49(9), 1849–1859.

Jin, H., Bai, M., Benitez, J., Liu, X. (2017). The generalized inverses of tensors and application to linear models. *Computers and Math. with Appl.*, 74(3), 385–397.

Khatri, C.G. and Rao, C.R. (1968). Solutions to some functional equations and their applications to characterization of probability distributions. *Sankhya, Indian J. Statistics, Series A*, 30, 167–180.

Khatri, C.G. and Rao, C.R. (1972). Functional equations and characterization of probability laws through linear functions of random variables. *J. of Multivariate Analysis*, 2, 162–173.

Kibangou, A.Y. and Favier, G. (2007). Blind joint identification and equalization of Wiener-Hammerstein communication channels using PARATUCK-2 tensor decomposition. *Proc. EUSIPCO'2007*, Poznan, Poland.

Kibangou, A.Y. and Favier, G. (2009a). Identification of parallel-cascade Wiener systems using joint diagonalization of third-order Volterra kernel slices. *IEEE Signal Processing Letters*, 16(3), 188–191.

Kibangou, A.Y. and Favier, G. (2009b). Non-iterative solution for PARAFAC with a Toeplitz matrix factor. *Proc. of EUSIPCO*, Glasgow.

Kibangou, A.Y. and Favier, G. (2010). Tensor analysis-based model structure determination and parameter estimation for block-oriented nonlinear systems. *IEEE Journal of Selected Topics in Signal Processing*, 4(3), 514–525.

Kiers, H.A.L. (2000). Towards a standardized notation and terminology in multiway analysis. *J. Chemometrics*, 14(2), 105–122.

Kilmer, M.E. and Martin, C.D. (2011). Factorization strategies for third-order tensors. *Linear Algebra and Its Applications*, 435, 641–658.

Kofidis, E. and Regalia, P.A. (2002). On the best rank-1 approximation of higher-order supersymmetric tensors. *SIAM J. Matrix Anal. Appl.*, 23(3), 863–884.

Kolda, T.G. and Bader, B.W. (2009). Tensor decompositions and applications. *SIAM Review*, 51(3), 455–500.

Kolda, T.G. and Mayo, J.R. (2014). An adaptive shifted power method for computing generalized tensor eigenpairs. *SIAM J. Matrix Anal. Appl.*, 35(4), 1563–1581.

Koning, R.H., Neudecker, H., Wansbeek, T. (1991). Block Kronecker products and the vecb operator. *Linear Algebra and Its Applications*, 149, 165–184.

Koukoulas, P. and Kalouptsidis, N. (1995). Nonlinear system identification using Gaussian inputs. *IEEE Tr. Signal Processing*, 43(8), 1831–1841.

Kroonenberg, P.M. and de Leeuw, J. (1980). Principal component analysis of three-mode data by means of alternating least squares algorithms. *Psychometrika*, 45(1), 69–97.

Kruskal, J.B. (1977). Three-way arrays: Rank and uniqueness of trilinear decompositions, with application to arithmetic complexity and statistics. *Linear Algebra Appl.*, 18(2), 95–138.

Kruskal, J.B. (1989). Rank, decomposition, and uniqueness for 3-way and N-way arrays. In *Multiway Data Analysis*, Coppi, R., Bolasco, S. (eds). Elsevier, Amsterdam.

Kruskal, J.B., Harshman, R.A., Lundy, M.E. (1989). How 3-MFA data can cause degenerate PARAFAC solutions, among other relationships. In *Multiway Data Analysis*, Coppi, R., Bolasco, S. (eds). Elsevier, Amsterdam.

Lacoume, J.-L., Amblard, P.-O., Comon, P. (1997). *Statistiques d'ordre supérieur pour le traitement du signal*. Masson, Paris.

Lahat, D., Adah, T., Jutten, C. (2015). Multimodal data fusion: An overview of methods, challenges and prospects. *Proceedings of the IEEE*, 103(9), 1449–1477.

Lancaster, P. and Tismenetsky, M. (1985). *The Theory of Matrices with Applications*. Academic Press, New York.

Landsberg, J.M. (2012). *Tensors: Geometry and Applications*. American Mathematical Society, Providence, RI.

de Launey, W. and Seberry, J. (1994). The strong Kronecker product. *Journal of Combinatorial Theory*, 66(2), 192–213.

de Lathauwer, L. (1997). Signal processing based on multilinear algebra. PhD. Thesis, KUL, Leuven.

de Lathauwer, L. (2006). A link between the canonical decomposition in multilinear algebra and simultaneous matrix diagonalization. *SIAM J. Matrix Anal. Appl.*, 28(3), 642–666.

de Lathauwer, L. (2008). Decompositions of a higher-order tensor in block terms. Part II: Definitions and uniqueness. *SIAM J. Matrix Anal. Appl.*, 30(3), 1033–1066.

de Lathauwer, L. (2011). Blind separation of exponential polynomials and the decomposition of a tensor in rank-$(L_r, L_r, 1)$ terms. *SIAM J. Matrix Anal. Appl.*, 32(4), 1451–1474.

de Lathauwer, L., de Moor, B., Vandewalle, J. (2000a). A multilinear singular value decomposition. *SIAM J. Matrix Anal. Appl.*, 21(4), 1253–1278.

de Lathauwer, L., de Moor, B., Vandewalle, J. (2000b). On the best rank-1 and rank-(R1, R2, ..., RN) approximation of higher-order tensors. *SIAM J. Matrix Anal. Appl.*, 21(4), 1324–1342.

Latorre, J.I. (2005). Image compression and entanglement [Online]. Available at: https://arxiv.org/abs/quant-ph/0510031.

Lawson, C.L. and Hanson, R.J. (1974). *Solving Least Squares Problems*. Prentice-Hall, Englewood Cliffs, NJ.

Lee, N. and Cichocki, A. (2014). Big data matrix singular value decomposition based on low-rank tensor train decomposition. In *Advances in Neural Networks-ISNN*, Zeng, Z., Li, Y., King, I. (eds). Springer, Cham.

Lee, N. and Cichocki, A. (2017). Fundamental tensor operations for large-scale data analysis using tensor network formats. *Multidim. Syst. Signal Process*, 29, 921–960.

Leonov, V.P. and Shiryaev, A.N. (1959). On a method of calculations of semi-invariants. *Theory Probab. Appl.*, IV(3), 319–328.

Lev Ari, H. (2005). Efficient solution of linear matrix equations with application to multistatic antenna array processing. *Communications in Information and Systems*, 5(1), 123–130.

Li, Y. and Ding, Z. (1994). A new nonparametric method for linear system phase recovery from bispectrum. *IEEE Tr. Circuits and Syst. II: Analog and Digital Signal Proc.*, 41, 415–419.

Li, W. and Ng, M. (2014). On the limiting probability distribution of a transition probability tensor. *Linear and Multilinear Algebra*, 62, 362–385.

Li, G., Qi, L., Yu, G. (2013). The Z-eigenvalues of a symmetric tensor and its application to spectral hypergraph theory. *Numerical Linear Algebra with Applications*, 20, 1001–1029.

Li, S., Dian, R., Fang, L., Bioucas-Dias, J.M. (2018). Fusing hyperspectral and multispectral images via coupled sparse tensor factorisation. *IEEE Tr. on Image Processing*, 27(8), 4118–4130.

Liang, M. and Zheng, B. (2018). Further results on Moore–Penrose inverses of tensors with application to tensor nearness problems. *Computers & Mathematics with Applications*, 77(5), 1282–1293.

Liang, M., Zheng, B., Zhao, R.-J. (2018). Tensor inversion and its application to the tensor equations with Einstein product. *Linear and Multilinear Algebra* [Online]. Available at: https://doi.org/10.1080/03081087.2018.1500993.

Lim, L.-H. (2005). Singular values and eigenvalues of tensors: A variational approach. *Proc. of the IEEE Int. Workshop on Computational Advances in Multi-Sensor Adaptive Processing (CAMSAP'05)*. Puerto Vallarta, Mexico.

Lim, L.-H. (2013). Tensors and hypermatrices. In *Handbook of Linear Algebra*, 2nd edition, Hogben, L. (ed.). Chapman & Hall/CRC Press, Boca Raton.

Lim, L.H. and Comon, P. (2009). Nonnegative approximations of nonnegative tensors. *J. of Chemometrics*, 23, 432–441.

Liu, S. and Trenkler, G. (2008). Hadamard, Khatri-Rao, Kronecker and other matrix products. *Int. J. Information and Syst. Sc.*, 4(1), 160–177.

Liu, J., Musialski, P., Wonka, P., Ye, J. (2013). Tensor completion for estimating missing values in visual data. *IEEE Tr. Pattern Analysis and Machine Intelligence*, 35(1), 208–220.

Liu, D., Li, W., Vong, S.-W. (2018). The tensor splitting with application to solve multilinear systems. *Journal of Computational and Applied Mathematics*, 330, 75–94.

Lu, H., Plataniotis, K.N., Venetsanopoulos, A.N. (2008). MPCA: Multilinear principal component analysis of tensor objects. *IEEE Trans. Neural Netw.*, 19(1), 18–39.

Lu, C., Feng, J., Chen, Y., Liu, W., Lin, Z., Yan, S. (2020). Tensor robust principal component analysis with a new tensor nuclear norm. *IEEE Tr. on Pattern Analysis and Machine Intelligence*, 42(4), 925–938.

Luo, X., Zhou, M., Xia, Y., Zhu, Q. (2014). An efficient non-negative matrix-factorization-based approach to collaborative filtering for recommender systems. *IEEE Tr. Industrial Informatics*, 10(2), 1273–1284.

Lyakh, D.I. (2015). An efficient tensor transpose algorithm for multicore CPU, Intel Xeon Phi, and NVidia Tesla GPU. *Computer Physics Communications*, 189, 84–91.

Magnus, J.R. (2010). On the concept of matrix derivative. *J. of Multivariate Analysis*, 101, 2200–2206.

Magnus, J.R. and Neudecker, H. (1979). The commutation matrix: Some properties and applications. *Annals of Statistics*, 7(2), 381–394.

Magnus, J.R. and Neudecker, H. (1985). Matrix differential calculus with applications to simple Hadamard, and Kronecker products. *J. Mathematical Psychology*, 29, 474–492.

Magnus, J.R. and Neudecker, H. (1988). *Matrix Differential Calculus with Applications in Statistics and Econometrics*. John Wiley & Sons, Chichester.

Mahoney, M.W. and Drineas, P. (2009). CUR matrix decompositions for improved data analysis. *Proc. of the National Academy of Sciences (PNAS)*, 106(3), 697–702.

Mahoney, M.W., Maggioni, M., Drineas, P. (2008). Tensor-CUR decompositions for tensor-based data. *SIAM J. on Matrix Analysis and Applications*, 30(3), 957–987.

Makantasis, K., Doulamis, A., Nikitakis, A. (2018). Tensor-based classification models for hyperspectral data analysis. *IEEE Tr. Geoscience and Remote Sensing*, 56(12), 6884–6898.

Marcus, M. and Khan, N.A. (1959). A note on the Hadamard product. *Canadian Math. Bull.*, 2, 81–83.

Mathews, V.J. and Sicuranza, G.L. (2000). *Polynomial Signal Processing*. John Wiley & Sons, New York.

McCullagh, P. (1987). *Tensor Methods in Statistics*. Chapman & Hall/CRC Press, Boca Raton.

Mendel, J.M. (1991). Tutorial on higher-order statistics (spectra) in signal processing and system theory: Theoretical results and some applications. *Proc. of the IEEE*, 79(3), 278–305.

Meyer, C.D. (2000). *Matrix Analysis and Applied Linear Algebra*. SIAM, Philadelphia.

Miao, Y., Qi, L., Wei, Y. (2021). T-Jordan canonical form and T-Drazin inverse based on the T-product. *Communications on Applied Math. and Comput.*, 3, 201–220.

Mika, S., Schölkopf, B., Smola, A.J., Müller, K.-R., Scholz, M., Rätsch, G. (1999). Kernel PCA and de-noising in feature spaces. In *Advances in Neural Information Processing Systems 11*, Kearns, M.S., Solla, S.A., Cohn, D.A. (eds). MIT Press, Cambridge.

Mitrovic, N., Asif, M.T., Rasheed, U., Dauwels, J., Jaillet, P. (2013). CUR decomposition for compression and compressed sensing of large-scale traffic data. *16th Int. IEEE Conf. on Intelligent Transportation Systems (ITSC'2013)*, 1475–1480, The Hague, The Netherlands.

Morozov, A. and Shakirov, S. (2010). New and old results in resultant theory. *Theoretical and Math. Physics*, 163, 587–617.

Morup, M. (2011). Applications of tensor (multiway array) factorizations and decompositions in data mining. *Wiley Interdisciplinary Reviews: Data Mining and Knowledge Discovery*, 1(1), 24–40.

Morup, M., Hansen, L.K., Madsen, K.H. (2011). Modeling latency and shape changes in trial based neuroimaging data. *Asilomar-SSC*, Monterey, USA.

Na, Y.J., Kim, K.S., Song, I., Kim, T. (1995). Identification of nonminimum phase systems using the third and fourth order cumulants. *IEEE Tr. on Signal Proc.*, 43(8), 2018–2022.

Nagy, J.G., Ng, M.K., Perrone, L. (2004). Kronecker product approximation for image restoration with reflexive boundary conditions. *SIAM J. Matrix Anal. Appl.*, 25(3), 829–841.

Nandi, A.K. (1999). *Blind Estimation Using Higher-Order Statistics*. Kluwer Academic Publishers, Boston, MA.

Nanopoulos, A., Rafailidis, D., Symeonidis, P., Manolopoulos, Y. (2010). MusicBox: Personalized music recommendation based on cubic analysis of social tags. *IEEE Tr. Audio, Speech and Language Processing*, 18, 407–412.

Neudecker, H. and Liu, S. (2001). Some statistical properties of Hadamard products of random matrices. *Statist. Papers*, 42, 475–487.

Neudecker, H., Liu, S., Polasek, W. (1995). The Hadamard product and some of its applications in statistics. *Statistics*, 26(4), 365–373.

Ni, G. (2019). Hermitian tensor and quantum mixed state [Online]. Available at: arXiv:1902.02640v4.

Nie, J. (2017). Generating polynomials and symmetric tensor decompositions. *Foundations of Comput. Math.*, 17, 423–465.

Nikias, C.L. (1988). ARMA bispectrum approach to nonminimum phase system identification. *IEEE Tr. on Acoustics, Speech and Signal Proc.*, 36, 513–524.

Nikias, C.L. and Mendel, J.M. (1993). Signal processing with higher-order spectra. *IEEE Signal Proc. Magazine*, 10–37.

Nikias, C.L. and Petropulu, A.P. (1993). *Higher-order Spectra Analysis. A Nonlinear Signal Processing Framework*. Prentice-Hall, Englewood Cliffs, NJ.

Nikias, C.L. and Raghuveer, M.R. (1987). Bispectrum estimation: A digital signal processing framework. *Proc. of the IEEE*, 75(7), 869–891.

Novikov, A., Podoprikhin, D., Osokin, A., Vetrov, D. (2015). Tensorizing neural networks. *Proc. of the 28th Int. Conf. on Neural Information Processing Systems (NIPS'15)*, 1, 442–450.

Olive, M. and Auffray, N. (2013). Symmetry classes for even-order tensors. *Mathematics and Mechanics Complex Systems*, 1(2), 177–210.

Oseledets, I. (2011). Tensor-train decomposition. *SIAM J. Sci. Computing*, 33(5), 2295–2317.

Oseledets, I. and Tyrtyshnikov, E. (2009). Breaking the curse of dimensionality, or how to use SVD in many dimensions,. *SIAM J. Sci. Computing*, 31, 3744–3759.

Oseledets, I. and Tyrtyshnikov, E. (2010). TT-cross approximation for multidimensional arrays. *Linear Algebra and Its Applications*, 432, 70–88.

Paatero, P. (2000). Construction and analysis of degenerate PARAFAC models. *J. Chemometrics*, 14, 285–299.

Padhy, S., Goovaerts, G., Boussé, M., De Lathauwer, L., Van Huffel, S. (2019). The power of tensor-based approaches in cardiac applications In *Biomedical Signal Processing. Advances in Theory, Algorithms and Applications*, Naik, G. (ed.). Springer, Singapore.

Paleologu, C., Benesty, J., Ciochina, S. (2018). Linear system identification based on a Kronecker product decomposition. *IEEE/ACM Tr. on Audio, Speech, and Language Proc.*, 26(10), 1793–1808.

Pan, R. (2014). Tensor transpose and its properties [Online]. Available at: arXiv:1411.1503v1.

Panagakis, Y., Kotropoulos, C., Arce, G.R. (2010). Non-negative multilinear principal component analysis of auditory temporal modulations for music genre classification. *IEEE Tr. on Audio, Speech, and Language Processing*, 18(3), 576–588.

Panigrahy, K. and Mishra, D. (2018). An extension of the Moore–Penrose inverse of a tensor via the Einstein product [Online]. Available at: arXiv:1806.03655v1.

Panigrahy, K., Behera, R., Mishra, D. (2020). Reverse order law for the Moore–Penrose inverses of tensors. *Linear and Multilinear Algebra*, 68(2), 246–264.

Papalexakis, E., Faloutsos, C., Siropoulos, N.D. (2016). Tensors for data mining and data fusion: Models, applications, and scalable algorithms. *ACM Tr. Intelligent Systems and Technology*, 8(2).

Papoulis, A. (1984). *Probability, Random Variables and Stochastic Processes*. McGraw-Hill, New York.

Papy, J.M., de Lathauwer, L., Van Huffel, S. (2005). Exponential data fitting using multilinear algebra: The single-channel and multi-channel case. *Numerical Algebra with Applications*, 12(8), 809–826.

Pearson, K. (1901). On lines and planes of closest fit to systems of points in space. *Philosophical Magazine*, 2(6), 559–572.

Pearson, K.J. (2010). Essentially positive tensors. *Int. J. of Algebra*, 4(9), 421–427.

Pearson, K.J. and Zhang, J. (2014). On spectral hypergraph theory of the adjacency tensor. *Graphs and Combinatorics*, 30, 1233–1248.

Phan, A.-H., Tichavsky, P., Cichocki, A. (2013). Low complexity damped Gauss–Newton algorithms for CANDECOMP/PARAFAC. *SIAM J. Appl. Math.*, 34(1), 126–147.

Phan, A.-H., Tichavsky, P., Cichocki, A. (2017). Blind source separation of single channel mixture using tensorization and tensor diagonalization. *VA/ICA 2017*, LNCS 10169, 36–46.

Picinbono, B. (1993). *Random Signals and Systems*. Prentice-Hall, Englewood Cliffs, NJ.

Picinbono, B. (1994). On circularity. *IEEE Tr. Signal Processing*, 42(12), 3473–3482.

Pisarenko, V.F. (1973). The retrieval of harmonics from a covariance function. *Geophysical J. Int.*, 33(3), 347–366.

Pitsianis, N.P. (1997). The Kronecker product in approximation and fast transform generation. PhD Thesis, Cornell University.

Pollock, D.S.G. (2011). On Kronecker products, tensor products and matrix differential calculus. Working paper 11/34, Department of Economics, University of Leicester [Online]. Available at: http://www.le.ac.uk/ec/research/RePEc/lec/leecon/dp11-34.pdf.

Qi, L. (2005). Eigenvalues of a real supersymmetric tensor. *J. Symbolic Computation*, 40, 1302–1324.

Qi, L. (2012). The spectral theory of tensors [Online]. Available at: arXiv:1201.3424v1.

Qi, L., Wang, F., Wang, Y. (2009). Z-eigenvalue methods for a global polynomial problem. *Mathematical Programming*, 118, 301–316.

Ragnarsson, S. and Van Loan, C.F. (2013). Block tensors and symmetric embeddings. *Linear Algebra and Its Applications*, 438, 853–874.

Raimondi, F., Cabral Farias, R., Michel, O., Comon, P. (2017). Wideband multiple diversity tensor array processing. *IEEE Tr. on Signal Processing*, 65(20), 5334–5346.

Ran, B., Tan, H., Wu, Y., Jin, P.J., (2016). Tensor based missing traffic data completion with spatial-temporal correlation. *Physica A: Statistical Mechanics and its Applications*, 446(15), 54–63.

Regalia, P.A. and Mitra, S.K. (1989). Kronecker products, unitary matrices and signal processing applications. *SIAM Review*, 31(4), 586–613.

Rendle, S. and Schmidt-Thieme, L. (2010). Pairwise interaction tensor factorization for personalized tag recommendation. *Proc. of the Third ACM Intern. Conf. on Web Search and Data Mining*, 81–90.

Rezghi, M. and Elden, L. (2011). Diagonalization of tensors with circulant structure. *Linear Algebra and its Applications*, 435, 422–447.

Rocha, D.S., Fernandes, C.E.R., Favier, G. (2019a). MIMO multi-relay systems with tensor space-time coding based on coupled nested Tucker decomposition. *Digital Signal Processing*, 89(3), 170–185.

Rocha, D., Favier, G., Fernandes, C.E.R. (2019b). Closed-form receiver for multi-hop MIMO relay systems with tensor space-time coding. *Journal of Communication and Information Systems*, 34(1), 50–54.

Roth, W.E. (1934). On direct product matrices. *Bulletin of the American Mathematical Society*, 40, 461–468.

Sanguansat, P. (2012). *Principal Component Analysis: Multidisciplinary Applications*. Intech, London.

Schetzen, M. (1980). *The Volterra and Wiener Theories of Nonlinear Systems*. John Wiley & Sons, New York.

Schölkopf, B., Smola, A.J., Müller, K.-R. (1998). Nonlinear component analysis as a kernel eigenvalue problem. *Neural Computation*, 10, 1299–1319.

Sidiropoulos, N.D. (2001). Generalizing Carathéodory's uniqueness of harmonic parameterization to N dimensions. *IEEE Tr. on Information Theory*, 47(4), 1687–1690.

Sidiropoulos, N.D. and Bro, R. (2000). On the uniqueness of multilinear decomposition of N-way arrays. *J. Chemometrics*, 14, 229–239.

Sidiropoulos, N.D. and Budampati, R. (2002). Khatri-Rao space-time codes. *IEEE Trans. Signal Process.*, 50(10), 2377–2388.

Sidiropoulos, N.D. and Kyrillidis, A. (2012). Multi-way compressed sensing for sparse low-rank tensors. *IEEE Signal Proc. Letters*, 19(11), 757–760.

Sidiropoulos, N.D. and Liu, X. (2001). Identifiability results for blind beamforming in incoherent multipath with small delay spread. *IEEE Trans. Signal Process.*, 49(1), 228–236.

Sidiropoulos, N.D., Bro, R., Giannakis, G.B. (2000a). Parallel factor analysis in sensor array processing. *IEEE Trans. Signal Process.*, 48(8), 2377–2388.

Sidiropoulos, N.D., Giannakis, G.B., Bro, R. (2000b). Blind PARAFAC receivers for DS-CDMA systems. *IEEE Trans. Signal Process.*, 48(3), 810–823.

Sidiropoulos, N.D., de Lathauwer, L., Fu, X., Huang, K., Papalexakis, E., Faloutsos, C. (2017). Tensor decomposition for signal processing and machine learning. *IEEE Tr. Signal Processing*, 65(13), 3551–3582.

Signoretto, M., Plas, R.V., De Moor, B., Suykens, J.A.K. (2011). Tensor versus matrix completion: A comparison with application to spectral data. *IEEE Signal Proc. Letters*, 18(7), 403–406.

de Silva, V. and Lim, L.-H. (2008). Tensor rank and the ill-posedness of the best low-rank approximation problem. *SIAM J. Matrix Anal. Appl.*, 30(3), 1084–1127.

Smilde, A.K., Bro, R., Geladi, P. (2004). *Multi-way Analysis. Applications in the Chemical Sciences*. Wiley, Chichester.

Söderström, T. (1994). *Discrete-time Stochastic Systems*. Prentice-Hall, Englewood Cliffs.

Song, Y. and Qi, L. (2015). Properties of some classes of structured tensors. *Journal of Optimization Theory and Applications*, 165(3), 854–873.

Sorensen, D.C. and Embree, M. (2015). A DEIM induced CUR factorization. Technical report [Online]. Available at: https://scholarship.rice.edu/handle/1911/102226.

Springer, P., Sankaran, A., Bientinesi, P. (2017). TTC: A tensor transposition compiler for multiple architectures. *Proc. of the 3rd ACM SIGPLAN Int. Workshop on Libraries, Languages, and Compilers for Array Programming*, 41–46, New York, USA.

Stanimirovic, P., Ciric, M., Katsikis, V., Li, C., Ma, H. (2020). Outer and (b, c) inverses of tensors. *Linear Multilinear Algebra*, 68(5) [Online]. Available at: https://doi.org/10.1080/03081087.2018.1521783.

Stegeman, A. (2006). Degeneracy in CANDECOMP/PARAFAC explained for $p \times p \times 2$ arrays of rank $p + 1$ or higher. *Psychometrika*, 71(3), 483–501.

Stegeman, A. (2008). On uniqueness conditions for CANDECOMP/PARAFAC and INDSCAL with full column rank in one mode. *Lin. Alg. Appl.*, 431(1–2), 211–227.

Stegeman, A. and Sidiropoulos, N.D. (2007). On Kruskal's uniqueness condition for the CANDECOMP/PARAFAC decomposition. *Lin. Alg. Appl.*, 420, 540–552.

Styan, G.P.H. (1973). Hadamard products and multivariate statistical analysis. *Linear Algebra Appl.*, 6, 217–240.

Sun, L., Zheng, B., Bu, C., Wei, Y. (2016). Moore–Penrose inverse of tensors via Einstein product. *Linear Multilinear Algebra*, 64, 686–698.

Swami, A. and Mendel, J.M. (1990). ARMA parameter estimation using only output cumulants. *IEEE Tr. on Acoustics, Speech and Signal Proc.*, 38, 1257–1265.

Symeonidis, P. and Zioupos, A. (2016). *Matrix and Tensor Factorization Techniques for Recommender Systems*. Springer, Cham.

Tan, H., Feng, G., Feng, J., Wang, W., Zhang, Y.-J., Li, F. (2013). A tensor-based method for missing traffic data completion. *Transportation Research Part C: Emerging Technologies*, 28, 15–27.

Therrien, C.W. (1992). *Discrete Random Signals and Statistical Signal Processing*. Prentice-Hall, Englewood Cliffs, NJ.

Tick, L.J. (1961). The estimation of transfer functions of quadratic systems. *Technom.*, 3, 563–567.

Tomasi, G. and Bro, R. (2005). PARAFAC and missing values. *Chemometrics and Intelligent Laboratory Systems*, 75(2), 163–180.

Tomasi, G. and Bro, R. (2006). A comparison of algorithms for fitting the PARAFAC model. *Computational Statistics & Data Analysis*, 50(7), 1700–1734.

Tracy, D.S. and Dwyer, P.S. (1969). Multivariate maxima and minima with matrix derivatives. *J. Amer. Statist. Assoc.*, 64(328), 1576–1594.

Tracy, D.S. and Singh, R.P. (1972). A new matrix product and its applications in matrix differentiation. *Statist. Neerlandica*, 26, 143–157.

Tucker, L.R. (1966). Some mathematical notes on three-mode factor analysis. *Psychometrika*, 31, 279–311.

Tugnait, J.K. (1990). Approaches to FIR system identification with noisy data using higher order statistics. *IEEE Tr. on Acoustics, Speech and Signal Proc.*, 38(7), 1307–1317.

Van Loan, C.F. (2000). The ubiquitous Kronecker product. *J. of Computational and Applied Math.*, 123, 85–100.

Van Loan, C.F. (2009). The Kronecker product. A product of the times. *SIAM Conf. on Applied Linear Algebra*, Monterey, CA.

Van Loan, C.F. and Pitsianis, N. (1993). Approximation with Kronecker products. In *Linear Algebra for Large Scale and Real-time Applications*, Moonen, M.S., Golub, G.H., de Moor, B.L. (eds). Kluwer Academic Publishers, Dordrecht.

Vasilescu, M.A.O. and Terzopoulos, D. (2002). Multilinear analysis of image ensembles: TensorFaces. *Proc. of the European Conf. on Computer Vision (ECCV '02)*, 447–460, Copenhagen.

Vasilescu, M.A.O. and Terzopoulos, D. (2005). Multilinear independent component analysis. *Proc. of the IEEE Conf. on Computer Vision and Pattern Recognition (CVPR '05)*, 1, 547–553, San Diego, USA.

Voronin, S. and Martinsson, P.-G. (2017). Efficient algorithms for CUR and interpolative matrix decompositions. *Adv. Comput. Math.*, 43, 495–516.

Wang, S. and Zhang, Z. (2013). Improving CUR matrix decomposition and the Nyström approximation via adaptive sampling. *J. of Machine Learning Res.*, 14, 2729–2769.

Wang, R., Li, S., Cheng, L., Wong, M.H., Leung, K.S. (2019). Predicting associations among drugs, targets and diseases by tensor decomposition for drug repositioning. *BMC Bioinformatics*, 20(26), 628.

Xie, J. and Qi, L. (2016). Spectral directed hypergraph theory via tensors. *Linear and Multilinear Algebra*, 780–794.

Ximenes, L., Favier, G., de Almeida, A. (2014). PARAFAC-PARATUCK semi-blind receivers for two-hop cooperative MIMO relay systems. *IEEE Tr. Signal Processing*, 62(14), 3604–3615.

Xu, L., Liang, G., Longxiang, Y., Hongbo, Z. (2012). PARALIND-based blind joint angle and delay estimation for multipath signals with uniform linear array. *EURASIP J. Advances in Signal Proc.*, 130.

Zarzoso, V. and Nandi, A.K. (1999). Blind source separation. In *Blind Estimation Using Higher-Order Statistics*, Nandi, A. (ed.), Kluwer Academic Publishers, Boston, MA.

Zhang, T. and Golub, G.H. (2001). Rank-one approximation of higher order tensors. *SIAM J. Matrix Anal. Appl.*, 23, 534–550.

Zhang, L., Qi, L., Zhou, G. (2014). M-tensors and some applications. *SIAM J. Matrix Anal. Appl.*, 35(2), 437–452.

Zhou, M., Liu, Y., Long, Z., Chen, L., Zhu, C. (2019). Tensor rank learning in CP decomposition via convolutional neural network. *Signal Processing: Image Communication*, 73, 12–21.

Zniyed, Y., Boyer, R., de Almeida, A., Favier, G. (2019a). High-order CPD estimation with dimensionality reduction using a tensor train model. *EUSIPCO*, Rome.

Zniyed, Y., Boyer, R., de Almeida, A., Favier, G. (2019b). Multidimensional harmonic retrieval based on Vandermonde tensor train. *Signal Processing*, 163(10), 75–86.

Zniyed, Y., Boyer, R., de Almeida, A., Favier, G. (2020). Tensor train representation of MIMO channels using the JIRAFE method. *Signal Processing*, 171, 107479.

Index

Other titles from

in

Digital Signal and Image Processing

2019

FAVIER Gérard
From Algebraic Structures to Tensors (Matrices and Tensors with Signal Processing Set – Volume 1)

MEYER Fernand
Topographical Tools for Filtering and Segmentation 1: Watersheds on Node- or Edge-weighted Graphs
Topographical Tools for Filtering and Segmentation 2: Flooding and Marker-based Segmentation on Node- or Edge-weighted Graphs

2017

CESCHI Roger, GAUTIER Jean-Luc
Fourier Analysis

CHARBIT Maurice
Digital Signal Processing with Python Programming

CHAO Li, SOULEYMANE Bella-Arabe, YANG Fan
Architecture-Aware Optimization Strategies in Real-time Image Processing

FEMMAM Smain
Fundamentals of Signals and Control Systems
Signals and Control Systems: Application for Home Health Monitoring

MAÎTRE Henri
From Photon to Pixel – 2nd edition

PROVENZI Edoardo
Computational Color Science: Variational Retinex-like Methods

2015

BLANCHET Gérard, CHARBIT Maurice
Digital Signal and Image Processing using MATLAB®
Volume 2 – Advances and Applications:The Deterministic Case – 2nd edition
Volume 3 – Advances and Applications: The Stochastic Case – 2nd edition

CLARYSSE Patrick, FRIBOULET Denis
Multi-modality Cardiac Imaging

GIOVANNELLI Jean-François, IDIER Jérôme
Regularization and Bayesian Methods for Inverse Problems in Signal and Image Processing

2014

AUGER François
Signal Processing with Free Software: Practical Experiments

BLANCHET Gérard, CHARBIT Maurice
Digital Signal and Image Processing using MATLAB®
Volume 1 – Fundamentals – 2nd edition

DUBUISSON Séverine
Tracking with Particle Filter for High-dimensional Observation and State Spaces

ELL Todd A., LE BIHAN Nicolas, SANGWINE Stephen J.
Quaternion Fourier Transforms for Signal and Image Processing

FERNANDEZ Christine, MACAIRE Ludovic, ROBERT-INACIO Frédérique
Digital Color Imaging

FERNANDEZ Christine, MACAIRE Ludovic, ROBERT-INACIO Frédérique
Digital Color: Acquisition, Perception, Coding and Rendering

NAIT-ALI Amine, FOURNIER Régis
Signal and Image Processing for Biometrics

OUAHABI Abdeljalil
Signal and Image Multiresolution Analysis

2011

CASTANIÉ Francis
Digital Spectral Analysis: Parametric, Non-parametric and Advanced Methods

DESCOMBES Xavier
Stochastic Geometry for Image Analysis

FANET Hervé
Photon-based Medical Imagery

MOREAU Nicolas
Tools for Signal Compression

2010

NAJMAN Laurent, TALBOT Hugues
Mathematical Morphology

2009

BERTEIN Jean-Claude, CESCHI Roger
Discrete Stochastic Processes and Optimal Filtering – 2nd edition

CHANUSSOT Jocelyn *et al.*
Multivariate Image Processing

DHOME Michel
Visual Perception through Video Imagery

GOVAERT Gérard
Data Analysis

GRANGEAT Pierre
Tomography

MOHAMAD-DJAFARI Ali
Inverse Problems in Vision and 3D Tomography

SIARRY Patrick
Optimization in Signal and Image Processing

2008

ABRY Patrice *et al.*
Scaling, Fractals and Wavelets

GARELLO René
Two-dimensional Signal Analysis

HLAWATSCH Franz *et al.*
Time-Frequency Analysis

IDIER Jérôme
Bayesian Approach to Inverse Problems

MAÎTRE Henri
Processing of Synthetic Aperture Radar (SAR) Images

MAÎTRE Henri
Image Processing

NAIT-ALI Amine, CAVARO-MENARD Christine
Compression of Biomedical Images and Signals

NAJIM Mohamed
Modeling, Estimation and Optimal Filtration in Signal Processing

QUINQUIS André
Digital Signal Processing Using Matlab

Printed and bound by CPI Group (UK) Ltd, Croydon, CR0 4YY